Industrial Enzyme Applications

Industrial Enzyme Applications

Edited by
Andreas Vogel and Oliver May

WILEY-VCH

Editors

Dr. Andreas Vogel
c-LEcta GmbH
R&D Enzyme Development
Perlickstr. 5
04103 Leipzig
Germany

Dr. Oliver May
DSM Nutritional Products Ltd
Wurmisweg 576
4303 Kaiseraugst
Switzerland

Library of Congress Card No.:
applied for

British Library Cataloguing-in-Publication Data
A catalogue record for this book is available from the British Library.

Bibliographic information published by the Deutsche Nationalbibliothek
The Deutsche Nationalbibliothek lists this publication in the Deutsche Nationalbibliografie; detailed bibliographic data are available on the Internet at <http://dnb.d-nb.de>.

Print ISBN: 978-3-527-34385-0
ePDF ISBN: 978-3-527-81375-9
ePub ISBN: 978-3-527-81377-3
oBook ISBN: 978-3-527-81378-0

Cover Design Adam-Design, Weinheim, Germany
Typesetting SPi Global, Chennai, India
Printing and Binding

Printed on acid-free paper

10 9 8 7 6 5 4 3 2 1

Contents

Preface *xiii*

Part I Overview of Industrial Enzyme Applications and Key Technologies *1*

1.1 Industrial Enzyme Applications – Overview and Historic Perspective *3*
Oliver May
1.1.1 Prehistoric Applications *3*
1.1.2 Growing the Scientific Basis *5*
1.1.3 The Beginning of Industrial Applications and the Emerging Enzyme Industry *12*
References *21*

1.2 Enzyme Development Technologies *25*
Andreas Vogel
1.2.1 Introduction *25*
1.2.2 Identification of Wild-Type Enzymes *26*
1.2.2.1 Selection Parameters for Starting Enzymes *28*
1.2.3 Enzyme Engineering *30*
1.2.3.1 Types of Enzyme Modifications *30*
1.2.3.2 General Engineering Strategies. Library Design and Generation *30*
1.2.3.3 Screening for Better Enzymes *37*
1.2.4 Impact of Enzyme Development Technologies Today and Tomorrow *38*
Acknowledgments *41*
References *41*

1.3 Eukaryotic Expression Systems for Industrial Enzymes *47*
Lukas Rieder, Nico Teuschler, Katharina Ebner, and Anton Glieder
1.3.1 Eukaryotic Enzyme Production Systems *47*
1.3.2 Special Considerations for Working with Eukaryotic Expression Systems *47*

1.3.2.1 Choice of Expression Host *47*
1.3.2.2 Comparison of Cell Structure and Their Influence on Molecular Biology *49*
1.3.3 Differences in Vector Design for Eukaryotic and Prokaryotic Hosts *51*
1.3.4 Differences in Regulation of Gene Expression in Eukaryotes and Prokaryotes *56*
1.3.4.1 Different Types of Promoters *58*
1.3.5 Industrial Enzyme Production *58*
1.3.6 Enzyme Production on Industrial Scale *61*
1.3.6.1 Homologous Protein Production *61*
1.3.6.2 Heterologous Protein Production *62*
References *63*

1.4 Process Considerations for the Application of Enzymes *71*
Selin Kara and Andreas Liese
1.4.1 Biocatalyst Types Used in Industrial Processes *71*
1.4.2 Enzyme Immobilization for Biocatalytic Processes *74*
1.4.3 Reaction Medium Applied in Enzymatic Catalysis *76*
1.4.3.1 Monophasic Systems – Organic Media *77*
1.4.3.2 Multiphasic Systems – Liquid/Liquid Mixtures *80*
1.4.3.3 Multiphasic Systems – Gas/Liquid Mixtures *83*
1.4.3.4 Multiphasic Systems – Solid/Liquid Mixtures *84*
1.4.4 Appropriate Reactor Types in Enzyme Catalysis *87*
1.4.5 Assessment Criteria for Enzymatic Applications *90*
References *92*

Part II Enzyme Applications for the Food Industry *95*

2.1 Enzymes Used in Baking *97*
Joke A. Putseys and Margot E.F. Schooneveld-Bergmans
2.1.1 Introduction *97*
2.1.2 The Baking Process – The Baker's Needs *98*
2.1.2.1 Flour Quality and Standardization *98*
2.1.2.2 Mixing and Dough Handling *100*
2.1.2.3 Fermentation and Dough Stability *105*
2.1.2.4 Baking and Oven Spring *109*
2.1.3 The Bread Quality – The Consumers' Needs *111*
2.1.3.1 Color and Flavor *111*
2.1.3.2 Shelf Life *112*
2.1.4 Trends and Opportunities for Baking Enzymes *116*
2.1.4.1 Fine Baking and Confectionary *116*
2.1.4.2 Consumer Preference: Health, Individual Values, and Convenience *117*
2.1.5 Conclusion *118*
References *119*

2.2 **Protein Modification to Meet the Demands of the Food Industry** *125*
Andrew Ellis
2.2.1 Food Proteins *125*
2.2.2 Processing of Food Protein *127*
2.2.3 Enzymes in the Processing of Food Proteins *127*
2.2.4 Food Protein Value Chain *130*
2.2.5 Recent Enzyme Developments *131*
2.2.5.1 Simple Protein Modification (Value Level 3) *131*
2.2.5.1.1 Developing Microbial Alternatives to Plant and Animal Enzymes *131*
2.2.5.2 Specialized Enzyme Modification (Value Level 4) *134*
2.2.5.2.1 Whey Protein Hydrolysates *134*
2.2.5.2.2 Plant Protein Hydrolysates *134*
2.2.5.3 Highly Specific Protein Modification (Value Level 5) *135*
2.2.5.3.1 Gluten Modification *135*
2.2.5.3.2 Acrylamide Reduction *135*
2.2.5.3.3 Bioactive Peptides *136*
2.2.6 Enzymes to Meet Future Needs *137*
Acknowledgments *139*
References *139*

2.3 **Dairy Enzymes** *143*
Peter Dekker
2.3.1 Introduction *143*
2.3.2 Coagulants *145*
2.3.2.1 Traditional Rennets *147*
2.3.2.2 Microbial Rennets *148*
2.3.2.3 Fermentation Produced Chymosin *151*
2.3.3 Ripening Enzymes *152*
2.3.3.1 Proteases/Peptidases *153*
2.3.3.2 Lipases/Esterases *154*
2.3.4 Lactases *154*
2.3.4.1 Neutral Lactase *156*
2.3.4.2 Acid Lactase *158*
2.3.4.3 GOS Production *158*
2.3.5 Miscellaneous Enzymes *161*
2.3.5.1 Oxidases/Peroxidases *161*
2.3.5.2 Phopholipases *162*
2.3.5.3 Cross-linking Enzymes *162*
2.3.5.4 Preservation *163*
2.3.6 New Developments *163*
References *163*

2.4 Enzymatic Process for the Synthesis of Cellobiose *167*
Birgit Brucher and Thomas Häßler
2.4.1 Enzymatic Synthesis of Cellobiose *167*
2.4.2 Cellobiose – Properties and Applications *168*
2.4.3 Existing Routes for Cellobiose Synthesis *170*
2.4.4 Enzyme Development *171*
2.4.5 Process Development *173*
2.4.5.1 Synthesis of Cellobiose *174*
2.4.5.2 Purification of Cellobiose *174*
2.4.6 Summary and Future Perspective *176*
References *176*

2.5 Emerging Field – Synthesis of Complex Carbohydrates. Case Study on HMOs *179*
Dora Molnar-Gabor, Markus J. Hederos, Sebastian Bartsch, and Andreas Vogel
2.5.1 Introduction to Human Milk Oligosaccharides (HMOs) *179*
2.5.1.1 Discovery and Function of HMOs *179*
2.5.1.2 Structure of HMOs *180*
2.5.1.3 HMO Production, Regulatory Authorizations, and Commercial Launch – Historical Overview *181*
2.5.2 Glycom A/S Technologies Toward Commercial HMO Production *184*
2.5.2.1 Whole Cell Microbial Fermentation to HMOs (*In Vivo* Process) *185*
2.5.2.2 The Glycom *In Vitro* Concept to Diversify HMO Blends *187*
2.5.2.3 Validation of the HMO Diversification Concept with Non-optimized Enzymes *187*
2.5.3 Enzyme Development *189*
2.5.3.1 Optimization of the α1-3/4 Transfucosidase *189*
2.5.3.2 Optimization of the α2-6 Transsialidase *192*
2.5.4 Applications of the Optimized Enzymes for the HMO Profiles *195*
2.5.4.1 Scale-Up of the Lacto-*N*-fucopentaose III (LNFP-III), Sialyl Lacto-*N*-neotetraose (LST-c), and Sialyl Lacto-*N*-tetraose (LST-a) HMO Profiles *195*
2.5.5 Conclusion and Perspective *197*
References *198*

Part III Enzyme Applications for Human and Animal Nutrition *203*

3.1 Enzymes for Human Nutrition and Health *205*
Yoshihiko Hirose
3.1.1 Introduction *205*
3.1.2 Current Problems of Enzymes in Healthcare Business *205*
3.1.3 Enzymes in Existing Healthcare Products *206*
3.1.3.1 Digestive Enzymes *206*

3.1.3.1.1 Digestive Enzymes in United States *206*
3.1.3.1.2 Therapeutic Digestive Enzymes *207*
3.1.3.2 Acid Lactase *207*
3.1.3.3 α-Galactosidase (ADG) *208*
3.1.3.4 Dextranase *208*
3.1.3.5 Glucose Oxidase *208*
3.1.3.6 Acetobacter Enzymes *210*
3.1.3.7 Laccase (Polyphenol Oxidase) *210*
3.1.4 New Enzyme Developments in Healthcare Products *211*
3.1.4.1 Transglucosidase *211*
3.1.4.2 Laccase *211*
References *215*

3.2 Enzyme Technology for Detoxification of Mycotoxins in Animal Feed *219*
Dieter Moll
3.2.1 Introduction to Mycotoxins *219*
3.2.2 Mycotoxin Mitigation Strategies *220*
3.2.3 Enzyme Applications *224*
3.2.4 FUM*zyme*® *225*
3.2.4.1 The Substrate: Fumonisins *225*
3.2.4.2 Enzyme Discovery *227*
3.2.4.3 Enzyme Selection *230*
3.2.4.4 Enzyme Activity Assays *232*
3.2.4.5 Enzyme Characterization and Evaluation *233*
3.2.4.6 Enzyme Feeding Trials and Biomarker Analysis *234*
3.2.4.7 Enzyme Engineering *237*
3.2.4.8 Enzyme Production *238*
3.2.4.9 Enzyme Registration *239*
3.2.5 Future Mycotoxinases *240*
3.2.6 Conclusions *242*
References *243*

3.3 Phytases for Feed Applications *255*
Nikolay Outchkourov and Spas Petkov
3.3.1 Phytase As a Feed Enzyme: Introduction and Significance *255*
3.3.2 Historical Overview of the Phytase Market Development *256*
3.3.3 From Phytate to Phosphorus: Step by Step Action of the Phytase *259*
3.3.3.1 Properties of Phytate *259*
3.3.3.2 Phytases Structural and Functional Classification *260*
3.3.3.2.1 Phytases from the Histidine Acid Phosphatases (HAP) Superfamily *261*
3.3.3.2.2 β-Propeller Phytase (BPP) *261*
3.3.3.2.3 Cysteine Phytase (CPhy) *263*
3.3.3.2.4 Purple Acid Phytases (PAPhy) *263*
3.3.3.2.5 Classification of the Phytases Based on Phytate Dephosphorylation Steps *263*

3.3.4 Nutritional Values of Phytase in Animal Feed *265*
3.3.5 Phytase Application As Feed Additive *265*
3.3.6 Effective Phytate Hydrolysis in the Upper Digestive Tract of the Animal *266*
3.3.7 Kinetic Description of Ideal Phytases *269*
3.3.8 Resistance to Low pH and Proteases *271*
3.3.9 Temperature Stability *271*
3.3.10 In lieu of Conclusion: Lessons from Phytase Super Dosing Trials *274*
References *275*

Part IV Enzymes for Biorefinery Applications *287*

4.1 Enzymes for Pulp and Paper Applications *289*
Debayan Ghosh, Bikas Saha, and Baljeet Singh
4.1.1 Refining and Fiber Development Enzyme *290*
4.1.1.1 Microscopic Evaluation *291*
4.1.1.2 Evaluation of Enzyme-Treated Handsheets *293*
4.1.1.2.1 Case Study 1 *293*
4.1.1.2.2 Case Study 2 *295*
4.1.2 Drainage Improvement Enzyme *296*
4.1.2.1 Case Study 3 *299*
4.1.2.2 Case Study 4 *300*
4.1.3 Stickies Control Enzyme *301*
4.1.3.1 Case Study 5 *303*
4.1.4 Deinking Enzymes *306*
4.1.4.1 Case Study 6 *307*
4.1.5 Hardwood Vessel Breaking Enzyme *308*
4.1.5.1 Fiber Tester Image Analysis *308*
4.1.6 Native Starch Conversion Enzyme *310*
4.1.7 Bleach Boosting Enzyme *312*
4.1.7.1 Common Bleaching Agents *312*
4.1.7.1.1 Case Study 7 *313*
4.1.7.2 Overcoming Challenges Faced by Bleaching Enzymes in Pulp and Paper industry *315*
4.1.8 Paper Mill Effluent Treatment Enzymes *315*
4.1.8.1 Case Study 8 *316*
4.1.9 Slushing Enzyme *317*
4.1.9.1 Case Study 9 *317*
4.1.9.2 Role of Enzymes in Pulp and Paper Industry – End Note! *318*
References *319*

4.2 Enzymes in Vegetable Oil Degumming Processes *323*
Arjen Sein, Tim Hitchman, and Chris L.G. Dayton
4.2.1 Introduction *323*
4.2.2 General Seed Oil Processes *324*
4.2.2.1 Phospholipids *325*

4.2.2.2 A Molecular View of the Degumming Process *327*
4.2.3 Enzymatic Degumming *330*
4.2.3.1 Phospholipase C *331*
4.2.3.2 Ways to Cope with Poor Conversion/Poor Quality Oils in PLC-Based Processes *333*
4.2.3.3 Phospholipase A *336*
4.2.4 Enzymatic Degumming in Industrial Practice *337*
4.2.4.1 Introduction Hurdles *341*
4.2.5 Other Applications of Enzymes in Oil – Outlook *343*
4.2.5.1 Enzymatic Interesterification of Triglyceride Oils *343*
4.2.5.2 Biodiesel *344*
4.2.5.3 Enzyme-Assisted Decoloring *344*
4.2.5.4 Enzyme-Assisted Oil Extraction *344*
4.2.6 Conclusion *345*
Acknowledgments *345*
References *345*

Part V Enzymes used in Fine Chemical Production *351*

5.1 KREDs: Toward Green, Cost-Effective, and Efficient Chiral Alcohol Generation *353*
Chris Micklitsch, Da Duan, and Margie Borra-Garske
5.1.1 Introduction *353*
5.1.2 Ketoreductases *355*
5.1.3 Cofactor Recycling *356*
5.1.4 CodeEvolver® Protein Engineering Technology *358*
5.1.5 Reduction of a Wide Range of Ketones/Aldehydes *358*
5.1.6 Critical Selectivity Tools for Enantiopure Asymmetric Carbonyl Reduction *364*
5.1.7 Examples of Improved KREDs for Improved Manufacturing *369*
5.1.8 KREDs: Going Green and Saving Green *373*
References *377*

5.2 An Aldolase for the Synthesis of the Statin Side Chain *385*
Martin Schürmann
5.2.1 Introduction – Biocatalysis *385*
5.2.1.1 Enzymes as Biocatalysts in Chemical Process *385*
5.2.1.2 Biocatalytic Routes to the Statin Side Chain *387*
5.2.2 The Aldolase DERA in Application *387*
5.2.2.1 DERA-Catalyzed Aldol Reactions *387*
5.2.2.2 Feasibility Phase of DERA-Enabled Statin Side Chain Process *390*
5.2.3 Directed Evolution and Protein Engineering to Improve DERA *392*
5.2.3.1 Rational Design *392*
5.2.3.2 Directed Evolution of DERA *394*
5.2.3.3 Other Approaches to Suitable or Improved DERAs *396*

5.2.3.4 Other Applications of Process Intermediates and the DERA Technology *397*
5.2.4 Conclusions *398*
Acknowledgments *400*
References *401*

Index *405*

Preface

Industrial enzyme applications are part of our everyday life since mankind discovered the benefits of transforming milk, grapes, and grains to durable, palatable, more tasteful products such as yoghurt, wine, beer, and bread. At that time, the microorganisms and enzymes that led to this transformation were used unconsciously. Since then, our knowledge about enzymes, and the development of technologies on how to adapt, produce, and apply them has improved dramatically, especially in the last few decades. This has led to a wide spectrum of enzyme applications that we come across in products from the food and drink, chemical, pharmaceutical, biorefinery, and the human and animal nutrition industry.

This book shall direct the perspective to a variety of enzyme types and their applications that are actually used in industrial processes. When we conceived this book, we asked the following questions: Which solutions can enzymes provide to address the needs of industries and consumers? Which enzymes provide which solutions, and why have these particular enzymes been chosen and are successfully competing with other solutions? What was the decisive advantage of the use of enzymes over the competitive process? What enzyme features were required and how were these obtained during enzyme development?

The book is written by experts from different industries who develop and apply enzymes in their teams to address various problems: from enabling "simple" cost reduction and process innovations for existing products (which often goes hand in hand with ecologic benefits) to product innovations, developing new products that provide additional value (e.g. sustainability, purity, nutritional) for consumers. We are very happy that we could attract authors from various industries that were willing to share insights into their work. Writing publications from an industrial position is often not the prime priority and is conflicted by company policies. Particularly for this reason, the contributions in this book provide unique and individual insights into industrial drivers and strategies for the implementation of enzymatic processes.

This textbook on Industrial Enzyme Application will serve as a reference guide for academic and industrial researchers and provides a unique, industrial perspective on the development and application of enzymes. It will help understand the industrial drivers in searching for and developing an enzymatic solution.

We also hope the described examples inspire students and practitioners to develop new applications, providing our world with more sustainable solutions that enzymes are perfectly capable of providing.

c-LEcta, Leipzig
DSM, Kaiseraugst
April 2019

Andreas Vogel
Oliver May

Part I

Overview of Industrial Enzyme Applications and Key Technologies

1.1
Industrial Enzyme Applications – Overview and Historic Perspective

Oliver May

DSM Nutritional Products, Wurmisweg 576, 4303 Kaiseraugst, Switzerland

The field of industrial enzyme applications has been extensively reviewed, including its historic background [1–4]. Therefore this chapter will not repeat the excellent work of others but provide historic context in the following chapters and give a general overview of the field, including topics that we have chosen not to cover in detail in this book. It is also meant as a tribute to the many pioneering scientists that enlightened us with their brilliant minds about the miracles of life and how enzymes work (see Section 1.1.2) and to the visionary entrepreneurs shaping a growing multibillion enzyme business (see Section 1.1.3).

While we have come a long way in unraveling how enzymes work, we have still not captured all the details on how they achieve the huge rate acceleration of the reactions they catalyze, nor have we explored exhaustively their full potential in existing as well as new application fields. This leaves room for us and future generations standing on the shoulders of giants some of which are mentioned in the following, or are hidden in the prehistoric world that applied enzymes, even without knowing they existed.

1.1.1 Prehistoric Applications

Enzymes played an important role early in the history and development of humanity. The Neolithic Revolution of around 12 500 years ago marked the transition of lifestyles from hunting and gathering to agriculture and settlement. With this transition, farming practices were invented, leading to domestication of plants and animals. While safe storage of hunted or gathered food was certainly already important during the Neolithic times, the unconscious use of microbes and the action of their enzymes allowed to preserve the food and supply it to others who could focus on other activities outside the primary production of food. Certainly, people also enjoyed improved palatability or a desirable taste of food that was in contact with microorganisms that contributed to these desired properties.

While no one can really determine the exact date when the first products were made in which enzymes from microbes, plant, or animal tissues played such a

Industrial Enzyme Applications, First Edition. Edited by Andreas Vogel and Oliver May.

beneficial role, the first scientifically proven evidence, based on residues found in pottery vessels for cheese making, dates back to 7500 years ago [5]. What a great invention to preserve milk without a fridge and make it palatable as well as enjoyable! Today's enzyme applications in dairy are covered in Chapter 2.3 and the achievements of Christian Hansen, an entrepreneur starting a very successful enzyme and starter culture business in 1874, are described in Section 1.1.3.

The first indication of enzyme-assisted grain processing to produce an alcoholic beverage was found in the Neolithic village of Jiahu in China and dates back to 7000 BCE [6]. Based on chemical analyses of organics absorbed into ancient pottery the authors have shown that a mixed fermented beverage of rice, honey, and fruit (hawthorn fruit and/or grape) was being produced. The earliest proof of wine production dated to 5400–5000 BCE, at the Neolithic site of Tepe in Mesopotamia [7] where tartaric acid was found in an old jar, and around 5000 BCE from grape juice residues found in Dikili Tash in Greece [8]. For over 2500 years *Aspergillus* strains have been extensively used in China as starter cultures in grain (soy, rice) fermentation, a traditional practice for production of rice wine (sake) or other distilled products (shochu), which were imported from Japan by Buddhist monks [9]. The Japanese word Koji still uses the Chinese character (麹) that means (wheat) grains fermented by fungi. Today, grain processing and malt production are separate industries whereas in the Western world malt production can be considered as the first sector that industrialized enzyme production (see also Section 1.1.3).

Next to being used for fermentative processes, Koji has been used as digestive aid, as first described 2500 years ago in a Chinese classic book entitled "Zuo-Zhuan," in the English Chronicle of Zuo or Commentary of Zuo [10]. In the description, wheat-based Koji was used to treat digestion problems. This tradition was later turned into the first industrial application of a fungal enzyme by Jokichi Takamine, one of the pioneering entrepreneurs discussed in Section 1.1.3 and who also inspired the application of enzymes for nutrition and health, reviewed by Yoshihiko Hirose in Chapter 3.1. The good old tradition of fermented food is carried through to date and is estimated to provide about 20–40% of our food supply today [11].

The Egyptians, who already used yeast to brew beer, began to employ microorganisms as well as malt to make bread for which samples were found in different archeological sites dating to 2000–1200 BCE [12]. This ancient malt application in baking developed into a major field of enzyme application, which is reviewed by Putseys and Schooneveld in Chapter 2.1.

However, it is not only food that has benefited from enzymes long before their existence was known. One of the first technical (materials) applications can be found in leather processing, which provided ancient civilization with leather for water skins, bags, boats, or shoes as early as 7000 to 3300 BCE [13]. In this traditional process, the bating step that softens the material was a fermentative process. It relied on enzymes produced by bacteria found in pigeons or dog dung, which was added in this step. The replacement of dung by enzymes was a huge improvement that was started by Otto Röhm, whose pioneering work and the foundation of his enzyme company are described in Section 1.1.3.

Enzyme applications in organic synthesis (see Chapters 5.1 and 5.2), feed (see Chapters 3.2 and 3.3), textiles, detergents, and in biorefineries (see Chapters 4.1–4.2) obviously do not root back to ancient times and are inventions of the twentieth century, which are further discussed in Section 1.1.3. More comprehensive reviews of detergent, textile, and biofuel applications can be found here [14–17].

1.1.2 Growing the Scientific Basis

Why does science matter if people already enjoyed the benefits of enzymes for thousands of years as described above? Well, while humankind indeed enjoyed the desired effects of enzymes such as reducing spoilage of valuable raw materials, creating an appealing taste, or supporting better digestion of food products, the desired effects sometimes just did not happen, or turned in highly undesired directions. Science was therefore needed to reduce or prevent such failures. Applying the insight science provided was not only making existing applications more robust. Science also enabled the discovery, development, and efficient production of new enzymes with desired properties, which opened up new opportunities to broaden their application field. The following section highlights key achievements of many brilliant minds in history that created the scientific foundation on which today's commercial success is built.

"Seeing is believing," and seeing microorganisms was not possible before the pioneering Dutch scientist and gifted craftsman Antoni van Leeuwenhoek developed an analytical instrument with so far unprecedented magnifying power. With his specially prepared lenses he created simple but still very powerful microscopes that could reach a 300-fold magnification. With these microscopes, he could not only see living microbes, which he originally referred to as animalcules (from Latin *animalculum* = "tiny animal"), but also already made observations in 1675 that brought him very close to recognizing which "magic forces" are behind some of the desired effects created by microorganisms and their enzymes. Figure 1.1.1 shows a copy of his famous letter to the Royal Society, which was translated in English by Henry Olden, the Editor of the Journal Philosophical Transactions of the Royal Society, where he not only disclosed the discovery of "living creatures" but also commented on the "bubbling water" comparing it with fermenting beer [18].

Leeuwenhoek decided not to share the details of preparing the lenses with the scientific community and took this secret with him when he died in 1723. Because he kept his knowledge secret, his discoveries were doubted or even dismissed during the following century as other scientists could not reproduce his results.

After the discovery of living microbes, another important achievement was reported by Anselme Payen and Jean-Francois Persoz. The French chemists were working at a sugar factory to improve the process of starch conversion. In 1833, Payen and Persoz reported that an alcohol precipitate of malt extract (probably the oldest and highest volume industrial enzyme product) contained a substance that converted starch into sugar [19]. They named the substance "diastase" after the Greek word διάστασις (*diastasis*) (a parting, a separation) as it caused

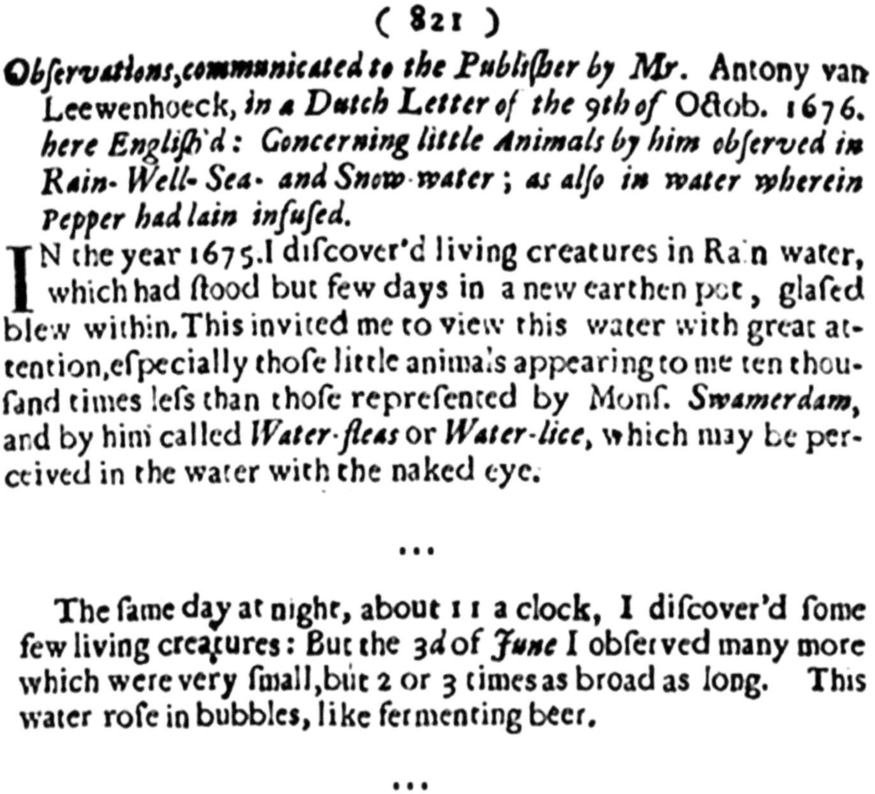

(821)

Obſervations, communicated to the Publiſher by Mr. Antony van Leewenhoeck, *in a Dutch Letter of the 9th of* Octob. 1676. *here Engliſh'd: Concerning little Animals by him obſerved in Rain- Well- Sea- and Snow-water; as alſo in water wherein Pepper had lain infuſed.*

IN the year 1675. I diſcover'd living creatures in Rain water, which had ſtood but few days in a new earthen pot, glaſed blew within. This invited me to view this water with great attention, eſpecially thoſe little animals appearing to me ten thouſand times leſs than thoſe repreſented by Monſ. *Swamerdam*, and by him called *Water-fleas* or *Water-lice*, which may be perceived in the water with the naked eye.

...

The ſame day at night, about 11 a clock, I diſcover'd ſome few living creatures: But the 3d of *June* I obſerved many more which were very ſmall, but 2 or 3 times as broad as long. This water roſe in bubbles, like fermenting beer.

...

Figure 1.1.1 Excerpt of Leeuwenhoek famous letter to the Royal Society, which was translated in English by Henry Olden, the Editor of the Journal Philosophical Transactions of the Royal Society.

the starch in the barley seed to transform quickly into soluble sugars and hence allowed separation of the husk from the rest of the seed. They were also the first to propose a nomenclature for such substances using the suffix "ase" after the root that indicates which substance was modified by the biomolecule. This was the beginning of a systematic enzyme nomenclature that is used to date, with the exception of proteolytic enzymes ending with "in," such as subtilisin. Today, "diastase" refers to any α-, β-, or γ-amylase (EC 3.2.1.1-3) that can break down carbohydrates and are applied in many different industrial applications. An overview of several enzymes based on the EC nomenclature and their application fields, some of which are further discussed below, is provided in Table 1.1.1. It is perhaps interesting to note that these fundamental discoveries, which inspired many innovations, were made in the context of industrial research.

Soon after the first preparation of diastase from a plant source, additional enzyme classes were isolated. Important examples were proteases (EC 3.4.X.X), which hydrolyze proteins and peptides. In 1836, Theodor Schwann studied the process of human digestion and succeeded with isolating an enzyme that he called pepsin [20]. This is the first enzyme prepared from animal tissue, which

Table 1.1.1 Overview of enzyme classes used in different applications.

EC number	Enzyme name	Application	Function
Oxidoreductases			
1.1.3.4	Glucose oxidase	Baking	Gluten modification/dough strengthening
		Brewing	Oxygen reduction/shelf life improvement
		Dairy	Milk preservation
		Textile	Bleaching
1.1.3.5	Hexose oxidase	Baking	Gluten modification/dough strengthening
1.10.3.2	Laccase	Pulp and paper	Bleaching
		Textile	Bleaching of dye to prevent backstaining
		Various	Cork treatment
		Various	Polymerization of lignin for production of wood fiberboards
1.11.1.6	Catalase	Brewing	Shelf live improvement
		Dairy	Milk preservation
		Textile	Hydrogen peroxide removal
		Various	Wastewater treatment
1.11.1.7	Peroxidase	Baking	Dough improvement
1.13.11.12	Lipoxygenase	Baking	Whitening of breadcrumb
Transferases			
2.3.2.13	Transglutaminase	Dairy	Texture improvement
2.4.1.5	Dextransucrase	Various	Production of dextrans
Hydrolases			
3.1.1.3	Triacylglycerol lipase	Baking	Bread improvement
		Dairy	Modification of cheese flavor
		Detergent	Greasy stain removal
		Pulp and paper	Pitch removal
3.1.1.11	Pectin methylesterase	Beverage	Yield increase for fruit (berry, apple) juices, citrus fruit peeling
3.1.1.26	Galactolipase	Baking	Emulsification
3.1.3.8	3-Phytase	Beverage	Mashing in brewing
		Feed	Phosphate release

(Continued)

Table 1.1.1 (Continued)

EC number	Enzyme name	Application	Function
3.1.3.26	6-Phytase	Beverage	Mashing in brewing
		Feed	Phosphate release
3.2.1.1	α-Amylase	Baking	Antistaling
		Beverage	Mashing (brewing), apple juice production
		Detergent	Removal of starch containing stains
		Starch processing	Starch hydrolysis for syrup production
		Feed	Degradation of starch
		Various	Viscosity reduction in oil drilling
		Textile	Desizing
3.2.1.2	β-Amylase	Beverage	Mashing (brewing)
3.2.1.3	Glucoamylase	Beverage	Mashing (brewing), apple juice production
		Starch processing	Hydrolysis of maltooligosaccharide for syrup production
		Various	Toothpaste
3.2.1.4	Endo-1,4-β-glucanase	Detergent	Softening, color improvement
		Textile	Cotton finishing, denim ageing
3.2.1.6	Endo-1,4(3)-β-glucanase	Feed	Increased feed efficiency
3.2.1.8	Endo-1,4-β-xylanase	Baking	Improved dough stability/handling
		Feed	Increased feed efficiency
		Textile	Flax retting
		Pulp and paper	Pulp bleaching
3.2.1.55	Arabinosidase	Beverage	Apple juice production
3.2.1.60	Endo-1,4-β-mannanase	Detergent	Removal of guar gum containing stains
		Textile	Flax retting
		Pulp and paper	Pulp bleaching
		Various	Viscosity reduction in oil drilling
3.2.1.91	Exo-cellobiohydrolase	Detergent	Softening, color improvement
		Textile	Cotton finishing
		Pulp and paper	Mechanical pulping

(Continued)

Table 1.1.1 (Continued)

EC number	Enzyme name	Application	Function
3.4.21.26	Prolyl oligopeptidase	Beverage	Improve stability of beer
		Supplement	Reduction of allergen
3.4.21.62	Subtilisin	Detergent	Removal of proteinaceous stains
		Feed	Improved feed efficiency
		Various	Membrane cleaning
3.4.21.63	Oryzin	Feed	Improved feed efficiency
3.4.23.4	Chymosin	Dairy	Cheese clotting
3.4.23.18	Aspergillopepsin I	Feed	Improved feed efficiency
3.4.23.22	Endothiapepsin	Dairy	Cheese clotting
3.4.23.23	Mucorpepsin	Dairy	Cheese clotting
3.4.24.28	Bacillolysin	Feed	Improved feed efficiency
Lyases			
4.1.1.5	α-Acetolactate decarboxylase	Beverage	Diacetyl removal for beer flavor enhancement
4.2.2.2	Pectate lyase	Textile	Cotton scouring
4.2.2.10	Pectin lyase	Beverage	Fruit juice production, citrus peeling
Isomerases			
5.3.1.5	Xylose isomerase	Starch processing	Conversion of glucose into fructose for high fructose corn syrup production
5.3.4.1	Protein disulfide isomerase	Various	Hair waving
Ligases			
6.3.2.28	Dipeptide ligase	Bioconversion	Production of dipeptides

Source: Aehle 2007 [1]. Adapted with permission of John Wiley & Sons.

found early applications in the leather industry to remove hair and residual tissue from animal hides prior to tanning (a process that produces leather from hides). It was also used in the recovery of silver from discarded photographic films by digesting the gelatin layer that holds the silver compound [21]. Another important contribution of Schwann was his recognition that "All living things are composed of cells and cell products," which became the foundation of what is known today as the cell theory [22].

Probably one of the most influential scientists of the nineteenth century was Luis Pasteur. He is best known to the public for his invention of the heat treatment technique of, for example, milk and wine to stop bacterial contamination, a process now called pasteurization. He was also very influential in the field of medicine by recognizing that partly inactivated microbes can result in immunity, which laid the foundation for vaccination that saved millions of people's lives [23]. Furthermore, Pasteur also laid the foundation for our understanding of molecular asymmetry (chirality). He was separating different crystal shapes from each other to form two piles of tartaric crystals: in solution, one form of these crystals rotated light to the left and the other to the right, while an equal mixture of the two forms canceled each other's effect and did not rotate the polarized light. He further observed that a solution of this tartaric acid derived from living things rotated the plane of polarization of light passing through it, whereas tartaric acid derived by chemical synthesis had no such effect although its elemental composition was the same. Today, we know that this phenomenon is attributed to the specific properties of enzymes. They can induce chirality of reacting with prochiral compounds or discriminate different enantiomers, which is a valuable property of enzymes for the synthesis of enantiomerically pure compounds as reviewed in Chapters 5.1 and 5.2.

Pasteur also studied alcoholic fermentation where he proved that living microorganisms (yeast) are indeed responsible for fermentation: he has shown that the skin of grapes was the natural source of yeasts and that sterilized grapes and grape juice never fermented. In 1862, when studying the fermentation of sugar to alcohol by yeast, Louis Pasteur along with Ferdinand Cohn and Robert Koch came to the conclusion that this fermentation was catalyzed by a "vital force" contained within the yeast cells, which were called "ferments" [24]. He then moved on to investigate the conversion of alcohol into vinegar and concluded that pellicle, which he called "the flower of vinegar," served as a method of transport for the oxygen in air to a multitude of organic substances, laying the foundation for applying biocatalysts in chemical syntheses [25]. In 1886, Brown confirmed Pasteur's findings and gave the causative agent in vinegar production the name *Bacterium xylinum*. Brown also found that this bacterium could oxidize propanol to propionic acid and mannitol to fructose [26]. Such oxidative biotransformation steps were later applied in the production of L-ascorbic acid (vitamin C) as described in Section 1.1.3.

Wilhelm Friedrich Kühne, a professor of physiology at the University of Heidelberg, was the first to propose the term "Enzyme" for "ferments" in a publication from 1876 as shown in Figure 1.1.2 [27]. Despite the fact that he was working on digestive enzymes (trypsin) from pancreas, the word enzyme is derived from a Greek word "ενζυμον" meaning "in yeast."

The connection between the function of fermenting yeast and enzymes was finally proved by Eduard Buchner in 1897. He studied the ability of yeast extracts to ferment sugar in the absence of living yeast cells. In a series of experiments at the University of Berlin, he found that sugar was fermented even when there were no living yeast cells in the mixture and named the enzyme (mix) that brought about the fermentation of sucrose "zymase" [28]. In 1907, he received the Nobel Prize in Chemistry "for his biochemical research and his discovery of cell-free fermentation."

Once the existence of enzymes was shown, research began to provide detailed insights into their unique properties and functions. The fact that enzymes are very specific can be concluded from Pasteur's discovery of chirality. While he observed the effect, Pasteur did not yet make the direct link between the phenomenon of chirality induced by living microorganisms and the specificity of enzymes that were discovered later, as just described. A starting point to

Ueber das Verhalten verschiedener organisirter und sog. ungeformter Fermente.

Sitzung am 4. Februar 1876.

Hr. W. Kühne berichtet über das Verhalten verschiedener organisirter und sog. ungeformter Fermente. Um Missverständnissen vorzubeugen und lästige Umschreibungen zu vermeiden schlägt Vortragender vor, die ungeformten oder nicht organisirten Fermente, deren Wirkung ohne Anwesenheit von Organismen und ausserhalb derselben erfolgen kann, als *Enzyme* zu bezeichnen. — Genauer untersucht wurde besonders das Eiweiss verdauende Enzym des Pankreas, für welches, da es zugleich Spaltung der Albuminkörper veranlasst, der Name *Trypsin* gewählt wurde. Das Trypsin vom

Figure 1.1.2 Excerpt of Kühne's original publication from 1876 where he introduced the word enzymes for the first time.

describe enzyme specificity was provided by Emil Fischer, who suggested in 1894 that enzymes and their substrates possess specific complementary geometric shapes that fit exactly into one another. This model is referred to as "the lock and key" model [29]. While this model can be used to explain enzyme specificity, it fails to explain one of the core functions of enzymes as a catalyst, which is the rate acceleration obtained by stabilization of a transition state. In 1958, Daniel Koshland suggested a modification to the lock and key model, which is known as the "induced fit" model: since enzymes have a rather flexible structure, the active site is continuously reshaped by interactions with the substrate [30]. This brought us closer to an explanation for how enzymes function but a real breakthrough in our understanding was achieved by enzyme structure determination.

Lysozyme was the first enzyme crystallized by Edward Penley Abraham and Robinson in 1937 [31] enabling many years later the elucidation of the three-dimensional structure of hen egg white lysozyme described by David Chilton Phillips in 1965. He obtained the first 2-Å resolution model via X-ray crystallography [32]. As a result of Phillips' determination of the structure of lysozyme, it was also the first enzyme to have a detailed, specific mechanism suggested, which provided an explanation for how enzymes speed up a chemical reaction.

Starting with the crystallographic data of Rosalind Franklin [33]; the elucidation of the DNA structure in 1953 [34] by, and the awarding of the Nobel Prize to, Watson, Crick, and Wilkins [35]; the unraveling of the genetic code for which Marshall Warren Nirenberg, Har Gobind Khorana, and Robert William Holley received the Nobel Prize in 1968 [36]; and the introduction of recombinant DNA technology in 1973 by Cohen et al. [37], the field of molecular biology revolutionized the cost-efficient production of enzymes, which is described in Chapter 1.4. It also provided efficient tools for enzyme discovery and engineering, which are further discussed in Chapter 1.2.

Building on the above-described work and many more scientific achievements, the following section highlights the pioneering work of entrepreneurs that marked the beginning of the industrial application of enzymes and the development of a multibillion Euro enzyme industry.

1.1.3 The Beginning of Industrial Applications and the Emerging Enzyme Industry

Looking back at the ancient enzyme applications and the scientific history described above, it is not surprising that the first industrial applications of enzymes did not make use of isolated enzymes but crude extracts (such as malt extracts), or even whole cell biocatalysts.

This is also true for the synthesis of chemical compounds in the so-called **biotransformation processes**. Vinegar production is perhaps the oldest and best known example of such an industrial scale biotransformation process in which oxidation of ethanol is catalyzed by acetic acid bacteria. Another pioneering example of an industrial biotransformation process is the production of L-(−)ephedrine, which was introduced by Knoll AG (today BASF) in the 1930s [38]. The process is based on the discovery of Neuberg and Hirsch who showed that in presence of yeast, benzaldehyde can condense with acetaldehyde to optically active 1-hydroxy-1-phenyl-2-propanone, which is then chemically converted to L-(−)ephedrine [39]. This was the first example of an enzymatic C–C coupling reaction at an industrial scale. More recent examples of industrial applications of aldolases catalyzing C–C coupling reactions are reviewed in Chapter 5.2.

Kluyver and de Leeuw, two Dutch scientists from Delft (NL), demonstrated in 1924 that *Acetobacter suboxydans* can oxidize D-sorbitol to L-sorbose [40]. This became an important intermediate in the Reichstein–Grüssner synthesis of L-ascorbic acid (vitamin C) [41]. This synthesis was turned into an industrial process by Roche (Roche Vitamins, now part of DSM) in the 1930s [42].

In 1953, Peterson et al. reported that *Rhizopus arrhius* can convert progesterone into 11-α-hydroxyprogesterone, which was used as an intermediate in the synthesis of cortisone [43]. This microbial hydroxylation simplified and considerably improved the efficiency of the multistep chemical synthesis of corticosteroid hormones and their derivatives. Although the chemical synthesis from deoxycholic acid developed at Merck (Germany) was in principle possible, it was too complicated and uneconomical: 31 steps were necessary to obtain 1 kg of cortisone acetate from 615 kg of deoxycholic acid. The introduction of microbial 11-α-hydroxylation of progesterone dramatically reduced the price of cortisone from \$200 to \$6 per gram. Further improvements have led to an estimated price of less than \$1 per gram [44].

Another remarkable example of applying a whole cell biocatalyst is the production of acrylamide. In the mid-1970s, Nitto Chemical Industries (now Mitsubishi Rayon) introduced its biocatalytic production using a nitrile hydratase in the form of *Rhodococcus* cells. This process is considered a milestone in industrial biocatalysis demonstrating the impressive efficiency of enzyme systems operating at product concentrations of up to 700 g/l for the production of a low-cost commodity chemical. It also demonstrates that enzymes can be applied in a very crude form as whole cell biocatalyst, providing a more cost-competitive solution. The enzyme process is also a "greener" solution by consuming less energy and avoiding heavy metal wastewater problems of the alternative chemical process [45].

At the end of the 1970s Toyo Jozo (Japan) in collaboration with Asahi Chemical Industry (Japan) pioneered the industrial production of 7-aminocephalosporanic acid (7-ACA) by a chemo-enzymatic two-step process starting from cephalosporin C. In the 1990s developments from Gist-Brocades applied metabolic as well as protein engineering, which enabled further breakthroughs not only for the production of cephalosporin C but also for the chemo-enzymatic synthesis of several derivatives as reviewed in Chapter 5.1.

In the early 1980s another pioneering process was introduced by Degussa (now Evonik Industries). In collaboration with Prof. Kula and Prof. Wandrey a racemic resolution process was developed for production of enantiomerically pure amino acids (e.g. L-methionine) applying an L-amino acylase (EC 3.5.1.14) from *Aspergillus oryzae* in an enzyme membrane reactor (EMR). Since then, many other racemic resolution processes have been developed [46]. A more recent example using a non-animal-derived pig liver esterase is reviewed in Chapter 5.1.

The same group from Jülich and Degussa that developed the abovementioned EMR process also introduced the first reductive amination process in the 1990s for the production of L-*tert*-leucine and L-neopentylglycine. In this process, a leucine dehydrogenase (EC 1.4.1.9) is used for reductive amination of a keto acid together with a formate dehydrogenase (EC 1.2.1.2) for recycling the cofactor NADH. More recent examples of applying dehydrogenases are presented in Chapter 5.2. A more comprehensive overview of enzymes used in organic synthesis and industrial biotransformations is provided here [2, 3]. While the above-described application field of enzymes delivers highly valuable solutions for challenging synthesis problems for the chemical, fine chemical, and the connected pharma field, its impact on growing the total enzyme business is still very limited, as shown in Figure 1.1.3.

In the early history of **brewing**, malt production was an integrated part of the brewing process. In the malting step, the water content and temperature of grains from malt, sorghum, or wheat were adjusted to allow them to germinate. During this germination process, grains that already contain enzymes (e.g. β-amylases) produce additional enzymes such as α-amylases. After stopping the germination process by a heating step, the malt product is obtained and contains a mix of enzymes that convert the starch of the grains into a fermentable sugar solution. During the last centuries, this malting step was separated from the brewing process, making malt probably the first commercial and to date the highest volume

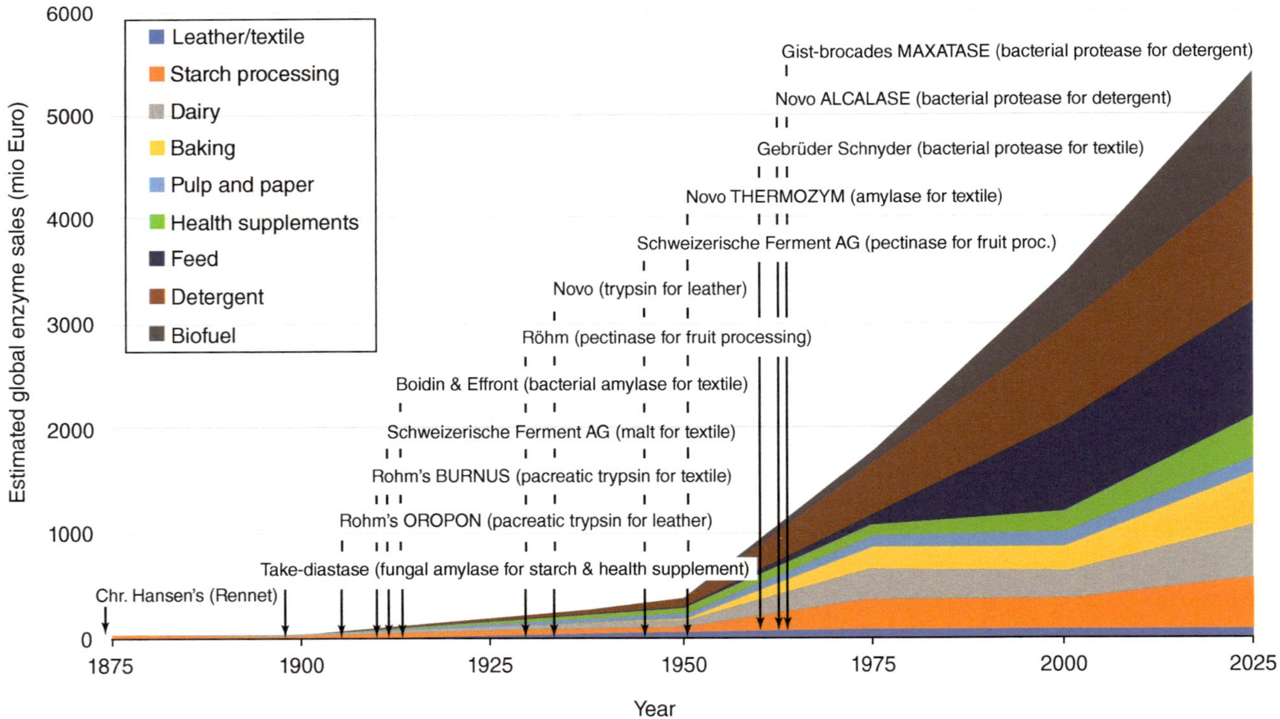

Figure 1.1.3 Historic events and estimated growth of enzyme business from 1875 until 2025, broken down into different application fields.

enzyme product. Its success can be attributed to the fact that it enabled an easier, faster, and more consistent brewing process. However, the malt enzymes, first isolated by Payen and Persoz in 1833 as described in the previous section, do have some limitations. They can only work at certain temperatures, pH values, or contain unwanted side activities giving rise to many enzyme innovations for the beverage industry, which is covered in Chapter 2.2. Access to standardized and relatively pure enzyme preparations was also an important factor to introduce amylases in the baking industries, which started to be applied in the beginning of the twentieth century with many following innovations covered in Chapter 2.1.

The history of modern enzyme technology really began when Christian Hansen, in a joint venture with pharmacist H.P. Madsen, opened the first enzyme factory in 1873. They started the first Danish production of pepsin preparations for **treating digestive problems**.

Christian Hansen also began working with rennet, the enzyme-rich substance extracted from the fourth stomach of ruminants. These extracts were used since thousands of years for the manufacture of cheese to make the milk coagulate. As demand rose sharply in the nineteenth century, Christian Hansen began a series of experiments on the production of rennet. In 1874 the Chr. Hansen's Teknisk-Kemiske Laboratorium opened, which is, today, known as the company Chr. Hansen. In a converted metal workshop in Copenhagen he began the commercial production of standardized quality-assured, liquid animal rennet for **dairy applications**, which are further described in Chapter 2.3. Just as in the previously described example for the brewing industry, simplification and standardization to reduce process/batch failures with a negative economic impact for the emerging cheese industry was again an important driver and success factor for the introduction of a commercial enzyme product.

In contrast to Western countries, which originally used malt in their brewing processes as described above, Asian countries used the traditional "Koji process." In this process *Aspergillus* strains are used as starter cultures to ferment rice or soy for production of rice wine (sake), soy sauce (shoyu), soybean paste (miso), and distilled spirits (shochu) [47]. Believing that Koji made from *A. oryzae* could revolutionize the American distillery industry and outcompete malt, Jokichi Takamine adapted the Koji process for the production of diastase (amylase), which he patented in 1894. This is reported to be the first patent on a microbial enzyme protecting the production of diastatic enzyme by growing fungi on bran and using aqueous alcohol to extract the enzyme as described in the patent application shown in Figure 1.1.4 [48].

While his initial attempts to introduce microbial diastase in the conservative American brewing industry was not commercially successful, he also realized that this enzyme preparation has potential medical applications. Parke, Davis & Company of Detroit received the license on the amylase and marketed it successfully under the brand name Taka-diastase as a digestive aid to treat dyspepsia, which is caused by incomplete digestion of starch. This example and the pepsin-based product developed by Chr. Hansen can be seen as starting points for the application field of **enzyme supplements**, which are further described in Chapter 3.1 by Hirose. Takamine expanded his business operations in both enzymes and pharmaceuticals and founded three major companies:

UNITED STATES PATENT OFFICE.

JOKICHI TAKAMINE, OF PEORIA, ILLINOIS.

PROCESS OF MAKING DIASTATIC ENZYME.

SPECIFICATION forming part of Letters Patent No. 525,823, dated September 11, 1894.

Application filed February 23, 1894. Serial No. 501,251. (No specimens.)

To all whom it may concern:

Be it known that I, JOKICHI TAKAMINE, a subject of the Emperor of Japan, residing at Peoria, in the county of Peoria and State of Illinois, have invented certain new and useful Improvements in Processes of Making Diastatic Enzyme; and I do hereby declare the following to be a full, clear, and exact description of the invention, such as will enable others skilled in the art to which it appertains to make and use the same.

The object of this invention is to prepare and manufacture diastatic enzyme, or soluble ferment in a concentrated form which possesses the power of transforming starch into sugar for use in various industries, by a process not hitherto practiced, and in a very economical and practical manner.

My invention is based upon the utilization of the property possessed by certain fungi during their growth on proper media of producing diastatic enzyme.

its weight of the spores of the fungus and the mass is then well mixed. Before mixing in the spores, the bran may be steamed with live steam, in order to sterilize the mass, with the object of preparing the enzyme in a purer form. When the bran is not steamed, the product afterward obtained is liable to be contaminated with more of the foreign elements, but the quantity of the enzyme obtained from the unsteamed bran on which the fungi has been grown is larger and consequently the diastatic power is stronger, from the same amount of bran than where steamed bran is used. After thoroughly mixing with the spores of the fungus, the bran is placed in a room with the temperature ranging from 25° to 30° centigrade, and the air saturated or nearly saturated with moisture, the mass is spread out into thin layers of from one to four inches in thickness on the floor or shelves. The thickness of the layers is determined by the nature of the bran, and

Figure 1.1.4 Jokichi Takamine's patent from 1894, the first enzyme patent protecting a process for production of diastase from a fungal source. From US patent office.

Sankyo Pharmaceutical Company of Tokyo, the International Takamine Ferment Company of New York, and the Takamine Laboratory of Clifton, New Jersey. With his death, the International Ferment Company of New York was dissolved. Takamine's son, Eben, continued running Takamine Laboratory. After his death in 1953, Eben's widow sold the company to Miles Laboratories ending up in today's enzyme business of DowDuPont as described in Figure 1.1.5.

Takamine's work on fungal amylases was succeeded by Boidin and Effront, who pioneered the work on bacterial enzymes using surface fermentation. One of their landmark patents was obtained in 1917 claiming the use of amylases in processes above 80 °C and under alkaline conditions, which is advantageous in industrial **starch processing** [49]. A few years later they founded the company Societé Rapidase in Seclin (France) commercializing not only enzymes important to the emerging starch industry but also pioneering enzymes for **textile applications** [50]. Prior to the introduction of enzymes (initially in the form of malt or pancreas extract) textiles were treated with acid, alkali, or oxidizing agents, or soaked in water for several days so that naturally occurring microorganisms could break down the starch that is used as sizing agent. However, both methods were difficult to control and sometimes damaged or discolored the material, which could be prevented by using enzymes. As shown in the overview of companies in Figure 1.1.5, Societé Rapidase was later acquired by Gist-Brocades (now part of DSM).

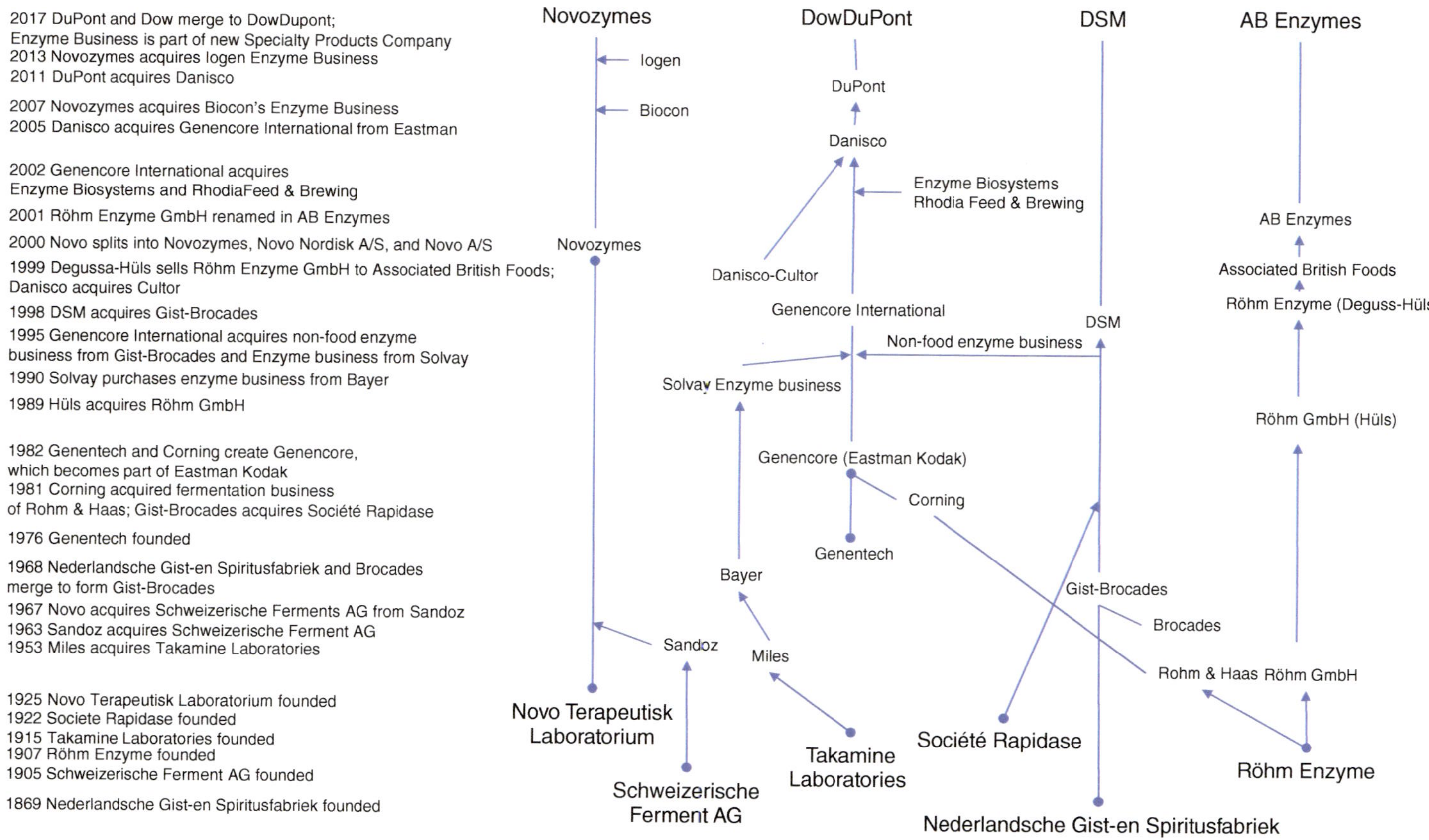

Figure 1.1.5 Overview of company roots starting with pioneering entrepreneurs to todays' world leading enzyme suppliers.

Another pioneering work contributing to important industrial enzyme applications was accomplished by Otto Röhm. According to traditional **leather processing**, bating required the excrement of dogs, a fact that did not improve the image of tanning, which was considered a stinking and unpleasant activity. Röhm's theory was that these excrements exerted their effect by residual amounts of the animals' digestive enzymes. If this was so, he concluded it might be possible to use extracts of the pancreas directly for bating. Such extracts containing the protease trypsin were tested and produced the expected positive results. Röhm accepted these results as confirmation of the correctness of his theory, but later experiments showed that it was not the animals' enzymes that were active, but rather enzymes of bacteria growing in the intestinal tract of the animals. He introduced the application of pancreatic proteases for bating of hides, replacing the use of less defined and less pleasant animal dung in the tanning process [51]. He standardized the enzyme activity and launched the product OROPON™. In 1907, OROPON went into production, and Röhm together with the merchant Otto Haas founded the company Röhm and Haas for commercialization of enzymes and other products. Later, they established a branch in Philadelphia (USA) and started to export enzymes to the United States and in 1911 to Japan. Within only a few years, Röhm replaced the manure bate with his enzyme product. Other successful applications of Röhm's pancreatic enzyme extracts were the removal of sericin gum from silk with the action of proteases and desizing of cotton fabrics at weaving mills with pancreatic amylase. The latter selectively removes the sizing agent (starch) that is added to yarn to improve the weaving process without affecting the fibers.

Probably the most famous application triggered by Röhm was the introduction of BURNUS® in 1914, the world's first enzyme product in **laundry applications**. BURNUS was not a detergent in the traditional sense, but a pure stain remover applied before the actual washing process. The protease-containing product released the dirt embedded by the protein, which would have penetrated the textiles through the high washing temperatures. The laundry was soaked in BURNUS for a few hours and then washed with only a little soap and much less harm to both the fibers and the person doing the washing.

Next to pioneering enzymes for leather, textile, and laundry applications, Röhm also introduced pectinase derived from *Aspergillus niger* for **fruit processing.** In the 1930s, the Swiss fruit juice industry discovered that pectinases increase the yield and quality of their products [52]. Supply issues during the war allowed Röhm's competitor Schweizerische Ferment AG founded in 1915 in Basel (Switzerland) to enter the market with their first microbial-derived product Pectinex in 1945. Today, the German branch of Röhm Enzymes belongs to AB-Enzymes and the United States-based Rohm and Haas ended up with DowDuPont. The Schweizerische Ferment AG, which also pioneered malt and pancreas-based enzyme products for **baking, textile, and laundry applications**, now belongs to Novozymes, as shown in Figure 1.1.5.

The root of Novozymes, today's enzyme business leader with enzyme sales of €1.9 billion in 2016, goes back to the foundation of Novo Terapeutisk Laboratorium in 1925, which produced insulin from pancreas. At the end of the 1930s, the company developed a process to extract both insulin and active trypsin, which was inactivated in the earlier process. This was the start of Novo's enzyme business with trypsin as the first product for tanning application. Around 1945 Novo started to work on bacterial amylase using submerged fermentation of *Bacillus subtilis*. In 1951, TERMOZYM® was launched as Novo's first microbial enzyme, followed by AQUAZYME® in 1954, which was the first liquid enzyme (amylase) formulation for the textile market. At that time, the textile industry was the largest market for industrial enzymes, with Novo offering pancreatic and bacterial amylases, Schweizerische Ferment AG malt-based amylase, and Societé Rapidase bacterial amylase, as described above. Over time, other enzymes have been successfully introduced in the textile industry next to amylases, such as cellulases to enhance the appearance and feel of cellulosic garments, catalases to eliminate residual hydrogen peroxide following bleaching or in dye house wastewater recycling, and proteases for the softening and anti-pilling of wool as well as for the degumming of silk [16].

As previously mentioned, amylases turned out to be very valuable products triggering innovations in the **starch processing** industry in the manufacture of specific types of syrup i.e. those containing a range of sugars that could not be produced by conventional acid hydrolysis. Another real breakthrough in this industry was reached early in the 1960s when an enzyme, glucoamylase, was launched for the first time, which could completely break down starch into glucose [53]. Within a few years, almost all glucose production changed over to enzymatic hydrolysis because of greater yield, higher degree of purity, and easier crystallization.

In 1959, Gebrüder Schnyder, a small detergent producer in Switzerland, launched BIO-40 and in 1962 the company Kortman & Schulte (Netherlands) launched BIOTEX containing bacterial proteases. Compared to the trypsin used in BURNUS the proteases in BIO-40 and BIOTEX were more active and stable under conditions used in **detergent applications**. The initial success triggered the other enzyme players to develop competitive products, with Novo launching ALCALASE® in 1963 and Gist-Brocades launching MAXATASE® in 1965. Within five years of their introduction into detergents, more than 50% of all heavy-duty laundry detergents for domestic use in Europe contained proteases, compared to only 15% in the United States. The rapid growth of the enzyme-containing detergents was temporarily discontinued in the early 1970s when industrial hygiene and safety problems became evident. Significant improvements in detergent manufacturing techniques, the development of low-dust encapsulated enzyme prills, and improvements in industrial hygiene practices resolved the issue. Since then, other enzyme classes have been successfully introduced in the detergent industry such as alkaline and temperature-tolerant α-amylases, lipases, and cellulases, making detergent enzymes the biggest business segment of industrial enzymes today, as shown in Figure 1.1.3.

Another major application field of **enzymes is in the feed** industry in which they help to improve the nutritional value of animal feed. One mode of action is to reduce the viscosity caused by non-starch polysaccharides. In their soluble form, these compounds can increase the viscosity of the digesta in the animals' small intestine. This leads to a reduction of the degree and rate of nutrient digestion and therefore a decrease of the animals' performance. Enzymes targeted against these antinutritional factors formed the initial basis of commercial products, with xylanases aimed at the arabinoxylan backbone of wheat-based diets and β-glucanases at the β-glucan backbone of barley. The first report of an enzyme product (PROTOZYME) derived from *A. oryzae* used in poultry diet dates back to the 1920s [54]. In the 1950s, additional scientific reports demonstrated the positive impact of enzymes for animal nutrition [55]. In 1988, commercial feed enzymes such as a β-glucanase (CORNZYME®) as well as some multienzyme solutions, including PORZYME® for barley-based pig starter diets and AVIZYME® for poultry diets, were launched by the Finnish company Cultor. Today, the global sales volume of feed enzymes consisting mainly of β-glucanases, cellulases, proteases, xylanases, and phytases is estimated at c. €900 million per year [56]. Commercially, still one of the most important feed enzyme class is phytase, which is reviewed in Chapter 3.3.

The global **Pulp and Paper** enzymes market size was estimated at over €100 million in 2016 [57]. Amylases are used in various applications including deinking, modification of starch coating, cleaning, and drainage improvement, particularly in the cardboard industry. The coating treatment makes the surface of paper smooth and strong to improve the writing quality of the paper. Cellulases are applied for fiber modification, improving machine run ability, reducing risk of unwanted deposits, and drainage improvement for recycled waste cardboard. In sulfite pulping lipases are used to reduce pitch problems and for removing of triglycerides whereas xylanases are applied for bleaching of Kraft pulps to attain a specified brightness. A more comprehensive overview of enzyme applications in the pulp and paper industry is provided by Saha in Chapter 4.5.

While enzymes have been applied for thousands of years for the production of potable alcohol as described above, their application in **ethanol production** for automotive fuel was seriously considered after the oil crisis in 1973. However, the lack of enzymes with required properties (high activity and stability) as well as efficient (recombinant) production systems (leading to acceptable costs) presented significant hurdles for the introduction of economically viable enzyme solutions. Enabled by the introduction of recombinant DNA technology as well as advanced tools of enzyme production and development (see Chapter 1.2), today grain-based fuel ethanol production in most of Europe and the United States applies enzymes. Thermostable α-amylases are used at a high-temperature (e.g. 105 °C) liquefaction step following a saccharification step in which starch dextrins are converted to glucose at lower temperature (e.g. 60 °C) using glucoamylases. More recently, much efforts have gone into the development of cellulase cocktails for the production of ethanol from cellulosic feedstocks as well as the development of lipases for the production of biodiesel, for which some recent reviews can be found here [17, 58, 59].

Several other applications have been developed during the last 50 years such as using enzymes as drugs, in personal care applications, for analytical purposes, or in molecular biology. While a lot is still to come and new enzyme companies are emerging, the roots of several of today's major enzyme companies can nicely be traced back to the few entrepreneurs highlighted in this section and in the overview shown in Figure 1.1.5.

References

1 Aehle, W. (ed.) (2007). *Enzymes in Industry: Production and Applications*, 3e. Weinheim: Wiley-VCH.

2 Drauz, K., Gröger, H., and May, O. (eds.) (2012). *Enzyme Catalysis in Organic Synthesis: A Comprehensive Handbook*, 3e. Weinheim: Wiley-VCH.

3 Liese, A., Seelbach, K., and Wandrey, C. (eds.) (2006). *Industrial Biotransformations*, 2e. Weinheim: Wiley-VCH.

4 Buchholz, K. and Collins, J. (2013). The roots—a short history of industrial microbiology and biotechnology. *Appl. Microbiol. Biotechnol.* 97: 3747–3762.

5 Salque, M., Bogucki, P.I., Pyzel, J. et al. (2013). Earliest evidence for cheese making in the sixth millennium BC in northern Europe. *Nature* 493 (7433): 522–525.

6 McGovern, P.E., Zhang, J., Tang, J. et al. (2004). Fermented beverages of pre- and proto-historic China. *Proc. Natl. Acad. Sci. U.S.A.* 101 (51): 17593–17598.

7 McGovern, P.E., Glusker, D.L., and Exner, L.J. (1996). Neolithic resinated wine. *Nature* 381: 480–481.

8 Valamoti, S.M., Mangafa, M., Koukouli-Chrysanthaki, C.H., and Malamidou, D. (2007). Grape-pressings from northern Greece: the earliest wine in the Aegean? *Antiquity* 81: 54–61.

9 Fukushima, D. (2004). Industrialization of fermented soy sauce production centering around Japanese shoyu. In: *Food Science and Technology* (ed. K.H. Steinkraus), 1–88. New York: Marcel Dekker.

10 Sailey, J. (1992). On bring the wonders of the Tso-Chuan to the English reader. *Monum Serica* 40: 363–368.

11 Campbell-Platt, G. (1994). Fermented food – a world perspective. *Food Res. Int.* 27: 253–257.

12 Samuel, D. (1996). Investigation of ancient Egyptian baking and brewing methods by correlative microscopy. *Science* 273: 488–490.

13 Possehl, G.L. (1996). Mehrgarh. In: *Oxford Companion to Archaeology* (ed. B. Fagan). Oxford University Press.

14 Ito, S., Kobayashi, T., Hatada, Y., and Horikoshi, K. (2005). Enzymes in modern detergents. In: *Microbial Enzymes and Biotransformations*, Methods in Biotechnology, vol. 17 (ed. J.L. Barredo), 151–161. Humana Press.

15 Maurer, K.-H. (2010). Enzymes, detergent. In: *Encyclopedia of Industrial Biotechnology* (ed. M.C. Flickinger), 1–16. New York: Wiley.

16 Shahid, M., Mohammad, F., Chen, G. et al. (2016). Enzymatic processing of natural fibres: white biotechnology for sustainable development. *Green Chem.* 18 (8): 2256–2281.

17 Lynd, L.R., Liang, X., Biddy, M.J. et al. (2017). Cellulosic ethanol: status and innovation. *Curr. Opin. Biotechnol.* 45: 202–211.

18 Leewenhoeck, A. (1677). Observation, communicated to the publisher by Mr. Antony van Leewenhoeck, in a Dutch letter of the 9 Octob. 1676 here English'd: concerning little animals by him observed in rain-well-sea and snow water; as also in water wherein pepper had lain infused. *Philos. Trans. R. Soc. London, Ser. A* 12: 821–831.

19 Payen, A. and Persoz, J.-F. (1833). Mémoire sur la diastase, les principaux produits de ses réactions et leurs applications aux arts industriels (Memoir on diastase, the principal products of its reactions, and their applications to the industrial arts). *Ann. Chim. Phys.* 2 (53): 73–92.
20 Schwann, T. (1836). Ueber das Wesen des Verdauungprocesses. *Arch. Anat. Physiol. Wiss. Med.* 1: 90–138.
21 Audran, R.G. and Marechal, C.M. (1967). Sensitization of photographic film in the ultraviolet. French Patent Application 1489146, filed 1 June 1966.
22 Schwann, T. (1839). Microscopical Researches into the Accordance in the Structure and Growth of Animals and Plants. Berlin. (English translation by Henry Smith, for the Sydenham Society, 1847).
23 Chevallier-Jussiau, N. (2010). Henry Toussaint and Louis Pasteur. Rivalry over a vaccine. *Hist. Sci. Med.* 44 (1): 55–64.
24 Vallery-Radot, R. (1923). *The Life of Pasteur*. Doubleday Page and Company.
25 Pasteur, L. (1862). Suite a une précédente communication sur les mycodermes; Nouveau procéde industriel de fabrication du vinaigre. *Compt. Rend.* 55: 28–32.
26 Sebek, O.K. (1982). Notes on the historical development of microbial transformations. In: *Microbial Transformations of Bioreactive Compounds*, vol. 1 (ed. J.P. Rosazza). Boca Raton, FL: CRC Press.
27 Kühne, W. (1877). Über das Verhalten verschiedener organisirter und sog. Ungeformter Fermente. *Verh. Heidelb. Naturhist.-Med. Ver., Neue Folge* 1 (3): 190–193.
28 Buchner, E. (1897). Alkoholische Gärung ohne Hefezellen. *Ber. Dtsch. Chem. Ges.* 30: 117–124.
29 Fischer, E. (1894). Einfluss der configuration auf die wirkung der enzyme. *Ber. Dtsch. Chem. Ges.* 27 (3): 2985–2993.
30 Koshland, D.E. (1958). Application of a theory of enzyme specificity to protein synthesis. *Proc. Natl. Acad. Sci. U.S.A.* 44 (2): 98–104.
31 Abraham, E.P. and Robinson, R. (1937). Crystallization of lysozyme. *Nature* 140: 24.
32 Blake, C.C., Koenig, D.F., Mair, G.A. et al. (1965). Structure of hen egg-white lysozyme. A three-dimensional Fourier synthesis at 2 Angstrom resolution. *Nature* 22 (206): 757–761.
33 Klug, A. (1974). Rosalind Franklin and the Double Helix. *Nature* 248: 787–788.
34 Watson, J.D. and Crick, F.H. (1953). Molecular structure of nucleic acids; a structure for deoxyribose nucleic acid. *Nature* 171 (4356): 737–738.
35 Nobelprize Organization (1962). The Nobel Prize in Physiology or Medicine 1962. http://www.nobelprize.org/nobel_prizes/medicine/laureates/1962/ (accessed 9 January 2018).
36 Nobelprize Organization (1968). The Nobel Prize in Physiology or Medicine 1968. http://www.nobelprize.org/nobel_prizes/medicine/laureates/1968/ (accessed 9 January 2018).
37 Cohen, S.N., Chang, A.C., Boyer, H.W., and Helling, R.B. (1973). Construction of biologically functional bacterial plasmids in vitro. *Proc. Natl. Acad. Sci. U.S.A.* 70 (11): 3240–3244.

38 Hildebrandt, G. and Klavehn, W. (1930) Verfahren zur Herstellung von 1-1-Phenyl-2-methylaminopropan-1-ol, Knoll AG Chemische Fabriken in Ludwigshafen. German Patent 548459, filed 09 April 1930 and issued 13 April 1932.

39 Neuberg, C. and Hirsch, J. (1921). Über ein Kohlenstoffketten knüpfendes Ferment (Carboligase). *Biochem. Z.* 115: 282–310.

40 Kluyver, A.J. and de Leeuw, F.J. (1924). *Acetobacter suboxydans*, een merkwaardige azijnbacterie. *Ned. Tijdschr. Geneeskd.* 10: 170.

41 Reichstein, T. and Grüssner, A. (1934). Eine ergiebige Synthese der l-ascorbinsäure (C-Vitamin). *Helv. Chim. Acta* 17 (1): 311–328.

42 Pappenberger, G. and Hohmann, H.P. (2014). Industrial production of L-ascorbic acid (Vitamin C) and D-isoascorbic acid. *Adv. Biochem. Eng. Biotechnol.* 143: 143–188.

43 Peterson, D.H., Murray, H.C., Epstein, S.H. et al. (1952). Microbiological oxygenation of steroids. I. Introduction of oxygen at carbon-11 of progesterone. *J. Am. Chem. Soc.* 74: 5933–5936.

44 Sebek, O.K. and Perlman, D. (1979). Microbial transformation of steroids and sterols. In: *Microbial Technology*, 2e, vol. 1 (ed. H.J. Peppler and D. Perlman), 484–488. New York: Academic Press.

45 Kobayashi, M., Nagasawa, T., and Yamada, H. (1992). Enzymatic synthesis of acrylamide: a success story not yet over. *Trends Biotechnol.* 10: 402–408.

46 Santaniello, E., Ferraboschi, P., Grisenti, P., and Manzocchi, A. (1992). The biocatalytic approach to the preparation of enantiomerically pure chiral building blocks. *Chem. Rev.* 92 (5): 1071–1140.

47 Zhu, Y. and Tramper, J. (2013). Koji – where East meets West in fermentation. *Biotechnol. Adv.* 31 (8): 1448–1457.

48 Takamine, J. (1894). Process of making diastatic enzyme. US Patent 5,25,823, filed 23 February 1894.

49 Boidin, A. and Effront, A. (1917). Process for treating amylaceous substances. US Patent 7,99,490, filed 6 November, granted 22 May 1917.

50 Boidin, A. and Effront, A. (1920). Treatment of textile fabrics or fibres. US Patent 1,505,534, filed 21 February 1914, granted 19 April 1920.

51 Roehm, O. (1908). Preparation of hides for the manufacture of leather. US Patent 8,86,411, filed 12 October 1907, granted 5 May 1908.

52 Mehlitz, A. (1930). Über die Pektasewirkung. *Biochem. Z.* 221: 217–231.

53 Kumar, P. and Satyanarayana, T. (2009). Microbial glucoamylases: characteristics and applications. *Crit. Rev. Biotechnol.* 29 (3): 225–255.

54 Clickner, F.H. and Follwell, E.H. (1926). Application of "protozyme" (*Aspergillus orizae*) to poultry feeding. *Poult. Sci.* 5: 241–247.

55 Jensen, L.S., Fry, R.E., Allred, J.B., and McGinnis, J. (1957). Improvement in the nutritional value of barley for chicks by enzyme supplementation. *Poult. Sci.* 36: 919–921.

56 Research and Makets (2014). Feed Enzymes Market by Livestock, by Types & by Geography - Global Trends & Forecasts to 2019.

57 Global Market Insights (2017). *Pulp and Paper Enzymes Market*. Global Market Insights. Report ID GMI1995.

58 Payne, C.M., Knott, B.C., Mayes, H.B. et al. (2015). Fungal cellulases. *Chem. Rev.* 115: 1308–1448.

59 Hama, S., Noda, H., and Kondo, A. (2018). How lipase technology contributes to evolution of biodiesel production using multiple feedstocks. *Curr. Opin. Biotechnol.* 50: 57–64.

1.2
Enzyme Development Technologies

Andreas Vogel

c-LEcta GmbH, R&D Enzyme Development, Perlickstr. 5, 04103 Leipzig, Germany

1.2.1 Introduction

Industrial applications of enzymes are indispensable in our everyday life. We come into contact with enzymes or products made with the aid of enzymes in nearly all areas of our daily routine. It started with the ancient and still highly significant biotechnological processes to make wine, beer, cheese and bread, which are still applied by using the same principles as thousands of years before. See Chapter 1.1 for an historical perspective of the use of enzymes. Today, enzyme applications are found in a number of products covering detergents, paper and pulp, feed, and the synthesis of bulk chemicals, vitamins, drugs, and food ingredients [1–4].

Two groundbreaking developments have brought the application of enzymes to a more advanced level: the development of recombinant DNA technology (Nobel Prize in 1980 to Paul Berg [5]) and directed evolution of enzymes (Nobel Prize in 2018 to Frances Arnold [6]). The first method, developed in the 1970s, opened up the natural sequence space and made wild-type enzymes available for investigation of their function and enabled standardized production processes (see Chapter 1.3). Recombinant DNA technology also allowed manipulation of the DNA sequence and thereby enabled inclusion of artificial changes in an enzyme sequence [7]. This paved the way to apply random mutagenesis methods, which, combined with selection and screening methods, mimic natural (Darwinian) evolution and were first applied in the early 1990s [8, 9]. These methods and their further development, which will be described in this chapter, have fundamentally changed our view on how enzymes can be applied in industrial processes [10]. Since then, a multitude of new applications have emerged from the knowledge of how enzymes can be adapted to meet industrial requirements, some of which are outlined in this book.

Industrial development does not differ much from academic research in the methods that are applied. The main driver is the focus. While academics are looking for novelty, the main driver in industry is to enable innovative new solutions to customer demands that can be successfully brought to market. Cost considerations play a pivotal role already in early stage development. Hence, industrial

Industrial Enzyme Applications, First Edition. Edited by Andreas Vogel and Oliver May.

development does ideally take the whole development program into account, which ranges from sourcing, optimization, and production of the enzyme to the application development under consideration of aspects such as scalability, raw material costs and availability, and finally the downstream processing of the product. Next to these technical considerations, intellectual property (IP) (freedom to operate, generation of new IP), regulatory aspects, and market situation (size of market, main competitors, price) are important key figures to decide whether a development program should be pursued or not. Enzyme development in industry is thus highly interdisciplinary and all development steps are guided by economics. All these aspects define the full feature profile of the enzyme. This results in the requirement to monitor multiple parameters in parallel throughout the enzyme development process, which very often limits the throughput that can be appliedto get reasonable measures for the application.

Enzyme development can be regarded as a mature technology, which is impressively proved by the multitude of current applications and the constantly growing number of new processes that are presented in the individual chapters of this book. This chapter outlines the enzyme development technologies with a focus on how they are applied in industry.

1.2.2 Identification of Wild-Type Enzymes

Every enzyme development needs a starting point. The only source used so far in industrial development is the repertoire provided by nature, which can be accessed by different approaches (Figure 1.2.1). This repertoire is vast! Researchers all over the world have access to nearly 1 billion sequence entries in NCBI databases (https://www.ncbi.nlm.nih.gov/genbank/statistics/), >368 000 strains in >103 culture collections [11], and published data on more than 370 000 enzymes in the BRENDA database [12]. This natural repertoire represents to a large extent a source of functional polypeptides that are utilized by nature to fulfill functions that the living cell needs.

This repertoire is so vast that the quest to make a reasonable selection of enzymes to be researched in the laboratory for a specific task is a serious issue. So far, selection of enzymes from sequence databases is done based on classifications by (mostly automated) annotations, phylogenetic criteria, homology groups, motif searches, and others. An automated annotation might be wrong. In order to get an answer to whether this annotated function is correct, the enzyme needs to be expressed in a desired host and investigated experimentally. This is also required to characterize the enzyme for further required properties such as stability, selectivity, and substrate scope. General applicable bioinformatic tools that would simplify this experimental approach and allow prediction of functions or expression yields from sequence analysis are rare. Notable exceptions include the PROSS approach for soluble expression [13], the "Hot Spot Wizard" [14], and catalophore approach [15] for substrate scope analysis, and the 3DM database [16] for homology information analysis.

Although GenBank® [17] is an important and rich source for new enzymes, the lack of experimental data associated with the majority of sequences is

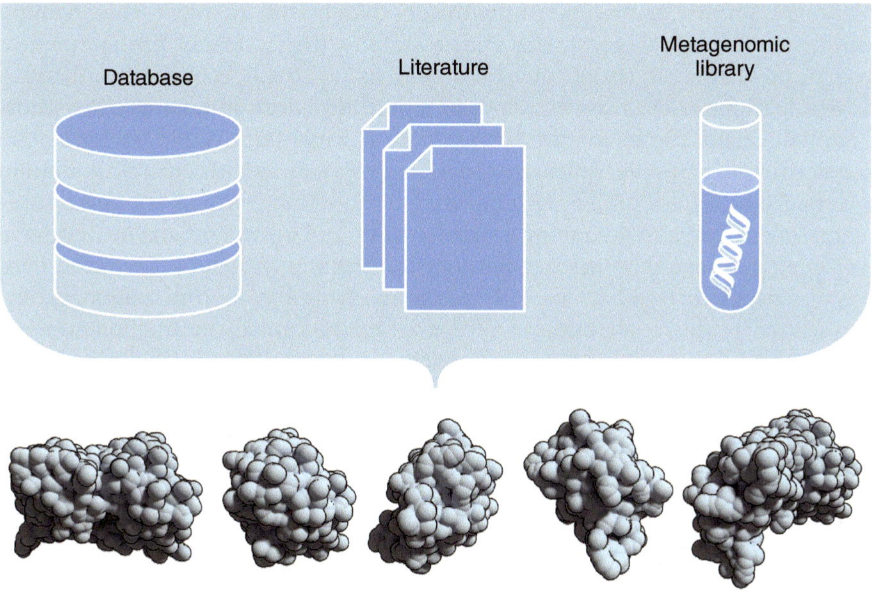

Figure 1.2.1 Sources of new enzymes. Wild-type enzymes from nature are made available for experimental testing mainly from three basic sources: sequence databases, literature, and metagenomic library screenings.

connected with an uncertainty, longer development times, and risks. Hence, the first choice often falls to an enzyme for which features have been already analyzed and published. Scientific literature from mostly academic groups about identification, characterization, and demonstrating the use of an enzyme for a synthetic application is the most important and invaluable source of information. The BRENDA databank is very efficient in collecting public data in a searchable database [12]. The identification and characterization of functions of new enzymes and their publication are particularly valuable to be able to go into an industrial development. This open-minded, driven by gain of knowledge, nonbiased research work is usually done by academics. For industry, the search for completely new enzymes is usually too risky and would always be biased toward a specific application. In that sense, successful industrial development is highly dependent on proof-of-concept knowledge in the public literature.

A third source, which is somewhat between those two previously mentioned, is the functional screening in metagenomic expression libraries [18]. Those libraries are made by random fragmentation of genomic DNA, which is transferred into an expression vector where the encoded enzymes are recombinantly expressed. So far, this technology works only for prokaryotic DNA and in prokaryotic expression hosts [19]. The microbiological source is usually unknown; only the functional parameters of the encoded open reading frames are of interest. The ability of the gene to be functionally expressed in the screening host might be regarded as a restriction, but is, however, a desired screening feature if the screening host shall also be the production host.

Genomic library screening is highly dependent on an efficient screening technology. The search resembles the "needle in the haystack" problem and a throughput of several 10 000 to 100 000 clones is usually required to obtain a reasonable number of target enzymes in a plasmid-based library [20, 21]. Thus, functional screens either require robotic high-throughput machinery, agar plate screens, or conceptually different technologies such as the cluster screening developed by c-LEcta [22]. Screening only by functions, which is independent of sequence constraints, enables access to new sequence landscapes that have never been connected before to the function that was screened. The latter case is very relevant to build up new IP positions. Sequence-based screening (e.g. using a DNA probe or oligo-based PCR methods) of metagenomic libraries will always be restricted to known sequence space and just add more information to sequences that already appear in databases and will thus be connected with the same limitations as described above. Still, new sequences can be sequestered by this approach and they can be valuable to build up new or to create freedom around existing IP.

Once a sequence has been chosen, irrespective of its source, its experimental investigation has been tremendously facilitated in recent years by the low cost of DNA synthesis. The price per base has fallen almost logarithmically during the last 15 years and that of a typical gene for an enzyme with 300 amino acids is currently ~300 € [23]. The number of genes that goes into the laboratory for experimental investigation is thus more limited by parallel cloning, expression, and assay procedures than by the costs of the DNA material.

1.2.2.1 Selection Parameters for Starting Enzymes

For an industrial application, there is usually a set of features that the enzyme needs to fulfill (Figure 1.2.2) [24, 25]. This set of features is very much dependent on the desired application. An increase in activity is always appreciated and is in most cases the limiting feature for an economic application. The second most frequent parameter that requires optimization is stability. This is usually associated with a high activity, on the one hand, due to intrinsic correlation of temperature and reaction rate. On the other hand, high enzyme stability is mandatory to run an industrial process, which very often requires an environment that is hostile for an enzyme. Protein stability also promotes evolvability [26] and thus is an important aspect to monitor in enzyme engineering programs. These two parameters are nowadays often included as standard in industrial enzyme development projects and are very often sufficient, especially if the native enzyme function is required for the application. If this is not the case, optimization of selectivity and substrate scope is required, which is the third most frequent parameter for enzyme optimization projects.

Activity yield of an enzyme production process is a product of specific activity and expression yield. This requires the consideration of the expression host in the selection of the starting enzyme. Stability is always an important feature that can have different meanings. The main focus is stability in the application, e.g. at 40 °C for 30 hours in the reaction medium. The reaction medium may contain further potential enzyme destabilization agents such as high concentration of substrates

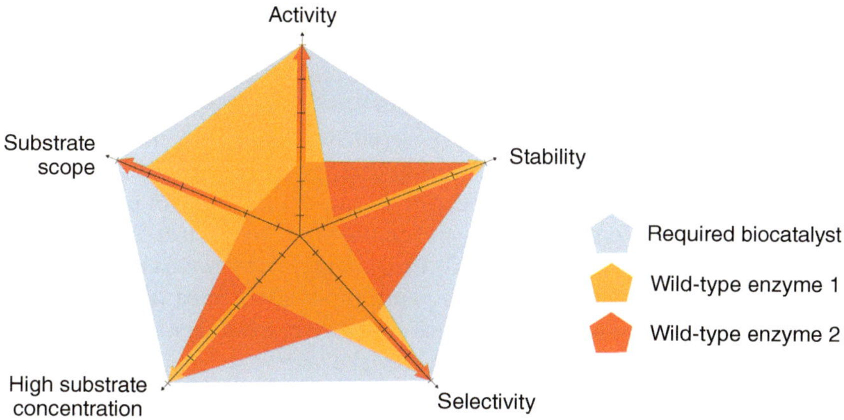

Figure 1.2.2 Selection criteria for industrial enzymes. Schematic illustration of a multi-parameter profile for the development of an industrial enzyme. Natural (wild-type) enzymes usually fulfill a subset of features and may be complementary. The optimization to obtain the full feature profile is done by enzyme engineering.

and product, specific pH values required for compound stability, or organic cosolvents. All these need to be taken into account in enzyme characterization.

Thus, stability is much more defined by the reaction conditions than by the thermal stability alone. However, there is often quite a good correlation between thermal stability and other stability measures [27, 28], and measuring the melting temperature of an enzyme is a very good and easily accessible indicator to compare different enzymes. At first sight, enzymes from thermophilic organisms appear to be a preferred choice when high stability is required [29]. Indeed, a thermophilic enzyme is a good choice when a high reaction temperature of >60 °C is envisioned. However, most processes are conducted at rather ambient temperatures of 30–50 °C, where the activity of enzymes from thermophilic organisms is often very low [30]. The selection thus most often ends up in enzymes from mesophilic organisms and (thermo)stability is a feature that wild-type enzymes sometimes inhere, like the lipase A from the cold adapted yeast *Candida antarctica* with a temperature optimum of 90 °C [31] or the fructose-6-phosphate aldolase from *Escherichia coli* with a half-life of 16 hours at 75 °C [32]. If stability of an enzyme is not sufficient, it is an adequate task for the optimization by enzyme engineering (see below).

Selectivity is another feature that is very often required from the enzyme. It can be in the meaning of regio-, stereo-, or chemical selectivity, which will dictate product yield and diminish side product formation [33]. Substrate scope is another feature that is demanded especially for pharmaceutical applications. This feature is rarely gained from wild-type enzymes and is a typical task for enzyme engineering [34] (see below).

When an enzyme development project shall be started, at a certain point the question arises as to which features can be obtained from nature and which features are better addressed by enzyme engineering (Figure 1.2.2). Industrial developers are committed to choosing the fastest and most effective route to an optimized enzyme and must take into account the prospects of success inherent in every development program. There is no general answer to this question and each application has its own boundary conditions. Next to the technical situation there are always other aspects that need consideration such as budget, timelines, IP situation, use of genetically modified material, and regulatory questions. For the technical site, the established and accessible methods and, more importantly, the experience of the researcher define the development program. A basic guideline is that wild-type enzymes have been "engineered" by nature and selected for the function that they need to fulfill in the organisms, whereas industrial applications need to work efficiently in a plant under conditions defined by the process. The bigger the differences are between the organism and the plant, the higher the need to engineer the enzyme in the laboratory.

Looking at nature is also relevant for setting upper limits in a development, which should be guided by numbers that wild-type enzymes achieve in nature. This means that in case a certain specific activity is required for the application and this number has not been reached by any wild-type enzyme utilizing the same reaction mechanism, the prospects of success should be set very carefully.

1.2.3 Enzyme Engineering

1.2.3.1 Types of Enzyme Modifications

For the optimization of enzymes for a desired process a series of different methods are used. On the one hand, a given polypeptide can be immobilized or chemically modified [35–37]. These methods do not modify the enzyme sequence and even enzymes that are isolated from natural sources can be modified in these ways. These methods mainly aim at stabilization of an enzyme or process relevant requirements (i.e. immobilized enzymes to be applied in a plug flow reactor; see Chapter 1.4). Usually they consist of a single step of modification and the effect can thus not be amplified further.

On the other hand, modification of the enzyme sequence by adding mutations allows for adaptation of the enzyme to a desired parameter within natural boundaries in an iterative way. The modification can be applied based on a rational idea, on pure random mutation, or by a combination of both approaches (semi-rational) [38–40]. All of these modifications will be referred to as "enzyme engineering" in this chapter.

1.2.3.2 General Engineering Strategies. Library Design and Generation

Beginning with the first elucidation of an enzyme structure in 1965 [41], scientists began to understand the role of individual amino acids, especially in the active site and substrate binding pocket. It was realized that the actual chemistry was executed by a few (2-4) amino acid side chains and further amino acids were positioned in a way to stabilize the transition state, thereby enabling the reaction to proceed faster. Research was very much focused on a static picture of the active site and the amino acid residues that interact with the substrate in the substrate binding pocket. Those amino acids are required to position the substrate in such a way as to enable perfect interaction with the active site. Scientists in the 1980s used this knowledge to make predictions on how a residue change might influence the enzyme actions [42, 43]. Since mutagenesis tools were at this time just starting to be developed and were time consuming, only a few mutants could be prepared with a reasonable effort. Thus, experimentalists were dependent on precise predictions by the structural biologist. The results from this "era" in the quest to optimize enzymes were limited to a few, but still impressive, studies [44, 45].

A real breakthrough in the optimization of enzymes for industrial applications came with the implementation of random mutagenesis methods by Frances Arnold and coworker [8] and Pim Stemmer [9]. Arnold used the error prone PCR (epPCR) [46, 47] to generate a library of genes with random point mutations and Stemmer shuffled homologous genes by random fragmentation and ligation. Those methods were clearly inspired by Darwinian principles of evolution and have their counterparts in asexual (point mutations) and sexual (shuffling, recombination) generation of variant genes in nature (Figure 1.2.3b) [48]. "Selection of the fittest" is in laboratory evolution either done by real selection systems, i.e. growth advantages of a host in which the improved variants are expressed, or mostly by high-throughput screening systems [49]. The evolutionary character is brought in by iterative cycles of mutation and selection (or screening) of the best variants (Figure 1.2.3a). The charm of this method is that no information about the structure or functional sites is required. These random methods are thus conceptually contradictory to rational design. Another charm is the evolutionary character, which allows subsequent step-by-step improvement of the evolved variants from a mutagenesis/screening round that is limited only by the effort, time, and budget (and clearly: chemical constraints) that can be put into a development program.

The epPCR method used in the initial work by Arnold allows random incorporation of point mutations, and by careful adaptation of the protocol, the average error rate can be varied. The distribution of mutations per gene follows a Gaussian distribution. The population thus contains at the low end a fraction of wild-type genes and at the high end genes with a high mutational load, which are much more susceptible for being non-folded and inactive. In addition, frame shift mutations may occur, which means that the desired full-length polypeptide will not be translated. Within a single round of epPCR a two or three base exchange within a codon is highly unlikely by statistical reasons. This limits the number of accessible amino acid exchanges to less than five on average due to the structure of the genetic code [50]. In the course of the repeated cycles of epPCR, mutations accumulate that do not contribute to the improved features of the variant. In the best case, the mutations are neutral, but they can be negative. This accumulation of non-beneficial mutations leads to a dead end in the "asexual" evolution [48]. Backmutation of the individual mutations or a random "sexual" mutation event by shuffling with the parental gene [51] will enable an improved variant with lower non-beneficial mutational load.

The epPCR, although anything other than perfect, allowed the scientific community to step in easily into directed evolution without the need for special

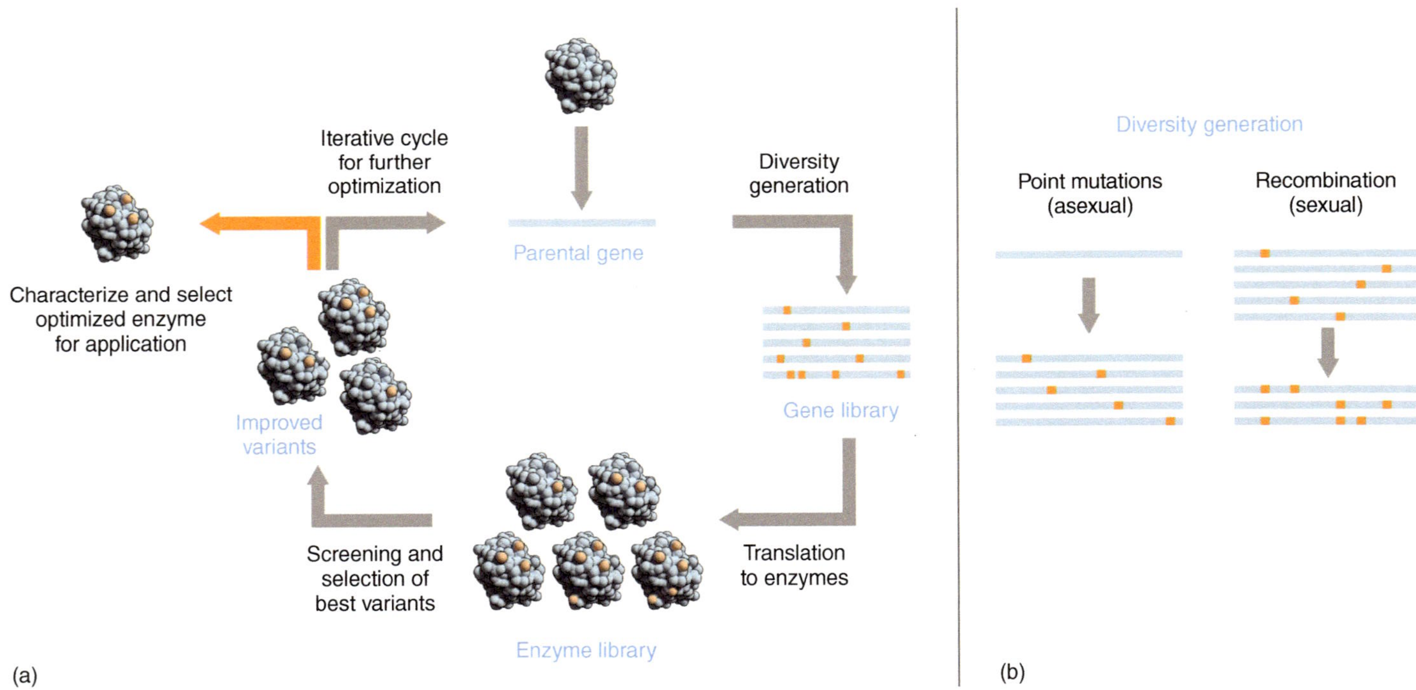

Figure 1.2.3 General strategies in directed evolution. (a) The evolutionary cycle starts with a parental gene, which is normally a wild-type sequence, and follows the basic cycle of diversity generation on the gene level, translation, and functional screening of an enzyme library. The best candidates (or the sequence information of the best candidates) can either enter another evolutionary cycle or an improved variant is selected for the application after characterization of the relevant features. (b) Diversity generation is done following basically two principles. Either mutations are incorporated in one parental gene (asexual) or the information of various mutations is recombined to new combinations of existing mutations (sexual).

expertise in structural biology or enzymology. In addition, given the restraints of epPCR, the scientific community was motivated to think about better strategies and methods for the creation of random mutational libraries [52, 53].

A boost to methodical developments started with a series of memorable acronyms coming up. Several reviews describe the different methods in more detail [53, 54]. Only some methods that were considered important in industrial enzyme engineering will be briefly touched upon.

The mutational load of an epPCR was a debate [55–57] as well as improvement on the epPCR protocol itself [58, 59], which addressed the limitations of codon exchanges to access a broader amino acid space. These methods, however, suffer from a complex laboratory protocol. The problem of accumulation of negative mutations was addressed with the ProSAR protocol [60], which is a statistical algorithm that dissects between negative, neutral, and beneficial mutations in later stages of evolution. It requires, however, large data sets of sequence–function–relationship.

The epPCR protocol has low relevance in industrial enzyme engineering today and has been nearly completely replaced by oligonucleotide-based library generation methods (see below) that allow better control over the mutation sites and substitutions. An early application of oligonucleotide methods was saturation mutagenesis [61–63]. In combination with epPCR, it makes all amino acid exchanges accessible once a beneficial position was identified. This strategy, epPCR in combination with saturation mutagenesis (and eventually shuffling), was in a certain way the dominant strategy in directed evolution in the late 1990s and early 2000s [63–66].

There was also a series of methodical development for the preparation and design of recombination libraries, which was reflected in numerous creative acronyms such as SHIPREC, SCHEMA, ITCHY, SCRATCHY, and RACHITT (reviewed in [53, 54]). In general, those methods suffer from a high fraction of inactive genes. One aspect is the inherent limitations in the protocols, which often contain a high fraction of frame shift mutations and deletions. Another explanation, which is especially valid for family shuffling, is considering the three-dimensional protein structure. The fragments that are mixed and derived from different genes, although coding for a complementary secondary structural element, have a lot of spatial interactions. A disturbance of this interaction influences the folding process and the stability within the three-dimensional structure and is thus very delicate. The SCHEMA algorithm [67] aiming to address this, however, did not find broader application. Looking at the current methods, DNA shuffling in the sense describe above has no more relevance in industrial development. Still, the concept of shuffling and recombination is realized by using the information of beneficial mutations with the use of defined oligonucleotide-based methods (see below).

Noteworthy are methods that are based on a synthetic construction of the entire gene by assembly of oligonucleotides [68–70]. By doping with oligonucleotides that contain mutations, these methods offer the ability for full control over the number, sites, and amino acid exchanges in the library. Looking nearly perfect on paper, in practice these methods are accompanied by a high unwanted error rate, which is in part inherent to the oligonucleotide assemble method, but mainly due to limited oligonucleotide quality. Although oligonucleotide quality may be very high, even a low number of erroneous oligonucleotides sums up in the library generation process, since in the end the entire library material consists of synthetic oligonucleotides.

It is easily seen that protein engineering methods were adopted and developed further in industry very early, driven by the battle to get main market shares in the detergent field. Subtilisin was the target protease on which several companies were very active, namely Genencor, Gist-Brocades, and Novo Nordisc. Starting in the mid-1980s several publications and patents were published, which document the application of rational engineering, saturation mutagenesis, and cassette mutagenesis for that target in industry [71–73]. It was thus no surprise that the initial epPCR work was performed on subtilisin [8] and that DNA shuffling [74] and complete site saturation mutagenesis [25] approaches were rapidly applied to subtilisin as well.

The so-called "consensus approach" is another example that was developed in industry in the quest to stabilize phytase for feed applications (see also Chapter 3.3). The concept used by researchers at F. Hoffmann-La Roche [75] is to analyze information from multiple sequence alignments and make an attempt to move the target sequence toward the most utilized amino acids in a certain position within the alignment (the consensus). This approach leads to a very small set of amino acid exchanges (in the range of 10–40), which often have an impressive impact on enzyme stability. The use of homology information for library design has been since extended and is used by several academic groups and industries [76–78].

It is notable how fast random methods have been installed and adopted in enzyme engineering and further developed by industry [79]. This is especially well illustrated by the list of companies for which Pim Stemmer and his DNA shuffling approach provided the basis [80]: Maxygen, a successful initial public offering in 2000, Verdia (sold in 2004 to DuPont), and Codexis, which is one of the forerunners in industrial enzyme engineering.

Another landmark in methodical development that resulted in broad acceptance by the community was the CASTing [34, 81] and iterative saturation mutagenesis (ISM) [82, 83] methods developed by the Reetz group. The CASTing strategy aimed at efficient shaping of the substrate binding site by systematic scanning of the residues surrounding the substrate for broadening substrate scope and adjustment of stereoselectivity. Several small libraries are designed considering spatial interactions and library size. Residues to be mutated are selected in groups of two or three positions where the amino acid side chains interact in space, thereby allowing simultaneous adaptation where direct interaction is expected. At the same time, the library is limited to a size that allows reasonable coverage with standard screening equipment (i.e. handling of 96-well plates). In the ISM concept, the information about hot spot residues and preferred amino acid exchanges is used in subsequent library designs, which ultimately cover the entire substrate binding pocket [84]. The basic idea is that subsequent screening of several small libraries is more effective than the screening of one large complex library. The concept turned out to be very successful and provided a systematic strategy that could be readily applied without the need for robotic high-throughput screening equipment. Enzyme engineering became a tool that could be applied in any laboratory with molecular biology expertise and manual 96-well plate liquid handling and analytics [85–87]. As stated above, the ISM concept focuses on the substrate binding pocket and thus is most effective for broadening the substrate scope and stereoselectivity.

For industrial enzyme development CASTing and ISM may represent a good first step. The limited number of amino acid side chains forming the substrate binding pocket can be comprehensively explored with this approach, but higher activity improvements will still be required. Distant mutations have proved to be important for activity that cannot necessarily be explained or even predicted on a structural basis [88, 89]. The dynamics of the enzyme is one explanation where distant mutations exert a remote effect on the shape and flexibility of the substrate binding pocket [90, 91]. The trajectory of the substrate into the active site, in a tunnel, a cleft, or on the surface, is another mechanism of direct interactions between substrate and enzyme that influences the on/off kinetics of the substrate and product with a subsequent impact on activity [14, 92]. Speaking about stability, the surface residues have often been discussed as the preferred places for beneficial mutations [93, 94]. Taking all these positions into account in the library design, a choice of well over 100 positions will be selected. The consequence is that strategies and screening equipment need to be considered again.

The most comprehensive way to investigate the impact of each amino acid on a certain property is the systematic scanning of each possible single amino acid mutation (Figure 1.2.4). A full site saturation library consisting of every possible single mutation will still yield a screenable number of several thousands of individual mutants, still having a high content of neutral and deleterious mutations (low hit rate). This type of comprehensive single mutation screening has been realized by several groups in the industry [25, 95]) and academics [96, 97]. Although being comprehensive, those libraries are done using synthetic oligonucleotides position by position and thus require substantial costs, time, and effort. In order to reduce time and effort, one needs to make a decision about a subset of positions and amino acid exchanges to reduce the library size to a few hundred to a thousand variants [98–100]. If the selection is made on a meaningful rational basis, the hit rate should increase as well. This strategy has been optimized by c-LEcta and a bioinformatics toolset was developed, which was called the multi-dimensional mutagenesis (MDM) platform. It uses structural, molecular dynamics, functional, and homology information and can be edited with rational ideas. It provides a routine for library design that can be gradually adjusted in size and thus aligned with the screening system or effort allocated to a project (see Chapters 2.4 and 2.5 for application examples of this approach). With this knowledge-driven approach, typically a set of 50–150 amino acid positions is selected and an individual set of exchanges for each position is provided (Figure 1.2.4). In-house molecular biology tools have been optimized for the generation of single site mutation libraries with an individual set of amino acid exchanges in 96-well plates to yield high-quality mutant genes (i.e. very low fraction of wild-type or frame shift mutations). The workaround time for the generation of these libraries is typically one week.

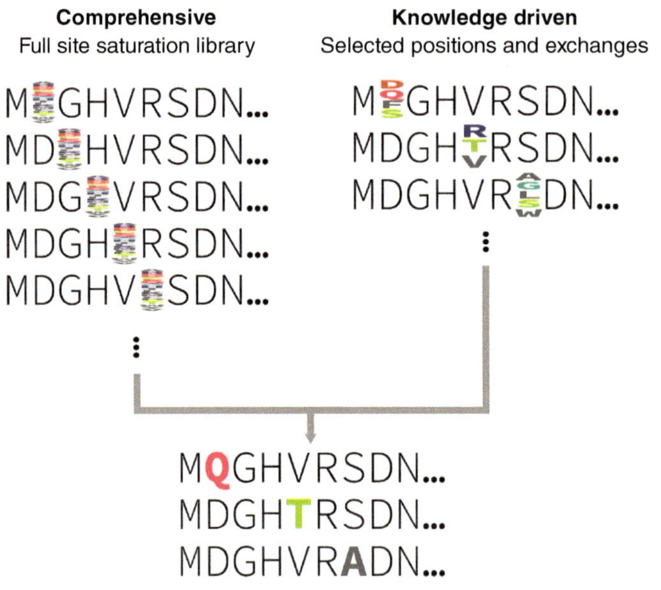

Figure 1.2.4 Strategies for single mutation libraries. In the full site saturation library, every single mutation is created separately, whereas in the knowledge-driven approach, the position and type of amino acid exchanges is chosen based on knowledge. Both strategies lead to a set of improved variants containing one amino acid exchange. This information is used for further optimization by recombination.

Using single site mutation libraries covering a broad sequence space has become a standard strategy as a first step for a number of companies (e.g. DuPont, BASF, Codexis, c-LEcta) and has been discussed in the scientific literature [101]. It compromises library size and the desire to cover a broad sequence space. The synergistic effect of multiple mutations will be overseen by this procedure as will be the negative effectors. However, such a screening usually provides a large number of beneficial single mutations and their combined effect will be subsequently analyzed in a second recombination step. This two-step approach has turned out to be very effective in terms of fast generation of multiple variants and still covers a large sequence space. The combination of beneficial single mutations may be additive, negative, or even synergistic [102]. This effect cannot be predicted yet and thus needs to be explored experimentally. Several strategies for the design of a recombination library based on a set of beneficial mutations can be envisioned: all beneficial positions may be recombined allowing the wild-type and preferred mutations (thus allowing two amino acids per position). More diversity in terms of amino acid exchanges may be added at this stage. The library size needs to be carefully taken into account as the numbers are increasing exponentially [100]. Thus, a dissection into several small libraries, basically following the ISM concept, can be realized at this stage as well. This strategy implements the abovementioned "sexual" recombination idea on a very precise basis by taking the information about individual positions into a recombination library, rather than randomly shuffling fragments of a pool of improved (multi-mutated) genes.

The best variants from a recombination library screening can go into single site screening, again driving the evolutionary character of the strategy. Even if a comprehensive full site saturation library screening was performed in the first place, new mutations will show up since the template has changed. The beauty of evolutionary enzyme engineering is that further improvements of the enzyme are always possible and is only limited by the effort to be put in and, obviously, the natural boundaries.

1.2.3.3 Screening for Better Enzymes

Screening analytics is clearly the key for success in evolutionary enzyme engineering. "You get what you screen for" has been formulated by Frances Arnold and is seen as the "first law in directed evolution" [48]. This law is still valid as proved for many examples. A negative example is illustrated by screening a lipase for the racemic resolution of chiral carboxylic acids. Owing to its simplicity, a screening system may be setup by a hydrolytic screening of *para*-nitrophenyl esters of separated *R*- and *S*-compounds [103]. Screening is done in parallel and the ratio of the kinetic rates is regarded as a measure of enantioselectivity. However, as shown in the literature [104] and by research at c-LEcta, the readout is likely not transferable to the racemic resolution of nonactivated esters (such as ethyl-esters). It was even not indicative of activity improvement when switching from hydrolysis of *para*-nitrophenol butyrate to hydrolysis of tributyrin (own data). Thus, the challenge is to implement the parameters required for application in the screening system. It is absolutely mandatory that the assay reflects the industrial needs and is meaningful for the application.

An illustration of this is the screening for alcohol dehydrogenases for chiral alcohol synthesis (see also Chapter 5.1). An initial and easy screening system is to follow the absorption of the NAD(P)H cofactor at 340 nm. This allows a fast, photometric screen for activity improvement at conditions that are compatible with a photometric measurement. This means that such screening conditions cannot implement cofactor regeneration systems; inhomogeneous phases (e.g. arising from high substrate conditions) and cofactor concentration must be set to be compatible with the readout. Such a screening is still useful to look for initial activity toward a substrate (e.g. widening the substrate scope) or to sort out a large fraction of inactive clones if the library design is set in such a way. However, the photometric NAD(P)H monitoring cannot be applied for screening at high conversion rates at high substrate concentration with the implementation of isopropanol for cofactor regeneration, which is in fact often the desired type of application in this case.

In the setup for an application-based screening system additional features beside activity need to be implemented. Staying with the example of an alcohol dehydrogenase (ADH) that shall be applied in isopropanol for substrate solubility and cofactor regeneration, the screening needs to cover besides activity, stability toward isopropanol, activity with isopropanol, activity (and stability) toward high substrate concentration, and ability to act toward high conversions. The latter feature does include high stability over the entire reaction time and low inhibition by either substrate or product. The different features may be screened individually, but if the desired application can be realized in a screening system, it is not required anymore to dissect the contributions of individual feature improvements. In fact, it is not of main interest for the industrial enzyme developer to understand and dissect the effects that were obtained by enzyme engineering on a molecular level. It is of much more importance to obtain an enzyme that works in the application, no matter why it does so.

According to what was said above, a lot of effort is going into screening analytics in industrial development before enzyme engineering will be started. The screening readout has to be meaningful for the application. HPLC analytics turned out to be the method of choice for most applications. A throughput of ~1000 clones a week can be easily achieved with standard HPLC systems, which is very much in line with the library designs applied at c-LEcta and described above. A conceptually different approach is the cluster screening developed at c-LEcta [105], which was also published in a similar form by the Bommarius group [106]. Sampling in a mixture of different individual clones in a well allows a significantly higher throughput of 10 000 to 100 000 variants per microtiter plate, and is still compatible with complex analytics. This allows the creation of new functionalities where single site mutations fail to show an effect and simultaneous mutations are required to create a first activity. This was exemplified by converting an aspartase into a β-amino acid lyase, which required three mutations to create a first detectable activity [107].

Care must be taken about the quality of the readouts since small improvements are of great interest in a single mutation library screening. Thus, cultivation of the expression strain, enzyme workup from the cell culture, reaction conditions, and sample setup for the analytics need to be carefully adjusted in a multi-well format for each screening. Statistical variations, which are inherent to each system, need to be monitored and taken into account in the data interpretation.

In the end, the implementation of a successful screening system for industrial enzyme development is not so much a result of special technologies. It is more a result of a deep understanding of the objectives, a well-thought-out directed evolution strategy, a solid work flow with optimized and standardized procedures for each step, and the choice of suitable methods for analytics. Finally, success depends highly on an experienced and well-trained team.

1.2.4 Impact of Enzyme Development Technologies Today and Tomorrow

Looking at the multitude of enzyme applications in the various fields as illustrated in the following chapters, it becomes clear that enzymes play an almost hidden, but indispensable, role in our everyday life. Owing to the boost in innovations in enzyme technology and its implementation in further processes expanding the application space, there is increasing acceptance and trust in the development and application of enzyme technology.

The potential of today's industrial enzyme development technologies opens up new application fields connected to regulatory challenges. Today, highly engineered enzymes produced recombinantly in *E. coli* are used to prepare food ingredients. The corresponding regulatory approval processes are challenging and require much resources and time. The approval authorities face the challenge of having to carry out safety assessments with these novel products and to set new standards. A more detailed description would go beyond the scope of this chapter. Nevertheless, it should be mentioned that the more examples play through this process, the simpler the approval procedures will become. The approval of recombinant enzymes from *E. coli* for the production of ingredients for baby food is a major step forward in that sense (see Chapter 2.5).

In terms of technologies for enzyme development, there has been a movement from specialized (high-throughput) methods that could only be realized by industries or academic institutes with high financial capabilities to an almost everyone's technology that can use smart library design for the initial steps (Figure 1.2.5). The development of a fully optimized enzyme for industrial use still requires expert teams with experience and efficient, standardized technologies in the whole development chain.

Developments in bioinformatics are becoming more and more important. Big data analysis and machine learning may open up faster routes to optimized enzymes [86, 108]. However, at present this is not realized and big data analysis will require the availability (hence, collection) of big data, which is inherently connected to a substantial effort in the experimental setup and data processing.

Methodologies based on structural analysis are still regarded as the most promising routes to manipulate an enzyme in a designed way. *De novo* design of enzymes is a recent development that inspires the imagination of new possibilities in enzyme application [109]. Although *de novo* design has made a big step forward in recent years, namely by the work of David Baker and the development of the Rosetta algorithm [110], it is still far away from creating industrially applicable enzymes immediately. Currently, it can provide new starting points that require further optimization by the directed evolution methods described above [111]. Enzyme dynamics do play an increasingly important role in explaining remote mutations obtained by random methods [94]. The view that enzyme dynamics is important for function extends the simple lock and key principle (which reduces the view to the substrate binding pocket) to the entire network of the polypeptide surrounding the active site. Enzyme dynamics can thus provide important design principles for future enzyme development. The constantly increasing calculation capacity of modern computers opens up wider access and more accurate simulations of enzyme action. Figure 1.2.5 illustrates some landmark inventions in enzyme technologies and their impact on industrial implementation. See also Figure 1.1.3 in Chapter 1.1 for enzyme product introductions before 1960.

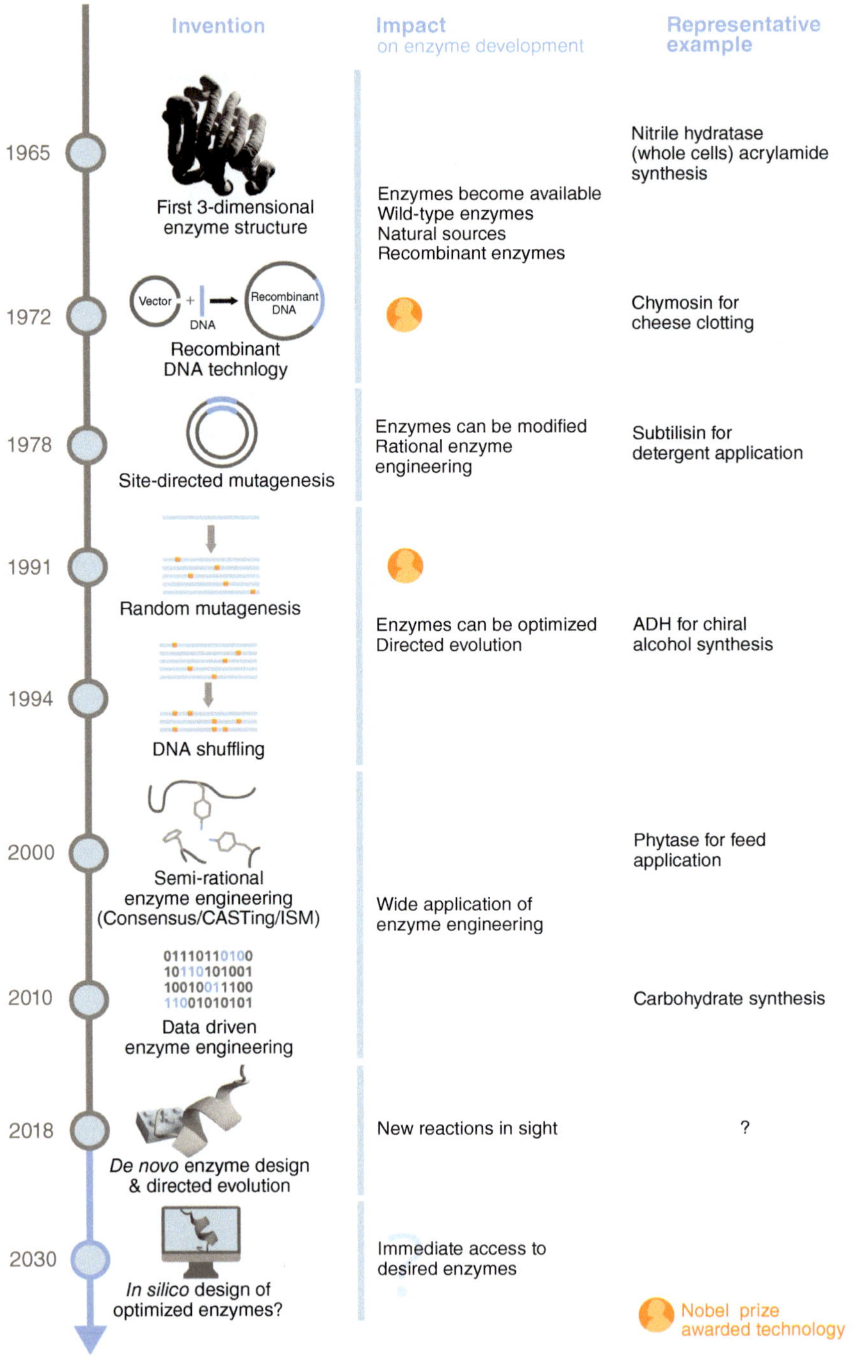

Figure 1.2.5 Landmark inventions in enzyme technology and their impact on industrial enzyme development.

If enzyme function, stability, and folding should be fully understood, then the immediate access of an enzyme for an industrial application from a computational design would become feasible and would make the generation and screening of variant libraries unnecessary. Still, we are far away from this today.

Acknowledgments

Thanks to Sebastian Bartsch (c-LEcta GmbH) for discussions about this topic, which continues to fascinate us, and for revising the manuscript.

References

1 Aehle, W. (ed.) (2007). *Enzymes in Industry: Production and Applications*, 3e. Weinheim: Wiley-VCH.
2 Kirk, O., Borchert, T.V., and Fuglsang, C.C. (2002). *Curr. Opin. Biotechnol.* 13: 345–351.
3 van Beilen, J.B. and Li, Z. (2002). *Curr. Opin. Biotechnol.* 13: 338–344.
4 Hughes, D.L. (2018). *Org. Process Res. Dev.* 22: 1063–1080.
5 Berg, P. and Mertz, J.E. (2010). *Genetics* 184: 9–17.
6 Arnold, F.H. (2018). *Angew. Chem. Int. Ed.* 57: 4143–4148.
7 Dalbadie-McFarland, G. and Richards, J.H. (1983). *Annu. Rep. Med. Chem.* 18: 237–245.
8 Chen, K.Q. and Arnold, F.H. (1991). *Biotechnology (NY)* 9: 1073–1077.
9 Stemmer, W.P. (1994). *Nature* 370: 389–391.
10 Bornscheuer, U.T., Huisman, G.W., Kazlauskas, R.J. et al. (2012). *Nature* 485: 185–194.
11 Wu, L., Sun, Q., Desmeth, P. et al. (2016). *Nucleic Acids Res.* 2017: D611–D618.
12 Schomburg, I., Jeske, L., Ulbrich, M. et al. (2017). *J. Biotechnol.* 261: 194–206.
13 Goldenzweig, A., Goldsmith, M., Hill, S.E. et al. (2016). *Mol. Cell* 63: 337–346.
14 Pavelka, A., Chovancova, E., and Damborsky, J. (2009). *Nucleic Acids Res.* 37: W376–W383.
15 Steinkellner, G., Gruber, C.C., Pavkov-Keller, T. et al. (2014). *Nat. Commun.* 5: 4150.
16 Kuipers, R.K., Joosten, H.J., van Berkel, W.J. et al. (2010). *Proteins* 78: 2101–2113.
17 Sayers, E.W., Cavanaugh, M., Clark, K. et al. (2018). *Nucleic Acids Res.* 47: D94–D99.
18 Handelsman, J., Rondon, M.R., Brady, S.F. et al. (1998). *Chem. Biol.* 5: R245–R249.
19 Ferrer, M., Beloqui, A., Timmis, K.N., and Golyshin, P.N. (2009). *J. Mol. Microbiol. Biotechnol.* 16: 109–123.

20 Steele, H.L., Jaeger, K.E., Daniel, R., and Streit, W.R. (2009). *J. Mol. Microbiol. Biotechnol.* 16: 25–37.
21 Gabor, E.M., Alkema, W.B., and Janssen, D.B. (2004). *Environ. Microbiol.* 6: 879–886.
22 Czaja, R. (2015). *BIOspektrum* 21: 344–346.
23 Carlson, R. (2009). *Nat. Biotechnol.* 27: 1091–1094.
24 Woodley, J.M. (2013). *Curr. Opin. Chem. Biol.* 17: 310–316.
25 Aehle, W. and Estell, D.A. (2008). Systematic evaluation of sequence and activity relationships in proteins using site evaluation libraries for engineering multiple properties. WO2008002472A2, 64pp. 3 January 2008. Danisco US Inc., Genencor Division, USA.
26 Tokuriki, N. and Tawfik, D.S. (2009). *Curr. Opin. Struct. Biol.* 19: 596–604.
27 Polizzi, K.M., Bommarius, A.S., Broering, J.M., and Chaparro-Riggers, J.F. (2007). *Curr. Opin. Chem. Biol.* 11: 220–225.
28 Reetz, M.T., Soni, P., Fernandez, L. et al. (2010). *Chem. Commun. (Camb.)* 46: 8657–8658.
29 Demirjian, D.C., Morís-Varas, F., and Cassidy, C.S. (2001). *Curr. Opin. Chem. Biol.* 5: 144–151.
30 Sterner, R. and Brunner, E. (2008). *Thermophiles, Biology and Technology at High Temperatures* (ed. F. Robb, G. Antranikian, D. Grogan and A. Driessen), 25–38. Boca Raton, FL: CRC Press LLC.
31 Kirk, O. and Christensen, M.W. (2002). *Org. Process Res. Dev.* 6: 446–451.
32 Schurmann, M. and Sprenger, G.A. (2001). *J. Biol. Chem.* 276: 11055–11061.
33 Reetz, M.T. (2016). *Directed Evolution of Selective Enzymes: Catalysts for Organic Chemistry and Biotechnology*. Weinheim: Wiley-VCH.
34 Reetz, M.T., Bocola, M., Carballeira, J.D. et al. (2005). *Angew. Chem. Int. Ed.* 44: 4192–4196.
35 Hanefeld, U., Gardossi, L., and Magner, E. (2009). *Chem. Soc. Rev.* 38: 453–468.
36 Means, G.E. and Feeney, R.E. (1990). *Bioconjug. Chem.* 1: 2–12.
37 Cowan, D.A. and Fernandez-Lafuente, R. (2011). *Enzyme Microb. Technol.* 49: 326–346.
38 Bornscheuer, U.T. and Pohl, M. (2001). *Curr. Opin. Chem. Biol.* 5: 137–143.
39 Chen, R. (2001). *Trends Biotechnol.* 19: 13–14.
40 Lutz, S. (2010). *Curr. Opin. Biotechnol.* 21: 734–743.
41 Blake, C.C., Koenig, D.F., Mair, G.A. et al. (1965). *Nature* 206: 757–761.
42 Brannigan, J.A. and Wilkinson, A.J. (2002). *Nat. Rev. Mol. Cell Biol.* 3: 964–970.
43 Ulmer, K.M. (1983). *Science* 219: 666–671.
44 Estell, D.A., Graycar, T.P., and Wells, J.A. (1985). *J. Biol. Chem.* 260: 6518–6521.
45 Wilkinson, A.J., Fersht, A.R., Blow, D.M. et al. (1984). *Nature* 307: 187–188.
46 Leung, D.W., Chen, E., and Goeddel, D.V. (1989). *Technique* 1: 11–15.
47 Cadwell, R.C. and Joyce, G.F. (1992). *PCR Methods Appl.* 2: 28–33.
48 Arnold, F.H. (1998). *Acc. Chem. Res.* 31: 125–131.

49 Reetz, M.T. (2016). Selection versus screening in directed evolution. In: *Directed Evolution of Selective Enzymes* (ed. M.T. Reetz). https://doi.org/10.1002/9783527655465.ch2.
50 Patrick, W.M., Firth, A.E., and Blackburn, J.M. (2003). *Protein Eng.* 16: 451–457.
51 Stemmer, W.P. (1994). *Proc. Natl. Acad. Sci. U.S.A.* 91: 10747–10751.
52 Arnold, F.H. and Georgiou, G. (eds.) (2003). *Directed Evolution Library Creation.* https://doi.org/10.1385/159259395X.
53 Reetz, M.T. (2016). Directed evolution of enzymes. In: *Enzyme Catalysis in Organic Synthesis* (ed. K. Drauz, H. Gröger and O. May). https://doi.org/10.1002/9783527639861.
54 Lutz, S. and Patrick, W.M. (2004). *Curr. Opin. Biotechnol.* 15: 291–297.
55 Drummond, D.A., Iverson, B.L., Georgiou, G., and Arnold, F.H. (2005). *J. Mol. Biol.* 350: 806–816.
56 Nov, Y. (2014). *Methods Mol. Biol.* 1179: 261–278.
57 Zhao, J., Kardashliev, T., Joelle, R.A. et al. (2014). *Biotechnol. Bioeng.* 111: 2380–2389.
58 Wong, T.S., Tee, K.L., Hauer, B., and Schwaneberg, U. (2004). *Nucleic Acids Res.* 32: e26.
59 Murakami, H., Hohsaka, T., and Sisido, M. (2002). *Nat. Biotechnol.* 20: 76–81.
60 Fox, R.J., Davis, S.C., Mundorff, E.C. et al. (2007). *Nat. Biotechnol.* 25: 338–344.
61 Estell, D.A., Graycar, T.P., Miller, J.V. et al. (1986). *Science* 233: 659–663.
62 Wells, J.A., Vasser, M., and Powers, D.B. (1985). *Gene* 34: 315–323.
63 Miyazaki, K. and Arnold, F.H. (1999). *J. Mol. Evol.* 49: 716–720.
64 Kuchner, O. and Arnold, F.H. (1997). *Trends Biotechnol.* 15: 523–530.
65 Reetz, M.T., Wilensek, S., Zha, D., and Jaeger, K.E. (2001). *Angew. Chem. Int. Ed.* 40: 3589–3591.
66 Liebeton, K., Zonta, A., Schimossek, K. et al. (2000). *Chem. Biol.* 7: 709–718.
67 Voigt, C.A., Martinez, C., Wang, Z.G. et al. (2002). *Nat. Struct. Biol.* 9: 553–558.
68 Coco, W.M., Encell, L.P., Levinson, W.E. et al. (2002). *Nat. Biotechnol.* 20: 1246–1250.
69 Ness, J.E., Kim, S., Gottman, A. et al. (2002). *Nat. Biotechnol.* 20: 1251–1255.
70 Zha, D., Eipper, A., and Reetz, M.T. (2003). *ChemBioChem* 4: 34–39.
71 Wells, J.A. and Estell, D.A. (1988). *Trends Biochem. Sci.* 13: 291–297.
72 von der Osten, C., Branner, S., Hastrup, S. et al. (1993). *J. Biotechnol.* 28: 55–68.
73 Bryan, P.N. (2000). *Biochim. Biophys. Acta* 1543: 203–222.
74 Ness, J.E., Welch, M., Giver, L. et al. (1999). *Nat. Biotechnol.* 17: 893–896.
75 Lehmann, M., Pasamontes, L., Lassen, S.F., and Wyss, M. (2000). *Biochim. Biophys. Acta* 1543: 408–415.
76 Lehmann, M., Loch, C., Middendorf, A. et al. (2002). *Protein Eng.* 15: 403–411.
77 Polizzi, K.M., Chaparro-Riggers, J.F., Vazquez-Figueroa, E., and Bommarius, A.S. (2006). *Biotechnol. J.* 1: 531–536.

78 Kourist, R., Jochens, H., Bartsch, S. et al. (2010). *ChemBioChem* 11: 1635–1643.
79 (2018). *Nat. Biotechnol.* 36: 1023.
80 Larrick, J.W., Schellenberger, V., and Barbas III, C.F. (2013). *Nat. Biotechnol.* 31: 584.
81 Reetz, M.T., Carballeira, J.D., Peyralans, J. et al. (2006). *Chemistry* 12: 6031–6038.
82 Reetz, M.T., Wang, L.W., and Bocola, M. (2006). *Angew. Chem. Int. Ed.* 45: 1236–1241.
83 Reetz, M.T., Carballeira, J.D., and Vogel, A. (2006). *Angew. Chem. Int. Ed.* 45 (46): 7745–7751.
84 Reetz, M.T. and Carballeira, J.D. (2007). *Nat. Protoc.* 2: 891–903.
85 Bommarius, A.S., Blum, J.K., and Abrahamson, M.J. (2011). *Curr. Opin. Chem. Biol.* 15: 194–200.
86 Badenhorst, C.P.S. and Bornscheuer, U.T. (2018). *Trends Biochem. Sci.* 43: 180–198.
87 Liu, Q., Xun, G., and Feng, Y. (2018). *Biotechnol. Adv.* (in press).
88 Oue, S., Okamoto, A., Yano, T., and Kagamiyama, H. (1999). *J. Biol. Chem.* 274: 2344–2349.
89 Morley, K.L. and Kazlauskas, R.J. (2005). *Trends Biotechnol.* 23: 231–237.
90 Agarwal, P.K., Billeter, S.R., Rajagopalan, P.T. et al. (2002). *Proc. Natl. Acad. Sci. U.S.A.* 99: 2794–2799.
91 Osuna, S., Jimenez-Oses, G., Noey, E.L., and Houk, K.N. (2015). *Acc. Chem. Res.* 48: 1080–1089.
92 Goldsmith, M. and Tawfik, D.S. (2017). *Curr. Opin. Struct. Biol.* 47: 140–150.
93 Strickler, S.S., Gribenko, A.V., Gribenko, A.V. et al. (2006). *Biochemistry* 45: 2761–2766.
94 Nick, P.C., Scholtz, J.M., and Grimsley, G.R. (2014). *FEBS Lett.* 588: 2177–2184.
95 DeSantis, G., Wong, K., Farwell, B. et al. (2003). *J. Am. Chem. Soc.* 125: 11476–11477.
96 Funke, S.A., Eipper, A., Reetz, M.T. et al. (2003). *Biocatal. Biotransform.* 21: 67–73.
97 Brissos, V., Eggert, T., Cabral, J.M., and Jaeger, K.E. (2008). *Protein Eng. Des. Sel.* 21: 387–393.
98 Reetz, M.T., Kahakeaw, D., and Lohmer, R. (2008). *ChemBioChem* 9: 1797–1804.
99 Kille, S., Acevedo-Rocha, C.G., Parra, L.P. et al. (2013). *ACS Synth. Biol.* 2 (2): 83–92.
100 Kazlauskas, R.J. and Bornscheuer, U.T. (2009). *Nat. Chem. Biol.* 5: 526–529.
101 Tracewell, C.A. and Arnold, F.H. (2009). *Curr. Opin. Chem. Biol.* 13: 3–9.
102 Reetz, M.T. (2013). *Angew. Chem. Int. Ed.* 52: 2658–2666.
103 Reetz, M.T., Zonta, A., Schimossek, K. et al. (1997). *Angew. Chem. Int. Ed. Engl.* 36: 2830–2832.
104 Wu, Q., Soni, P., and Reetz, M.T. (2013). *J. Am. Chem. Soc.* 135: 1872–1881.
105 Greiner-Stöffele, T. and Struhalla, M. (2003). Method from the selection of biomolecules from biomolecule variant libraries. WO2005040376.

106 Polizzi, K.M., Parikh, M., Spencer, C.U. et al. (2006). *Biotechnol. Prog.* 22: 961–967.

107 Vogel, A., Schmiedel, R., Hofmann, U. et al. (2014). *ChemCatChem* 6: 965–968.

108 Saito, Y., Oikawa, M., Nakazawa, H. et al. (2018). *ACS Synth. Biol.* 7: 2014–2022.

109 Zanghellini, A. (2014). *Curr. Opin. Biotechnol.* 29: 132–138.

110 Huang, P.S., Boyken, S.E., and Baker, D. (2016). *Nature* 537: 320–327.

111 Zeymer, C., Zschoche, R., and Hilvert, D. (2017). *J. Am. Chem. Soc.* 139: 12541–12549.

1.3

Eukaryotic Expression Systems for Industrial Enzymes

Lukas Rieder, Nico Teuschler, Katharina Ebner, and Anton Glieder

Graz University of Technology, Institute of Molecular Biotechnology, Petersgasse 14, 8010 Graz, Austria

1.3.1 Eukaryotic Enzyme Production Systems

The main focus of this chapter is to provide insights into the field of "industrial biotechnology," a biotechnological approach using living cells or pure enzymes for chemical manufacturing with the aim of reducing energy consumption and waste [1].

While prokaryotic expression hosts are still commonly used in industry (Figure 1.3.1), this work mainly focuses on eukaryotic microbial expression systems, the resulting specific considerations that have to be made, and the consequences for modern industrial enzyme production. The chapter also provides insights into current industrial production processes: "Which enzymes are currently produced by eukaryotic microbes?" and "What are the state-of-the-art techniques used for industrial enzyme production and application?" Since the introduction of the first recombinantly produced enzyme both the diversity of expression systems and the diversity and quality of enzymes used in our daily life have exploded.

Sales of industrially relevant enzymes today show an estimated value of about €4.0 billion per year [2] compared to about €100 billion in therapeutic protein production in the year 2015 [3]. With a share of almost 50% of the total enzyme market, Novozymes was the leading company in this field in 2015 and, together with DuPont and DSM, held about 75% of the total market value in this field [4]. See also Chapter 1.1 for an historic perspective and an overview of the industrial enzyme market.

1.3.2 Special Considerations for Working with Eukaryotic Expression Systems

1.3.2.1 Choice of Expression Host

All available expression hosts have advantages and drawbacks for the production of individual target proteins. Major criteria to be considered to find the best

Industrial Enzyme Applications, First Edition. Edited by Andreas Vogel and Oliver May.

suited system are, for example, total yield, space time yield (STY), specific productivity, quality, type and efficiency of folding and post-translational processing, cost of media and downstream processing (DSP), and compatibility of the enzyme product with the desired application. Especially for food, feed, and pharma applications regulatory aspects for legal product approval need to be evaluated as well.

The filamentous fungus *Trichoderma reesei (syn. Hypocrea jecorina)* and the nonconventional yeast *Komagataella phaffii* (*Pichia pastoris*), which will be discussed in the following paragraphs, are only two hosts within the large portfolio of expression systems commonly used in industry but typical and representative examples. Figure 1.3.1 gives an overview about the broad range of organisms applied for industrial protein expression including animal cells, plants, and bacteria, as listed by the Association of Manufacturers and Formulators of Enzyme Products (AMFEP) (https://amfep.org). Following this listing, 37% of the enzymes are produced in prokaryotic organisms while enzymes made in fungi dominate the eukaryotic domain of life.

AMFEP is a nonprofit European industry association and provides information related to the regulatory status of enzymes in the European Union. Basically, all enzyme producing companies providing enzymes for the European market are members of this association and provide the AMFEP with a complete list of the manufactured enzymes. As a result, an overview about all enzymes for industrial applications in Europe is available.

In addition, information provided by the AMFEP also includes the exact species used for the respective production process. The most common eukaryotic microorganisms are listed in Table 1.3.1 (https://amfep.org).

In Table 1.3.2, adapted from Gomes et al., the advantages and disadvantages of some commonly used prokaryotic and eukaryotic host organisms for industrial applications are shown [5].

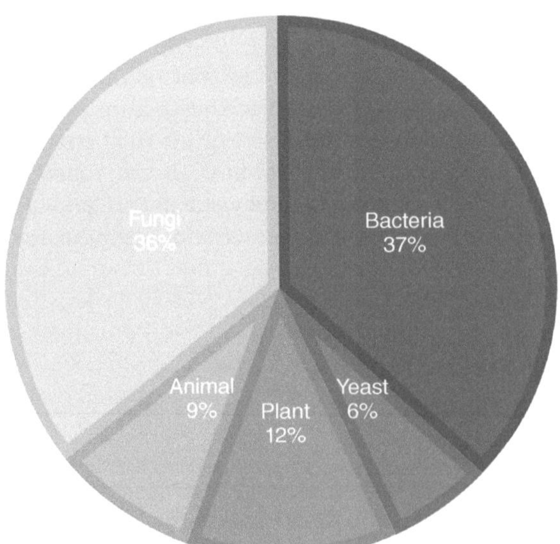

Figure 1.3.1 Production hosts for industrially relevant enzymes classified by the kingdoms of life. Source: Data provided on the AMFEP website: https://amfep.org.

Table 1.3.1 List of common eukaryotic host microorganisms for the production of industrially relevant enzymes.

Aspergillus niger	*Pichia angusta* or *Hansenula polymorpha*
Aspergillus oryzae	*Pichia pastoris*
Cryphonectria or *Endothia parasitica*	*Saccharomyces cerevisiae*
Fusarium venenatum	*Schizosaccharomyces pombe*
Kluyveromyces lactis	*Trichoderma reesei* or *longibrachiatum*
Penicillium camemberti	

Source: Data provided on the AMFEP website: https://amfep.org.

Since target-specific challenges define the requirements for an adequate host organism, no "universal production host" suitable for all applications can be depicted. Generally speaking, the best suited choice for a specific production host depends on the targeted application. For example, the specific requirements for proteins produced for pharmaceutical applications are fundamentally different from those produced for consumer applications, food or feed purposes, or as biocatalysts for chemical production. In case of pharmaceutical products it is absolutely necessary to have flawless, preferably homogenous, posttranslational modifications, low error rates in translation, and perfect folding of the protein [6]. On the other hand, for laundry enzymes the price, specific productivity, cost of media, complexity in DSP, and hurdles in commercial product approval will be the main effectors for the decision on a specific production host.

The following paragraphs focus on yeasts and fungi as expression hosts as these are the most commonly used eukaryotic organisms in industrial biotechnology and complementary to frequently used bacterial host systems.

1.3.2.2 Comparison of Cell Structure and Their Influence on Molecular Biology

The molecular organization of cells is a major issue that influences the potential use as "cell factories" for the production of enzymes and chemicals. Eukaryotic cells structures show many significant differences to prokaryotic ones, which has consequences for their respective fields of application.

The subcellular organelles of eukaryotes have varying substrate permeability, redox conditions, and capabilities for posttranslational modifications. Targeting of enzymes specifically to these organelles provides biological challenges, as well as new opportunities by employing eukaryotic hosts. For example, protein secretion involves specific targeting to the endoplasmic reticulum (ER) directed by specific signal peptides such as the CBH1 secretion signal of the exoglucanase 1 (CBH1) of *T. reesei*, or the alpha mating factor secretion signal of *Saccharomyces cerevisiae*. In this compartment, specific posttranslational modifications such as N- and O-glycosylation take place. Additionally, the quality of (re-)folding of the disulfide bridge containing enzymes in this oxidative compartment is important before the respective enzymes follow their path of further transport and modification through different Golgi compartments prior to fusing with the cell membrane and secretion to the culture supernatant [7, 8].

Table 1.3.2 List of industrial host organisms for protein production and the respective advantages and disadvantages, according to Gomes et al. [5].

		Organism	Advantages	Disadvantages
Prokaryotic	Bacteria	*Bacillus subtilis*	GRAS, efficient secretion	Secreted proteases, sometimes bad expression of heterologous genes, low stability of plasmids, sporulation of WT strains
		Escherichia coli	Easiest, fast, and cheap system, well-defined genetic and metabolic models, no glycosylation, different platform strains	No N-, and O-, glycosylation, pathogenic strains exist, no efficient secretion
		Pseudomonas	High cell densities	Limited plasmid stability, pathogenic strains exist
Eukaryotic	Yeast	*Pichia pastoris*	High cell density, easy scale up, high levels of functional proteins	Hypermannosylation of mammalian proteins
		Saccharomyces cerevisiae	Well established, can be used for therapeutic application	Hyperglycosylation and branched mannosylation of secreted proteins
	Fungi	*Aspergillus niger*	GRAS, secretion potential, typical eukaryotic posttranslational modifications	Complex manipulation, low transformation rates
		Myceliophthora thermophila	Efficient secretion, typical eukaryotic posttranslational modifications, interesting spectrum of own secreted enzyme cocktail for lignocellulose degradation	Complex manipulation, low transformation rates
		Trichoderma reesei	Efficient secretion, typical eukaryotic posttranslational modifications, interesting spectrum of secreted endogenous enzyme cocktail for lignocellulose degradation	Complex manipulation, secreted proteases

In prokaryotic systems, all processes connected to the DNA, such as replication and transcription, take place in "one pot," the cytoplasm, while in eukaryotes these processes are compartmentalized and located to the nucleus. As a consequence, in a prokaryotic system such as *Escherichia coli* the DNA introduced during genetic engineering only needs to pass over the outer and inner cell membrane to be replicated, transcribed, and translated.

In case of microbial eukaryotes such as *T. reesei* or *P. pastoris,* foreign DNA has to get through the cell wall as well as the pores of the nucleolus to reach the place of transcription or integrate into the host genome before it gets degraded in the host cytoplasm. Following transcription as well as mRNA processing within the nucleus, the mature mRNA is translocated to the cytoplasm for translation and, subsequently, potential further translocation of the protein, e.g. secretion of protein to the extracellular space [9, 10].

Such host-specific differences also influence the design of expression vectors, expression cassettes, and finally the overall expression strategy.

1.3.3 Differences in Vector Design for Eukaryotic and Prokaryotic Hosts

In order to enable simple cloning and DNA amplification, most expression constructs are designed as double-stranded DNA molecules called plasmids, which can easily be propagated using *E. coli*. Per definition, plasmids are circular or linear extrachromosomal replicons that can be used for the introduction, modification, or removal of target genes [11]. Generally, one major goal of expression construct design is to keep the size of the vector as small as possible, since rising construct size increases metabolic load on the organism and decreases plasmid stability [12, 13]. One of the largest plasmids found in nature originates from *Azospirillum brasilense* CBG497 and has a size of 1.59 megabases [14]. However, natural *E. coli* plasmids are typically 10–100 kb large, and commercially available and commonly used expression vectors are usually not bigger than 3–5 kb [10].

The exact design of an expression vector depends on the host organism of choice. Nevertheless, it is possible to distinguish between autonomously replicating plasmids, also called episomal vectors, artificial chromosomes, which are large chromosomes such as DNA molecules with (an) endogenous origin(s) of replication, and integrative vector systems, which do not replicate autonomously but are integrated into the genome and maintained during cell division due to chromosomal replication origins [15].

Plasmid-based expression systems are more frequently used in prokaryotic organisms but they can also be used in lower eukaryotes. In higher eukaryotes their application is limited to transient expression [12]. Plasmid-based systems for both types of organisms, prokaryotic and eukaryotic, share some necessary genetic parts, but also show significant differences (Figure 1.3.2). For eukaryotes, typically shuttle plasmids with regulatory elements and selection markers working in both prokaryotes and eukaryotes are used, as cloning is much easier and faster in *E. coli* [16]. In both systems, an origin of replication (ORI) for

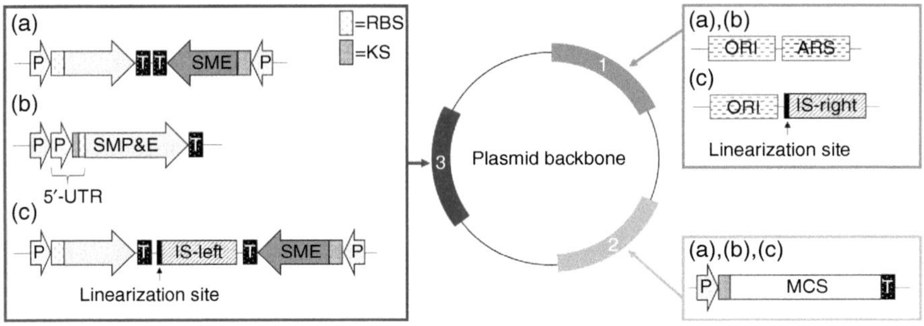

Figure 1.3.2 Different possibilities for the construction of a eukaryotic expression vector. Generally, three crucial regions can be discer'ned: A region for autonomous replication that works either only in prokaryotes (ORI) or in the eukaryotic host as well (ORI and ARS) (1), a region for inserting the GOI (2), and a region containing selection marker(s) (3). Depending on the combination of the regions one can create (a) an episomal plasmid with two marker genes, (b) one marker gene that works in the prokaryotic propagation host and the final eukaryotic expression host, or (c) a typical eukaryotic integrative plasmid with two integration sites (IS) flanking the multiple cloning site (MCS).

independent replication of the DNA within the cell, a selection marker (SM-) for use in prokaryotes (P) and/or eukaryotes (E), depending on the system, and a multiple cloning site (MCS, Figure 1.3.2, 2a–c) for the integration of the gene of interest (GOI) into the vector are needed. Typically, expression vectors are assembled from individual genetic parts either by classical restriction cloning or by state-of-the-art scarless DNA assembly using recombinant cloning technologies such as Gibson isothermal assembly. Special considerations that have to be made in regard to the promoter and terminator regions, which drive expression of the selection marker, as well as the GOI, are discussed later in more detail.

Generally, vectors for expression in eukaryotes (episomal as well as chromosomal) need in addition to the prokaryotic selection system a marker for selection in the final eukaryotic host (E). These cassettes mostly consist of an antibiotic resistance gene or a gene for complementation of an auxotrophy, with the corresponding promoter and terminator regions (Figure 1.3.2, 3a,c). Keeping in mind the vector size an alternative approach uses resistance genes as selection markers (SMs) that work in both prokaryotic (P) and eukaryotic (E) organisms (SMP and E). Here, the promoter for prokaryotic transcription is integrated in the 5′ untranslated region (5′-UTR) of the eukaryotic core promoter region (Figure 1.3.2, 3b).

For autonomous replication in eukaryotic hosts a specific feature called ARS (autonomously replicating sequence) is indispensable [17]. It is shown in Figure 1.3.2 (1a,b) and is responsible for the replication of DNA molecules in eukaryotes, similar to the ORI in prokaryotes.

While episomal vectors are still frequently used for protein expression in prokaryotic hosts and baker's yeast, integrative vectors are more commonly used and most times preferred for many other yeasts (e.g. *P. pastoris, Hansenula polymorpha, Kluyveromyces lactis, Schizosaccharomyces pombe*) and even standard for filamentous fungi [18, 19].

One of the two major benefits of genomic integration is the high chance to obtain a stable production strain, in contrast to most plasmid-based autonomously replicating vectors [20]. The second reason why production strains with stably integrated expression cassettes are preferred in industrial applications is that no further selection pressure (e.g. by antibiotics) is needed after transformation and selection [21, 22].

With integrative systems, expression strength usually depends on the specific integration locus as well as differences in numbers of integrated copies of the expression cassette (copy number) [23]. To control and predict locus effects, integration can be targeted to a specific locus using extensions to the expression cassette, which are identical (homologous ends) to a specific preferred targeting site (Figure 1.3.2, 1c and 3c). Such specific loci can show special properties such as inducible or especially strong expression. In addition, some preferred loci show low or neglectable epigenetic effects. An example of this strategy is the directed integration into the *AOX1* locus in the methylotrophic yeast *P. pastoris*. Expression from the *AOX1* promoter is strongly inducible by methanol and as an added benefit, disruption of the *AOX1* gene reduces biomass production from methanol to a minimum; therefore, uncoupling biomass formation from target gene expression is possible [24]. In addition, less toxic hydrogen peroxide and heat are produced as by-products of the methanol metabolism in less tightly controlled feeding schemes such as high-throughput cultivation on a small scale. Site-specific integration can be identified due to the slow growth on methanol as the sole carbon source caused by the disruption or replacement of the gene open reading frame coding for the alcohol oxidase 1 (AOX1).

After the introduction of DNA into an organism selection pressure has to be applied to force the host to retain the foreign DNA or to identify cells with such DNA integrated into their genome. In prokaryotes and also to some part in eukaryotes selection of transformants is mostly done using marker systems based on antibiotics. The prerequisite here is that the expression vector carrying the GOI also carries an expression cassette for the expression of a protein conferring resistance to a specific normally toxic substance. Table 1.3.3 lists the most commonly used antibiotic-based selection systems.

In order to keep the expression construct small in size, selection markers working in prokaryotes as well as in eukaryotes can be used. The double functionality of some selection markers is based on the interaction with highly conserved and vital cell functions. Resistance to the broadly active antibiotic Zeocin, a widely used selection system and probably most commonly used for *P. pastoris*, for example, leads to a cleavage of the host's genomic DNA and the gene product of the selection marker binds and thereby inactivates the antibiotic. Hygromycin B, which is often used with *T. reesei*, inhibits the translocation of mRNA and tRNA on the ribosome. For marker gene expression in prokaryotes as well as in eukaryotes, dual promoters are useful. For this purpose, the prokaryotic promoter is usually put to a nonfunctional and non-translated area of the eukaryotic core promoter or the 5′-UTR.

While many experiments in laboratory scale use antibiotic resistance markers due to their efficiency and simplicity, markers relying on growth deficiency such as auxotrophy or lack of competencies for special carbon source utilization are

Table 1.3.3 Frequently used selection markers based on antibiotics for prokaryotic and eukaryotic organisms, based on data from Addgene.

	Substrate	Resistance gene	Interaction	Concentration (μg/ml)
Prokaryotic	Kanamycin	*nptII*	Binds 30S ribosomal subunit; causes miss-translation	50–100
	Ampicillin	*bla*	Inhibits cell wall synthesis	100–200
	Bleomycin (Zeocin)[a]	*ble*	Induces DNA breaks	5–100
	Carbenicillin	*bla*	Inhibits cell wall synthesis	100
	Chloramphenicol	*cat*	Binds 50S ribosomal subunit; inhibits peptidyl translocation	5–25 (EtOH)
	Erythromycin	*erm*	Blocks 50S ribosomal subunit; inhibits aminoacyl translocation	50–100 (EtOH)
	Spectinomycin	*aadA14*	Binds 30S ribosomal subunit; interrupts protein synthesis	7.5–50
	Streptomycin	*aadA14*	Inhibits initiation of protein synthesis	25–100
	Tetracycline	*tet*	Binds 30S ribosomal subunit; inhibits protein synthesis (elongation step)	10
Eukaryotic	Blasticidin	*bsd, bls, bsr*	Inhibits termination step of translation	2–10
	G418/Geneticin[a]	*neo*	Blocks polypeptide synthesis at 80S; inhibits chain elongation	100–800
	Hygromycin B[a]	*hygB*	Blocks polypeptide synthesis at 80S; inhibits chain elongation	50–500
	Puromycin	*pac*	Inhibits protein synthesis; premature chain termination	1–10

a) Markers working in prokaryotic and eukaryotic cells [25].
Source: Addgene 2017 [25].

preferred for industrial applications. This strategy allows to avoid the presence of antibiotic resistance genes in biomass of large scale processes and is especially beneficial in cases in which the selection pressure has to be maintained on production scale to keep the plasmids stable [25].

Table 1.3.4 Frequently used auxotrophic systems based on data from Addgene [26] and Nett et al. [27].

Gene	Target
arg1	L-Arginine
arg2	L-Arginine
his3	L-Histidine
leu1+	L-Leucine
leu2	L-Leucine
lys2	L-Lysine
met15	L-Methionine and overproduces hydrosulfide ions
trp1	L-Tryptophan
ura3	Pyrimidine (uracil)
ura4+	Pyrimidine (uracil)

Auxotrophic systems work as combinations of strains lacking functionality of certain essential proteins or enzymes (auxotrophic strains) and the complementation of such deficiency, by introducing a functional copy of the corresponding gene via plasmids or linearized vector fragments for genomic integration [26]. Auxotrophic systems are mostly based on complementation of missing functionality in the synthesis of amino acids or nucleotides. Table 1.3.4 is based on data from Addgene [25] and Nett et al. [27] and lists the frequently used autotrophic markers.

Likewise, the deficiencies in carbon source utilization can also be used for the selection. For example, the deletion of the *GUT1* (glycerol kinase 1) gene in yeasts leads to an inability to use glycerol as the sole carbon source. The advantage of carbon source utilization deficiencies for selection lies in the simplicity of media preparation and strain cultivation as well as the possibility to apply complex media for growth, which is not the case for amino acid auxotrophy based selection [28]. Dying cells during cultivation are another problem as this may lead to sufficient provision of the missing amino acids to the growing cells, leading to a loss of selection pressure.

The triose phosphate isomerase (TPI1) system, reported for the production of insulin by Novo Nordisk, is a prominent example for the industrial application of a carbon utilization marker. The *TPI1* gene was deleted in the genome of the host *S. cerevisiae* and by introducing a plasmid with the corresponding gene from *S. pombe (POT1),* the pathway was complemented, leading to a very stable production strain. Cells that have no plasmid hardly grow independent of the growth media, as a major enzyme for glycolysis is missing [29].

Counterselection markers, a further possibility to select for transformants, are especially applicable for the generation of multi knockout strains. This method uses positive and negative selection strategies. During the first round of selection to select for cells that took up the foreign DNA during transformation, such markers lead to the survival of the transformed cells (positive) while the same marker gene can be used to kill the host cell during the second round of selection (negative), as applied, for example, to identify cells that have lost the expression cassette for marker recycling [30]. An important counterselection marker for use in yeast and fungi is the amdS marker system. This marker was, for example, used for introducing several mutations to the genome of *S. cerevisiae* as marker recycling is possible and straightforward. In the first round of selection the amdS gene is introduced to the cell, which allows the use of acetamide as the sole nitrogen source. After selection, the amdS gene is removed from the genome. This process is called marker recycling as it allows use of the same marker gene in several successive mutation rounds. Once the marker is removed, the cells can be counterselected on media containing fluoroacetamide. In case the marker is successfully removed, no toxic fluoroacetic acid is produced anymore by an amdS-catalyzed amidase reaction, leading to the survival of the cells [31].

However, all the considerations for maintaining selection pressure only apply to expression strains employing episomal plasmids. In case of stable integration of the expression cassettes into the genome selection pressure is only necessary after the initial transformations process. After genomic integration the selection marker is stably maintained in the host genome and no selection is necessary anymore in the production process or scale-up. Thus, the problem of applying genes that confer antibiotic resistance for selection is more or less reduced to an approval and public acceptance issue.

1.3.4 Differences in Regulation of Gene Expression in Eukaryotes and Prokaryotes

Promoters are special regulatory elements of the DNA that coordinate the binding of the RNA polymerase to initiate transcription. It was observed that these A/T-rich DNA elements, which form the transcription start site, have unusual structures and low stability [32].

Promoter regions from prokaryotic cells (Figure 1.3.3a) show structural differences compared to eukaryotic organisms (Figure 1.3.3b).

In prokaryotic promoters, the sigma factor binding sites, the −35 and −10 regions, are highly conserved and are the most important part of the whole sequence. These conserved hexameric motifs are centered on or near to the positions −35 and −10. The term −10 indicates the position 10 base pairs upstream of the transcription start, which is marked as +1 in Figure 1.3.3a [32]. Although this region is necessary for the binding of the RNA polymerase it is known that inducible and constitutive promoters have different consensus sequences and use different sigma factors. For the translation of the GOI the 16s rRNA of the ribosome recognizes the ribosome binding site (RBS), also called Shine–Dalgarno (SD) sequence, and initiates translation. Against common assumption, the RBS is not strongly conserved within the whole kingdom of prokaryotes and is located 5–10 base pairs upstream of the start codon, the starting point of the translation [33], but their respective composition and location influence the efficiency of translation.

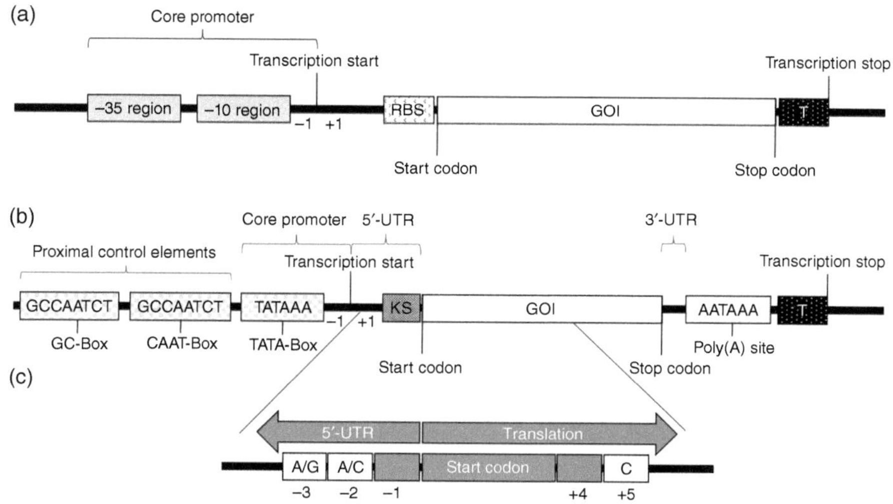

Figure 1.3.3 Comparison of prokaryotic and eukaryotic promoter regions (a, b) showing severe differences in regulation motifs and complexity. The close-up of the Kozak sequence (c) in white boxes shows the most important positions and bases for efficient translation in eukaryotes.

Eukaryotic promoters are more complex in structure (Figure 1.3.3b). The promoter sequence can be divided into a core promoter, which is usually rich in adenosine and thymine bases (A/T rich), and other regulatory elements such as the GC-Box and the CAAT-Box, upstream of the core promoter. The most characteristic element in the core promoter is the TATA-Box, located upstream of the transcription start and generally known as a positive factor for strong transcription. The exact location is dependent on the organism and promoter. Although this sequence is quite conserved and is a standard building block for synthetic promoters, some mismatches can occur and in many promoters it is totally absent (TATA less promoters). While 80% of *S. cerevisiae* promoters are TATA less, TATA containing promoters are frequently used for recombinant gene expression due to their high levels of transcriptional activation [34]. Next to transcription factors, which support the binding of the polymerase at the core promoter region, high expression levels in eukaryotic systems are also strongly dependent on the 5′UTR and 3′UTR [35].

In general, translation is more tightly regulated in eukaryotic hosts. The Kozak Consensus Sequence (KS) facilitates initiation of translation at the start codon AUG of the mRNA to start protein synthesis. This consensus sequence differs from species to species but generally the following base pairs are conserved amongst most species: −3A/G, −2A/C, and +5C, although −3A/G is the most crucial position for good translation (Figure 1.3.3c) [36].

Although the huge impact of the 5′-UTR on the expression efficiency is known, this DNA element is often ignored [37]. Kozak showed already in 1986 that hairpin structures within the 5′UTR are crucial for the expression level and that the expression level decreases with the strength of the hairpin [38]. A new approach of Weenink et al. showed that it is possible to predict the downregulation in expression by introducing hairpin structures to the 5′UTR [39].

1.3.4.1 Different Types of Promoters

Regardless of the differences between the different kingdoms of life, all organisms have constitutive and inducible promoters.

Although constitutive promoters are per definition active all the time, they usually show different regulation patterns in different phases of growth. In *T. reesei* the constitutive pyruvate decarboxylase promoter (P_{PDC}) was used in industrial applications for the production of xylanase II [40, 41]. Another example would be the glyceraldehydes-3-phosphate dehydrogenase promoter (P_{GAP}) in *P. pastoris*, which was described to lead to high STYs in cases where the recombinant protein causes no physical disadvantages to the host [31].

Inducible promoters can be divided into positive and negatively controlled ones. The promoter of the alcohol oxidase 1 gene (P_{AOX1}) is a typical tightly regulated promoter for applications with the host *P. pastoris*. The wild-type promoter can be induced by methanol, but mutant variants can be used without this toxic and flammable substance [24]. Similarly, the methanol oxidase promoter (P_{MOX}) of *H. polymorpha* is not repressed by glycerol in *Hansenula* and can be induced by simple derepression in *H. polymorpha* while it is typically repressed by glucose [42]. However, this derepression effect seemed to be host specific rather than promoter specific, since the *P. pastoris AOX1* promoter was also derepressed in the presence of glycerol, when used for protein expression by *H. polymorpha* [43].

The promoter of the exoglucanase 1 (P_{CBH1}) is an example for a strong inducible promoter of *T. reesei*. P_{CBH1} is induced by the presence of complex sugars, such as cellulose, and repressed when glucose is the sole carbon source [44]. P_{AOX1}, P_{MOX}, and P_{CBH1} are classical and frequently used inducible promoters for industrial applications using *P. pastoris, H. polymorpha,* and *T. reesei*, respectively.

Besides the well-known monodirectional promoters (MDPs) bidirectional promoters (BDPs) or dual promoters can also be found in nature. While MDPs only direct transcription in one direction BDPs work in both directions. This application brings an enormous advantage, as co-expression of different genes is important in many applications. Applications are, for example, co-expression of a selection marker with the GOI, or the simultaneous co-expression of a chaperon or a redox partner [45]. A well-known natural example of such dual promoters of *S. cerevisiae* is the strong inducible *GAL1/GAL10* promoter, which is strong in both directions [46]. A first library approach using BDPs to improve co-expression by *P. pastoris* was reported recently by Vogl et al. [45]

1.3.5 Industrial Enzyme Production

The omnipresence and versatility of enzymes was already discussed in Chapter 1.1. In order to be able to be in line with the high volumetric demand on enzymes several improvements are usually necessary to increase yields by the preferred host. For example, multiple integration of expression cassettes is one of the most frequently used strategies to increase expression yields [47]. Multicopy integration can be achieved by different strategies. Two very easy approaches to achieve multicopy integration are the use of either high amounts of DNA for transformation or the use of media with high antibiotic concentration for selection after the transformation procedure [48, 49]. Next to these common approaches another possibility is to use a weak promoter regulating the transcription of the resistance gene, resulting in transformants only in case of multiple integration events. Alternatively, multiple copies can be integrated simultaneously or consecutively using multiple expression constructs with different selection markers and also by using two gene copies in bidirectional expression vectors. The probably most successful method generating strains with multiple copies of the expression cassette is to use a mix of the different methods. However, it has to be considered that the optimal conditions can vary depending on the vector and host strains. For industrial applications also, the genetic stability of the obtained strains has to be considered, since multiple copy strategies can lead to stability issues in scale-up experiments and long-term cultivation procedures.

In 2008, Sunga et al. published the so-called posttransformational vector amplification (PTVA) method to generate up to 6% strains that have more than 10 constructs within the genome [48]. The method is based on transferring the cells from agar plates with low concentrations of antibiotics to those with high concentrations. Clones, which survive on the highest antibiotic concentration, are more likely to have several copies of the expression construct. Deriving from the original method a method called liquid PTVA was devised using liquid culture media for the cultivation of the cells. Therefore, media are changed at regular time periods with increasing amount of antibiotic. Only the last selection round is done by using plates, which leads to a decrease in costs [50].

However, frequently similar effects as in the case of multicopy integration can be obtained by codon optimization (more efficient translation), or stronger promoters, which leads to high abundance of mRNA such as the multiple expression cassettes but in some cases leading to more stable production strains. In general, combinations of these strategies lead to the most efficient production strains where a maximum of folded and processed active enzyme should be obtained while keeping the gene copy number at a minimum in view of potential genetic stability issues. Owing to homologous recombination, a loss of expression cassettes might cause trouble during long-time cultivation and large-scale production in industry [51, 52].

Aside from vector design another strategy for enhancing protein production is the development and engineering of industrial production strains. The *T. reesei* strain QM6a discovered during World War II on the Solomon Island was the basis for the most popular *T. reesei* production strains QM9414 and RutC30 (Figure 1.3.4), which are used due to their high capacity of cellulase secretion by this filamentous fungus [54–56].

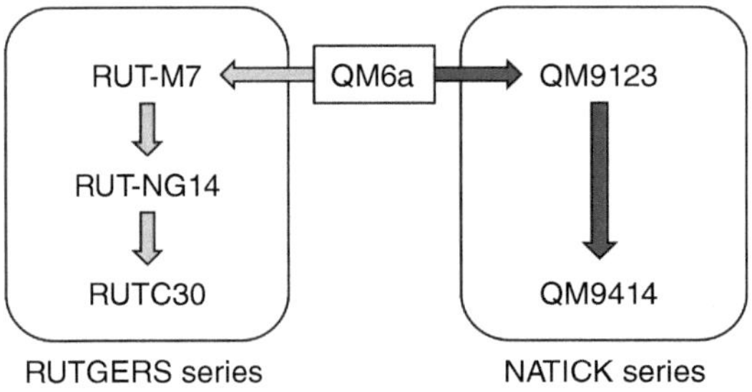

Figure 1.3.4 Genealogical tree of *T. reesei*. *T. reesei* QM6a as origin for modern production strains QM9414 and RutC30 made in several rounds of random mutagenesis, according to Kubicek [53]. Source: Kubicek 2013 [53]. Reproduced with permission of Elsevier.

Similarly, most industrial *P. pastoris* production strains were derived from the strain NRRL-Y11430 [57]. Frequently used derivatives of this wild type (WT) strain are the GS115 and the X-33 strains, available from Thermo Fisher Scientific & Research Corporation Technologies, Inc (RCT) or the CBS7435 and BG10 based platforms, described more recently by Näätsaari et al. and Sturmberger et al., respectively [28, 58].

In the beginning strain engineering was done mainly by using by UV-light or chemical mutagenesis, which is simple but causes much screening effort and multiple mutations in the genome [56, 59]. Most recent approaches use targeted methods such as CRISPR/Cas9-based technologies to inactivate or knock out specific genes e.g. proteases in order to eliminate side product formation or to avoid protein degradation due to protease effects, leading to a higher production of the target protein [60].

A well-known strategy to screen for clones with the desired gene disruption is the use of media components that become toxic (suicide substrates) by enzymatical turnover of the knockout target. The suicide substrate ensures that only organisms having the knockout will survive, as they are not able to form the toxic product [61]. Fluorouracil is one of such typically used compound. Similarly, suicide substrates were also applied to isolate protease-resistant mutants for improved heterologous protein production [51].

Next to modifying the production host or testing different vector and host combinations the enzyme itself can be tailored. One of the most common tailoring tools, especially if there is not much known about the enzyme, is random mutagenesis, combined with screening. In most cases, the aim of enzyme engineering for the purpose of increased production efficiency is to increase the solubility of an enzyme, its folding properties, or simply its stability in the culture supernatant during expression or during DSP. These are often neglected challenges and such approaches can influence enzyme yields in production processes significantly. Simple "self-cloning" of the host's own genes, for example with point mutations and/or in combination with alternative own promoters of the host genome, as well as the amplification of native expression cassettes, can also increase expression yields while avoiding the introduction of genetic changes, which cannot simply take place by nature, too.

1.3.6 Enzyme Production on Industrial Scale

The first enzyme produced in a genetically modified microorganism (GMM) introduced to the EU market was a lipase with the name Lipolase® by Novo Nordisk (today three companies: Novo Nordisk A/S, Novozymes A/S, and NNIT in 2000) in 1987 [62].

As mentioned above, the AMFEP collects information about the enzymes for industrial use in Europe. The latest update contained 238 enzymes belonging to 69 different categories according to the EC-number classification (EC, enzyme class). The exact vector/host combination and the scale of cultivation for industrial enzyme production are mostly not publicly known. Examples of industrially important enzymes and the respective host system and expression yields are shown in Table 1.3.5. As an example, the expression system used by DSM to produce a product called Brewers Clarex® is described. The endoprotease hydrolyzes proteins causing the formation of haze during the storage of beer. The enzyme is expressed in a glycoamylase and protease negative *Aspergillus niger* strain. Transformants were screened using an amdS marker system. For transcriptional control the *GLAA* (glucoamylase) promoter from *A. niger* was used [70]. This is a strong promoter inducible by starch and its products from hydrolysis and is frequently used in industrial fermentation processes [71].

1.3.6.1 Homologous Protein Production

Using "waste" material from food production (e.g. soy straw) containing cellulose, hemicellulose, and lignin, the most abundant polymers on our planet, for production of enzymes is a major goal of current industrial and political efforts. However, such renewable resources mostly have to be pretreated, which is either done chemically or physically, to function as efficient accessible carbon source [72].

One organism that can use such natural polymers after pretreatment for growing without any genetic engineering is *T. reesei*. The organism is well

Table 1.3.5 Selected examples of organisms with corresponding enzyme and yields for industrial homologous protein production.

Expression host	Enzyme	Yield	Activity	References
Aspergillus oryzae	Glycomylase	20 g/l	—	Finkelstein [63]
Aspergillus niger	Amylase	20 g/l	—	Wang et al. [64]
Aspergillus niger	Acid protease	—	3600 U/l	O'Donnell et al. [65]
Aspergillus niger	Pectinase	2.15 g/l	—	Suhaimi et al. [66]
Aspergillus niger	Catalase	16.5 g/l	—	O'Donnell et al. [65]
Aspergillus flavus	Glucosidase	21 mg/l	0.82 IU/ml	Dutt and Kumar [67]
Penicillium notatum	Glucose oxidase	—	112 U/ml	Sabir et al. [68]
Trichoderma reesei	Cellulase	100 g/l	—	Cherry and Fidantsef [69]

known for the effective production of biomass degrading enzymes such as cellobiohydrolases (CBH), endo-β-1,4 gluconases (EG), and β-glucosidases. This cocktail of enzymes, referred to as cellulase cocktail, has a high value for modern industry, waste treatment, and renewable energy production. They are used in the textile industry, paper industry, for laundry applications and, last but not least, in biorefineries [73]. Another feature that makes *T. reesei* an outstanding production host is the excellent protein secretion properties; modern strains can secrete up to 100 g/l of total protein. For this reason, *T. reesei* is assumed to be one of the most efficient organisms for protein production.

Myceliophthora thermophila C1 is another lignin cellulose degrading strain, which was established for industrial enzyme production by Dyadic (now DuPont/Genencor). Like *T. reesei,* the secretion potential of *M. thermophila* is enormous and reaches about 100 g/l with a purity of up to 80% [74]. In contrast to *T. reesei, M. thermophila* showed low protease activity during continuous cultivation processes, which made it a good production strain for recombinant protein production [75].

1.3.6.2 Heterologous Protein Production

Table 1.3.6 lists examples for heterologous recombinant proteins expressed in eukaryotic hosts. Nevertheless, the secretion of such heterologous proteins by filamentous fungi is often relatively low compared to native proteins [69]. Two main reasons for the low secretion yield are the high proteases activity of the fungi and the incorrect folding of recombinant proteins. To overcome this problem many different strategies, such as optimizing the codon usage, multiple random integration, and the use of engineered promoters, have been developed over the years [8, 90].

Another approach to increase the amount of secreted recombinant protein is the optimization of the cultivation conditions as it was observed that the pH during the cultivation has a major influence on the proteolytic degradation of the secreted proteins. Two major proteases are secreted depending on the pH. Acid aspartic protease is the main protease secreted at pH 5 or lower and a trypsin-like alkaline serine protease is secreted at pH 6 [91, 92]. The resulting strategy, which showed improved stability for recombinantly produced proteins, was to control the pH (pH 6) and to knock out the alkaline protease gene [93].

At last, it is important to mention that in addition to the topics discussed in this chapter, such as expression hosts, vector design, and strain construction, another important aspect of "industrial biotechnology" is the process development for bioreactor cultivation and induction, as well as DSP. Although these industrial aspects are less discussed in scientific literature, they are essential for highly efficient enzyme production, as they cause major cost contributions on the total manufacturing process and the different proteins expressed show different influences on the host strain physiology and specific productivity. Finally, also product recovery, polishing, and formulation are important factors in commercial enzyme manufacturing [94].

Table 1.3.6 Exemplifying list of organisms with corresponding enzymes produced, production yields, and promoter for heterologous protein production.

Expression host	Enzyme	Origin of gene	Promoter	Yields	References
Aspergillus oryzae	Aspartyl protease	*Rhizomucor miehei*	*AMYA*	3.3 g/l	Christensen et al. [76]
Aspergillus oryzae	Laccase	*Scytalidium thermophilum*	*AMYA*	1.8 g/l	Berka et al. [77]
Aspergillus oryzae	Xylanase	*Thermomyces lanuginosus*	—	—	X2753 – Novozymes
Aspergillus niger	Lysozyme	*Chicken*	*GLAA*	0.18 g/l	Wongwicharn et al. [78]
Aspergillus niger	Phytase	*Aspergillus ficuum*	*GLAA*	2.8 g/l	van Paridon et al. [79]
Aspergillus niger	Human interleukin-6	*Human*	*GLAA*	0.3 g/l	Punt et al. [80]
Aspergillus niger	Lipase	*Candida antarctica*	—	—	L3170 – Novozymes
Aspergillus awamori	Chymosin	*Calf*	*GLAA*	1.3 g/l	Dunn-Coleman et al. [81]
Acremonium chrysogenum	Alkaline protease	*Fusarium* sp.	*ALP*	4 g/l	Morita et al. [82]
Myceliophthora thermophila	Hydrolytic enzymes	*Trichoderma reesei*	*CBH2*	—	Gusakov et al. [83]
Neurospora crassa	HT186-D11 (scFv)	*Human antibody fragment*	*PCCG1NR*	3 mg/l	Havlik et al. [84]
Pichia pastoris	Glucose oxidase	*Penicillium notatum*	*AOX1*	2.5 g/l	Gao et al. [85]
Pichia pastoris	Cellobiohydrolase II	*Trichoderma reesei*	*AOX1*	6.55 g/l	Mellitzer et al. [86]
Trichoderma reesei	Lignin oxidase	*Phlebia radiata*	*CBH1*	0.02 g/l	Saloheimo et al. [87]
Trichoderma reesei	Acid phosphatase	*Aspergillus niger*	*CBH1*	0.5 g/l	Miettinen-Oinonen et al. [88]
Trichoderma reesei	Chymosin	*Calf*	*CBH1*	0.04 g/l	Uusitalo et al. [89]
Trichoderma reesei	Xylanase	—	*PDC*	1.52 g/l	Nevalainen and Peterson [40]

References

1 Glaser, J.A. (2005). White biotechnology. *Clean Technol. Environ. Policy* 7 (4): 233–235.

2 Singh, R., Kumar, M., Mittal, A., and Mehta, P.K. (2016). Microbial enzymes: industrial progress in 21st century. *3 Biotech* 6(2): 174.

3 Bandaranayake, A.D. and Almo, S.C. (2014). Recent advances in mammalian protein production. *FEBS Lett.* 588 (2): 253–260.
4 Marja Paloheimo, T.H., Mäkinen, S., and Vehmaanperä, J. (2016). *Production of Industrial Enzymes in Trichoderma reesei.* Springer International Publishing Switzerland.
5 Gomes, A.R., Byregowda, S.M., Veeregowda, B.M., and Balamurugan, V. (2016). An overview of heterologous expression host systems for the production of recombinant proteins. *Adv. Anim. Vet. Sci.* 4 (7): 10.
6 Lakowitz, A., Godard, T., Biedendieck, R., and Krull, R. (2018). Mini review: recombinant production of tailored bio-pharmaceuticals in different *Bacillus strains* and future perspectives. *Eur. J. Pharm. Biopharm.* 126: 27–39.
7 Lippincott-Schwartz, J., Roberts, T.H., and Hirschberg, K. (2000). Secretory protein trafficking and organelle dynamics in living cells. *Annu. Rev. Cell Dev. Biol.* 16: 557–589.
8 Wang, W., Meng, F., Liu, P. et al. (2014). Construction of a promoter collection for genes co-expression in filamentous fungus *Trichoderma reesei. J. Ind. Microbiol. Biotechnol.* 41 (11): 1709–1718.
9 Scherrer, K. (2018). Primary transcripts: from the discovery of RNA processing to current concepts of gene expression - review. *Exp. Cell Res.* 373 (1–2): 1–33.
10 Lodish, H., Berk, A., Zipursky, S.L. et al. (2000). *Molecular Cell Biology,* 4e. New York: W. H. Freeman.
11 Shintani, M., Sanchez, Z.K., and Kimbara, K. (2015). Genomics of microbial plasmids: classification and identification based on replication and transfer systems and host taxonomy. *Front. Microbiol.* 6: 242.
12 Palomares, L.A., Estrada-Mondaca, S., and Ramirez, O.T. (2004). Production of recombinant proteins: challenges and solutions. *Methods Mol. Biol.* 267: 15–52.
13 Baneyx, F. (1999). Recombinant protein expression in *Escherichia coli. Curr. Opin. Biotechnol.* 10 (5): 411–421.
14 Acosta-Cruz, E., Wisniewski-Dye, F., Rouy, Z. et al. (2012). Insights into the 1.59-Mbp largest plasmid of *Azospirillum brasilense* CBG497. *Arch. Microbiol.* 194 (9): 725–736.
15 Van Craenenbroeck, K., Vanhoenacker, P., and Haegeman, G. (2000). Episomal vectors for gene expression in mammalian cells. *Eur. J. Biochem.* 267 (18): 5665–5678.
16 Chen, Y.Y., Shieh, H.R., Lin, C.T., and Liang, S.Y. (2011). Properties and construction of plasmid pFW213, a shuttle vector with the oral *Streptococcus* origin of replication. *Appl. Environ. Microbiol.* 77 (12): 3967–3974.
17 Deshpande, A.M. and Newlon, C.S. (1992). The ARS consensus sequence is required for chromosomal origin function in *Saccharomyces cerevisiae. Mol. Cell. Biol.* 12 (10): 4305–4313.
18 Nevalainen, K.M., Te'o, V.S., and Bergquist, P.L. (2005). Heterologous protein expression in filamentous fungi. *Trends Biotechnol.* 23 (9): 468–474.
19 Jensen, N.B., Strucko, T., Kildegaard, K.R. et al. (2014). EasyClone: method for iterative chromosomal integration of multiple genes in *Saccharomyces cerevisiae. FEMS Yeast Res.* 14 (2): 238–248.

20 Cereghino, J.L. and Cregg, J.M. (2000). Heterologous protein expression in the methylotrophic yeast *Pichia pastoris*. *FEMS Microbiol. Rev.* 24 (1): 45–66.
21 Olson, P., Zhang, Y., Olsen, D. et al. (1998). High-level expression of eukaryotic polypeptides from bacterial chromosomes. *Protein Expr. Purif.* 14 (2): 160–166.
22 Zaman, S.B., Hussain, M.A., Nye, R. et al. (2017). A review on antibiotic resistance: alarm bells are ringing. *Cureus* 9 (6): e1403.
23 Felber, M., Pichler, H., and Ruth, C. (2014). Strains and molecular tools for recombinant protein production in *Pichia pastoris*. *Methods Mol. Biol.* 1152: 87–111.
24 Krainer, F.W., Dietzsch, C., Hajek, T. et al. (2012). Recombinant protein expression in *Pichia pastoris* strains with an engineered methanol utilization pathway. *Microb. Cell Fact.* 11: 22.
25 Addgene (2017). *Plasmids 101: A Desktop Resource*, 3e. Addgene.
26 Mulleder, M., Campbell, K., Matsarskaia, O. et al. (2016). *Saccharomyces cerevisiae* single-copy plasmids for auxotrophy compensation, multiple marker selection, and for designing metabolically cooperating communities. *F1000Research* 5: 2351.
27 Nett, J.H., Hodel, N., Rausch, S., and Wildt, S. (2005). Cloning and disruption of the *Pichia pastoris* ARG1, ARG2, ARG3, HIS1, HIS2, HIS5, HIS6 genes and their use as auxotrophic markers. *Yeast* 22 (4): 295–304.
28 Näätsaari, L., Mistlberger, B., Ruth, C. et al. (2012). Deletion of the *Pichia pastoris* KU70 homologue facilitates platform strain generation for gene expression and synthetic biology. *PLoS ONE* 7 (6): e39720.
29 Nielsen, J. (2013). Production of biopharmaceutical proteins by yeast: advances through metabolic engineering. *Bioengineered* 4 (4): 207–211.
30 Reyrat, J.M., Pelicic, V., Gicquel, B., and Rappuoli, R. (1998). Counterselectable markers: untapped tools for bacterial genetics and pathogenesis. *Infect. Immun.* 66 (9): 4011–4017.
31 Solis-Escalante, D. et al. (2013). *amdSYM*, a new dominant recyclable marker cassette for *Saccharomyces cerevisiae*. *FEMS Yeast Res.* 13 (1): 126–139.
32 Kanhere, A. and Bansal, M. (2005). Structural properties of promoters: similarities and differences between prokaryotes and eukaryotes. *Nucleic Acids Res.* 33 (10): 3165–3175.
33 Omotajo, D., Tate, T., Cho, H., and Choudhary, M. (2015). Distribution and diversity of ribosome binding sites in prokaryotic genomes. *BMC Genomics* 16: 604.
34 Yang, L. et al. (2014). Characterization of TATA-containing genes and TATA-less genes in *S. cerevisiae* by network topologies and biological properties. *Genomics* 104 (6): 562–571.
35 Pedersen, A.G., Baldi, P., Chauvin, Y., and Brunak, S. (1999). The biology of eukaryotic promoter prediction--a review. *Comput. Chem.* 23 (3–4): 191–207.
36 Nakagawa, S., Niimura, Y., Gojobori, T. et al. (2008). Diversity of preferred nucleotide sequences around the translation initiation codon in eukaryote genomes. *Nucleic Acids Res.* 36 (3): 861–871.
37 Sonenberg, N. and Hinnebusch, A.G. (2009). Regulation of translation initiation in eukaryotes: mechanisms and biological targets. *Cell* 136 (4): 731–745.

38 Kozak, M. (1986). Influences of mRNA secondary structure on initiation by eukaryotic ribosomes. *Proc. Natl. Acad. Sci. U.S.A.* 83 (9): 2850–2854.
39 Weenink, T., McKiernan, R.M., and Ellis, T. (2017). *Rational Design Of RNA Structures That Predictably Tune Eukaryotic Gene Expression*. bioRxiv.
40 Nevalainen, H. and Peterson, R. (2014). Making recombinant proteins in filamentous fungi- are we expecting too much? *Front. Microbiol.* 5: 75.
41 Li, J., Wang, J., Wang, S. et al. (2012). Achieving efficient protein expression in *Trichoderma reesei* by using strong constitutive promoters. *Microb. Cell Fact.* 11: 84.
42 Raschke, W.C., Neiditch, B.R., Hendricks, M., and Cregg, J.M. (1996). Inducible expression of a heterologous protein in *Hansenula polymorpha* using the alcohol oxidase 1 promoter of *Pichia pastoris*. *Gene* 177 (1–2): 163–167.
43 Gellissen, G., Kunze, G., Gaillardin, C. et al. (2005). New yeast expression platforms based on methylotrophic *Hansenula polymorpha* and *Pichia pastoris* and on dimorphic *Arxula adeninivorans* and *Yarrowia lipolytica* - a comparison. *FEMS Yeast Res.* 5 (11): 1079–1096.
44 Portnoy, T., Margeot, A., Linke, R. et al. (2011). The CRE1 carbon catabolite repressor of the fungus *Trichoderma reesei*: a master regulator of carbon assimilation. *BMC Genomics* 12: 269.
45 Vogl, T., Kickenweiz, T., Pitzer, J. et al. (2018). Engineered bidirectional promoters enable rapid multi-gene co-expression optimization. *Nat. Commun.* 9 (1): 3589.
46 West, R.W. Jr.,, Chen, S.M., Putz, H. et al. (1987). GAL1-GAL10 divergent promoter region of *Saccharomyces cerevisiae* contains negative control elements in addition to functionally separate and possibly overlapping upstream activating sequences. *Genes Dev.* 1 (10): 1118–1131.
47 Lopes, T.S., Klootwijk, J., Veenstra, A.E. et al. (1989). High-copy-number integration into the ribosomal DNA of *Saccharomyces cerevisiae*: a new vector for high-level expression. *Gene* 79 (2): 199–206.
48 Sunga, A.J., Tolstorukov, I., and Cregg, J.M. (2008). Posttransformational vector amplification in the yeast *Pichia pastoris*. *FEMS Yeast Res.* 8 (6): 870–876.
49 Scorer, C.A., Clare, J.J., McCombie, W.R. et al. (1994). Rapid selection using G418 of high copy number transformants of *Pichia pastoris* for high-level foreign gene expression. *Biotechnology (NY)* 12 (2): 181–184.
50 Aw, R. and Polizzi, K.M. (2016). Liquid PTVA: a faster and cheaper alternative for generating multi-copy clones in *Pichia pastoris*. *Microb. Cell Fact.* 15: 29.
51 Zhou, Z., Dang, Y., Zhou, M. et al. (2016). Codon usage is an important determinant of gene expression levels largely through its effects on transcription. *Proc. Natl. Acad. Sci. U.S.A.* 113 (41): E6117–E6125.
52 Choi, Y.J., Bourque, D., Morel, L. et al. (2006). Multicopy integration and expression of heterologous genes in *Methylobacterium extorquens* ATCC 55366. *Appl. Environ. Microbiol.* 72 (1): 753–759.
53 Kubicek, C.P. (2013). Systems biological approaches towards understanding cellulase production by *Trichoderma reesei*. *J. Biotechnol.* 163 (2): 133–142.

54 Schuster, A. and Schmoll, M. (2010). Biology and biotechnology of *Trichoderma*. *Appl. Microbiol. Biotechnol.* 87 (3): 787–799.

55 Peterson, R. and Nevalainen, H. (2012). *Trichoderma reesei* RUT-C30--thirty years of strain improvement. *Microbiology* 158 (Pt. 1): 58–68.

56 Ghosh, A., Ghosh, B.K., Trimino-Vazquez, H. et al. (1984). Cellulase secretion from a hyper-cellulolytic mutant of *Trichoderma reesei* Rut-C30. *Arch. Microbiol.* 140 (2–3): 126–133.

57 Kurtzman, C.P. (2009). Biotechnological strains of *Komagataella (Pichia) pastoris* are *Komagataella phaffii* as determined from multigene sequence analysis. *J. Ind. Microbiol. Biotechnol.* 36 (11): 1435–1438.

58 Sturmberger, L., Chappell, T., Geier, M. et al. (2016). Refined *Pichia pastoris* reference genome sequence. *J. Biotechnol.* 235: 121–131.

59 Rowlands, R.T. (1984). Industrial strain improvement: mutagenesis and random screening procedures. *Enzyme Microb. Technol.* 6 (1): 3–10.

60 Yuen, G., Khan, F.J., Gao, S. et al. (2017). CRISPR/Cas9-mediated gene knockout is insensitive to target copy number but is dependent on guide RNA potency and Cas9/sgRNA threshold expression level. *Nucleic Acids Res.* 45 (20): 12039–12053.

61 Kim, S.B. and Timmusk, S. (2013). A simplified method for gene knockout and direct screening of recombinant clones for application in *Paenibacillus polymyxa*. *PLoS ONE* 8 (6): e68092.

62 Novo Nordisk A/S (2010). *Novo Nordisk History*, 3e. Novo Nordisk A/S.

63 Finkelstein, D.B. (1987). Improvement of enzyme production in *Aspergillus*. *Antonie Van Leeuwenhoek* 53 (5): 349–352.

64 Wang, S., Lin, C., Liu, Y. et al. (2016). Characterization of a starch-hydrolyzing alpha-amylase produced by *Aspergillus niger* WLB42 mutated by ethyl methanesulfonate treatment. *Int. J. Biochem. Mol. Biol.* 7 (1): 1–10.

65 O'Donnell, D., Wang, L., Xu, J. et al. (2001). Enhanced heterologous protein production in *Aspergillus niger* through pH control of extracellular protease activity. *Biochem. Eng. J.* 8 (3): 187–193.

66 Suhaimi, N., Ramli, S., Malek, R.A. et al. (2016). Optimization of pectinase production by *Aspergillus niger* using orange pectin based medium. *J. Chem. Pharm. Res.* 8 (2): 259–268.

67 Dutt, D. and Kumar, A. (2014). Optimization of cellulase production under solid-state fermentation by *Aspergillus flavus* (AT-2) and *Aspergillus niger* (AT-3) and its impact on stickies and ink particle size of sorted office paper. *Cellul. Chem. Technol.* 48 (3): 285–298.

68 Sabir, S., Bhatti, H.N., Zia, M.A., and Sheikh, M.A. (2007). Enhanced production of glucose oxidase using *Penicillium notatum* and rice polish. *Food Technol. Biotechnol.* 45 (4): 443–446.

69 Cherry, J.R. and Fidantsef, A.L. (2003). Directed evolution of industrial enzymes: an update. *Curr. Opin. Biotechnol.* 14 (4): 438–443.

70 Food Standards Australia & New Zealand (2019). Application A1057 – Endo-protease as a Processing Aid (Enzyme). http://www.foodstandards.gov.au/code/applications/Pages/applicationa1057endo5114.aspx (accessed 03 March 2019).

71 Ganzlin, M. and Rinas, U. (2008). In-depth analysis of the *Aspergillus niger* glucoamylase (glaA) promoter performance using high-throughput screening and controlled bioreactor cultivation techniques. *J. Biotechnol.* 135 (3): 266–271.
72 Kricka, W., Fitzpatrick, J., and Bond, U. (2014). Metabolic engineering of yeasts by heterologous enzyme production for degradation of cellulose and hemicellulose from biomass: a perspective. *Front. Microbiol.* 5: 174.
73 Meng, F., Wei, D., and Wang, W. (2013). Heterologous protein expression in *Trichoderma reesei* using the cbhII promoter. *Plasmid* 70 (2): 272–276.
74 Emalfarb, M.A. (2015). There's a saying that "Expression systems aren't everything, but everything is nothing without a good expression system". *J. Bioanal. Biomed.* 7 (5): 27.
75 Karnaouri, A., Topakas, E., Antonopoulou, I., and Christakopoulos, P. (2014). Genomic insights into the fungal lignocellulolytic system of *Myceliophthora thermophila*. *Front. Microbiol.* 5: 281.
76 Christensen, T., Woeldike, H., Boel, E. et al. (1988). High level expression of recombinant genes in *Aspergillus oryzae*. *Nat. Biotechnol.* 6: 419–1422.
77 Berka, R., Thompson, S., Brown, S. et al. (1995). Cloning of laccases from the thermophilic fungi *Myceliophthora thermophila* and *Scytalidium thermophilum* and their heterologous expression in *Aspergillus oryzae*. In: *Abstracts of Papers of the American Chemical Society*, vol. 209, 196-BIOT. Washington, DC: American Chemical Society.
78 Wongwicharn, A., McNeil, B., and Harvey, L.M. (1999). Effect of oxygen enrichment on morphology, growth, and heterologous protein production in chemostat cultures of *Aspergillus niger* B1-D. *Biotechnol. Bioeng.* 65 (4): 416–424.
79 van Paridon, P.A., Veenstra, A.E., Luiten, R.G.M. et al. (1989). *Cloning and Expression of Microbial Phytase*. G.-B. NV, BASF SE. EP0779037A1.
80 Punt, P.J., van Biezen, N., Conesa, A. et al. (2002). Filamentous fungi as cell factories for heterologous protein production. *Trends Biotechnol.* 20 (5): 200–206.
81 Dunn-Coleman, N.S., Bloebaum, P., Berka, R.M. et al. (1991). Commercial levels of chymosin production by *Aspergillus*. *Biotechnology (NY)* 9 (10): 976–981.
82 Morita, S., Kuriyama, M., Nakatsu, M., and Kitano, K. (1994). High level expression of *Fusarium alkaline* protease gene in *Acremonium chrysogenum*. *Biosci. Biotechnol., Biochem.* 58 (4): 627–630.
83 Gusakov, A.V., Sinitsyn, A.P., Salanovich, T.N. et al. (2005). Purification, cloning and characterisation of two forms of thermostable and highly active cellobiohydrolase I (Cel7A) produced by the industrial strain of *Chrysosporium lucknowense*. *Enzyme Microb. Technol.* 36 (1): 57–69.
84 Havlik, D., Brandt, U., Bohle, K., and Fleissner, A. (2017). Establishment of *Neurospora crassa* as a host for heterologous protein production using a human antibody fragment as a model product. *Microb. Cell Fact.* 16 (1): 128.
85 Gao, Z., Li, Z., Zhang, Y. et al. (2012). High-level expression of the *Penicillium notatum* glucose oxidase gene in *Pichia pastoris* using codon optimization. *Biotechnol. Lett.* 34 (3): 507–514.

86 Mellitzer, A., Ruth, C., Gustafsson, C. et al. (2014). Synergistic modular promoter and gene optimization to push cellulase secretion by *Pichia pastoris* beyond existing benchmarks. *J. Biotechnol.* 191: 187–195.

87 Saloheimo, M., Niku-Paavola, M.L., and Knowles, J.K. (1991). Isolation and structural analysis of the laccase gene from the lignin-degrading fungus *Phlebia radiata*. *J. Gen. Microbiol.* 137 (7): 1537–1544.

88 Miettinen-Oinonen, A., Torkkeli, T., Paloheimo, M., and Nevalainen, H. (1997). Overexpression of the *Aspergillus niger* pH 2.5 acid phosphatase gene in a heterologous host *Trichoderma reesei*. *J. Biotechnol.* 58 (1): 13–20.

89 Uusitalo, J.M., Nevalainen, K.M., Harkki, A.M. et al. (1991). Enzyme production by recombinant *Trichoderma reesei* strains. *J. Biotechnol.* 17 (1): 35–49.

90 Te'o, V.S., Cziferszky, A.E., Bergquist, P.L., and Nevalainen, K.M. (2000). Codon optimization of xylanase gene xynB from the thermophilic bacterium *Dictyoglomus thermophilum* for expression in the filamentous fungus *Trichoderma reesei*. *FEMS Microbiol. Lett.* 190 (1): 13–19.

91 EneyskayaA, E.V., KulminskayaA, A., Savel'evN, N. et al. (1999). Acid protease from *Trichoderma reesei*: limited proteolysis of fungal carbohydrases. *Appl. Microbiol. Biotechnol.* 52 (2): 226–231.

92 Haab, D., Hagspiel, K., Szakmary, K., and Kubicek, C.P. (1990). Formation of the extracellular proteases from *Trichoderma reesei* QM 9414 involved in cellulase degradation. *J. Biotechnol.* 16 (3–4): 187–198.

93 Zhang, G., Zhu, Y., Wei, D., and Wang, W. (2014). Enhanced production of heterologous proteins by the filamentous fungus *Trichoderma reesei* via disruption of the alkaline serine protease SPW combined with a pH control strategy. *Plasmid* 71: 16–22.

94 Vittaladevaram, V. (2017). Fermentative production of microbial enzymes and their applications: present status and future prospects. *J. Appl. Biol. Biotechnol.* 5 (4): 090–094.

1.4

Process Considerations for the Application of Enzymes

Selin Kara[1] *and Andreas Liese*[2]

[1] *Aarhus University, Department of Engineering, Biological and Chemical Engineering Section, Biocatalysis and Bioprocessing Group, Gustav Wieds Vej 10, 8000 Aarhus, Denmark*
[2] *Institute of Technical Biocatalysis, Hamburg University of Technology, Denickestr. 15, 21073 Hamburg, Germany*

1.4.1 Biocatalyst Types Used in Industrial Processes

The application of enzymes, including their use for the synthesis of chemicals, which is the focus of this chapter, has been a key emerging field to meeting the current and future needs of our society. Other parts of this book introduce the enzyme applications for the production of bulk chemicals, their application in food and feed sectors (Chapters 2 and 3), and fine chemicals in the pharmaceutical industry (Chapter 5). In addition to these, the role of enzymes for the conversion of renewable feedstocks is discussed in Chapter 4. Nowadays, challenges of using enzymes in biotransformation processes can be overcome by means of modern molecular biotechnology techniques (Chapter 1.2) for designing biocatalysts with the desired characteristics such as

- high activity towards a nonnatural substrate,
- high stability at elevated substrate and product concentrations,
- high stability at elevated temperatures,
- high stability in the presence of organic solvents, and
- high selectivity (e.g. chemo-, regio-, and enantioselectivity).

On the other hand, the number of enzymes still hidden in nature and which are yet to be discovered also indicates that we can potentially find new biocatalysts that may be highly effective for specific chemical transformations. In this context, newly discovered extremophiles would be of high interest for industrial applications.

In a conventional way, catalysis can be classified as application driven (homo- and heterogeneous catalysis as well as biocatalysis) and origin based (chemo- and biocatalysis). However, as also done in the case of chemocatalysis, biocatalysis can be categorized as homogeneous and heterogeneous [1]. Homogeneous biocatalysis covers the use of isolated enzymes (purified or semi-purified crude extract), whereas heterogeneous biocatalysis covers the use of whole cells (resting or growing), immobilized enzymes, or immobilized cells. The choice

Industrial Enzyme Applications, First Edition. Edited by Andreas Vogel and Oliver May.

of biocatalyst type (enzymes or whole cells) and the biocatalyst preparation (homogeneous or heterogeneous) depend on the

- type of reaction,
- need for an external cofactor, e.g. nicotinamide,
- scale of the bioprocess, and
- operation mode, e.g. batch or continuous.

In the following parts of this section the use of whole cells (growing or resting) and isolated enzymes (*in vitro* approach) are introduced whereas the immobilization of biocatalysts is discussed in Section 1.4.2.

The main advantages of using whole cells are that

- biocatalyst costs is lower as no enzyme isolation or purification is required,
- enzyme stability can be higher as enzymes are found in their natural environment, and
- there is no need for the addition of external cofactors as they are directly provided from the cell metabolism.

However, there are some major disadvantages of using whole cells, such as

- molecular design of microorganisms (i.e. "designer bugs") can be material and time intensive,
- competitive reactions catalyzed by other intracellular enzymes can lead to low selectivity and productivity,
- high substrate and product concentrations – which are prerequisites at industrial scales – can be toxic to the cells,
- adjusting the enzymes' expression levels is not straightforward, and
- controlling and scaling up the bioprocess can be difficult.

Unlike the *whole cell catalysis* approach, reaction systems with isolated enzymes can be easily controlled and optimized, and high product yields and purities can be achieved due to the absence of competing side reactions. In principle, the aforementioned advantages of the *in vivo* approach can be regarded as the disadvantages of the *in vitro* strategy. However, with the current technology and knowledge, it is possible to optimize *in vitro* systems by means of methods such as

- protein purification methods resulting in high product yields and low material costs,
- reuse of expensive cofactors with a broad spectrum of *in situ* regeneration techniques,
- alteration of enzyme pH and temperature profiles,
- enzyme stabilization via immobilization [2] (Section 1.4.2), protein engineering [3] (Chapter 1.2), use of additives [4] or cosolvents [5], and
- compartmentalization of incompatible enzymes and/or their reaction conditions, as nature does in cellular organelles or compartments.

A crucial aspect of developing a biocatalytic synthesis route is the selection of readily available biocatalysts with sufficient activity, selectivity, and stability.

Table 1.4.1 Classification of enzymes with respect to their class, availability, and utility.

Enzyme class (EC)	Number		Utility[a]
	Classified	Available	
EC1, Oxidoreductases	~700	~100	+ + +
EC2, Transferases	~750	~100	++
EC3, Hydrolases	~650	~180	+ + +
EC4, Lyases	~300	~40	++
EC5, Isomerases	~150	6	+
EC6, Ligases	~80	5	±
EC7, Translocases	new established class by the International Union of Biochemistry in 2018		

a) Utility ranges from very useful (+ + +) to of little use (±).
Source: Faber 2011 [6]. Reproduced with permission of Springer Nature.

Table 1.4.2 Characteristics of resting and growing cells.

	Resting cells	Growing cells
Number of reaction steps	Few	Many
Number of enzymes involved in the biocatalysis route	Few[a]	Many
Starting material	Substrate	C + N source
Product	Natural or non-natural	Natural
Substrate resembles the product	Yes	No
Suitable for high substrate/product concentrations	Yes	No
Product isolation	Easy	Tedious
Side reactions	Few	Many
Possible biocatalyst form	Homogeneous or heterogeneous	Homogeneous
Contamination risks	No	Yes

a) Multienzymatic catalysis offers the combinations of enzymes in a cascade fashion; however, still the number of reactions are less than in the fermentative processes.
Source: Faber 2011 [6]. Reproduced with permission of Springer Nature.

Table 1.4.1 lists the biocatalysts based on their enzyme classes (ECs), commercial availability, and their utility for the conversion of nonnatural substrates.

Fermentative processes, i.e. the use of growing cells, have significant differences compared to the use of resting cells as summarized in Table 1.4.2. The recent achievements in the development of production strains to reach high product yields and specific productivities, to have high substrate and product tolerance, and to use diverse substrates allow more efficient optimization of fermentative processes. Renewable resources can be efficiently converted into valuable products such as amino acids, polymer precursors, and active pharmaceutical ingredients (APIs). Next to the diverse strategies to design highly productive strains, the rapid and reliable screening of the designed variants also plays a crucial role in bioprocess development.

1.4.2 Enzyme Immobilization for Biocatalytic Processes

The use of immobilized enzymes has become a standard procedure in industrial applications. The need for enzyme immobilization might arise from different reasons, e.g. to (i) increase enzyme stability, (ii) enable a facile enzyme separation from the reaction medium, (iii) recycle the enzyme for economic feasibility, (iv) run continuous processes, and (v) design robust multienzymatic cascades. Diverse immobilization techniques are available for the preparation of heterogeneous biocatalysts. Enzyme immobilization techniques can be categorized into three groups: (i) binding to a solid support [2c, d, 7], (ii) cross-linking [8], and (iii) encapsulation or entrapment within a matrix [9] (Figure 1.4.1). There is no universal method that can be applied for the immobilization of any enzyme and hence each individual immobilization method and the enzyme of interest should be evaluated separately under the immobilization conditions (i.e. treatment of enzyme during immobilization) as well as process conditions (i.e. agitation type and rate, reaction medium, etc.).

The requirements of a practical and efficient immobilization method, including the carrier properties, can be listed as

- a simple immobilization procedure,
- low costs of the immobilization method and carrier,
- no or low enzyme activity lost during the immobilization,
- high enzyme stability,
- no or low enzyme leaching (desorption) during the reaction,
- (ideally) recoverable solid support,
- sufficiently large carrier particles for ease of retention and separation of the immobilized enzymes,

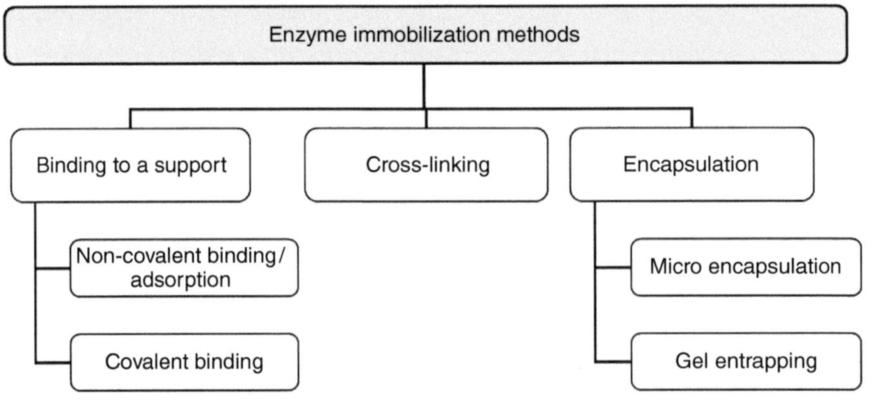

Figure 1.4.1 Enzyme immobilization methods.

- sufficiently large carrier pore size for low internal diffusion limitations, and
- mechanically stable carriers under shear stress.

For details on the specific characteristics of different immobilization methods, excellent reviews are available [2c, d, 7, 8]. In the following parts we shall focus on the abovementioned motivations for enzyme immobilization, respectively.

Increased stability: When attached on/into a carrier material, enzymes lose their molecular flexibility and hence become less sensitive toward storage and operational conditions, which would otherwise denature the enzymes. Enzyme immobilization in turn enables the use of highly efficient solvent-free systems and a variety of organic solvents (when needed), as it restricts conformational mobility and avoids unfolding. Hence, enzyme immobilization offers efficient biocatalytic applications in nonaqueous media.

Enzyme recovery: From the industrial application point of view protein contamination of the product can be minimized or eliminated by using heterogeneous biocatalysts, which can be easily removed. This is especially important in personal care products [10] since the final product should be protein free; otherwise allergenic reactions may occur.

Enzyme recycle: Immobilization facilitates efficient recovery and reuse of enzymes, which consequently makes the biocatalytic process economically viable. This is especially crucial when the enzyme is expensive. In these cases, the cost contribution of the enzyme to per kilogram of product can be reduced by recycling the enzymes. This is realized by batches running in repetitive mode whereby the immobilized enzyme is filtrated to recover and then reused for the next batch. Moreover, immobilized enzymes can be employed in a continuous process, e.g. in a stirred tank reactor – equipped with a filtration unit – or a packed bed reactor [11] (see Section 1.4.4).

Multienzyme reactions: The use of heterogeneous enzyme preparations has been a highlight in the design and implementation of multienzyme cascade reactions with the required spatial separation [12]. In this context, enzyme immobilization is a key method that offers possibilities for both "compartmentalization" [13] and "co-localization" [14] of biocatalysts in order to maximize productivities and economics. Enzymes are compartmentalized to ensure that each biocatalytic step runs at optimal conditions and to eliminate the deactivation effects of the reaction medium and/or any substrates or products on the enzymes. If the cascade reactions can run under similar conditions, a spatial separation is not necessary. On the contrary, a spatial co-localization of the enzymes will be advantageous in this case. The spatial proximity of the enzymes is intended to increase the effective concentration of the reaction intermediates and thus facilitate the transfer of these intermediates from one enzyme to another, the so-called "substrate channeling" [15].

To achieve industrially required productivities on a technical scale, heterogeneous biocatalysts should be optimized by reaction engineering for the design of the processes based on the (i) catalytic properties of the newly developed immobilized enzymes (e.g. k_{cat}, K_m, K_i for substrate and product), (ii) mass-transfer characteristics (internal and external diffusion, etc.), and (iii) material properties (void volume, porosity, swelling, etc.) [16]. Hence, the performance of the immobilized enzymes has to be evaluated to elucidate the limitations and hence to overcome those.

1.4.3 Reaction Medium Applied in Enzymatic Catalysis

Biocatalysis may be considered as "green chemistry" as it takes place in aqueous media, runs under environmentally benign conditions, and as enzymes are biodegradable catalysts. Among these characteristics, the use of water as reaction medium, however, has been a common discussion topic in the biocatalysis community. Water as a reaction medium may be considered *per se* as nontoxic. Unfortunately, this statement is not true as "contaminated" wastewater is generated and, therefore, pretreatment is necessary before its release or reuse [17]. With respect to downstream processing (dsp), the high boiling point of water, high energy consumption for vaporization, and yield-lowering emulsion formations make the separation of product(s) cost-intensive and tedious, making water often the least desired solvent in organic syntheses.

Enzymes are amongst the catalysts of choice for synthetic chemistry if a high selectivity (regio-, chemo-, and stereoselectivity) is desired. However, besides excellent selectivity, high product concentrations are also necessary for technical and economic feasibility. Recent reports on establishing efficient chemical transformations have revealed important parameters. Among the evaluated factors (e.g. atom economy, environmental factors, product yield, etc.), volume–time output (i.e. volumetric productivity) was determined to be the most significant parameter for process assessment [18]. However, owing to the limited solubility of hydrophobic reagents and water-induced side reactions (hydrolysis of labile groups, polymerization of quinones, racemization of cyanohydrins, acyl migration, etc.), high volumetric productivities could not typically be achieved in solely aqueous media. Another disadvantage of using water as reaction media is substrate and product inhibition issues, which can be suppressed in organic media owing to the partitioning of substrates and products between the bulk organic phase and the microenvironment around the enzyme. In addition, microbial contamination can be problematic when using water as a reaction medium. The use of nonaqueous media is especially helpful in shifting thermodynamic equilibria in the synthesis direction and away from the hydrolysis side in hydrolase-catalyzed reactions. Finally, yet importantly, many of the reactions leading to enzyme denaturation are induced by water [19]. Owing to the above given unfavorable aspects of using water as reaction media, the application of enzymes in low water media is of great interest in the biocatalysis community.

So far, different types of reaction media have been evaluated for biocatalytic systems [20]. The ones that are commonly used in an industrial environment can be categorized as (i) monophasic and (ii) multiphasic media (Figure 1.4.2) [6].

The first category, i.e. the use of monophasic aqueous media, covers the application of enzymes in solely aqueous media as well as in a mixture of water and a water-miscible organic solvent (e.g. isopropanol, methanol, dimethyl sulfoxide). The water-miscible solvent is commonly used at 10–20% (v/v) to solubilize

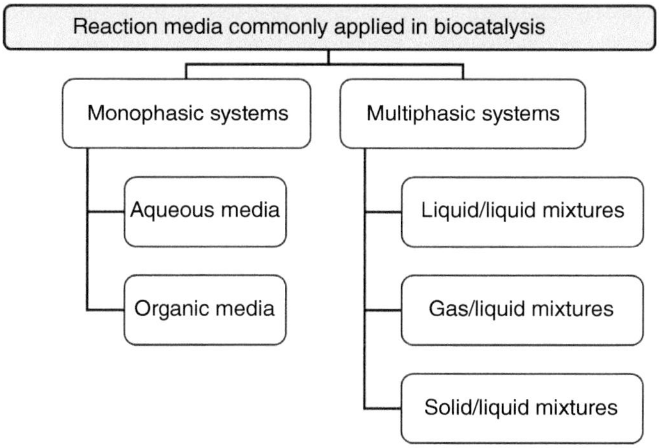

Figure 1.4.2 An overview of the applied reaction media in biocatalytic transformations.

hydrophobic substrates and/or to enhance enzyme catalytic performance. Owing to the aforementioned disadvantages of using aqueous media, application of alternative reaction media plays a vital role from an industrial application perspective and hence shall be discussed in the following parts of this chapter.

1.4.3.1 Monophasic Systems – Organic Media

The use of a water-immiscible organic solvent instead of aqueous medium as a bulk phase for an enzymatic reaction has been the focus of research since the 1960s. Development of an efficient biocatalytic process in predominantly organic media necessitates careful handling of several factors from the solvent and enzymatic point of views. Figure 1.4.3 gives an overview of the factors that might influence the performance of such enzymatic systems. When a solid

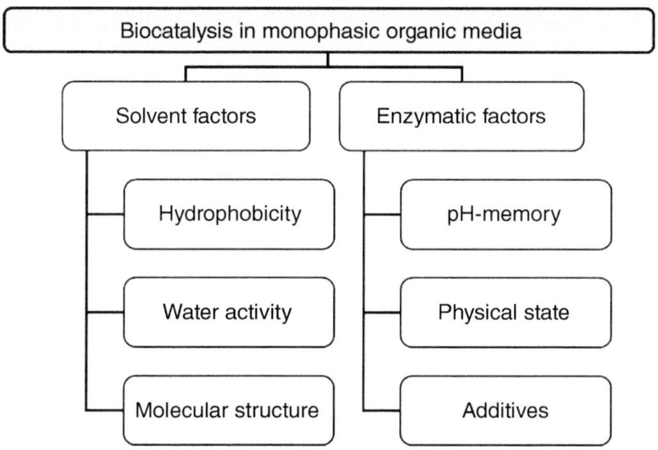

Figure 1.4.3 The schematic representation of parameters affecting the efficiency of a biocatalytic reaction in monophasic organic media.

enzyme – lyophilized, precipitated, or crystallized – is used in a monophasic organic medium, a suspension is formed. Therefore, these kinds of reactions would resemble a heterogeneous catalysis. Solvent factors that may affect the efficiency of a biocatalytic reaction in monophasic organic media can be listed as (i) hydrophobicity, (ii) water activity, and (iii) molecular structure (Figure 1.4.3).

To start with the first solvent factor, several parameters were used to describe solvents' hydrophobicity such as the Hildebrandt solubility parameter (δ), Dimroth–Reichardt polarity parameter (E_T, transition energy), the dielectric constant (ε), the dipole moment (μ), and the logarithm of the partition coefficient ($\log P$) of a given solvent between 1-octanol and water [21]. Among the parameters mentioned above, the $\log P$ value has been accepted as the most reliable in expressing solvent hydrophobicity. In general, it is documented that solvents with $\log P$ values of ≥ 2 cause negligible enzyme distortion, thereby ensuring high enzymatic activity. Conversely, solvents with $\log P$ values < 2 lead to enzyme deactivation [6, 22]. This is mainly due to the fact that polar solvents with a $\log P$ of < 2 can strip off the "structural water" bound to the enzyme and consequently lead to limited molecular flexibility of the enzyme in performing catalysis. However, several studies have also proved that $\log P$ cannot be the only criterion, since high enzyme activities were documented in organic solvents with $\log P$ values of < 2, such as in methyl-tert-butylether (MTBE) ($\log P = 1.0$), in diisopropylether (DIPE) ($\log P = 1.4$), in 2-methyltetrahydrofuran (2-MeTHF) ($\log P = 0.7$), and in cyclopentylmethylether (CPME) ($\log P = 1.6$) [23].

The second solvent factor to be considered is the amount of water bound to the enzyme, or the structural water. It represents the minimum amount of water molecules required for the enzyme to follow conformational changes during the formation of an enzyme–substrate complex. The amount of the bound water can be determined and adjusted indirectly. An indirect quantification is possible as the amount of bound water is influenced by the thermodynamic "water activity" (a_W) of the media, which has been used as the most suitable parameter to measure and control the amount of water in organic media [22]. By using some assumptions, the water activity can be calculated by $a_W = \frac{P_i}{P_0}$, where P_i is the partial pressure of water in the sample and P_0 is the partial pressure of pure water. Hence, a_W of pure water is 1.0. Different enzymes have different water requirements for optimal activity and hence the a_W values have to be adjusted to meet the enzyme's requirements. For example, in case of lipases (EC 3) a_W values of 0.0–0.2 are sufficient, whereas oxidoreductases (EC 1) require a_W values of 0.1–0.7 for catalytic activity [22]. A water activity of zero would mean water free; indeed, some lipases show considerable catalytic activity at water activities as low as 0.0001 [22]. In principle, as long as the enzyme's structural water is preserved, the remaining bulk water can be replaced with an organic media without significant enzyme deactivation. As previously mentioned, the number of water molecules per protein molecule may differ significantly depending on the enzyme. For example, in lysozyme the formation of a monolayer hydration shell covering the protein surface required 300 water molecules per protein molecule [24].

The use of reagent grade salts (e.g. $LiCl_2$, KAc, $MgCl_2$, K_2CO_3, $Mg(NO_3)_2$, NaBr, NaCl, KCl, K_2SO_4, etc.) with defined water activities is a standard method used for adjusting the a_W values [23b, 25]. In this context, the ability of an organic solvent to strip off the structural water from an enzyme's surface, which leads to enzyme deactivation, depends on both the solvent's hydrophobicity ($\log P$) and water activity (a_W). In polar solvents, the critical a_W values to reach maximum enzymatic activity were determined to be significantly higher depending on the solvent used. This can be attributed to the direct effects of polar solvents on the enzyme [26]. When comparing polar and nonpolar solvents, the a_W and solvent availability in the latter are almost independent because the water content is low, even at an a_W close to 1.0 [26]. For example, at an a_W value of 0.8 the water content (v/v, %) of acetone is 11.7%, whereas that of toluene is 0.02%.

In addition, the molecular structure of the solvent (i.e. cyclic vs. long chain, bulkiness, and functional groups) may also play a role in the catalytic performance of the enzymes in different organic solvents. However, this third solvent factor has been considered to a lesser extent compared to the first two parameters described above, which is most probably because molecular dynamic simulations are required for a deep understanding of the interactions between the solvent molecules and the protein. Hence, evaluation of this solvent factor can be regarded as a "trial and error" approach.

Besides the above-described solvent factors, enzymatic factors also play an important role in biocatalytic reactions in monophasic organic media, which can be listed as (i) pH-memory, (ii) physical state, and (iii) use of additives (Figure 1.4.3). The pH of an organic media, which has only a small percentage of water in it, cannot be measured easily. As previously noted, solid enzymes form aggregates when placed in a water-immiscible organic solvent. In this case, the pH-memory [27] of the enzyme is a parameter to be considered. The pH-memory accounts for the pH of the aqueous medium used for the preparation of the solid enzyme, and it determines the enzyme's catalytic performance in an organic media.

The second factor is the enzyme's physical state, which can be either in lyophilized, precipitated, or crystalline form. Solid enzymes suspended in an organic medium form a soft aggregate with cavities, which contains the organic solvent and the reactants. Mass transfer of substrate(s) and product(s) into and out of these void structures can be enhanced by means of efficient agitation in order to achieve high productivities. By doing so, it can be guaranteed that not only the enzyme molecules on the outer part of the aggregate catalyze the reaction but also the enzyme molecules embedded deep inside the aggregate perform the catalysis. In this context, enzyme immobilization (Section 1.4.2) on a macroscopic carrier (inorganic or organic) can provide a better distribution of enzymes on a solid surface so that higher productivities can be achieved in organic media. It is worth mentioning that adsorptive enzyme immobilization, which has a high risk of enzyme leaching in aqueous media, can be applied efficiently for the reactions running in organic media. The absence of a bulk aqueous phase would guarantee that the enzyme stays attached on the carrier surface although the interactions between the enzyme and the carrier surface are weak non-covalent interactions (e.g. van der Waals forces, hydrogen bonds, ionic interactions) in such an adsorptive immobilization [2c]. It has to be noted that, while using immobilization carriers in nonaqueous media, the polarity of the carrier should be taken into account as the carrier itself can also adsorb water.

The last enzymatic factor is the use of additives, which has been commonly applied for biocatalysis in organic media for enzyme stabilization and activation. So far, several types of additives have been used such as polymers (polyethylene glycol [PEG], polyvinyl alcohol, etc.), polyalcohols (e.g. carbohydrates, sugar alcohols or glycerol), and salts (NaCl, KCl, etc.) [28]. Lyophilization has been a common method of choice for the preparation of solid enzymes. Certain types of compounds can be added into the aqueous solution before lyophilization. Substrate analogs may be used as additives for co-lyophilization of the enzyme, which can prevent the collapse of the active site of the enzyme during lyophilization via the so-called "molecular imprinting" effect [28d]. Polyols and PEG have been commonly applied as lyoprotectants to preserve the overall enzyme structure during dehydration, which gave up to 100-fold increase in enzymatic activity [28d]. Whether the effects of different lyoprotectants are additive or not cannot be generalized and hence should be analyzed for each individual enzyme and the type of additives used.

As previously mentioned, enzymes may show enhanced stability in organic solvents compared to aqueous media. For example, porcine pancreatic lipase was reported to be active at 100 °C in predominantly organic medium; however, it denatured at the same temperature when applied in solely aqueous medium [29]. Enzymes become rigid when suspended in low water media and are less prone to lose the catalytically active conformation. Moreover, deactivating interactions between protein and water molecules, e.g. OH^-- or H^+-catalyzed hydrolysis of peptide bonds and reduction of disulfide bonds, are reduced in the absence of bulk water.

Overall, establishing an efficient biocatalytic process in organic media (i.e. use of organic solvents or solvent-free systems) requires a deep understanding and a careful evaluation of several parameters from both solvent and enzyme sides and interactions thereof in order to derive a conclusion about the enzymes' performance in different organic solvents.

1.4.3.2 Multiphasic Systems – Liquid/Liquid Mixtures

The use of a nonpolar organic solvent as a second phase (the so-called "two-liquid-phase system," 2LPS) [30] has been a common approach to overcome a range of limitations in biocatalysis such as enzyme inhibition by the substrate and/or product and unfavorable thermodynamic equilibria. In 2LPSs, the aqueous phase contains the enzyme where the biocatalysis takes place, and substrate(s) and product(s) are mostly found in the organic phase. Hence, the second phase acts as a substrate reservoir and a product sink, which keeps the concentration of the substrate(s) and product(s) at ideal levels in the aqueous medium, thereby avoiding enzyme inhibition. In addition, due to the *in situ* product removal (ISPR) from the aqueous medium, where the catalysis takes place, reactions can proceed to completion (Figure 1.4.4). The *in situ* product extraction, which already takes place during the biocatalysis, makes the dsp straightforward.

In order to develop an efficient and sustainable 2LPS, criteria for the selection of the organic phase can be listed as

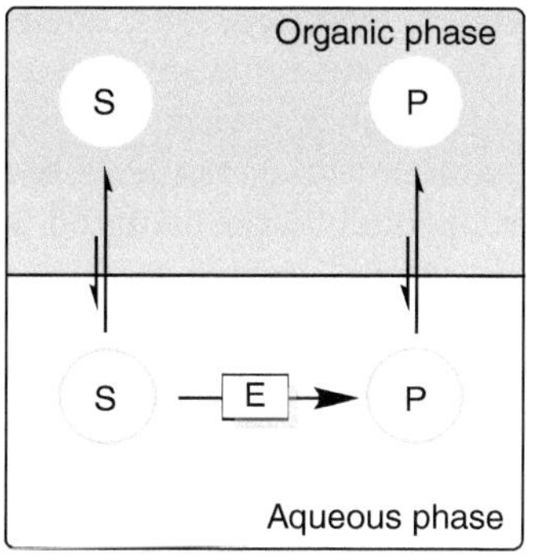

Figure 1.4.4 A 2LPS for the use of an organic phase as a substrate reservoir and a product sink to avoid enzyme inhibition, and to drive the reaction toward completion by *in situ* product removal. The enzyme is dissolved only in the aqueous phase. S, substrate; P, product; and E, enzyme.

- the capacity of the organic solvent to solubilize the reactants,
- the partitioning of a reactant between two phases (see below),
- possible denaturing effects of the organic solvents (see below), and
- the environmental compatibility, or so-called "greenness," of the organic solvent.

The number of partition coefficients to be considered in a biphasic reaction system depends on the number of substrates and products involved. It is important to mention that an efficient 2LPS requires a high interfacial area to avoid poor mass transfer between the two phases. Therefore, high power input by stirrers (or static mixers) is needed to obtain a sufficient mass transfer. The productivity of a biocatalytic 2LPS is mainly determined by a partition coefficient – a thermodynamic parameter – and by a mass transfer coefficient – a kinetic parameter. The partitioning effect becomes dominating when the dissolved substrate concentration in the aqueous medium, where the reaction takes place, becomes limited ($\leq$10-times K_m value). This means that by using an appropriate organic solvent that selectively dissolves the reactants (thermodynamics) and by applying effective agitation (kinetics), high productivities can be achieved. As a screening tool, partition coefficients of the reactants can be simulated in different solvents. For this purpose, mainly the Conductor-like Screening Model for Realistic Solvation (COSMO-RS) has been applied [31]. COSMO-RS is based on quantum chemical calculations and statistical thermodynamics, and can be used for the prediction of molecular interactions of molecules in a solution. Solvents identified to be promising by computational models can be then verified with experimental data.

Another aspect to be considered in biphasic systems is enzyme deactivation. Two types of deactivation are encountered in 2LPSs, namely, (i) molecular deactivation and (ii) interfacial deactivation. Molecular deactivation results from direct interactions between enzyme and solvent molecules dissolved in the aqueous phase, whereas interfacial deactivation results from the interaction of the protein structure with the aqueous–organic interface. It is generally assumed that interfacial deactivation plays a major role in enzyme deactivation occurring in 2LPSs; however, a rigorous analysis to differentiate these effects is necessary to elucidate which of these deactivation types is more detrimental to the biocatalyst. From the environmental and toxicity point of views, various methodological studies have been reported on the selection of solvents [32] for (bio)catalysis, while the main focus has become their "greenness" [33]. Hence, the conventional and environmentally unpreferable solvents can be replaced with greener alternatives to be used in an industrial environment.

Emulsions having either an aqueous phase dispersed in a continuous organic phase or *vice versa* have been applied in biocatalysis to achieve high volumetric productivities. A good dispersion in a continuous phase resulting in small droplets (diameter of approximately 1 μm) can be achieved by rigorous agitation. The dispersed droplets with small sizes may ensure the stability of the emulsion so that the two phases stay separated. However, due to the differences in the fluid densities and interfacial tension phase separation may occur when the emulsion mixture is stored for a long time. To alleviate this, there should be either a continuous energy input through agitation or surface-active agents, the so-called surfactants, present in the system to stabilize the emulsion.

Microemulsions and reverse micelles are classified as special types of an emulsion. In the continuous aqueous and dispersed organic phase, surfactants form micelles where the hydrophobic tails of the surfactants form the core and the hydrophilic heads are immersed in the surrounding aqueous medium (Figure 1.4.5a), whereas when the continuous phase is an organic medium, the hydrophilic heads form the core in the dispersed aqueous medium (Figure 1.4.5b). The latter, the so-called reversed micelles, allow the enzyme to be found in its optimal microenvironment. In this case the dispersed droplets' sizes are in general significantly smaller (diameter of 1.5–2 nm) and hence a very high interfacial area (surface area per droplet volume) of up to 100 m^2/ml can be achieved. Depending on the nature of the enzyme, it can be embedded in the aqueous droplet, located in the aqueous–organic interphase, or found partly in the organic medium [34]. It is worth mentioning here that there is no clear differentiation between the microemulsions and reversed micelles with respect to the droplet size [35].

The advantages of reversed micelles compared to the normal emulsions can be listed as follows:

- Reversed micelles are thermodynamically stable; the dispersed phase is distributed in the continuous phase as tiny droplets and hence the emulsion may keep its form for a longer period of time.

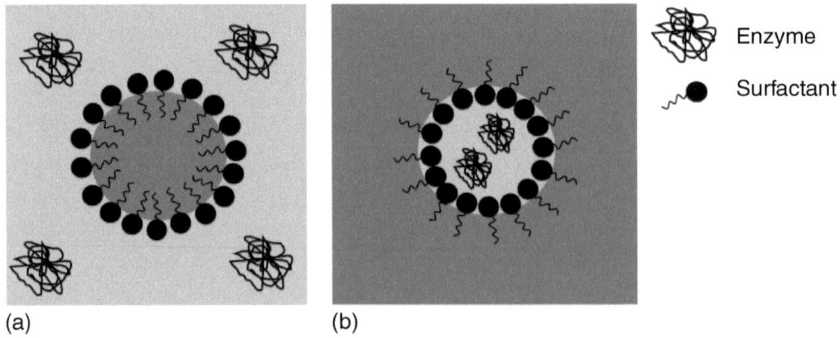

Figure 1.4.5 (a) Dispersed phase: organic medium and continuous phase: aqueous medium; (b) dispersed phase: aqueous medium and continuous phase: organic medium.

- The high interfacial area ensures that the mass transfer between the two phases is efficient.
- Owing to the presence of a surfactant the enzyme is spatially separated from the organic solvent, which avoids enzyme deactivation especially when the enzyme is sensitive against organic medium.

The size of the dispersed droplets in reversed micelles depends on

- the ratio of the water concentration to surfactant concentration,
- temperature,
- ionic strength, and
- composition of the mixture.

Overall, the use of surfactants is advantageous for the stabilization of the emulsions; however, often the complete separation of the surfactant from the product is very challenging and limits its application. In those cases, where the macroscopic separation of the dispersed and the continuous phase is difficult, an additional unit operation such as membrane filtration, electrocoalescence, or froth flotation process would be required. Small droplets dispersed in a continuous phase are useful from the mass transfer point of view; however, dsp would be easier when the droplets are sufficiently large. Hence, there is often a trade-off between an efficient biocatalytic process and a cost-efficient product isolation. When the difference in the fluid densities is very large, sedimentation would be enough for phase separation; otherwise, centrifugation would be needed. Depending on the individual multiphase system, separation of the two phases can also be possible through temperature changes.

Owing to the challenges related to the phase separation, microemulsions and reverse micelles have been applied in a limited manner in enzymatic catalysis. However, depending on the physical characteristics of the product, in some cases phase separation might not be necessary. For example, *in situ* removal of highly volatile products is possible through e.g. pervaporation using a membrane. By doing so, the target product can be isolated as "surfactant-free" since the free surfactant molecules – which are not attached to micelle aggregates – cannot pass through the pervaporation membrane.

1.4.3.3 Multiphasic Systems – Gas/Liquid Mixtures

Among the enzymes applied for redox catalysis, oxidases and monooxygenases are of high importance as they catalyze oxidation reactions selectively to high added-value compounds. Since molecular oxygen is a substrate needed at stoichiometric amounts in these reactions, efficient supply of oxygen plays a crucial role for the enzymatic performance. The gas/liquid mixtures are used in enzymatic catalysis for different purposes such as (i) substrate supply (as mentioned above), (ii) gentle mixing, and (iii) *in situ* by-product removal, e.g. removal of water in esterification reactions [36] to drive the reaction to completion. The last two aspects in which a bubble column reactor (bcr) is introduced are further discussed in Section 1.4.4. While using gas as a substrate the decisive parameter is the mass transfer from the gas phase to the liquid phase, where the enzymatic catalysis takes place. The mass transfer coefficient "k_L" (m/s or m/h) and the interfacial area "a," i.e. volume per specific area (m^{-1}), have a significant effect on the efficiency of enzymatic catalysis. Consequently, "k_La" (s^{-1} or h^{-1}) as a combined term is commonly used as a mass transfer coefficient. The "k_L" depends on (i) flow velocity, (ii) geometry of mass transfer system, (iii) fluid viscosity, and (iv) fluid diffusivity. Those parameters determining k_L can be experimentally easily determined, whereas it is generally difficult to estimate "a." When the gas is sparged into the liquid, the interfacial area will depend on the size and the number of the bubbles present, whereby these depend on medium composition, stirrer speed, and gas flow rate [37].

In a fermentative as well as enzyme-catalyzed process, factors affecting the oxygen transfer are therefore

- bubbles size,
- sparging, stirring, and medium properties,
- antifoam agents,
- temperature,
- gas pressure and oxygen partial pressure, and
- presence of cells and macromolecules [37].

1.4.3.4 Multiphasic Systems – Solid/Liquid Mixtures

The use of very high amounts of substrates in a liquid media might lead to solid–liquid mixtures, the so-called suspensions. Indeed, these multiphasic systems have been applied in enzymatic catalysis for the synthesis of peptides, β-lactam antibiotics, glycosides, glycamides, and esters [38]. Biocatalytic reactions that involve precipitated products in the aqueous medium would avoid the use of an organic solvent as a second phase (Section 1.4.3.2) to shift the thermodynamic equilibrium toward the synthesis direction. Hence, high productivities can be achieved directly in suspensions involving pure solids and a liquid phase (Figure 1.4.6). There are different classifications for solid–liquid mixtures in biocatalytic reactions, whereby as mentioned above a certain amount of water is needed for the enzymatic catalysis, and sometimes an organic phase is added to enhance the formation of the liquid phase where the enzymatic reaction takes place.

Table 1.4.3 summarizes the types of suspension systems involving solid substrate(s) and product(s) in the presence of a liquid phase (organic or aqueous).

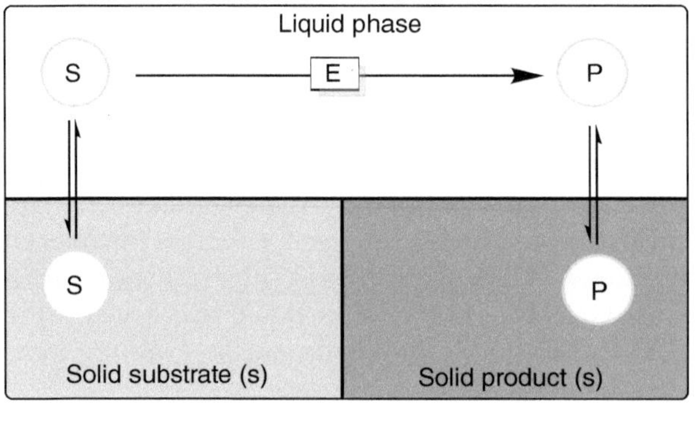

Figure 1.4.6 Solid–liquid biocatalysis involving solid substrate(s) and precipitated product(s) with a liquid phase needed for the enzymatic catalysis. S, substrate; P, product; and E, enzyme.

Table 1.4.3 Classification of biocatalytic systems with solid substrates and products.

Presence of	Solid substrate	Solid product	Liquid fraction	Organic solvent (O) or aqueous media (A)	Characteristics of the system
Suspended substrates					**One or both substrates are present mainly as solid suspended in the solvent. Products do not precipitate and yields tend to be low, unless the equilibrium lies in the synthetic direction (e.g. in the presence of organic solvents).**
	Yes	No	Low[a]	A	Mainly solid substrate with little aqueous liquid between particles
	Yes	No	Low[a]	O	Mainly solid substrates with little organic liquid between particles
	Yes	No	High	A	Excess solid substrate suspended in an aqueous phase
	Yes	No	High	O	Excess solid substrate suspended in an organic phase
Substrate suspended in a liquid substrate					**One solid substrate is suspended in a second substrate, which is liquid at the reaction temperature and which acts also as liquid phase. Yields can vary depending on several parameters.**
	Yes	Yes	High	O	One solid substrate is suspended in a second liquid substrate that acts as solvent
	Yes	No	High	O	One solid substrate is suspended in a second liquid substrate that acts as solvent
Suspended substrates and precipitating products					**Both, substrates and products, are present mainly as solids suspended in a liquid phase. Very good yields are commonly observed.**
	Yes	Yes	Low[a]	A/O	Mainly solid substrates yielding precipitating products
	Yes	Yes	High	A/O	The so-called "suspension-to-suspension", with excess solids

(Continued)

Table 1.4.3 (Continued)

Presence of	Solid substrate	Solid product	Liquid fraction	Organic solvent (O) or aqueous media (A)	Characteristics of the system
Product precipitated from a reaction in solution					**A homogeneous solution from which the product precipitates. These systems have been used for e.g. peptide synthesis since the 1930s.**
	No	Yes	High[b)]	O	Precipitation-driven synthesis of peptides in organic solvent
	No	Yes	High[b)]	A	Precipitation-driven synthesis of peptides in aqueous phase
Eutectic melting					**Two solid substrates form a liquid phase by eutectic melting**
	No	No	High	No/A[c)]	The two substrates melt and no additional solvent needs to be added
	No	Yes	High	No/A[c)]	The two substrates melt and no additional solvent needs to be added. When product precipitates the reaction mixture solidifies
	Yes	Yes	High	No/A[c)]	Also possible, can have one solid substrate and eutectic melt of mixture.

a) Working with suspended substrates, another component, normally termed an "adjuvant," has been added in a small amount (usually 5– 30%), to cause or enhance formation of a liquid phase.
b) Might be L at the end of reaction.
c) No solvent.
Source: Basso et al. [38]. Adapted from John Wiley & Sons.

Overall, the development of an efficient enzymatic catalysis in a suspension depends on the reaction system, whereby reaction thermodynamics, enzyme kinetics, and enzyme stability should be assessed for each individual case. In suspensions, rigorous agitation would play a crucial role in enhancing the mass transfer between the phases. In addition, if an organic solvent is preferred, the choice of this solvent would play an important role in the enzymatic performance, as mentioned in Section 1.4.3.1. Different approaches with respect to reactor engineering, as given in Section 1.4.4, should be followed to achieve high productivities in suspensions.

1.4.4 Appropriate Reactor Types in Enzyme Catalysis

The batch reactor, with addition of all reactants at time point zero, was for sure the first reactor setup used in enzymatic biotransformations. However, the first advanced setup in view of continuous operation was a packed bed reactor (pbr) with invertase adsorbed on bone char used for the splitting of sucrose. This fundamental process developed by Nelson and Griffin was to be used commercially for the production of Golden Syrup by Tate and Lyle during World War II, because sulfuric acid as the preferred reagent was unavailable at that time [39]. Later on, this concept of continuous processing was further utilized for the production of L-amino acid to overcome the disadvantages of soluble enzymes such as higher labor costs, complicated product separation, low yields, and high enzyme costs because of non-reusability of the enzyme. Tanabe Seiyaku Co., Japan, started in 1969 the industrial production of L-methionine by aminoacylase immobilized on DEAE-Sephadex in a pbr [40]. To overcome diffusion limitations this process was transferred in 1980 by Degussa-Hüls AG, Germany, in a continuous process utilizing homogeneously soluble aminoacylase, being retained by an ultrafiltration membrane in a continuously operated stirred tank reactor (cstr) to produce enantiomerically pure L-amino acids [41].

Basics for these developments was the detailed knowledge about the principles of (bio)chemical reaction engineering.

In principle, there are three different fundamental reactor types (Figure 1.4.7), where any other configuration is either a variation or combination of those three [35, 42]. With detailed knowledge of kinetics (e.g. simple Michaelis–Menten, ordered bi–bi, ping-pong mechanisms) as well as thermodynamics of the biotransformation, the most appropriate reactor setup can be chosen [43].

In the **batch reactor** the reactant concentrations change with time. At a given time point the concentrations are the same in every place in the reactor, assuming ideal conditions. Derivatizations are the **fed batch**, where one reactant is step-wise or continuously fed to the reactor as a function of time, or the **repetitive batch**, where the enzyme is recycled at the end of the reaction by filtration with an ultrafiltration membrane in the case of homogeneously soluble enzyme, or via standard filtration if an enzyme immobilisate is utilized. An alternative approach is the application of an aqueous/organic two-phase system, where the enzyme-containing aqueous phase is separated and recycled.

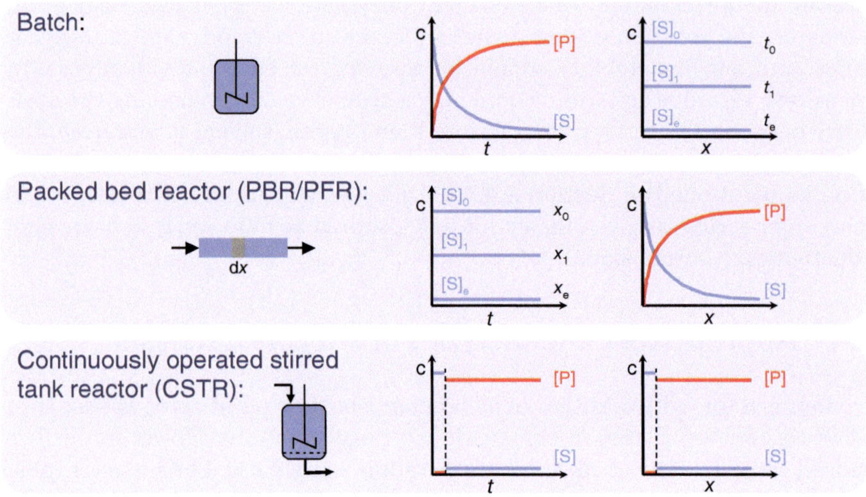

Figure 1.4.7 Fundamental types of reactors (t = time; x = place).

The **plug flow reactor** (**pfr**) or **pbr**, if used with immobilized enzymes, is the continuous variant of the batch reactor. From a reaction engineering point of view, in comparison to the batch reactor, only the x-axis, time, and place are interconverted. The reactant concentrations change over the length of the reactor. At a given place the concentrations are the same at every time point in the reactor, assuming ideal conditions.

The **cstr** demonstrates a totally different reaction engineering profile. Under ideal conditions it is operated under the so-called "outflow conditions," meaning that in contrast to the pfr, the reactant concentration at the outlet is identical to any place in the cstr and does not change with time. The cstr is always operated at maximum conversion, respectively, minimum substrate concentration in the reactor volume, even if the inlet substrate concentration is very high. Therefore, its reaction engineering characteristics are totally different from those of the batch reactor, or pfr.

The reaction time in continuously operated reactors is called **residence time** τ describing the average time a fluid element remains in the reactor, and is defined as the quotient of volume V $[L]$ and flow rate F $[L/h]$.

$$\tau = \frac{V}{F}$$

Additionally, a **dimensionless operation time** is introduced to determine the required concentration of the enzyme preparation m_{cat} (g/L) in the reactor to obtain the desired conversion X from a determined initial substrate concentration $c_{S,i}$ after a given time. The specific activity a_{sp} (μmol/$g_{\text{E,prep}}$/min) relates to the applied substrate under the given reaction conditions. This equation is true for a single substrate reaction. For more complex reactions, see Illanes et al. [44]. From the three operational variables (m_{cat}, t, and $c_{S,i}$), two can be established separately.

$$\frac{m_{cat}a_{sp}}{K_m}t = f(c_s, \chi)$$

For a given X and $c_{S,I}$ the mathematical product of m_{cat} times t will be constant. In the case of continuously operated reactors t will be replaced with the residence time τ. Consequently, the same performance will be obtained at a high enzyme concentration and a short reaction time, or at a low enzyme concentration and a long reaction time. The appropriate conditions will then be determined by economics.

The following **fundamental principles** (**rules of thumb**) apply for the **selection of the appropriate reactor** type:

1. In the case of significant product inhibition, batch and pfr will be the reactors of choice.
2. In the case of substrate surplus inhibition, fed batch or cstr will be the reactor of choice.
3. The enzyme consumption in a cstr will be generally higher than that in a pfr/pbr under comparable reaction conditions.
   Since the cstr always operates at the lowest substrate concentration and maximum conversion in the whole reactor volume, kinetically the lowest enzyme activity results from this operation point (close to origin of Michaelis–Menten curve). In batch or pfr, the whole concentration range of the Michaelis–Menten curve is utilized.
4. Kinetic resolutions yield higher enantioselectivities in batch and pfr, in contrast to cstr, since in the latter the substrate racemate is continuously fed during one residence time, reducing the resolution efficiency of the process (limiting the final enantiomeric excess of the product).
5. A pfr/pbr can be transformed to cstr characteristics, if additionally a circulation loop integrating the pbr is established, where the flow rate of the cycle is at least 20 times faster than the flow rate establishing the residence time.

Based on the three fundamental types of reactors, special reactor configurations are applied for special tasks:

Immobilized enzymes:

- *Bubble column reactor (bcr)*: This reactor is of special use if one reactant is gaseous, or if a component needs to be removed (stripped) with the gas phase *in situ*, e.g. water removal in an enzyme-catalyzed esterification [36, 45]. Additionally, it is also of advantage if biotransformations are carried out in highly viscous media. Here, the injected gas bubbles reduce the macroscopic viscosity. A pbr is then not applicable, since the pressure drop over the length of the reactor is too high.
- *Rotating bed reactor (rbr)*: The simplest possible setup with immobilized enzymes is the batch reactor with suspended immobilisate. In contrast to the pbr, a low mass transfer on the particle will be observed due to a low convective flow (resulting in diffusion limitation) on the particles. To overcome this limitation, the enzyme immobilisate is contained in a packed

bed inside a rotating cylinder [46]. As the rbr spins, a continuously circulating flow develops that sucks in the liquid from the bottom of the reactor and percolates it through the rotating packed bed, establishing a convective flow. This is especially useful for immiscible liquids as well as those of high viscosity.

Homogeneously soluble enzymes:

- *Enzyme membrane reactor (emr)*: Homogeneously soluble enzymes are retained with appropriate ultrafiltration membranes, whereby the reactants of low molecular weight, in contrast to the enzymes, can pass the membrane [47]. In laboratory versions often flat sheet membranes are applied. For scale-up convective flow membrane modules such as spiral wound, tubular, or hollow fibers are used to prevent membrane fouling caused by concentration polarization.

1.4.5 Assessment Criteria for Enzymatic Applications

In the following the most significant criteria to assess enzymatic transformations are described. The base case is a simple irreversible single substrate reaction with a possible side product *NP*.

$$|\nu_S|\, S \rightarrow |\nu_P|\, P + |\nu_{NP}|\, NP$$

Conversion χ describes the number of moles of reacted substrate ($n_{S,0} - n_S$) in relation to the number of moles of substrate at time point zero n_S. Conversion does not consider if any target product at all and/or mainly side product is produced.

$$\chi = \frac{n_{S,0} - n_S}{n_{S,0}}$$

Selectivity σ describes the number of moles of product n_P that was formed in relation to the reacted substrate.

$$\sigma = \frac{n_P - n_{P,0}}{n_{S,0} - n_S} \cdot \frac{|\nu_S|}{|\nu_P|}$$

Yield η is the mathematical product of selectivity and conversion. It describes the number of moles of product being synthesized in relation to the substrate at time point zero.

$$\eta = \chi \cdot \sigma = \frac{n_P - n_{P,0}}{n_{S,0}} \cdot \frac{|\nu_S|}{|\nu_P|}$$

Additionally, analytical and isolated yields are differentiated. Analytical yield describes the yield generated in the synthesis itself, omitting any steps of isolation/dsp. In contrast, isolated yield describes the yield at the end of the dsp integrating the analytical yield and the yield of the individual unit operations being part of the dsp process. If multiple steps are combined the individual yields are mathematically multiplied to give the overall isolated yield. The highest possible yield is targeted to increase the economic efficiency.

In enzyme-catalyzed reactions the catalytic activity is given in "**Unit**" (1 U = 1 μmol/min), which is calculated by substrate consumption. The reaction conditions such as temperature, pH, buffer, and reactant concentrations need to be given under which the Units were determined. In relation to the applied amount of enzyme preparation, the mass specific activity per milligram of enzyme μmol/min/mg is differentiated from the volume specific activity μmol/min/ml. In the latter case, the enzyme concentration in terms of mg/ml has to be given additionally.

Alternatively, **turnover frequency** (**TOF**) is used (s^{-1}). The number of moles of converted substrate per time is calculated on basis of the number of moles of catalyst n_{cat}.

$$\mathrm{TOF} = \frac{n_\mathrm{S}}{t \cdot n_\mathrm{cat}}$$

If the Michaelis–Menten kinetics is known, TOF can be described with the kinetic parameter as k_{cat}, describing the rate-limiting step of product formation from the enzyme–substrate complex.

The dimensionless **turnover number** (**TON**) describes the number of moles of product formed in relation to the applied number of moles of catalyst. In industry, sometimes also the mass m to yield the specific ratio (kg/kg) is used, requiring additionally the individual molecular weight M.

$$\mathrm{TON} = \frac{n_\mathrm{P}}{n^*_\mathrm{cat}} = \frac{m_\mathrm{P}}{m_\mathrm{cat}} \cdot \frac{M_\mathrm{cat}}{M_\mathrm{P}}$$

Comparison of different catalysts on the basis of the TON gives indirect information on the enzyme stability and efficiency. The higher the TON, the more attractive is the given process for industry. In Table 1.4.4 typical ranges of TON for different industrial application areas ranging from pharma to bulk are given. In all cases, the different reaction conditions and unit operations of dsp need to be considered, which might differ significantly from application to application.

In industry, an even more simple measure is the **biocatalyst consumption** (**bc**) just relating the consumed mass of the biocatalyst preparation to the mass of

Table 1.4.4 Catalytic productivity (=turnover number, TON) for different products and related costs of products [48].

Product	Costs of product (€/kg)	Range of catalytic productivity (= TON)		
		Cell dry mass	Isolated enzyme	Immobilized enzyme
Pharma	>100	10–35	250	50–100
Fine chemical	>15	70–230	670–1700	330–670
Special chemical	0.25	140–400	1000–4000	400–2000
Bulk	0.05	700–2000	5000–20000	2000–10000

Source: Tufvesson et al. 2011 [48]. Reproduced with permission of ACS.

product (g/kg):

$$bc = \frac{m_{cat}}{m_P}$$

In the preparation of pharmaceutical products of high value the biocatalyst is often discarded without recycling. The biocatalyst stability might be affected during the course of conversion by thermal deactivation, or alternatively, by deactivation caused by reactants. In those cases, rather the differential biocatalyst consumption is investigated to identify an optimal conversion point for terminating the reaction and separating the reaction solution from the biocatalyst.

Space–time yield (**sty**) describes the time-dependent volume productivity for continuous processes in the steady state expressed in mass of product per reactor volume V_R and residence time.

$$sty = \frac{m_P}{\tau \cdot V_R}$$

Depending on the costs of product, typical *sty* values are >100 g/L/d for expensive pharma products and >500 g/L/d for low-cost bulk chemical products.

The simplest way to analyze the sustainability, "greenness" of an enzymatic transformation, or even a respective reaction sequence, is the **atom economy** (**atom efficiency**). This term describes the efficient use of all atoms of the starting materials and additionally in reaction sequences added ones in the desired products [49].

$$\text{atom economy} = \frac{\text{molecular weight of desired product}}{\text{molecular weight of all reactants}}.$$

References

1 Yuryev, R. and Liese, A. (2010). *ChemCatChem* 2: 103–107.

2 (a) Klibanov, A.M. (1976). *Anal. Biochem.* 93: 1–25. (b) Mateo, C., Abian, O., Fernandez-Lafuente, R. et al. (2000). *Enzyme Microb. Technol.* 26 (7): 509–515. (c) Hanefeld, U., Gardossi, L., and Magner, E. (2009). *Chem. Soc. Rev.* 38: 453–468. (d) Cantone, S., Ferrario, V., Corici, L. et al. (2013). *Chem. Soc. Rev.* 42: 6262–6276.

3 (a) Bornscheuer, U.T. (2003). *Angew. Chem. Int. Ed.* 42: 3336–3337. (b) Bornscheuer, U.T., Huisman, G.W., Kazlauskas, R.J. et al. (2012). *Nature* 485: 185–194.

4 Spickermann, D., Kara, S., Barackov, I. et al. (2014). *J. Mol. Catal. B: Enzym.* 103: 24–28.

5 Vagenende, V., Yap, M.G.S., and Trout, B.L. (2009). *Biochemistry* 48: 11084–11096.

6 Faber, K. (2011). *Biotransformations in Organic Chemistry*, 6e. Berlin Heidelberg, Heidelberg, Dordrecht, London, New York: Springer-Verlag.

7 Magner, E. (2013). *Chem. Soc. Rev.* 42: 6213–6222.

8 (a) Sheldon, R.A. (2007). *Adv. Synth. Catal.* 349: 1289–1307. (b) Sheldon, R.A. and van Pelt, S. (2013). *Chem. Soc. Rev.* 42: 6223–6235.

9 (a) Gómez Baraibar, Á., Reichert, D., Mügge, C. et al. (2016). *Angew. Chem. Int. Ed.* 55: 14823–14827. (b) Temiño, D.M.-R.D., Hartmeier, W., and Ansorge-Schumacher, M.B. (2005). *Enzyme Microb. Technol.* 36: 3–9.

10 Ansorge-Schumacher, M.B. and Thum, O. (2013). *Chem. Soc. Rev.* 42: 6475–6490.

11 Xue, R. and Woodley, J.M. (2012). *Bioresour. Technol.* 115: 183–195.

12 Kara, S. and Schrittwieser, J.H. (2017). *BIOspektrum* 23: 468–470.

13 Wedde, S., Rommelmann, P., Scherkus, C. et al. (2017). *Green Chem.* 19: 1286–1290.

14 Rocha-Martín, J., Rivas, B.d.l., Muñoz, R. et al. (2012). *ChemCatChem* 4: 1279–1288.

15 (a) France, S.P., Hepworth, L.J., Turner, N.J., and Flitsch, S.L. (2017). *ACS Catal.* 7: 710–724. (b) Wheeldon, I., Minteer, S.D., Banta, S. et al. (2016). *Nat. Chem.* 8: 299–309. (c) Zhang, Y.H.P. (2011). *Biotechnol. Adv.* 29: 715–725.

16 Liese, A. and Hilterhaus, L. (2013). *Chem. Soc. Rev.* 42: 6236–6249.

17 (a) Ni, Y., Holtmann, D., and Hollmann, F. (2014). *ChemCatChem* 6: 930–943. (b) Domínguez de María, P. and Hollmann, F. (2015). *Front. Microbiol.* 6: 1257.

18 Dach, R., Song, J.J., Roschangar, F. et al. (2012). *Org. Process Res. Dev.* 16: 1697–1706.

19 Adlercreutz, P. and Mattiasson, B. (1987). *Biocatalysis* 1: 99–108.

20 Bommarius, A.S. and Riebel, B.R. (2008). *Protein Science Encyclopedia*. Wiley-VCH.

21 Laane, C., Boeren, S., Vos, K., and Veeger, C. (1987). *Biotechnol. Bioeng.* 30: 81–87.

22 Adlercreutz, P. (2008). *Organic Synthesis with Enzymes in Non-Aqueous Media* (ed. G. Carrea and S. Riva), 3–34. Weinheim: Wiley-VCH.

23 (a) Kara, S., Spickermann, D., Weckbecker, A. et al. (2014). *ChemCatChem* 6: 973–976. (b) Petrenz, A., María, P.D.d., Ramanathan, A. et al. (2015). *J. Mol. Catal. B: Enzym.* 114: 42–49.

24 Careri, G., Gratton, E., Yang, P.H., and Rupley, J.A. (1980). *Nature* 284: 572–573.

25 (a) Greenspan, L. (1977). *J. Res. Natl. Bur. Stand. Phys. Chem.* 81A: 89–96. (b) Wehtje, E., Svensson, I., Adlercreutz, P., and Mattiasson, B. (1993). *Biotechnol. Tech.* 7: 873–878.

26 Bell, G., Janssen, A.E.M., and Halling, P.J. (1997). *Enzyme Microb. Technol.* 20: 471–477.

27 Constantino, H.R., Griebenow, K., Langer, R., and Klibanov, A.M. (1997). *Biotechnol. Bioeng.* 53: 345–348.

28 (a) Adlercreutz, P. (1993). *Biochim. Biophys. Acta, Protein Struct. Mol. Enzymol.* 1163: 144–148. (b) Khmelnitsky, Y.L., Welch, S.H., Clark, D.S., and Dordick, J.S. (1994). *J. Am. Chem. Soc.* 116: 2647–2648. (c) Dabulis, K. and Klibanov, A.M. (1993). *Biotechnol. Bioeng.* 41: 566–571. (d) Dai, L. and Klibanov, A.M. (1999). *Proc. Natl. Acad. Sci. U.S.A.* 96: 9475–9478.

29 Klibanov, A.M. and Zaks, A. (1984). *Science* 224: 1249–1251.

30 Lilly, M.D. (1982). *J. Chem. Technol. Biotechnol.* 32: 162–169.

31 (a) Ritter, E., Yordanova, D., Gerlach, T. et al. (2016). *Fluid Phase Equilib.* 422: 43–55. (b) Mohammad, S., Held, C., Altuntepe, E. et al. (2016). *Fluid Phase Equilib.* 416: 83–93.
32 (a) Peters, C., Kölzsch, R., Kadow, M. et al. (2014). *ChemCatChem* 6: 1021–1027. (b) Prat, D., Hayler, J., and Wells, A. (2014). *Green Chem.* 16: 4546–4551.
33 (a) Capello, C., Fischer, U., and Hungerbühler, K. (2007). *Green Chem.* 9: 927. (b) Hernaiz, M.J., Alcantara, A.R., Garcia, J.I., and Sinisterra, J.V. (2010). *Chemistry* 16: 9422–9437. (c) Jessop, P.G. (2011). *Green Chem.* 13: 1391–1398.
34 Khmelnitsky, Y.L., Levashov, A.V., Klyachko, N.L., and Martinek, K. (1988). *Enzyme Microb. Technol.* 10: 710–724.
35 Briechle, S., Howaldt, M., Röthig, T., and Liese, A. (2011). *Bioprozesstechnik*, 3e (ed. H. Chmiel), 427–476. Heidelberg: Spektrum Akademischer Verlag.
36 Hilterhaus, L., Thum, O., and Liese, A. (2008). *Org. Process Res. Dev.* 12: 618–625.
37 Doran, P.M. (2013). *Bioprocess Engineering Principles*, 2e, 379–444. London: Academic Press.
38 Basso, A., Cantone, S., Ebert, C. et al. (2008). *Organic Synthesis with Enzymes in Non-Aqueous Media* (ed. G. Carrea and S. Riva), 279–301. Wiley-VCH.
39 Cheetham, P.S.J. (1995). The applications of enzymes in industry. In: *Handbook of Enzyme Biotechnology* (ed. A. Wiseman), 493–498. Ellis Harwood.
40 Vasic-Racki, D. (2006). History of industrial biotransformations dreams and realities. In: *Industrial Biotransformations*, vol. 40 (ed. A. Liese, K. Seelbach and C. Wandrey), 1–36. Wiley-VCH.
41 Bommarius, A.S., Drauz, K., Groeger, H., and Wandrey, C. (1992). *Chirality in Industry* (ed. A.N. Collins, G.N. Sheldrake and J. Crosby), 372–397. Wiley.
42 (a) Hilterhaus, L. and Liese, A. (2010). *Encyclopedia of Industrial Biotechnology* (ed. D. Flickinger), 1–14. Wiley. (b) Levenspiel, O. (1998). *Chemical Reaction Engineering*. Wiley.
43 Segel, I. (1993). *Enzyme Kinetics: Behavior and Analysis of Rapid Equilibrium and Steady-State Enzyme Systems.* Wiley.
44 Illanes, A., Wilson, L., and Vera, C. (2014). *Problem Solving in Enzyme Biocatalysis*. Wiley.
45 Kern, S., Himmelspach, A., Grammann, K. et al. (2016). *Org. Process Res. Dev.* 20 (11): 1930–1936.
46 (a) Mallin, H., Muschiol, J., Bystrom, E., and Bornschauer, U.T. (2013). *ChemCatChem* 5: 3529–3532. (b) Aguilera, A.F., Tolvanen, P., Heredia, S. et al. (2018). *Ind. Eng. Chem. Res.* 57 (11): 3876–3886.
47 (a) Wichmann, R., Wandrey, C., Bückmann, A.F., and Kula, M.-R. (1981). *Biotechnol. Bioeng.* 23 (12): 2789–2802. (b) Haberland, J., Hummel, W., Dausmann, T., and Liese, A. (2002). *Org. Process Res. Dev.* 6 (4): 458–462. (c) Greiner, L., Laue, S., Wöltinger, J., and Liese, A. (2007). *Top. Curr. Chem.* 276: 111–124. (d) Yuryev, R., Strompen, S., and Liese, A. (2011). *Beilstein J. Org. Chem.* 7: 1449–1467.
48 Tufvesson, P., Lima-Ramos, J., Nordblad, M., and Woodley, J.M. (2011). *Org. Process Res. Dev.* 15: 266–274.
49 Sheldon, R.A. (2000). *Pure Appl. Chem.* 72 (7): 1233–1246.

Part II

Enzyme Applications for the Food Industry

2.1

Enzymes Used in Baking

Joke A. Putseys and Margot E.F. Schooneveld-Bergmans

DSM Biotechnology Centre, Alexander Fleminglaan 1, 2613 AX Delft, the Netherlands

2.1.1 Introduction

Baked cereal products have been prepared as an everyday food by mankind for thousands of years, because of their nutritional value, mainly coming from the starch and protein content, the ease of growing cereals at different conditions, and the possibility to store them without serious quality loss. The first evidence of wild barley, wheat, and oat grains being ground for human consumption dates back to 23 000 years ago [1, 2]. Most likely, these crushed grains were originally eaten as such, and later mixed with water and cooked to make it, as a porridge, more palatable. The next steps in the development of bread, the most commonly consumed baked cereal product known today, was to bake a thicker dough into a flat bread on hot stones or, eventually, in an oven, and finally, to use the leftover dough of the day before to transform the rather solid flat bread into a wild-yeast leavened product [3]. The art of further refining particularly the ground grains of wheat into white flour by development of milling and sifting techniques, having proper mixing and fermentation equipment, and enriching the basic recipe of flour, water, and leaven with fat, egg, and many other ingredients has resulted in a broad range of baked cereal foods, such as bread, biscuits, cakes, and pastry [4, 5].

The major asset of wheat in comparison to other cereals is its specific protein composition that allows to make leavened baked products. The industrial production of baker's yeast in the nineteenth century and that of amylase, a starch degrading enzyme, in the twentieth century initiated the use of biotechnologically produced components in bread making. Standardized, relatively pure fungal amylase capable of replacing malt flour and providing bread of constant quality [6] paved the way for the wide variety of microbially produced enzymes that are applied in baking nowadays (see also Chapter 1.1).

In this chapter, the focus will be on the bread making process, the final bread quality, and how the various baking enzymes play a role considering the baker's and consumer's needs. The chapter will be concluded with an outlook on opportunities and trends, including a few specific enzymes and baking applications other than bread.

Industrial Enzyme Applications, First Edition. Edited by Andreas Vogel and Oliver May.

2.1.2 The Baking Process – The Baker's Needs

2.1.2.1 Flour Quality and Standardization

A myriad of aspects is to be considered in defining the quality of flour. Starting from the quality of the final product to be prepared, i.e. a light loaf of bread, **wheat** is clearly the cereal of choice. Wheat is unique in its ability to form an elastic and cohesive mass when mixed with water, and hold the gas, produced by yeast, during the whole process of fermentation, handling, and baking [7]. This is attributable to the **protein** content and, even more importantly, to the specific protein composition, being a combination of the gluten storage proteins gliadin and glutenin. Rye also contains these storage proteins, but the lower protein content of rye flour and its specific protein composition create a less elastic protein network in the dough, causing loaves to be heavy and dense. Often a mix of rye and wheat flour is used to make such bread lighter [7, 8].

With the focus on wheat for its unique bread making properties, it is relevant to realize that different **species of wheat** are grown for different purposes: durum wheat (*Triticum durum*) is typically grown for pasta production, common wheat (*Triticum aestivum*) for use in bakery products such as breads and rolls, and club wheat (*Triticum compactum*) for cakes and cookies. Within the common wheat species, many **different varieties** are known. These can be classified according to kernel hardness (hard or soft wheat), grain color (red or white wheat), and season of planting (spring or winter wheat) [4, 8]. Grain color is of lesser importance for standard white flour production, as the red pigment in the outer layers of the kernel is removed during milling. In whole wheat flour, however, red wheat tends to result in darker bread than white wheat, leading to an increased healthiness perception. It also produces a slightly different flavor [9].

An important factor determining flour quality is kernel hardness, which relates to the force required to crush the kernels. This also links to the particle size of the final flour, since soft wheats yield flour with a finer particle size than hard wheats. In addition, hard wheats generally have a higher protein content, which has implications for the final use of these flours. Soft wheat flour is mainly used for biscuits, muffins, pastries, and cakes, while hard wheat flour with higher protein content is primarily used for yeast-leavened products, such as bread and rolls [4, 10].

Environmental conditions during growth and storage such as soil type, climate conditions, pests and diseases, sprout damage, and moisture content affect the quality of the flour also, and are of primary importance to the miller. The miller transforms the wheat kernels in different grades of flour, suited for a specific purpose. The **milling process** is aimed at the separation of the endosperm, making up over 80% of the kernels' weight and containing the starch and gluten, from the germ and bran. This is done by roller milling, in which breaking, reducing, and sieving are the main processes, following the cleaning and tempering of the kernels. The bran and germ contain enzymes and oil that affect the storage stability and baking quality of the flour negatively. On the other hand, removing these parts of the kernel also reduces the nutritive value, as the bran contains fibers, vitamins, and minerals (= ash). Different grades of flour are indicated by their **extraction rate**: whole wheat flour has 100% extraction rate, while straight-grade, or "all purpose," flour has an extraction rate of approximately 75%. Patent flour can have an extraction rate of 45–60%, and also contains the lowest level of ash [4, 10, 11]. Milling inevitably leads to the formation of **damaged starch** in which part of the starch granules in the endosperm are cracked due to friction. In hard wheat, this level is generally higher than in soft wheat. Damaged starch absorbs more water in the dough mixing stage than the non-damaged starch, which is beneficial in this stage of the bread making process. During dough fermentation, on the other hand, this higher water absorbing capacity can limit the expansion of the dough, which adversely affects the final bread quality [4, 11]. Also, **endogenous enzymes** that might be present in the wheat kernel can interact with damaged starch. Endogenous α-amylase and β-amylase can readily degrade the damaged starch to dextrins and maltose, hence providing additional sugars to the yeast, while also influencing the dough rheological properties [4, 12]. Many other endogenous enzymes have been reported to be present in wheat, such as arabinoxylan-degrading enzymes, proteases, lipases, esterases, phytase, lipoxygenase, superoxide dismutase, polyphenol oxidases, peroxidase, catalase, protein disulfide isomerase, phosphorylase, ascorbic acid oxidase, and dehydrogenases. They mainly originate from the outer bran-associated layers of the kernels [13, 14]. Primarily the lipolytic enzymes cause flour quality to deteriorate during storage, which runs faster for whole wheat flour than for white flour. The changing ratio of tri-, di-, monoacylglycerols, and free fatty acids, resulting from lipolytic enzyme activity, is generally thought to have an overall negative impact on the flour's bread making quality. In addition, wheat lipoxygenase may oxidize the polyunsaturated fatty acids and cause oxidative rancidity and co-oxidation of the gluten [9, 15, 16]. Flour quality is also influenced by preharvest sprouting, leading to increased levels of amylases, proteases, and arabinoxylan-degrading enzymes, of which the endoxylanases appear to be primarily from microbial origin, whereas the α-amylase, β-amylase, and protease are mainly endogenous [17, 18].

Main wheat flour quality factors determining the overall mixing, proofing, and baking quality for bread making are quantity and quality of the protein, color and ash content (as indicators for the extraction rate), water absorption capacity, and level of enzyme activity. Many empirical rheological methods are available to assess wheat quality, such as the farinograph and mixograph for mixing quality, rheofermentometer, alveograph, and extensograph for proofing quality, and falling number, amylograph, and rapid visco analyzer for endogenous α-amylase activity. Near infrared reflectance spectroscopy is used to determine the protein or moisture content and other compositional features of flour, while correlation of rheological and compositional data now gives opportunities for prediction of baking performance [4, 11, 19].

Flour standardization is practiced by millers to provide a bread making flour of consistent quality. Oxidizing agents are added to age and bleach the flour and to strengthen the dough, the latter mainly by oxidation of sulfhydryl groups of the gluten to disulfides. Potassium bromate used to be the preferred oxidant for dough strengthening, but is replaced nearly completely by ascorbic acid [4, 8, 20]. Although ascorbic acid as such is a reducing agent, it acts as an oxidant in the formation of disulfide bridges in the wheat gluten, as it is converted by the wheat endogenous ascorbate oxidase to dehydroascorbic acid in the presence of oxygen [21].

Amylase is added to standardize the variable and generally low level of endogenous α-amylase of flour to provide sufficient maltodextrins, which are converted by the wheat endogenous β-amylase into fermentable sugars for the yeast, to optimize the dough handling properties and the loaf volume, crumb structure, flavor, and crust color of the final bread [4, 6, 22]. The types and mechanism of action of amylases is explained in Section 2.1.2.3, as the enzymes have a major role in that stage of the bread making process. Initially, barley malt was used for amylase supplementation. In the 1950s and 1960s, barley malt started to be gradually replaced by fungal α-amylase, originating from *Aspergillus oryzae (see also Chapter 1.1)*. The advantage of this fungal α-amylase (also known as TAKA) compared to the α-amylase of malt flour can be attributed to its lower proteolytic activity, lower temperature stability reducing the risk of overdosing, and decreased sensitivity toward the major α-amylase inhibitors in wheat [6, 22]. Addition of other baking enzymes, such as cellulase and xylanase, at the mill is possible to enable a proper distribution of enzymes in the flour and obtain flour with improved rheological and baking properties, but this is not a common practice [23].

The wheat species and variety used, the concentration of damaged starch and the levels of endogenous enzymes of the wheat flour, and the bread making process selected can vary tremendously. All these factors are important determinants for the choice and dosage of enzymes required to result in a baked product of the desired quality and acceptance.

2.1.2.2 Mixing and Dough Handling

Bread making starts with the basic ingredients flour, yeast, salt, and water. Flour brings the two main structural components of bread: protein and starch. The protein consists of glutenin and gliadin, which, upon hydration with water while mixing, form the viscoelastic gluten network able to hold the gas produced by the yeast. Damaged starch hydrates upon mixing and non-damaged starch during baking and then transform into a gel that sets upon cooling. Flour also contains minor amounts of non-starch polysaccharides, which compete with the protein and starch for water, and lipids that have a role in dough stability. In addition to the production of carbon dioxide, yeast also provides flavor precursors. Salt enhances flavor, toughens the gluten network, and controls the rate of fermentation. Other components can be added to this basic recipe for specific functionalities as well, such as shortening (baking specific formulations of solid fats) or emulsifiers for softness or dough strengthening, sugar for volume increase and sweetness, oxidants for dough strengthening, enzymes as processing aids, and preservatives to inhibit molds. Most hard-crusted or soft breads and rolls are low in sugar and fat and are known as lean dough products [7, 8].

Mixing is not only required to evenly distribute the ingredients into a homogeneous dough, but it is also crucial for the incorporation of air bubbles, required to transform the dough during fermentation into a foam, and for the development of the gluten network. An optimally developed gluten network provides extensibility to allow expansion during fermentation, while simultaneously resisting collapse. This is achieved by administering sufficient mixing intensity and total energy input, which is concomitantly brought about by friction of the mass in the mixer. Initially, the gluten is present as distinct clusters, which later become extended and form a continuous network, with starch granules evenly dispersed through it, acting as a filler. Stretching of the hydrated protein is primarily caused by uncoiling of the high molecular weight glutenins, which form and reform intra- and intermolecular disulfide bonds upon energy input (see Figure 2.1.1). Too extensive mixing, however, makes the dough lose its consistency and become sticky, most likely due to rupture of essential disulfide bonds. Dough stickiness is also observed when adding reducing agents, such as cysteine and sodium bisulfite, demonstrating the value of the disulfide bonds. The quality of the gluten network is determined by the protein content, the composition, the molecular size distribution, and the location of the cysteine residues in the glutenin and gliadin. Flours with high levels of protein require higher mixing intensity and longer development, while being more resistant to overmixing, providing higher elasticity and better bread making potential than flours with protein levels below 12%. Besides the gluten network, the presence of a liquid phase, containing surface-active components such as proteins and lipids that stabilize the expanding gas bubbles during fermentation, is pivotal. Therefore, the amount of water required to produce a dough of proper consistency and stability needs to be balanced, and is generally in the range of 55–64% of the weight of the flour. Since the energy input during mixing increases the dough temperature to some extent, the temperature of the flour and water needs to be managed in relation to the fermentation temperature of the final dough [4, 8, 20].

Dough mixing methods vary around the world and can be categorized into three basic processes. The "straight dough" process consists of mixing all ingredients into a developed dough in one step. This is followed by fermentation for a period of minutes up to hours, which may or may not be interrupted for punching, and followed by a final proof. Depending on the fermentation time, the dose of yeast and the temperature of the dough can vary, both being generally higher for shorter fermentation times. The second method, often used in the United States, is known as the "sponge and dough" process. Approximately two-thirds of the flour, part of the water, and the yeast are mixed to form a slack dough, or "sponge," which is allowed to ferment for a few hours at relatively low temperature. Subsequently, it is mixed into a developed dough with the remaining ingredients, and is either scaled immediately or given additional proof. This process normally results in softer bread, having a finer cell structure and more flavor than bread produced according to the "straight dough" method. Additionally, the dough is less sensitive to overfermentation, which is beneficial for large-scale production. The third method, known as the "Chorleywood bread process," is mainly used in the United Kingdom and Australia, and consists of high-speed mixing of all ingredients in 2–4 minutes under partial vacuum. Owing to this high energy input, the dough temperature can rise by 14 °C during mixing, which requires cooling or the use of ice water. Moreover, this process generally requires more water, as less damaged starch is hydrolyzed in the relatively short mixing time, which also requires a higher yeast dose. In the "Chorleywood bread process,"

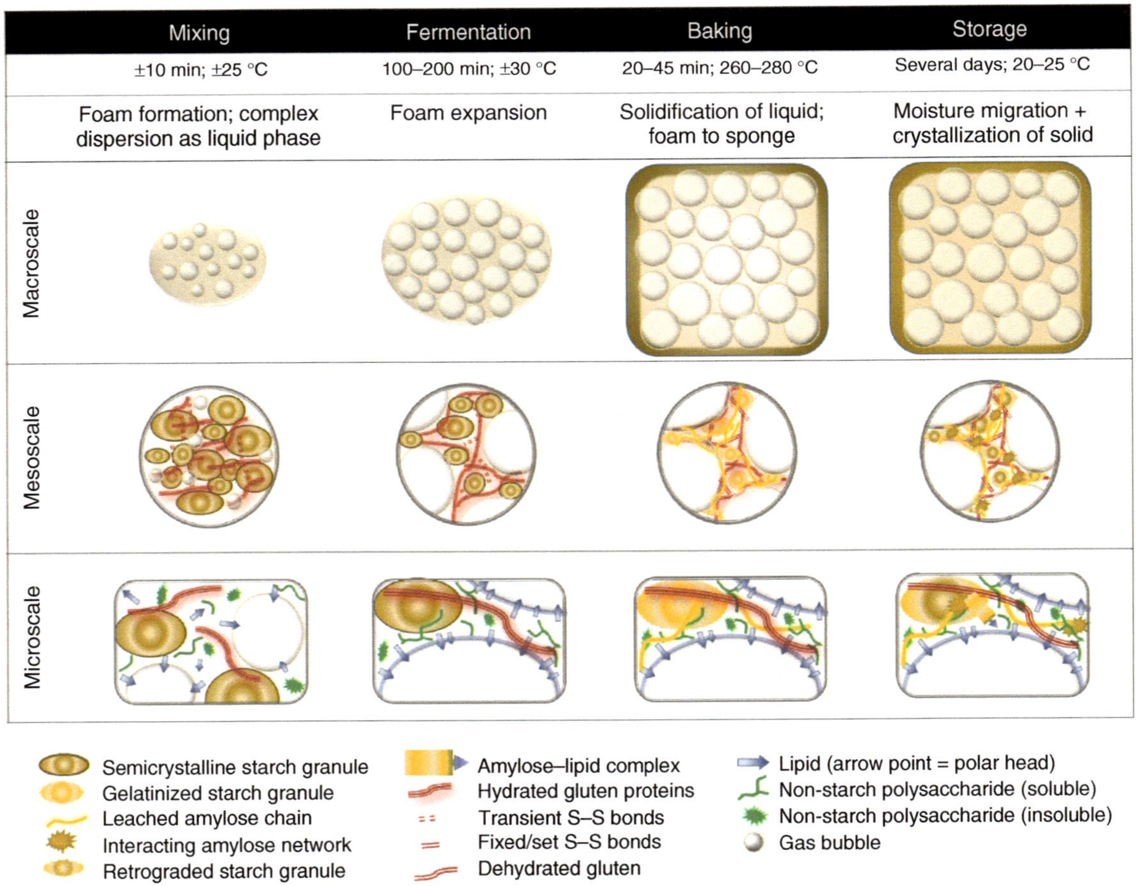

Figure 2.1.1 Schematic presentation of the various stages of mixing, fermentation, baking, and storage in the bread making process; conceptually showing the main events taking place at macro-, meso- and microscale from visually observable physical transformations in the foam to molecular interactions between starch, protein, lipids, and non-starch polysaccharides.

dough development largely depends on the presence of sufficient oxidant. Alternative methods, such as liquid preferment and sourdough, as variations on the "sponge and dough" method, are also applied, while variations on the "Chorleywood bread process" were explored in the United States, but were not well accepted [4, 7, 8].

Enzymes have a significant contribution in the rheological properties of dough, even more than the main ingredients and the mixing conditions. The ease of dough handling after mixing is important, and is primarily governed by its bulk consistency and surface properties. Slack and sticky dough is particularly disastrous in industrial bakeries as it interrupts the continuous transfer of dough in the equipment for resting, dividing, fermentation, scaling, and baking, and consequently also affects the next batches of dough as longer processing times cause defects in the final product.

Hemicellulases are a diverse group of enzymes capable of hydrolyzing hemicellulose, which in wheat flour primarily consists of arabinoxylans. Hemicellulose and pentosan are alternative terms used in baking for non-starch polysaccharides, and are intended to indicate the water-unextractable arabinoxylan (WU-AX) and water-extractable arabinoxylan (WE-AX), respectively. Although arabinoxylans only constitute approximately 2% of the flour, of which two-thirds is water-unextractable, they are able to absorb large quantities of water, and thus affect dough rheology. Arabinoxylans consist of a backbone of β-(1,4)-linked xylo-pyranosyl residues, to which single α-arabinofuranoside units are linked via the C(O)-3 and C(O)-2 of the xylose. Some of the arabinose units are esterified with ferulic acid. The difference between WE-AX and WU-AX could not be attributed to the structure and distribution of the arabinose residues along the xylan backbone, and was speculated to be mainly caused by minor differences in composition in which ferulic acid may play an important role [24, 25]. The hemicellulase of primary importance to modify the rheological properties of dough is endoxylanase (E.C.3.2.1.8), classified according to protein structure in the glycosyl hydrolase (GH) families 5, 7, 8, 10, 11, and 43, and catalyzing the endo-hydrolysis of the 1,4-β-xylosidic linkage. The GH11 endoxylanases are most often used in bread making, and differ from the other main group of GH10 xylanases by releasing longer fragments from the arabinoxylan polymer, and by having a higher affinity for WU-AXs. This mode of action of the GH11 endoxylanases matches well with the observed solubilization of WU-AXs during the bread making process, with concomitant increase in the viscosity of the dough liquid [24, 26]. A recent study with endoxylanases with modified binding sites showed that enhanced preference for WU-AXs further improved the effect of the xylanase in bread making [27]. It is also postulated that the (partial) conversion of the needle-like WU-AXs into solubilized arabinoxylans decreases the interference with the gluten network, resulting in more stable dough. The benefits of the use of xylanase in dough handling consequently give rise to improved oven spring, larger loaf volume, and improved structure and softness of the crumb [24, 26]. Xylanases became commercially available for baking applications in the 1980s, and are produced by fungi, such as *Aspergillus, Trichoderma, Humicola,* and *Thermomyces,* and bacteria, such as *Bacillus* [28, 29]. They can be either classical products, containing accessory enzymes such as arabinofuranosidase, which can also enhance the xylanase effect, or they can be expressed in rather pure form in another microbial host. Combinations of xylanases and **cellulases** were shown to have beneficial and synergistic effects on dough handling and final baked product quality despite the very low level of cellulose in wheat flour [28]. Academic studies also showed the potential of psychrophilic xylanases, especially GH8 xylanases, to increase bread volume at lower dose than the mesophilic xylanases [26, 30]. A disadvantage of xylanases is that they can cause dough stickiness, particularly when highly dosed or not sufficiently specific for the WU-AXs [22].

Glucose oxidase (E.C.1.1.3.4) is an enzyme that can strengthen and dry a slack and sticky dough caused by excessive xylanase action or just by improving the gluten network of a relatively weak flour having low gluten quality or quantity [26, 31]. The enzyme oxidizes glucose in the presence of oxygen into hydrogen peroxide and glucono-δ-lactone, and the latter is chemically instable and converted into gluconic acid. The hydrogen peroxide can subsequently oxidize SH groups of the gluten proteins into S—S bonds in the presence of wheat endogenous peroxidase. Next to that, oxidative cross-linking of ferulic acid units of neighboring arabinoxylans can take place. It is also hypothesized that cross-linking of two tyrosine residues or of feruloylated arabinoxylan to tyrosine moieties in the gluten can occur in dough [14, 32, 33]. As the enzyme requires both oxygen and glucose, its strengthening effect on the bulk rheology of the dough occurs mainly during mixing, where oxygen is not limiting. Its effect on drying of the surface of the dough can proceed in later stages of the bread making process, as long as glucose is available. This enzyme is typically from fungal origin (e.g. from *Penicillium* or *Aspergillus*) and is used more and more to replace chemical additives that are banned for health reasons, such as potassium bromate. Its commercialization started at the end of the 1980s. Supplementation with the optimal dosage of glucose oxidase not only results in stronger and less sticky dough but the bread volume is also increased and its crumb structure improved [22]. Here as well, it is possible to overdose the enzyme, which would create too stiff a gluten network that hampers rise during fermentation and baking, resulting in a very dense dough and bread. Besides glucose oxidase, other redox enzymes, such as sulfhydryl oxidase, hexose oxidase, pyranose oxidase, laccase and peroxidase, were explored for their effects in bread making, but have not been equally commercially successful as glucose oxidase yet [21, 26, 33].

Proteases (E.C.3.4.x.x) hydrolyze peptide bonds in proteins, and are applied in bread making in a limited manner. They are classified according to their endo- or exo-type mode of action, catalytic mechanism, and structural homology [34, 35]. For optimization of dough mixing and handling properties of particularly strong wheat flours endoproteases are applied. They decrease mixing time by faster water absorption and increase dough extensibility, leading to easier dough handling. As endoproteases interfere with the covalent linkages in the gluten network, their action is not reversible, as is the case with reducing agents. For that reason, only very low amounts and/or relatively mild acting proteases are applied in bread making, such as endoproteases produced by *Aspergillus* or *Bacillus* [22, 34].

Amylases can also affect the rheological properties of the dough, primarily the wheat endogenous amylases, malt flour, and the fungal amylase produced by *A. oryzae*, as already discussed in the previous section on flour standardization.

2.1.2.3 Fermentation and Dough Stability

After mixing, the dough is allowed to rest such that the formed gluten network can recover from the mechanical deformation to which it was subjected. In this step, the dough is placed at slightly higher temperatures and humidity (typically 30 °C and 85% relative humidity) to also create optimal conditions for fermentation. During fermentation, the *Saccharomyces cerevisiae* yeast is typically used to convert the fermentable sugars present in the dough into, predominantly, carbon dioxide (CO_2). Alcohols and aromatic compounds are generated as well. CO_2 thus produced cannot create new gas cells from scratch, as the pressure required to accomplish this would be infinitely high. The CO_2 partially dissolves in the aqueous phase and diffuses to the air pockets in the dough that were incorporated during mixing, thereby expanding the gas cells and increasing the overall dough volume tremendously.

Fermentation or proofing can take place in one or multiple steps, partly in bulk at the start and/or continuing after shaping of the dough toward the end of fermentation. In case of multiple steps, the dough is punched in between the various fermentation phases. This is done to release some of the gas already formed earlier on in the fermentation, on the one hand, and to redistribute the already formed gas cells and alcoholic compounds, on the other hand [36]. Moreover, the deformation of the dough during punching and shaping also adds to the elasticity and extensibility of the gluten, making the walls of the gluten network surrounding the gas cells thinner, which will result in a finer crumb grain after baking [36]. This way, the fermentation step in the baking process transforms the dough into a foam-like material with the flour–water phase acting as the liquid layer that surrounds the air–CO_2 gas cells (as schematically shown in Figure 2.1.1).

Initially, the **air bubbles** in the dough grow independently and freely of one another. After a while, they come into contact with adjacent bubbles, which could lead to bubble coalescence [37]. The homogeneity and stability of the fermenting dough depends on the diameter of the growing bubbles and thickness and strength of the intermittent dough cell walls [37]. As the air bubbles in the dough are subjected to thermodynamic phenomena such as Ostwald ripening, the internal pressure makes large bubbles grow even larger and small bubbles disappear [38, 39]. This, in combination with significant expansion of the bubbles during fermentation, makes the dough a very unstable foam. Redistribution of the gas cells, e.g. by punching the dough, is essential to obtain a homogeneous distribution of gas cells in the bread crumb.

Moreover, having good **extensibility and surface-active properties** is critical for the dough to be able to retain the extra gas produced during fermentation. These essential dough properties can be attributed to an interplay between bulk viscosity and interfacial features, such as interfacial viscosity, surface activity, and concentration of amphiphilic molecules at the interface [40].

The dough's bulk viscosity is determined by the various flour components. Gluten of good quality and quantity is instrumental in obtaining a dough with acceptable (biaxial) extensibility and elasticity. Next to gluten, starch and lipids are also present in the flour, as well as non-starch polysaccharides such as arabinogalactan peptides or arabinoxylans [22, 37, 40, 41]. WU-AXs can be regarded as needles that might damage the foam structure by puncturing/deflating the gas cells. WE-AXs, on the other hand, are beneficial [22, 24]. They bind much water and contribute significantly to the bulk viscosity. This creates a more stable foam, as higher viscosities of the liquid phase prevent the bubbles from moving upwards and disappearing into the atmosphere [41].

The gas–liquid interface is occupied by amphiphilic flour components, such as proteins and lipids. Already early on in fermentation, disruptions in the protein network occur that do not lead to a collapsing dough. This is attributed to the presence of a liquid film surrounding the air bubbles [41, 42]. This liquid film is of essence to obtain a stable dough and to prevent gas cell disruption or coalescence during fermentation and baking. To gain more insight into the main contributors of this phase, the supernatant after centrifugation of the dough can be analyzed. This so-called dough liquor is rich in proteins, lipids, and non-starch polysaccharides. The lipids are probably the most important fraction, impacting the gas cell stability either indirectly by interacting with proteins at the interface and/or directly by their surface-active nature, positioning them at the interface [41, 42].

Depending on the type of bread that is being made, the importance and extent of the various attributes just described can vary. For tin-baked bread, the final step of fermentation takes place in the mold. In this case, improved dough stability leads, for example, to larger loaf volumes and a more homogeneous crumb structure. Floor-baked bread, on the other hand, benefits from good shape retention and oven spring, both determined by the stability of the dough as well.

Enzymes are of utmost importance during this phase of baking. Not only can enzymes boost fermentation by increasing the total amount of fermentable sugars they can also influence the bulk viscosity and increase dough stability by generating more surface-active compounds. The most important causes for dough instability during fermentation are lack of extensibility, on the one hand, and inhomogeneities in the foam structure, on the other hand [37]. Enzymes can help in overcoming these problems.

Amylases (of which the action mechanism is schematically drawn in Figure 2.1.2) can have a significant impact on the fermentation step in the baking process, next to the dough rheological properties already discussed in the previous sections. The amylase enzymes are categorized into GH families (E.C.3.2.1.x). *Endo-acting α-amylases* (E.C. 3.2.1.1), belonging to GH13, can be from bacterial, fungal, or plant origin, with this biological variation also being reflected in differences in binding sites and substrate specificity [22, 26]. These enzymes cleave internal α-1,4 bonds between glucose units inside the starch polymers, amylose, and amylopectin, resulting in polymers with drastically reduced chain lengths. Especially those α-amylases able to act on native or damaged starch are of relevance during fermentation, as starch is not yet transformed into a more enzyme-accessible form at this stage. That only happens once starch gelatinizes during baking (see Section 2.1.2.4 on baking and oven spring). Exo-acting amylases cut glucose or maltose units off from the end of a glucose polymer or oligomer. *β-Amylase* (E.C.3.2.1.2), belonging to GH14, cuts maltose off in its β-anomeric form [22, 26]. *Amyloglucosidase* (E.C. 3.2.1.3), also called

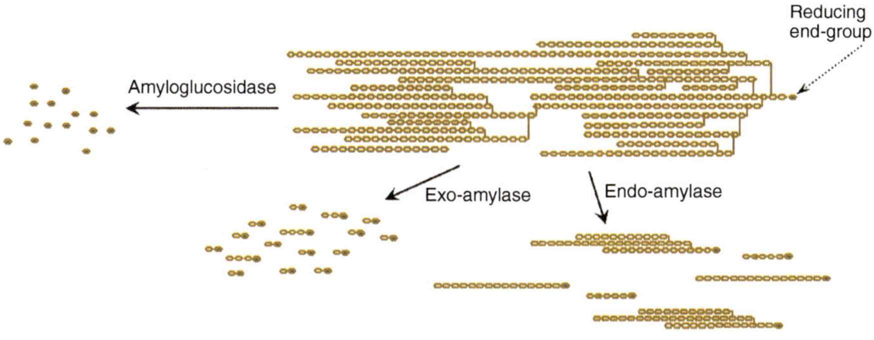

Figure 2.1.2 Mode of action of various starch degrading baking enzymes. Source: Adapted from [22, 31].

glucoamylase, is an exo-acting amylase as well, but one that cleaves glucose monomers off instead of maltose. This enzyme of GH family 15 is typically from fungal origin (e.g. from *Aspergillus niger*). It prefers the conditions generally found in bread dough during mixing and fermentation, with a pH around 4.5–5.5 and temperatures between 25 and 40 °C. Once in the oven, however, amyloglucosidase quickly inactivates at temperatures of 65 °C or higher.

These amylases as such, but especially their combination, result in the production of a higher amount of low molecular weight fragments, oligosaccharides, and short sugars such as maltose or glucose. The latter two can be used by yeast as extra source of fermentable sugars [43]. This way, endogenous or added amylases can prolong fermentation time and increase the total CO_2 produced. Next to that, endo-acting amylases can also alter the viscosity of the starch, especially once it is gelatinized (*cfr. infra*). The lower bulk viscosity thus obtained allows for a longer oven rise and results in a higher loaf volume of the bread [43]. Moreover, addition of amyloglucosidase generates extra glucose units that can contribute in the Maillard reaction, resulting in a darker crust color after baking [43]. Similar to most baking enzymes, amylases can have a negative effect when overdosed, which would lead to too sticky doughs or gummy breads.

Lipases are enzymes that act on the endogenous lipid fraction in the dough. A typical white wheat flour contains up to 2% lipids, of which the apolar fraction, containing *a.o.* triacylglycerols, diacylglycerols, and free fatty acids, comprises about half. The first generation lipases (E.C.3.1.13) hydrolyzed these neutral or apolar triacylglycerol lipids, resulting in free fatty acids and diacyl- or monoacylglycerols. The improvements that these enzymes brought were mainly related to crumb structure and softening and rather limited with regard to improving dough stability. To address the latter, more efforts were directed toward optimization of more beneficial lipases. This resulted in second generation lipases that were active on a broader range of lipids, including polar lipids such as phospholipids and galactolipids [26, 31]. Addition of these enzymes resulted in doughs with better stability and a more appealing crumb grain. New generations of lipases followed, mainly with improved or diversified specificity. Those acting primarily on polar phospholipids endogenously present in the flour are called *phospholipases*. These can be further subdivided, based on the position they prefer to cleave, into phospholipase A1 (E.C.3.1.1.32), phospholipase A2 (E.C.3.1.1.4), phospholipase B (E.C.3.1.1.5), phospholipase C (E.C.3.1.4.3), and phospholipase D (E.C.3.1.4.4). The last three are of less importance for baking application [26]. Galactolipases (E.C.3.1.1.26), which function similar to phospholipases albeit on galactolipids, exist as well [31]. By enzymatically cleaving off one of the fatty acid groups, an even more polar lysolipid is obtained (shown in Figure 2.1.3) having improved surface-active and emulsifying properties. The concentration and organization of these polar amphiphilic lipids at the air–liquid interface improve dough stability significantly [42], helping the dough to withstand or recover from shocks or disruptions it might encounter, especially in industrial plants. The lipid layer also forms a barrier against coalescence during fermentation and especially during baking. Next to the *in situ* formation of emulsifiers, the reaction products of these lipases give rise to products that can interact with the gluten proteins in the dough, thus contributing to the rheological properties of the dough as well [31]. When too much of this type of enzyme is dosed, however, a detrimental impact on the dough and bread properties is observed. An overdose would break down the (beneficial) lysolipids (see Figure 2.1.3), losing all emulsifying properties and resulting in a weaker dough and lower bread volume [41, 42]. The optimal dosage of this type of enzyme typically of fungal origin (e.g. from *Aspergillus* or *Fusarium*), however, creates emulsifying compounds from endogenous flour components. The first generation of lipases were commercialized in the 1980s, and originated from fungi such as *Rhizopus* and *Thermomyces*. The second generation of lipases only came to the market in the 1990s [31], presumably because – till that point – the baking industry was satisfied with the chemical emulsifiers commonly used in bread, such as monoacylglycerols (MAG), sodium stearoyl lactylate (SSL), and diacetyl tartaric esters of mono- and diglycerides (DATEM). Since the 1990s, however, the call for clean(er) label products and addition of less synthetic chemicals in food is becoming louder and louder. Lipases bring an alternative for these chemical emulsifiers, by converting lipids that are present in the flour or recipe into more surface-active components, and creating similar benefits such as improved dough stability, increased loaf volume, and finer crumb structure [31, 42].

Glucose oxidase has a positive effect on dough rheological properties, as was already indicated in Section 2.1.2.2, which ultimately leads to improved dough stability during fermentation as well. The oxidation of SH groups of the gluten proteins into S—S bonds by hydrogen peroxide strengthens the gluten network in the dough, while the cross-linking of arabinoxylans also leads to an improvement

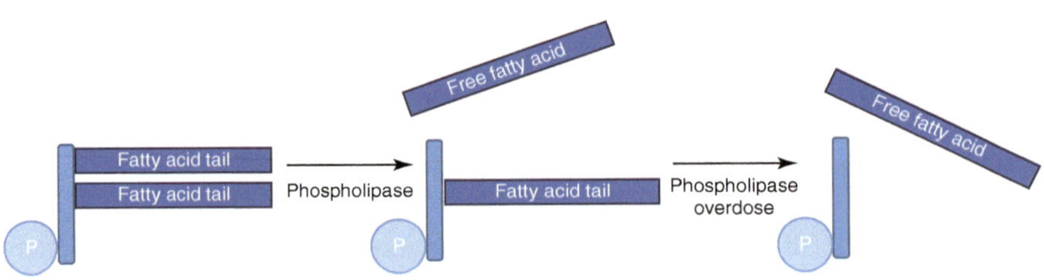

Figure 2.1.3 Schematic presentation of the action of phospholipases in bread making.

in dough stability by increasing the bulk viscosity of the dough [32]. As stated before, it is important to balance the glucose oxidase dosage, as overdosing leads to a stiff dough, low loaf volume, and dense crumb structure.

Endo-acting **xylanases** (E.C. 3.2.1.8) turning WU-AX into WE-AX are beneficial as they contribute to the bulk viscosity, thus increasing the dough's gas cell stability. This enables the dough to better cope with the mechanical stress encountered during expansion, allowing for longer oven rise during and larger loaf volume after baking [22]. The combination of this type of enzymes with glucose oxidase is commonly used as an alternative to chemical oxidizing agents. Potential problems occurring by overdosing xylanases, such as stickiness, can be counterbalanced by the dough drying and strengthening properties of the glucose oxidase enzyme [22].

2.1.2.4 Baking and Oven Spring

Once the proofed or fermented dough is put in the oven, the baking phase starts. During this important step in the bread making process, structural, physical, and chemical changes take place that transform the foam-like dough into sponge-like bread (as schematically depicted in Figure 2.1.1). These changes involve volume expansion, crumb setting, and crust formation and browning.

The higher temperatures in the oven initially boost fermentation in the dough until the yeast is killed at temperatures around 55–60 °C [43]. Next to that, the thermal expansion of the gas inside the air bubbles makes the dough rise in volume. This continues until the gas cell walls rupture under the pressure, resulting in a continuous porous open structure. The gas molecules can now escape to the environment, and further expansion of the dough is halted [44]. Also, crust formation (*cfr. infra*) puts a stop to this volume increase inside the oven, which is known as **oven rise** [44].

Crumb setting results from the combination of starch gelatinization and heat setting of the protein fraction, both predominantly taking place at relatively high moisture content and temperatures around 60–85 °C [37, 44, 45].

The gluten proteins unfold in response to the increasing temperatures in the oven. Initially, mainly the high molecular weight glutenin proteins denature and form a network, but at higher temperatures, the lower molecular weight gliadin proteins are also incorporated in the gluten network. This way, a thermoset network with permanent disulfide bonds as cross-links is formed [45].

Gelatinization is an irreversible process that happens when starch granules are heated above a certain temperature in the presence of water. During heating, the granules swell and take up more water. Once above its gelatinization temperature, starch's semicrystalline amylopectin crystals melt, and its linear amylose chains gain mobility and can leach out of the granule and associate with other amylose chains [45]. This leads to the formation of a transient amylose network connecting disrupted granule remnants rich in amorphous amylopectin. These irreversible changes to the starch granules make both amylose and amylopectin more prone to enzyme attack. As gelatinization occurs in the presence of at least 25–30% water, it might not (or only partly) take place in the outer crust where the dough is driest due to more extensive water evaporation.

The combination of protein denaturation and starch gelatinization ensures that at the end of baking a single protein network with discontinuous starch granules is present in the outer crust, whereas in the center of the bread a double network of both protein and starch molecules is responsible for solidification or setting of the crumb [44, 45]. During cooling of the freshly baked bread, the leached-out amylose chains interact with each other into double helices or twine around lipids to form amylose–lipid complexes. Both associations can potentially function as nuclei for further crystallization and as junction points for a more permanent semicrystalline amylose network. Although this network does not contribute to the setting of the crumb as such, it is important to be mentioned as this network is said to be accountable for the initial firmness and sliceability of freshly baked bread [45].

Browning of the crust is the third phenomenon happening during baking. Browning of the bread crust during baking can be described as a two-step process. In a first lag phase, the outer surface of the dough seems to become lighter in color. This could be due to either drying of the surface and/or smoothening of the dough surface by volume expansion, resulting in more reflection of the light. After that, the crust becomes darker as a result of heat and mass transfer at the dough surface, e.g. due to water evaporation. The very low water activity thus created leads to conditions that are favorable for first Maillard reaction and then caramelization to occur [46]. These two types of non-enzymatic chemical reactions, Maillard reaction and caramelization, not only generate a brown crust but also result in the formation of aroma or flavor compounds creating the well-known smell and taste of freshly baked bread [44]. Maillard reactions occur when reducing sugars and compounds containing a nitrogen group, such as amino acids, are heated to temperatures above 110–120 °C in a low moisture environment (with water activities <0.6) [46]. This type of reaction, however, can also generate toxic or carcinogenic reaction products, such as acrylamide. Caramelization occurs when sucrose or (reducing) sugars are heated to temperatures above 150 °C. Thermal degradation of sugars and amino acids into volatile compounds, such as acetic acid and furfural, occurs around 220 °C [47].

Although many enzymes have an indirect impact on the baking phase in bread making, no enzyme targets an improvement in this phase of the baking process alone. Therefore, the enzymes will not be dealt with elaborately in this section. As mentioned in the previous sections, many types of enzymes can be beneficial for oven rise. Lipases enhance the interfacial properties of the gas cells, enabling them to retain the expanding gas longer without collapsing or coalescing with neighboring cells. Xylanases and glucose oxidase can improve the bulk viscosity properties of the gluten and arabinoxylan network in the dough, which also allows for better gas retention and higher loaf volume after baking. Next to that, glucose oxidase creates a dough that is less sticky and more dry, giving rise to better, crispier crust formation during baking. Amylases cutting large starch chains internally impact the bulk viscosity of the dough, resulting in a prolonged oven rise and larger loaf volume. On the other hand, exo-acting amylases, such as β-amylase and amyloglucosidase, produce a large amount of mono- and disaccharides, which not only boost fermentation and – as such – give rise to a larger loaf volume but also contribute to an enhanced caramelization and Maillard reaction, resulting in a darker crust color.

2.1.3 The Bread Quality – The Consumers' Needs

2.1.3.1 Color and Flavor

The appearance, as exemplified by color, shape, and structure, and the sensory aspects, represented by taste, smell, and mouthfeel, are key attributes for consumers' acceptance of bread. For a long time, white wheat bread, starting from a luxury product for the rich and becoming available for everyone through the milling techniques developed, was the main bread product consumed [5]. Whiteness of the inner crumb and a golden-brown crust were major quality aspects. Shape, structure, and mouthfeel of the bread are primarily determined by the ingredients, processing aids, and type of bread making process, as discussed in the previous sections, while shelf life will be discussed in the next section.

Crust color is primarily determined by the baking process, as described in Section 2.1.2.4. Since monosugars and amino acids play a major role in the browning reactions, an enhanced crust color is obtained by addition of enzymes releasing such compounds.

Amyloglucosidase is particularly interesting in this respect, as the enzyme releases glucose at relatively high rate from damaged starch during mixing and fermentation [43] and at the initial stages of starch gelatinization, until its inactivation temperature is reached. Amyloglucosidase inactivates at slightly higher temperatures than yeast, resulting in remaining non-fermented glucose, which enhances crust browning. Although fungal amylase inactivates prior to starch gelatinization, addition of this enzyme is also reported to result in a browner crust color [22, 43]. Additionally, proteases, particularly exo-acting ones, can be applied for color and flavor enhancement [14, 22].

Crumb color is optically lightened by a very fine crumb structure, which can be achieved by the mixing method or by addition of some enzymes, as already indicated in previous sections. The most effective crumb whitening is achieved by bleaching oxidants such as chlorine and benzoyl peroxide, although not ubiquitously permitted anymore, and by enzyme-active soy flour. Both methods degrade the carotenoids present in the wheat flour, which causes the slight yellowish crumb [21, 48].

Lipoxygenases (E.C.1.13.11.12) are the active enzymes in the soy flour, catalyzing the oxidation of polyunsaturated fatty acids (particularly linoleic and linolenic acid) into hydroxyperoxides. They require oxygen for the formation of intermediate (free) radicals, which oxidize SH-containing proteins and pigments, such as carotenoids. Since oxidation of SH-groups in the gluten network is also important for the dough's rheological properties, applying enzyme-active soy flour in bread making logically also improves dough handling and increases loaf volume, next to the bleaching effect [14, 22, 48]. With all the biotechnological techniques available nowadays, it is somehow surprising that lipoxygenase is not commercially available as a microbially produced baking enzyme. Plant- and cyanobacterium-derived lipoxygenases have been heterologously expressed, but without commercial success in baking. Presumably the synergistic action of the various isozymes present in soy flour could not be reproduced in a similar way by a single lipoxygenase, despite the cyanobacterium-derived lipoxygenase having comparable effects on bleaching and rheology as benzoyl peroxide [48, 49].

Flavor is the combination of taste and smell. The flavor of wheat bread is determined by a long list of volatile compounds, including alcohols, aldehydes, esters, ethers, ketones, acids, hydrocarbons, pyrazines, pyrrolines, furans, lactones and/or sulfur compounds, and non-volatile compounds such as phenolics, amino acids, peptides, sugars, and fatty acids, present in the crumb and/or the crust [5, 47]. Besides the ingredients and applied process conditions, enzymatic and fermentative reactions play a key role in determining the flavor of the crumb, as well as lipid oxidation. In the crust, mainly the non-enzymatic thermal reactions described above govern flavor generation. The origin of the most prominent compounds is ascribed to fermentation (40%), Maillard reaction (33%), and lipid oxidation (27%). In the crumb, the most pleasant bread aromas are acetoin and 3-methyl-1-butanol, derived from fermentation of leucine by yeast. Hexanal, resulting from lipid oxidation by lipoxygenase action, and butyric acid are the most detested aromas in the bread crumb. Next to leucine, valine, isoleucine, phenyl-alanine, and methionine are important precursors for flavor generation by yeast. Moreover, sweet and fruity odors such as ethyl acetate and ethyl hexanoate are produced by acyltransferases in yeast. Lactic acid bacteria, particularly in sourdough, also influence the flavor profile in bread crumb. In the crust, 2,3-butanedione and 2-acetyl-1-pyrroline, with the latter having a roasty odor, are appreciated. During the Maillard reaction, lysine, leucine, and isoleucine are known to give pleasant odors, whereas methionine's products are unpleasant. Apart from the various amino acids, the reactivity of the sugars, with xylose being more reactive than glucose, and the conditions of baking influence the final odor production [47]. With the decreasing fermentation times employed by the industrialization of bread making, a reduction in the generation of flavor compounds is inevitable. However, this is compensated to some extent by the use of baking enzymes able to provide the desired flavor precursors, such as sugars and amino acids. Enzymes listed for their effect on flavor development are amylases, amyloglucosidase, proteases, oxidases, lipases, and xylanases [47, 50]. Yet, research into bread aroma profiles generated by different commercial baker's yeasts or nonconventional yeast strains also opens ways to influence the flavor of bread [51, 52].

2.1.3.2 Shelf Life

Baked bread is not a stable system. It changes continuously during storage, resulting in a lower quality product with decreasing consumer acceptance. Not only texture and flavor deterioration but also microbial spoilage is a limiting factor for the bread's shelf life.

Staling is the term used to describe the undesired physicochemical changes taking place during storage of the bread, leading to a product with decreasing freshness perception. Besides losing the typical "freshly baked bread" aroma, textural attributes also change: the bread crumb becomes firmer and its resilience decreases, while the crust loses its crispiness and becomes leathery [45, 53].

Retrogradation (or recrystallization) of amylopectin is one of the main contributors to bread staling. This slow crystallization process takes place during the first week or two after the bread is baked [54]. As many amylose chains leach out of the granules during baking, amylopectin side chains are more concentrated in the granule remnants of the baked bread, enabling easier interaction and (re)association between adjacent side chains into double helices. These double helices can then get ordered further into (the so-called B-type) crystals. This interaction can take place inside and in between granule remnants, creating a fixed semicrystalline network. During storage, the amylopectin crystals in the bread crumb continue to grow in size and number [45]. This process takes place more extensively at cold temperatures. Amylopectin nucleation is the limiting step of this rather slow crystallization process. Between 4 and 14 °C, however, nucleation happens more readily [54]. This means that bread crumb firms faster at these low temperatures, despite a slower water diffusion, decreasing the rate of moisture migration or the rate of evaporation. The B-type amylopectin crystals that are formed during retrogradation hold a lot of water. This water is taken from the gluten (or from other amorphous regions) in the bread crumb. As water is a plasticizer, its removal increases the glass transition temperature of the gluten network. This makes the amorphous areas in the crumb increasingly more rigid and brittle, which is noticed by the decreasing resilience of the bread crumb with storage time [45, 53].

Next to water migration from the amorphous gluten to the crystallizing amylopectin, water also transfers from crumb to crust during storage [54, 55]. Freshly baked bread has a dry and crispy crust and a moist crumb. With storage time, this inequilibrium is restored to some extent, as water evaporates in the crumb and condenses in the crust [56]. This way, the crumb loses up to 25% of its moisture while the crust gains water. This is perceived during consumption of the bread as an increasingly dry crumb and tougher crust with storage time. Also on an even larger time scale, moisture migration takes place with water evaporating from the bread to the environment. However, avoiding this moisture loss, e.g. by special packaging, still does not prevent bread staling. As such, this factor is not the most important contributor to firming of the bread.

Many efforts to prevent or retard bread staling have been described in the literature. Some focus on adjusting the process by baking for shorter times or at lower temperatures [54, 57], whereas others try to adjust the recipe by the addition of extra gluten or water [55]. The most effective measure to decrease the extent of staling, however, is the use of **anti-staling enzymes**. Any amylase able to act on amylopectin's side branches can be beneficial to prevent bread staling to some extent. Microbial amylases targeted for this functionality have to be more heat stable than the fungal amylases used for flour correction, as the latter cannot act on crystalline starch and are denatured by the time starch gelatinizes during baking. They are not able to cut the amylopectin side branches once they are molten. A more heat stable amylase that can act on gelatinized starch is thus required to interfere with amylopectin retrogradation. Already since the 1970s and 1980s bacterial α-amylase was marketed and used in this respect. However, this enzyme appeared to be too active and heat stable, resulting in a sticky crumb during storage [31, 45].

The best known and most effective type of anti-staling enzyme is **maltogenic amylase** (E.C.3.2.1.133). This enzyme from microbial origin (e.g. from *Bacillus stearothermophilus*) belongs to GH family 13 [26]. It was first commercialized in the 1990s. Maltogenic amylase has an intermediate thermostability and typically works on gelatinized starch during the baking phase. This is possible as starch gelatinizes from 60 °C onwards during baking and the enzyme is still active up to temperatures of 80–90 °C. Its action pattern is not fully understood, but it is known that – although it has a predominant exo-activity, cutting off maltose units – this enzyme is also able to cleave the α-1,4 bonds within the starch chains internally, especially at higher temperatures [22, 45, 58]. This way, amylopectin branches can be significantly reduced in chain length, making them too short to form double helices and crystallize. This reduces the level of retrogradation during storage as schematically depicted in Figure 2.1.1, and as such limits the amount of water being withdrawn from the gluten, resulting in softer bread crumb with better resilience during storage [45]. Maltogenic amylases do not work on amylopectin alone but also cut amylose chains. The shorter linear amylose chains thus formed have higher mobility, can more readily leach out of the granules during baking, and contribute in network formation. This results in the formation of a more extensive amylose network during cooling and a slightly higher initial firmness of the bread crumb [45]. Another anti-staling enzyme that seems to work somewhat similarly to maltogenic amylase is the **maltotetraose-producing amylase** (E.C. 3.2.1.60) from *Pseudomonas* sp. [58]. It differs from maltogenic amylase in that it primarily cleaves maltotetraose off from the main chain instead of maltose units. It also has a more pronounced and temperature-independent endo-activity than maltogenic amylase [58].

The combination of maltogenic amylases with lipases is often suggested to improve and prolong freshness of the bread during storage. The amylase can hamper starch retrogradation, while the free fatty acids or monoacyl lipids produced by the lipases can interact with amylose (or even the remaining amylopectin side branches) to form lipid complexes. This prevents the chains involved in complex formation from participating in starch (re)crystallization and contributing to starch retrogradation.

Bread's shelf life is not only limited by the textural attributes as a result of staling but also by **microbial spoilage**. The high temperatures reached in the oven during baking kill the fungal spores that might have been present in the flour or dough [59]. Mold growth occurring in the bread is thus mainly caused by contamination after baking. The lower the water activity in the baked product, the less prone the product is to microbial infestation. That explains the positive contribution of, for example, salt to the extension of the shelf life of bread. Also, some bacterial cultures used in sourdough recipes can prolong the mold-free shelf life of bread by lowering the pH in the bread and by producing antifungal molecules such as organic acids [59]. Calcium propionate, a chemical interfering with the metabolic pathway of fungi, thus inhibiting their growth, is commonly used as additive to retard mold growth in baked products, extending the bread's mold-free shelf life by a significant one to three days.

Figure 2.1.4 depicts a conceptual overview of the action of the main baking enzymes during the bread making process, as discussed in the previous sections. To emphasize the impact of the respective enzymes, the changes to its substrate(s) are indicated in color, whereas the other components are represented in greyscale. The figure clarifies that the structural changes the enzymes bring

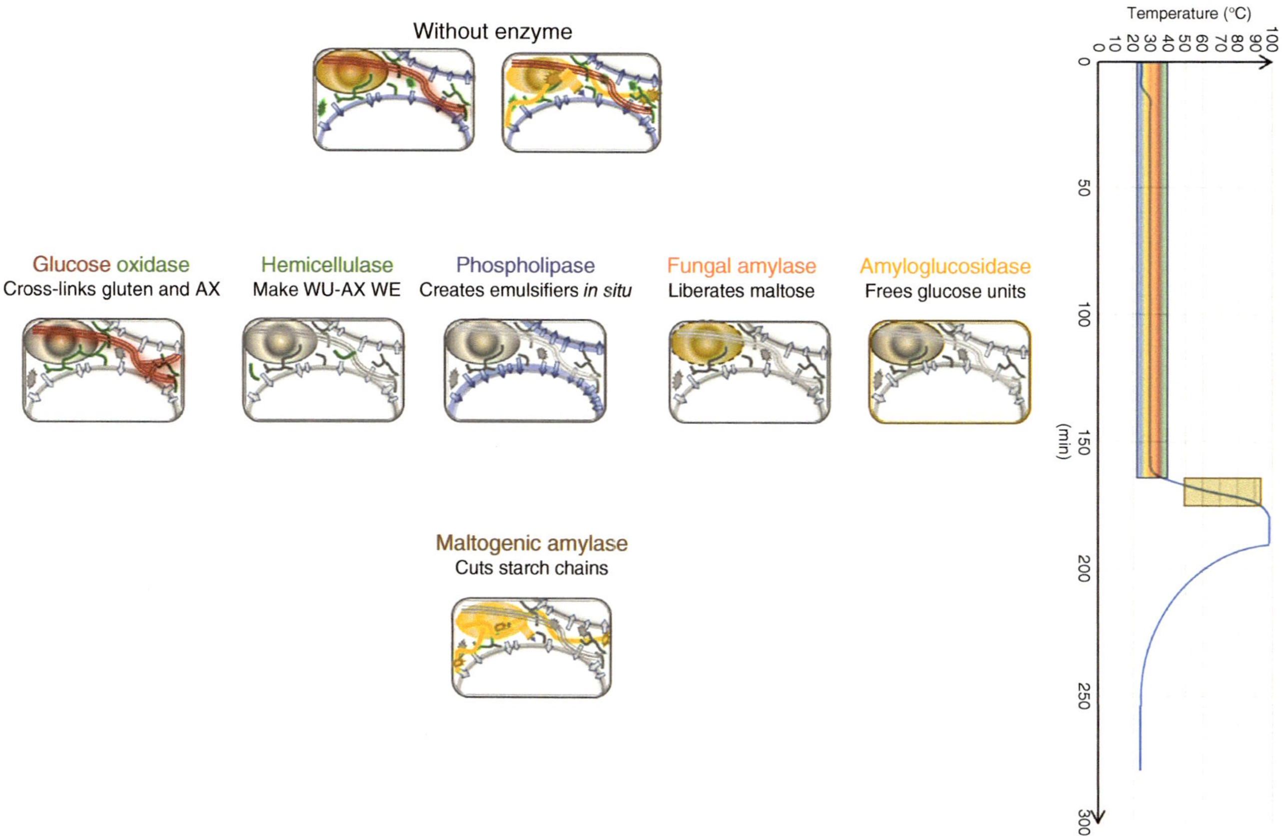

Figure 2.1.4 Conceptual visualization of the structural consequences of the activity of the most important baking enzymes in the bread making process with the microscale changes in comparison to the enzyme-free situation highlighted in color: glucose oxidase cross-links gluten (red) and arabinoxylans (green); hemicellulase converts the water-unextractable arabinoxylans (WU-AX) to water-extractable (WE) (green); phospholipase hydrolyzes an acyl chain from the phospholipids (blue), creating a more stable liquid film surrounding the gas cells; and the starch degrading enzymes modify starch and/or release glucose and maltose (orange-yellow). On the right-hand side of the figure, the temperature profile during the bread making process is represented, with the colored bars indicating when each of the represented enzymes is active. Overall, this figure clarifies that in the bread making process, subtle changes on molecular/microscale have a large impact on meso- and macroscale. The same legend as in Figure 2.1.1 applies.

about seem minor on microscale. However, these small adaptations have a strong impact on meso- and/or macroscale in each step of the bread making process (as described throughout the chapter). To bring the desired changes, each of the enzymes needs to be added in the correct dose, as overdosing is possible and leads to undesirable effects. The figure also indicates the timeframe of an average bread making process and the stage at which the selected enzymes are active.

2.1.4 Trends and Opportunities for Baking Enzymes

2.1.4.1 Fine Baking and Confectionary

Fine baking products include a large and increasingly growing range of products, such as muffins, cupcakes, pound cake, sponge cake, cookies, biscuits, wafers, and so on. These baked products are typically not consumed as staple food, but rather for indulgence. Next to wheat flour, also sugar, fat, and egg can be important in these bakery products, not only to give flavor and mouthfeel to the products but also from a structural point of view. For each of these fine baking products, the list of ingredients can vary tremendously and, consequently, the overall baking process as well. It would lead us too far to discuss every process in detail, but some more general remarks will be made.

In **cake**, egg and sugar are abundantly present besides wheat flour. Depending on the type of cake, fat can be a main ingredient as well. The cake baking process typically uses chemical leavening for CO_2 production. The high egg or fat content makes the viscosity of the cake batter too low for an extensive gluten network to develop. However, in cake as well, air is whipped in during mixing, and the gas cell stability of the batter is of importance to obtain a light and aerated cake after baking. During baking, the cake batter is predominantly set by the complex combination of starch swelling and gelatinization on the one hand, and egg denaturation on the other [60]. Enzymes can also be used to help in this baking process and improve the cake quality. The main enzymes are phospholipases, improving gas cell stability in the batter and volume in the cake, and anti-staling enzymes such as maltogenic amylase, to keep the cake soft and moist over a longer storage period.

In **cookies**, wheat flour, sugar, and fat are the main ingredients. Their ratio may vary depending on the type of cookie or biscuit, and thus on the baking process. Some of the main quality attributes for cookies are their flow or spread in the oven, and – related to that – cookie diameter, stack height, and color [61]. Cookies generally have very low water content after baking. During storage, they become softer or soggy by taking up water from the environment. The low water activity and high sugar levels make cookie dough prone to excessive Maillard reaction during baking. This can lead to too high levels of acrylamide in some type of cookies. To overcome this, the recipe or processes can be adjusted [62], or asparaginase can be used. This enzyme converts asparagine, the main precursor for acrylamide, into aspartic acid, thus reducing the level of acrylamide in the final baked cookie.

Wafers are sweet, thin, and dry biscuits, often used with ice cream or in chocolate bars. Lightness and crispiness are essential quality parameters in a wafer. Besides flour, salt, sugar, and the chemical leavening agent, water is also added to make the wafer batter more liquid [63]. To reduce the level of water in the recipe, protease can be added as a processing aid. By cleaving the gluten proteins in the flour, the chances of gluten proteins forming an elastic network are reduced, and a lower viscosity is obtained. This subsequently requires the addition of a lower amount of water to reach the desired low viscous wafer batter and also reduces the energy required to bake the wafers.

2.1.4.2 Consumer Preference: Health, Individual Values, and Convenience

White wheat bread has been losing consumers' interest for some time already, as its relative low level of vitamins, minerals, and fiber affect its nutritional value, its digestibility, rate of absorption, and many other health-related factors, which may be linked to the development of chronic diseases associated with obesity. Improved understanding of nutrition, increased income, and enhanced expectations on the role of the diet in the lifestyle of consumers create opportunities for bakery products to fit specific needs. Such **dietetic specialty products** include food adapted for intolerances (gluten-free), for healthier lifestyle (high-fiber), for specialized dietary requirements (obesity), for religious diet requirements (Kosher, Halal), for lifestyle choices (sports nutrition, vegetarian, vegan), and for various demographics (children, women, seniors) [5]. Two examples will be discussed here in more detail.

Wholemeal bread, or bread with added fibers, is known to be beneficial for intestinal microbiota and is more slowly digested. The production of these types of bread requires some adaptations in the bread making procedure, as wholemeal flour absorbs more water, the gluten is diluted, and the bran particles affect the final crumb structure and loaf volume by disrupting the gluten network and preventing regular expansion of the gas cells [11]. Furthermore, the presence of higher levels of endogenous enzymes, such as amylases and proteases, may directly influence the dough handling. Often, these adverse effects are corrected by the addition of strong white wheat flour or some vital wheat gluten [7]. More recent studies on increasing and improving the use of bran in bread have shown positive effects of particle size reduction, presoaking, thermal or enzymatic treatment (amylase, phytase, xylanase, glucose oxidase and cellulase), and fermentation of the bran [16].

For gluten-free bread a major challenge is to prepare bread without wheat, as the gluten proteins of wheat have an essential function in the quality characteristics of the final product. Breads are often made with plant proteins and cross-linking enzymes, such as oxidases, in addition to starch and hydrocolloids, in order to create a cohesive network in the dough or batter in the complete absence of gluten [14]. Alternative options for gluten-sensitive individuals are digestive aids with a prolyl endoprotease able to degrade the immunogenic peptides in the stomach after digestion [64], or potential consumption of wheat bread prepared with predigested gluten with the same specific prolyl endoprotease [65].

In **organic** breads, all ingredients should be of natural origin, including the enzymes that might be used as processing aids. This means that the flour comes from organically grown wheat crops, i.e. without the use of chemical pesticides on the field or during storage of the kernels. Moreover, no chemical additives can be used during baking. Common emulsifiers such as DATEM are not allowed in the production of organic bread. Although the quality attributes in these breads are different, as well as the customer's expectation thereof (e.g. an extended shelf life is neither desired nor necessary), enzymes offer attractive solutions to improve some attributes in the organic breads. The enzymes that are widely accepted, however, are the so-called "classical" or non-protein engineered enzymes. This restricts the number and type of enzymes that can be used.

Parbaked/frozen storage is becoming more common for supermarkets (and even some bakeries) to reduce food waste and increase convenience, by doing only the last part of the baking process in house. They bake off the bread when needed, delivering a freshly baked bread to the customers. When this is the case, the baking process as was described above is interrupted. The bread doughs or partially baked, also referred to as par(t)baked, breads are delivered to the stores where only the final bake-off step takes place. The dough or parbaked bread can be stored and transported frozen, at cold or even room temperatures. Care has to be taken that the dough is frozen at the right stage of fermentation; otherwise, the yeast can get damaged due to the low storage temperatures and lose functionality, resulting in lower quality bread regarding volume and taste. When bread is parbaked, the baking phase is split up into two parts, with the interruption occurring after the crumb is set, but before crust formation takes place. The latter is the aim of the second, final baking step: creating a brown and crispy crust via Maillard reaction [66]. This also results in the formation of aroma compounds, giving rise to the smell of freshly baked bread. The temperature at which the parbaked breads are stored intermittently between the two baking steps is an important factor determining the quality of the fully baked bread. When stored frozen, on the one hand, amylopectin retrogradation is halted, but, on the other hand, the bread can suffer from mechanical damage caused by the ice crystals [67, 68]. Although the second heating step, if sufficiently long to reach high enough temperatures in the center of the crumb, is able to melt the amylopectin crystals that might have formed in the parbaked bread, other bread characteristics, such as moistness, are not reversible. So, even though it has its economic benefits, adaptation of the baking process in this way has consequences, not only for the recipe and the process steps but potentially also for the quality parameters of the baked product.

2.1.5 Conclusion

In summary, many different types of enzymes are being used in baking, and each of them has a very specific molecular action. Yet, such action that only affects the microscale can contribute to macroscopic effects recognizable in processing properties, such as dough handling and dough stability, and in product quality characteristics, such as bread volume, crumb structure, crust color, and flavor. An overview of the role of each of the baking enzymes in all these aspects is given in Table 2.1.1.

Table 2.1.1 Overview of the role of the enzymes in all processing and product quality aspects of baking (black box indicating the key role of the enzyme in this baking aspect; grey box indicating the supporting role of the enzyme in this baking aspect; grey font indicates the unavailability of the enzyme as relatively pure commercial product).

	Fungal amylase	Amyloglucosidase	Bacterial amylase	Maltogenic amylase	Hemicellulase	Cellulase	(Glucose)oxidase	Protease	(Phospho)lipase	Lipoxygenase	Asparaginase
Process parameters											
Flour standardization	black	grey			grey						
Dough fermentability	grey	black	grey								
Dough handling	grey				black	grey	black	grey	grey		
Dough stability	grey	grey	grey		grey	grey	grey	grey	black		
Product quality parameters											
Taste and Flavor	grey	grey						grey			
Crust color	grey	black	grey					grey			
Crust crispiness							grey	grey			
Bread volume	grey	grey	grey		grey	grey	grey	grey	grey		
Crumb structure	grey	grey	grey	grey	grey	grey	grey		grey		
Crumb color					grey				grey	black	
Freshness/shelflife	grey		grey	black	grey				grey		
Health						grey	grey				black

References

1 Nadel, D., Piperno, D.R., Holst, I. et al. (2012). New evidence for the processing of wild cereal grains at Ohalo II, a 23 000-year-old campsite on the shore of the Sea of Galilee, Israel. *Antiquity* 86: 990–1003.

2 Piperno, D.R., Weiss, E., Holst, I., and Nadel, D. (2004). Processing of wild cereal grains in the Upper Palaeolithic revealed by starch grain analysis. *Nature* 430: 670–673.

3 Toussaint-Samat, M. (2008). The history of bread and cakes. In: *A History of Food*, 201–222. Oxford, UK: Blackwell Publishing Ltd.

4 Sievert, D., Hoseney, R.C., and Delcour, J.A. (2007). Bread and other baked products. In: *Ullmann's Encyclopedia of Industrial Chemistry*, 259–306. Wiley Online Library.

5 Zhou, W., Therdthai, N., and Hui, Y.H. (2014). Introduction to baking and bakery products. In: *Bakery Products: Science and Technology*, 2e (ed. W. Zhou, Y.H. Hui, I. De Leyn, et al.), 3–16. Chichester, West-Sussex, UK: Wiley.

6 Amos, A.J. (1955). The use of enzymes in the baking industry. *J. Sci. Food Agric.* 6: 489–495.

7 Hoseney, R.C. (1986). Yeast-leavened products. In: *Principles of Cereal Science and Technology. A General Reference on Cereal Foods*, 203–244. St Paul, Minnesota, USA: American Association of Cereal Chemists Inc.

8 Lai, H.M. and Lin, T.C. (2006). Bakery products: science and technology. In: *Bakery Products: Science and Technology* (ed. Y.H. Hui), 3–65. Ames, Iowa, USA: Blackwell Publishing.

9 Doblado-Maldonado, A.F., Pike, O.A., Sweley, J.C., and Rose, D.J. (2012). Key issues and challenges in whole wheat flour milling and storage. *J. Cereal Sci.* 56: 119–126.
10 Hoseney, R.C. (1986). Dry milling of cereals. In: *Principles of Cereal Science and Technology. A General Reference on Cereal Foods*, 133–152. St Paul, Minnesota, USA: American Association of Cereal Chemists Inc.
11 Pagani, M.A., Marti, A., and Bottega, G. (2014). Wheat milling and flour quality evaluation. In: *Bakery Products: Science and Technology*, 2e (ed. W. Zhou, Y.H. Hui, I. De Leyn, et al.), 19–53. Chichester, West-Sussex, UK: Wiley.
12 Struyf, N., Van der Maelen, E., Hemdane, S. et al. (2017). Bread dough and baker's yeast: an uplifting synergy. *Compr. Rev. Food Sci. Food Saf.* 16: 850–867.
13 Poutanen, K. (1997). Enzymes: an important tool in the improvement of the quality of cereal foods. *Trends Food Sci. Technol.* 8: 300–306.
14 Prasada Rao, U.J.S. and Hemalatha, M.S. (2014). Enzymes. In: *Bakery Products: Science and Technology*, 2e (ed. W. Zhou, Y.H. Hui, I. De Leyn, et al.), 276–294. Chichester, West-Sussex, UK: Wiley.
15 Fierens, E., Helsmoortel, L., Joye, I.J. et al. (2015). Changes in wheat (*Triticum aestivum* L.) flour pasting characteristics as a result of storage and their underlying mechanisms. *J. Cereal Sci.* 65: 81–87.
16 Hemdane, S., Jacobs, P.J., Dornez, E. et al. (2016). Wheat (*Triticum aestivum* L.) bran in bread making: a critical review. *Compr. Rev. Food Sci. Food Saf.* 15: 28–42.
17 Olaerts, H., Roye, C., Derde, L.J. et al. (2016). Impact of preharvest sprouting of wheat (*Triticum aestivum*) in the field on starch, protein and arabinoxylan properties. *J. Agric. Food Chem.* 64: 8324–8332.
18 Olaerts, H., Roye, C., Derde, L.J. et al. (2016). Evolution and distribution of hydrolytic enzyme activities during preharvest sprouting of wheat (*Triticum aestivum*) in the field. *J. Agric. Food Chem.* 64: 5644–5662.
19 Dobraszczyk, B.J. and Morgenstern, M.P. (2003). Rheology and the breadmaking process. *J. Cereal Sci.* 38: 229–245.
20 MacRitchie, F. (2016). Seventy years of research into breadmaking quality. *J. Cereal Sci.* 70: 123–131.
21 Joye, I.J., Lagrain, B., and Delcour, J.A. (2009). Use of chemical redox agents and exogenous enzymes to modify the protein network during breadmaking – a review. *J. Cereal Sci.* 50: 11–21.
22 Goesaert, H., Gebruers, K., Courtin, C.M. et al. (2006). Enzymes in breadmaking. In: *Bakery Products: Science and Technology* (ed. Y.H. Hui), 337–364. Ames, Iowa, USA: Blackwell Publishing.
23 Haros, M., Rosell, C.M., and Benedito, C. (2002). Improvement of flour quality through carbohydrases treatment during wheat tempering. *J. Agric. Food Chem.* 50: 4126–4130.
24 Courtin, C.M. and Delcour, J.A. (2002). Arabinoxylans and endoxylanases in wheat flour bread-making. *J. Cereal Sci.* 35: 225–243.

25 Gruppen, H., Hamer, R.J., and Voragen, A.G.J. (1992). Water-unextractable cell wall material from wheat flour. II. Fractionation of alkali-extractable polymers and comparison with water extractable arabinoxylans. *J. Cereal Sci.* 16: 53–67.
26 Miguel, A.S.M., Martins-Meyer, T.S., Verissimo de Costa Figueirido, E. et al. (2013). Enzymes in bakery: current and future trends. In: *Future Trends, Food Industry* (ed. I. Muzzalupo), 287–321. Intech Open Access Publisher.
27 Leys, S., Pauly, A., Delcour, J.A., and Courtin, C.M. (2016). Modification of the secondary binding site of xylanases illustrates the impact of substrate selectivity on bread making. *J. Agric. Food Chem.* 64: 5400–5409.
28 Hille, J.D.R. and Schooneveld-Bergmans, M.E.F. (2004). Hemicellulases and their synergism in breadmaking. *Cereal Foods World* 49: 283–286.
29 Polizeli, M.L.T.M., Rizzatti, A.C.S., Monti, R. et al. (2005). Xylanases from fungi: properties and industrial applications. *Appl. Microbiol. Biotechnol.* 67: 577–591.
30 Dornez, E., Verjans, P., Arnaut, F. et al. (2011). Use of psychrophilic xylanases provides insight into the xylanase functionality in bread making. *J. Agric. Food Chem.* 59: 9553–9562.
31 Bock, J.E. (2015). Enzymes in breadmaking. In: *Improving and Tailoring Enzymes for Food Quality and Functionality* (ed. R.J. Yada), 181–197. Cambridge, UK: Elsevier and Woodhead Publishing.
32 Decamps, K., Joye, I.J., Courtin, C.M., and Delcour, J.A. (2012). Glucose and pyranose oxidase improve bread dough stability. *J. Cereal Sci.* 55: 380–384.
33 Decamps, K., Joye, I.J., Rakotozafy, L. et al. (2013). The bread dough stability improving effect of pyranose oxidase from *Trametes multicolor* and glucose oxidase from *Aspergillus niger*: unraveling the molecular mechanism. *J. Agric. Food Chem.* 61: 7848–7854.
34 Rao, M.B., Tanksale, A.M., Ghatge, M.S., and Deshpande, V.V. (1998). Molecular and biotechnological aspects of microbial proteases. *Microbiol. Mol. Biol. Rev.* 62: 597–635.
35 Rawlings, N.D., Waller, M., Barrett, A.J., and Bateman, A. (2014). MEROPS: the database of proteolytic enzymes, their substrates and inhibitors. *Nucleic Acids Res.* 42: 503–509.
36 Takano, H., Ishida, N., Koizumi, M., and Kano, H. (2002). Imaging of the fermentation process of bread dough and the grain structure of baked breads by magnetic resonance imaging. *J. Food Sci.* 67: 244–250.
37 Della Valle, G., Chiron, H., Cicerelli, L. et al. (2014). Basic knowledge models for the design of bread texture. *Trends Food Sci. Technol.* 36: 5–14.
38 Kokelaar, J.J., van Vliet, T., and Prins, A. (1995). Surface and bulk properties in relation to bubble stability in bread dough. In: *Food Macromolecules and Colloids* (ed. E. Dickinson and D. Lorient), 277–284. Cambridge, UK: Royal Society of Chemistry.
39 Kokelaar, J.J. and Prins, A. (1995). Surface rheological properties of bread dough components in relation to gas bubble stability. *J. Cereal Sci.* 22: 53–61.
40 Turbin-Orger, A., Della Valle, G., Doublier, J.L. et al. (2015). Foaming and rheological properties of the liquid phase extracted from wheat flour dough. *Food Hydrocoll.* 43: 114–124.

41 Primo-Martin, C., Hamer, R.J., and de Jongh, H.H.J. (2006). Surface layer properties of dough liquor components: are they key parameters in gas retention in bread dough? *Food Biophys.* 1: 83–93.
42 Gerits, L., Pareyt, B., Masure, H.G., and Delcour, J.A. (2015). Native and enzymatically modified wheat (*Triticum aestivum* L.) endogenous lipids in bread making: a focus on gas cell stabilization mechanisms. *Food Chem.* 172: 613–621.
43 Struyf, N., Verspreet, J., Verstrepen, K.J., and Courtin, C.M. (2017). Investigating the impact of α-amylase, α-glucosidase and glucoamylase action on yeast-mediated bread dough fermentation and bread sugar levels. *J. Cereal Sci.* 75: 35–44.
44 Mondal, A. and Datta, A.K. (2008). Break baking – a review. *J. Food Eng.* 86: 465–474.
45 Goesaert, H., Slade, L., Levine, H., and Delcour, J.A. (2009). Amylases and bread firming – an integrated view. *J. Cereal Sci.* 50: 345–352.
46 Purlis, E. and Salvadori, V.O. (2009). Modelling the browning of bread during baking. *Food Res. Int.* 42: 865–870.
47 Pico, J., Bernal, J., and Gomez, M. (2015). Wheat bread aroma compounds in crumb and crust: a review. *Food Res. Int.* 75: 200–215.
48 Hayward, S., Cilliers, T., and Swart, P. (2017). Lipoxygenases: from isolation to application. *Compr. Rev. Food Sci. Food Saf.* 16: 199–211.
49 Zhang, C., Zhang, S., Lu, Z. et al. (2013). Effects of recombinant lipoxygenase on wheat flour, dough and bread properties. *Food Res. Int.* 54: 26–32.
50 Martinez-Anaya, M.A. (1996). Enzymes and bread flavor. *J. Agric. Food Chem.* 44: 2469–2480.
51 Aslankoohi, E., Herrera-Malaver, B., Rezaei, M.N. et al. (2016). Non-conventional yeast strains increase the aroma complexity of bread. *PLoS ONE* 11: e0165126. https://doi.org/10.1371/journal.pone.0165126.
52 Birch, A.N., Petersen, M.A., Arneborg, N., and Hansen, A.S. (2013). Influence of commercial baker's yeasts on bread aroma profiles. *Food Res. Int.* 52: 160–166.
53 Bollain, C., Angioloni, A., and Collar, C. (2005). Bread staling assessment of enzyme-supplemented pan breads by dynamic and static deformation measurements. *Eur. Food Res. Technol.* 220: 83–89.
54 Bosmans, G.M., Lagrain, B., Fierens, E., and Delcour, J.A. (2013). The impact of baking time and bread storage temperature on bread crumb properties. *Food Chem.* 141: 3301–3308.
55 Curti, E., Carini, E., Tribuzio, G., and Vittadini, E. (2014). Bread staling: effect of gluten on physico-chemical properties and molecular mobility. *LWT Food Sci. Technol.* 59: 418–425.
56 Monteau, J.Y., Purlis, E., Besbes, E. et al. (2017). Water transfer in bread during staling: physical phenomena and modelling. *J. Food Eng.* 211: 95–103.
57 Besbes, E., Jury, V., Monteau, J.Y., and Le Bail, A. (2014). Effect of baking conditions and storage with crust on the moisture profile, local textural properties and staling kinetics of pan bread. *LWT Food Sci. Technol.* 58: 658–666.

58 Derde, L.J., Gomand, S.V., Courtin, C.M., and Delcour, J.A. (2012). Characterisation of three starch degrading enzymes: thermostable β-amylase, maltotetraogenic and maltogenic α-amylases. *Food Chem.* 135: 713–721.

59 Belz, M.C.E., Mairinger, R., Zannini, E. et al. (2012). The effect of sourdough and calcium propionate on the microbial shelf-life of salt reduced bread. *Appl. Microbiol. Biotechnol.* 96: 493–501.

60 Wilderjans, E., Luyts, A., Brijs, K., and Delcour, J.A. (2013). Ingredient functionality in batter type cake making. *Trends Food Sci. Technol.* 30: 6–15.

61 Pareyt, B., Wilderjans, E., Goesaert, H. et al. (2008). The role of gluten in sugar-snap cookie system: a model approach based on gluten-starch blends. *J. Cereal Sci.* 48: 863–869.

62 Gökmen, V., Acar, Ö.C., Köksel, H., and Acar, J. (2007). Effects of dough formula and baking conditions on acrylamide and hydroxymethylfurfural formation in cookies. *Food Chem.* 104: 1136–1142.

63 Dogan, I.S. (2006). Factors affecting wafer sheet quality. *Int. J. Food Sci. Technol.* 41: 569–576.

64 Osorio, C., Wen, N., Gemini, R. et al. (2012). Targeted modification of wheat grain protein to reduce the content of celiac causing epitopes. *Funct. Integr. Genomics* 12: 417–438.

65 Rees, D., Holtrop, G., Chope, G. et al. (2018). A randomised, double-blind, cross-over trial to evaluate bread, in which gluten has been pre-digested by prolyl endoprotease treatment, in subjects self-reporting benefits of adopting a gluten-free or low-gluten diet. *Br. J. Nutr.* 119: 496–506.

66 Ronda, F., Caballero, P.A., Quilez, J., and Roos, Y.H. (2011). Staling of frozen partly and fully baked breads. Study of the combined effect of amylopectin recrystallization and water content on bread firmness. *J. Cereal Sci.* 53: 97–103.

67 Barcenas, M.E. and Rosell, C.M. (2006). Effect of frozen storage time on the bread crumb and aging of par-baked bread. *Food Chem.* 95: 438–445.

68 Bosmans, G.M., Lagrain, B., Ooms, N. et al. (2014). Storage of parbaked bread affects shelf life of fully baked end product: a ^{1}H NMR study. *Food Chem.* 165: 149–156.

2.2

Protein Modification to Meet the Demands of the Food Industry

Andrew Ellis

Biocatalysts Ltd, 1 Cefn Coed, Parc Nantgarw, Cardiff CF15 7QQ, Wales, UK

2.2.1 Food Proteins

Proteins were discovered by Jakob Berzelius in 1838 and are among the most actively studied molecules in food research and development [1]. The word "protein" is derived from the Greek word "protas" meaning "of primary importance" because of the fundamental role of proteins in sustaining life. Protein is essential for growth of the human body and maintenance of good health. Different foods vary in their protein content and in the properties of those proteins; the protein content in various food products is listed in Table 2.2.1. In addition to contributing nutritional properties, proteins impart various functional properties of foods such as solubility, water holding, fat absorption, emulsifying, foaming, and gelation. These properties make food proteins extremely versatile and useful. Each of the different protein types (from different sources, with and without different processing methods) possesses a unique balance of functional properties versus nutritional properties. This balance is governed by amino acid composition, ratios of essential amino acids, susceptibility to hydrolysis during digestion, source, and the effects of processing [3]. Other food macromolecules, primarily polysaccharides and lipids, are equally inherently functional by virtue of their individual molecular structures. Collectively, these food macromolecules possess the ability to interact to form complexes with each other or with other molecules.

In addition to proteins occurring intrinsically in many food types, protein is also consumed as protein ingredients – ingredients rich in protein or its derivatives that are specifically added in the manufacture of food products to impart functional and nutritional properties. Typically, such ingredients have been subjected to processing to release, change, or fix the valuable properties. The demand for protein ingredients continues to rise driven by several factors: (i) the continued growth of the human population and increasing standard of living in developing countries increases the overall demand for food protein (refer to Figure 2.2.1); (ii) global consumers are seeking lower cost, more flavorful, and nutritionally enhanced food products, and protein ingredients can play a key role here; (iii) food and food ingredient manufacturers are continuously looking

Industrial Enzyme Applications, First Edition. Edited by Andreas Vogel and Oliver May.

Table 2.2.1 The information in this table was obtained from [2].

Food product	Protein (g/100 g)
Almonds	21.15
Almond milk, sweetened, vanilla flavor	0.42
Apples, raw, with skin	0.26
Bananas, raw	1.09
Beef, grass-fed, strip steaks, lean only, raw	23.07
Bread, white	8.85
Broccoli, raw	2.82
Cheese, cheddar	22.87
Cheese, feta	14.21
Chicken, breast, skinless, boneless, meat only, raw	22.50
Chocolate, dark, 70–85% cacao solids	7.79
Corn flakes, cereals, ready-to-eat	7.50
Egg, whole, raw	12.56
Ice creams, vanilla	3.50
Lentils, raw	24.63
Milk, whole, 3.25% milk fat, with added vitamin D	3.15
Oranges, raw, navels	0.91
Pasta, whole wheat, dry	13.87
Potatoes, russet, flesh and skin, raw	2.14
Quinoa, uncooked	14.12
Rice, white, glutinous, unenriched, uncooked	6.81
Salmon, Atlantic, wild, raw	19.84
Soy milk, original and vanilla, with added calcium, vitamins A and D	2.60
Tofu, raw, firm, prepared with calcium sulfate	17.27
Tuna, light, canned in water, drained solids	19.44
Yogurt, plain, low fat	5.25
Yogurt, Greek, plain, non-fat	10.19

for innovative ways to maximize the value provided by protein including the use of some interesting new protein sources such as algae and insects. It is for these reasons that the market for protein ingredients, in terms of value, is projected to reach US $58.49 billion by 2022, at a compound annual growth rate of 6.0% from 2017 [5].

Enzymes are incredibly useful tools that can be used to make exquisite modifications to food proteins to further enhance their functionality and nutritional value. This chapter serves to demonstrate the role that enzymes can play in the modification of diverse types of food proteins and provides recent examples of innovative enzyme developments to meet specific market and consumer needs.

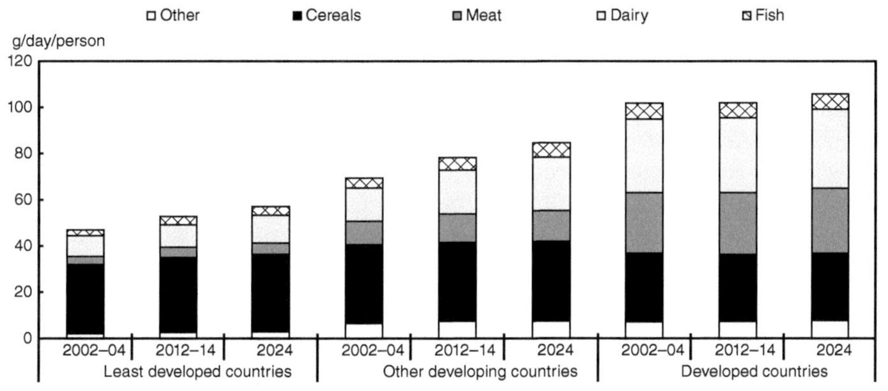

Figure 2.2.1 Protein intake per capita in least developed, other developing, and developed countries. *Note:* The category "other" includes sugar, vegetable oil, eggs, roots, and tubers. Sugar and vegetable oil represent negligible shares of total protein consumption. Vegetables, fruits, pulses, and other food items are not included in this figure. Source: OECD/FAO 2015 [4]. Reproduced with permission of OECD.

2.2.2 Processing of Food Protein

Proteins comprise of long chains of amino acid residues that fold into unique globular (tertiary) structures. These folds include one or more specific spatial conformations determined by covalent and non-covalent interactions. In aqueous environments these globular structures can self-associate to form multimers. Physiochemical properties such as types of amino acid groups, hydrophobicity, hydrophilicity, electrostatic interactions, and hydrogen bonding influence protein folding and self-association. These intrinsic properties together with extrinsic (environmental) factors such as protein concentration and composition of the surrounding solution can all influence the properties of proteins [6].

Food processing has evolved over the past two decades from basic objectives such as transformation, destruction of microorganisms, stabilization, and extension of shelf-life to more complex aims such as altering functional, nutritional, and chemosensory properties (taste, flavor, and texture). Food processing ranges from minimal to extensive modifications using physical, chemical, thermal, and enzymatic methods, which can be operated alone or in combinations. These modifications affect the structure and physiochemical properties of proteins, which in turn affect the performance of the proteins in the food manufacturing process and within the finished food product. An extensive review of thermal, chemical, and mechanical food processing methods and their effects on functionality of different protein types is provided by Aryee et al. [7].

2.2.3 Enzymes in the Processing of Food Proteins

Enzymes can play an extremely valuable role in food processing. There are various enzyme types that can modify food protein to provide enhanced functional, nutritional, and chemosensory properties. By far the most significant, in terms of volume of enzyme used for modifying food protein, are the proteases that act by breaking down (hydrolyzing) the protein structure. In the process of hydrolyzing these large molecules, functional amino groups and hydrophobic patches on the globular surface of the protein may be exposed. Furthermore, the molecular weight of the resulting protein or peptides is reduced. The properties of the hydrolyzed proteins (hydrolysates) are largely dependent on the type and specificity of enzyme used and the degree of hydrolysis (DH), with the degree of hydrolysis being affected by the temperature, pH, process time, substrate concentration, and enzyme dosage.

Endopeptidases (endoproteases) hydrolyze peptide bonds by acting on non-terminal amino acids (they tend to act away from the ends of the protein chain). They include trypsin, pancreatin, bromelain, papain, subtilisin, and fungal neutral endopeptidase. The hydrolytic efficiency of endopeptidase is dependent on the level of exposure of internalized peptide bonds; endopeptidase will first hydrolyze the exposed outer peptides, producing an increased availability of smaller peptides serving as substrates for further hydrolysis. Furthermore, the hydrolytic effect and efficiency of each endopeptidase is determined by its intrinsic specificity. Commercial enzyme preparations can be highly specific. For example, porcine trypsin hydrolyzes only at the carboxy terminal side of arginine or lysine [8]. That said, most commercial preparations of porcine trypsin are contaminated with chymotrypsin, which selectively catalyzes the hydrolysis of peptide bonds on the C-terminal side of tyrosine, phenylalanine, tryptophan, and leucine; this "additional" specificity can be a benefit or disadvantage depending on the application. Papain, containing four different cysteine endopeptidases, exhibits broad specificity [9]. It is interesting to note that, generally speaking, commercial preparations from fungal sources, e.g. neutral protease from *Aspergillus oryzae*, have broader specificity than bacterial protease preparations; during the manufacture of these preparations fungal fermentations tend to produce (secrete) a wider range of protease types. Furthermore, fungal enzyme preparations are generally more expensive (based on cost in use).

Exopeptidases catalyze the cleavage of the terminal (last) or next-to-last peptide bond from either the amino or the carboxy terminus of a peptide or protein, releasing a single amino acid or dipeptide. Examples of exopeptidases include dipeptidase and leucine amino peptidase from *A. oryzae*. In most cases, for exopeptidases to work on their target peptide bond the intact globular protein requires pre- or parallel treatment with endopeptidases (and/or thermal, chemical, or physical processing) to open up the globular structure and expose the termini.

Often, synergistic effects can be sought through the use of combinations of proteases from different biological sources used sequentially or in parallel [10, 11]. Such synergism can reduce the processing time (producing improved properties faster, thereby leading to cost reduction) and yield different or enhanced properties (relative to the performance of enzymes derived from a single biological source). For certain starting materials, the protein component is readily available for the action of proteases (or other types of protein-modifying enzymes). For example, fish protein in fish waste has previously been exposed by mechanical treatment, rendering the residual protein readily available for further protease digestion, e.g. in the production of protein hydrolysate for use in pet food palatability enhancers. However, in many food processing strategies, other enzyme types are used to assist with the release of protein from complex networks of macromolecules in the starting materials. For example, in the production of certain yeast extracts (flavor ingredient), protein is locked within the cell walls and within the yeast cells. Here, endo-β-glucanase e.g. Depol® 667P from Biocatalysts Ltd or Beta Glucanase BP Conc from Dupont Industrial Biosciences can be used to augment endogenous enzymes to improve the efficiency of cell lysis to yield valuable protein (and nucleotide) flavor molecules for further enzymatic hydrolysis.

In addition to proteases several other enzyme types can be used to modify proteins and amino acids; a summary can be found in Table 2.2.2. Several companies provide enzymes for protein modification including, Novozymes

Table 2.2.2 Examples of enzymes active towards food proteins.

Enzyme	Mode of action	Example sources	Effects (example applications)
Endopeptidase	Hydrolyze protein molecule by acting on non-terminal amino acids	Papaya (papain) Porcine pancreas (trypsin) *Bacillus subtilis* *Aspergillus* spp.	Hydrolysis of yeast cell protein for yielding yeast extract (flavor) Digestion of meat and fish proteins to produce protein (flavor) ingredients
Exopeptidase	Cleavage of the terminal (last) or next-to-last peptide bond from a, releasing a single amino acid or dipeptide	*Aspergillus oryzae*	Debittering of whey proteins and peptides Debittering of fish proteins and peptides
Transgutaminase	Catalyzes the formation of covalent bonds between the amino acid residues Lysine and Glutamine in proteins	*Streptomyces* spp.	Improves texture in bakery products Improves texture and connectivity in process meat products
Asparaginase	Hydrolyzes asparagine to aspartic acid and ammonia, thereby reducing free asparagine available for Maillard reactions to form, e.g. acrylamide	*Aspergillus niger* *Aspergillus oryzae*	Reduces acrylamide in foods
Glutaminase	Amidohydrolase enzyme that generates glutamate from glutamine	*Bacillus amyloliquefaciens*	Flavor enhancement of protein hydrolysates

A/S, Amano Enzyme Inc, Biocatalysts Ltd, Dupont Industrial Biosciences, AB Enzymes GmbH, and DSM Food Specialties.

2.2.4 Food Protein Value Chain

So far in this chapter we have described the importance of proteins for delivering nutritional and functional benefits in food. Protein can be present in food stuffs in their intrinsic state or present as typically processed food ingredients added during the manufacture of food products. Protein ingredients can be processed using various technologies to enhance the availability, form, or functionality of the protein. Enzymes can be used to assist in the extraction of proteins from their natural substrate; once extracted the protein can be further processed using enzymes that specifically target proteins. Different enzymes possess different modes of action, thereby enhancing the functionality and nutritional and chemosensory properties. Some of these modifications can be highly exquisite, leading to highly valuable protein or protein-derived ingredients or components. Hence, and generally speaking, the extent of modification of the protein correlates with the intrinsic (market) value of the protein; here, we describe this as the "protein value chain."

In Figure 2.2.2, the protein value chain is described by five distinct levels: Level 1 describes the intrinsic source of proteins. As described in Table 2.2.1, these

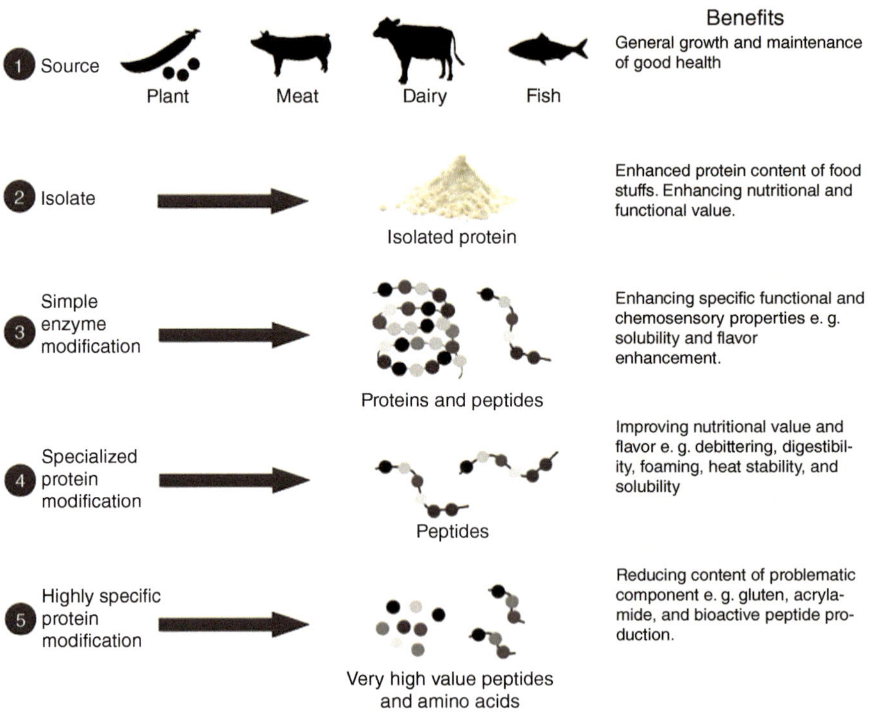

Figure 2.2.2 The food protein value chain according to the author's classification scheme.

proteins are often consumed with no or only minimal processing, e.g. cooking. Level 2 describes proteins that have been isolated from their source often through physical, chemical, and thermal processing methods. Membrane technology is often used in such processing to separate the protein of interest from other components [12]. As mentioned above for yeast lysis, in some applications, non-protease enzymes can be used to help improve the extraction of the isolated protein.

Whey protein is a primary example of a protein that has continued to develop through the protein value chain over the past three decades. Whey is a by-product of the cheese making process; when milk is clotted (which often uses an enzyme called rennet), the clotted protein ("curd") is used as the basis for cheese; the remaining liquid fraction ("unprocessed whey") contains mainly water and 0.4–0.8% whey proteins, 4.5–5% lactose, and 0.5–0.8% minerals and ash [13]. Traditionally, unprocessed whey was, and still is, used as an animal feed (pigs, sheep, cattle) or used as a fertilizer. Unprocessed whey has also provided value in the production of whey cheeses such as Ricotta and Mysost and in whey beverages, e.g. Taksi (Netherlands) and Rivella (Switzerland). However, with worldwide production of whey estimated at around 190×10^6 ton/yr and growing, production exceeds its demand for food and non-food applications and whey continues to be an environmental burden [14]. That said, over the past two decades major research efforts to further exploit the nutritional and functional properties of this protein have resulted in significant expansion of the uses of whey and its concomitant use through the protein value chain. Whey protein concentrates (WPC) are produced by removing a certain percentage of non-protein constituents from pasteurized whey. The finished WPC products are termed WPC34, which contains more than 34% whey protein, and WPC80, which contains more than 80% protein. WPC is processed by physical separation techniques such as precipitation, filtration, or dialysis. Whey protein isolate (WPI) is a further refined product containing >90% whey protein [14]. During the processing of whey a varying amount of lactose is removed and is used widely in the food and confectionary industries due to its low sweetness and can be further modified by enzyme treatment (β-galactosidase) to produce the prebiotic galacto-oligosaccharide. WPC and WPI are examples of proteins that fit in value level 2.

In the rest of this chapter we describe recent advances in enzyme-assisted protein modification aligned to levels 3, 4, and 5 of the protein value chain.

2.2.5 Recent Enzyme Developments

2.2.5.1 Simple Protein Modification (Value Level 3)

2.2.5.1.1 Developing Microbial Alternatives to Plant and Animal Enzymes

There is a long history of using proteases from animal and plant sources, many of which are used for relatively simple modification of food proteins. Calf rennet, containing the milk-clotting enzyme chymosin, was used by the ancient Egyptians for preserving milk (see also Chapter 1.1). The application of papain, a protease preparation extracted from papaya, for tenderization of meat was described in detail by Gottschall and Kies as far back as 1942 [15]. Pancreatin, also known as pancrelipase or pancreatic enzymes, is a commercial mixture of amylase, lipase, and protease derived from porcine pancreas. The US patent US2115505, dating from 1943, describes an early use of pancreatic enzymes in the making of cheese [16]. Both pancreatin and papain have broad specificity during protein hydrolysis and hence they are typically used where crude protein digestion is preferred.

Over the past 10 years, there has been a general trend away from the use of animal derived enzymes in food processing. This change has been driven, in part, by the increasing demand for dairy products and ingredients from emerging markets in Asia and India, with a requirement for such products to carry Kosher and Halal status. Furthermore, diseases such as BSE (bovine spongiform encephalopathy), which can be transmissible to humans through eating infected meat, have tainted the perception of animal enzymes. Consolidation in supply chains driven by the demand for logistical and financial efficiencies means that producers of such ingredients tend toward manufacturing a single product type to fit global preferences. Pancreatic enzymes have been used in the manufacture of enzyme modified cheese – a concentrated dairy flavor ingredient produced by the crude enzymatic hydrolysis of dairy fat and protein. Since the 1980s, Biocatalysts Ltd has provided microbial and animal derived enzymes for the manufacture of several types of enzyme modified cheese. In 2012, Biocatalysts Ltd launched Lipomod™957MDP, a product containing a combination of enzymes derived from microbial strains, specifically designed as a Kosher and Halal alternative to animal pancreatin (www.biocatalysts.com). The design of this product required a precise combination of different microbial proteases from different microbial strains that act synergistically to modify the globular structure of casein (in immature cheese, the starting material for enzyme modified cheese) producing a flavor profile similar to that obtained by pancreatin (which is itself a complex mixture of different endo- and exo-peptidase enzymes). The similar flavor profiles for enzyme modified cheeses produced by Lipomod957MDP and pancreatin can be seen in Figure 2.2.3.

The primary plant-derived enzyme used in food processing is papain. Papain is a cysteine endopeptidase that is stable and active under a wide range of pH and temperature conditions. The enzyme is derived from the tropical plant *Carica papaya*. The latex of *C. papaya* is a rich source of four cysteine endopeptidases including papain, chymopapain, glycyl endopeptidase, and caricain. The commercial preparation of papain contains a mixture of these endopeptidases [9]. It is used in food processing applications, primarily for crude digestion in applications such as meat tenderization, yeast lysis, and fish processing. From a commercial perspective, the main challenge with papain is its seasonal availability. Over recent years, the primary geographical locations from where papain is sourced, namely India and Congo, have been subjected to year-by-year fluctuating crop levels. This was the driver behind the design and launch of Promod™ 950L from Biocatalysts Ltd in 2016. Following similar design principles as those used for the microbial alternative to pancreatin,

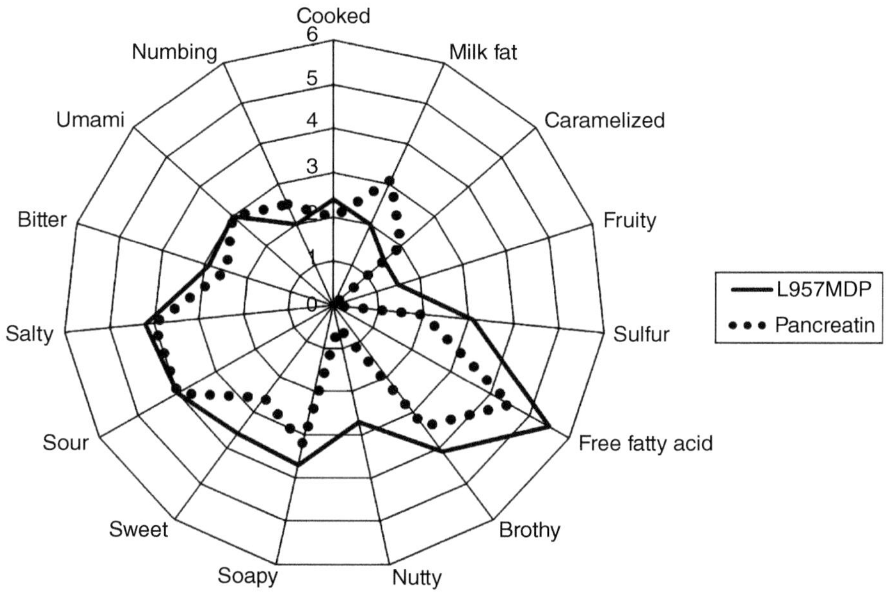

Figure 2.2.3 Flavor profiles for enzyme modified cheeses produced using either pancreatin or the microbial alternative to pancreatin, Lipomod™957MDP. Source: Courtesy of Professor Mary Anne Drake, North Carolina State University, US.

scientists at Biocatalysts Ltd deconstructed the performance of papain and matched it with a combination of individual enzymes from microbial sources (www.biocatalysts.com). Figure 2.2.4 depicts the relative performance of papain and Promod 950L in the digestion of yeast as determined by the release of free amino groups.

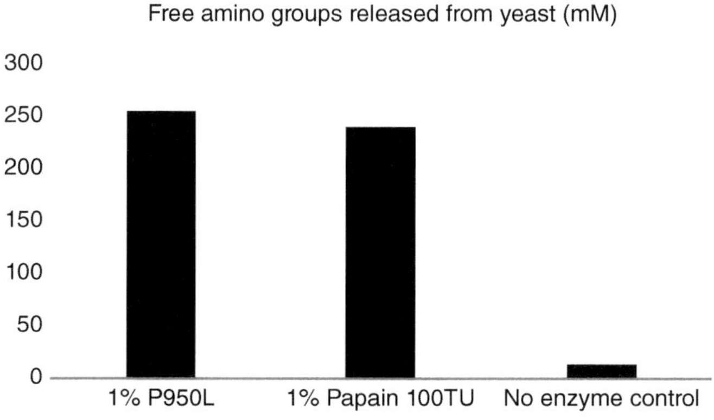

Figure 2.2.4 The relative performance of papain and Promod™ 950L in yeast hydrolysis expressed as millimole free amino acids released after 10 hours enzyme incubation (20% w/v dried baker's yeast in water with 1% w/w enzyme/yeast).

2.2.5.2 Specialized Enzyme Modification (Value Level 4)

2.2.5.2.1 Whey Protein Hydrolysates

Whey protein hydrolysate (WPH) is manufactured through enzymatic hydrolysis of WPC or WPI. The functional properties of WPH depend on numerous factors including the source of whey (rennet-derived, which is typically known as sweet whey, or acid-derived, typically known as acid whey); concentration of whey; the physiochemical state of the whey protein (influenced by the pre-treatment method e.g. heat denaturation); temperature, time, and pH of hydrolysis; and type and dosage of the enzyme used. The enhanced functional properties associated with WPH include improved foaming, emulsification, and gelling. The nutritionally valuable amino acids in WPH are more easily absorbed compared to WPC and WPI, and hence its increasing use in sports nutrition products such as drinks and bars [17].

Porcine trypsin is used extensively in the production of WPH. When used with heat denatured whey, trypsin can be used to produce WPH with a degree of hydrolysis of 3–8% with enhanced solubility, emulsifying properties, and foaming ability [18]. In 2017, Novozymes A/S launched a trypsin-like microbial enzyme (Formea™ TL) originating from *Fusarium oxysporum* and manufactured using *Fusarium venenatum*. This product possesses the same regioselectivity as porcine trypsin, i.e. hydrolyzes at the carboxy terminal side of arginine or lysine and has the added benefit of Kosher and Halal status. Formea TL has been shown to produce comparable peptide profiles when compared to porcine trypsin [19]. The main drawback with trypsin-produced WPH is its inherent bitterness owing to the release of bitter peptides. Biocatalysts Ltd specifically developed Flavorpro™ 750P for the manufacture of low bitterness WPH. For complete removal of bitterness, Flavorpro 750P is used in combination with the debittering enzyme Promod™ 782P (www.biocatalysts.com).

2.2.5.2.2 Plant Protein Hydrolysates

Over the past two decades, the global shortage of animal protein, driven by increasing demand for protein from a rapidly growing human population, has driven the increasing use of protein ingredients derived from plants [5]. Protein concentrates and isolates derived from beans/legumes, vegetables, and grains are now common ingredients in food products. The nutritional benefits of these plant proteins, relative to their animal counterparts, are complex and beyond the scope of this chapter; an interesting review is provided by Malav et al. [20]. Suffice to say, vegetable-derived proteins are generally less soluble and can benefit from enzymatic processing to improve this aspect. Although not yet as prevalent as WPHs, many ingredient companies now manufacture hydrolyzed plant proteins, including Hill Pharma Inc (Pea), Friesland Campina Domo (Wheat), ADM (Soy), A. Costantino & C. S.P.A. (Multiple types). dos Santos Aguilar and Sato (2018) describe the broad application of *Bacillus* alkaline protease for the hydrolysis of numerous plant-derived proteins including pea, soy, rice, chick pea, and wheat gluten [21]. Analogous to the progress of whey through the protein value chain over the past three decades, it would be expected to see the demand for plant hydrolysate increasing in the future. This may, in part, be driven by some of the unique nutritional features that result in certain vegetable proteins being a potential source of bioactive peptides (see further discussion below).

2.2.5.3 Highly Specific Protein Modification (Value Level 5)

2.2.5.3.1 Gluten Modification

Gluten is a protein complex found in wheat, barley, and rye. It gives breads and other grain products their shape, strength, and texture. Intolerance (including allergic responses) to gluten is expressed in different ways, but the main trigger is the same: specific gluten fragments that are exceptionally rich in proline amino acids (www.dsm.com/corporate/science/challenges/health-wellness/tolerase-g.html). In 2015, DSM Food Specialties introduced a prolyl endopeptidase (AN-PEP) under the brand name Tolerase® G. The enzyme is not intended for individuals with coeliac disease or who are gluten intolerant, but is aimed at the rising numbers of gluten-sensitive consumers who are already following a gluten-free diet and want help breaking down any residual gluten consumed in their diets. The enzyme was developed to function optimally in the acidic conditions within the stomach, specifically hydrolyzing peptides containing proline thereby enhancing gluten digestion. The enzyme is derived from the production strain *Aspergillus niger* and in addition to its application in gluten hydrolysis, it is used for prevention of chill-haze in beer and debittering of certain proline-rich protein hydrolysates.

The results of an *in vitro* study show that the AN-PEP enzyme degrades gluten molecules in the stomach more effectively than other commercially available supplements [22]. Another publication on the gluten-degrading effect of the AN-PEP enzyme describes a randomized, double-blind, placebo-controlled study that concluded that AN-PEP enzyme enhanced gluten digestion of healthy volunteers within a one hour period, irrespective of the caloric content of the meal [23].

2.2.5.3.2 Acrylamide Reduction

Acrylamide is a potential carcinogen that is formed when starchy foods are baked or fried at high temperatures. The main mechanism for acrylamide formation in starchy food involves reducing sugars and the amino acid asparagine, both common to these foods. The reducing sugars react with asparagine when heated and through a cascade of Maillard reactions, the side chain of asparagine is converted into acrylamide. Maillard reactions, also known as non-enzymatic browning, typically occur at temperatures above 100 °C and are desirable for important color and flavor development in fried and baked starch products. In 2007, Novozymes A/S launched Acrylaway®, an asparaginase enzyme that reduces acrylamide in baked and fried starchy foods without altering the appearance and flavor of the final product (www.novozymes.tv/video/6444601/acrylaway). Asparaginase converts asparagine into another amino acid, aspartic acid. Asparaginase can reduce acrylamide levels by up to 90%. It can be used in a broad range of foods such as biscuits, cookies, crackers, crisps, and toasted bread. Acrylaway is derived from the product strain *A. oryzae*. DSM Food Specialties also manufacture an asparaginase enzyme, derived from *A. niger*, sold under the brand name PreventAse® (www.dsm.com/markets/foodandbeverages/en_US/products/enzymes/baking/preventase).

c-LEcta GmbH developed an enzyme with unique properties that is suitable for food production processes involving higher temperatures. With a temperature optimum around 90 °C, c-LEcta's asparaginase derived from *Pyrococcus furiosus* reduces acrylamide formation by more than 90% in a broad range of foods [24]. The product has been out-licensed on an exclusive basis to Novozymes A/S. In 2013, this asparaginase was launched by Novozymes A/S under the brand name of Acrylaway HighT. During the first few years of commercial use of asparaginase, the removal of acrylamide was viewed as a high value (Value Level 5) application. Over the past decade, the routine use of asparaginase in many food applications has seen its use become commoditized.

2.2.5.3.3 Bioactive Peptides

Bioactive peptides are fragments that are present in the primary sequences of proteins and that confer functions beyond basic nutritional benefits. Examples of bioactive peptides and their claimed benefits are depicted in Table 2.2.3. The approach to the production of bioactive peptides involves first identifying a suitable protein source and then releasing peptide fragments through hydrolysis of peptide bonds by the action of endopeptidases. The resulting crude protein hydrolysate may undergo fractionation processes to yield an enriched bioactive peptide preparation or additional purification steps to isolate single

Table 2.2.3 Examples of bioactive peptides.

Product	Source	Claimed benefits	Type of fraction	Manufacturer
Lactium®	Milk	Relaxing	Peptide (YLGYLEQLL)	Ingredia, Arras Cedex, France
Sato Marine Super P	Sardine	Antihypertensive	Peptide (VY)	Sato Pharmaceutical Co., Ltd., Tokyo, Japan
Ameal S	Milk casein	ACE inhibition	Peptides (IPP and VPP)	Calpis, Japan
Vasotensin®	Bonito	Antihypertension	Peptide (LKPNM)	Metagenics, US
Peptide Nori S	Porphyra yezoensis	Antihypertension	Peptide (AKYSY)	Riken Vitamin, Japan
Seishou-sabou	Bovine and porcine blood	Anti-obesity	Peptide (VVYP)	Moringa & Co., Ltd., Japan
Marine peptide	Sardine	ACE inhibition	Peptides	SenmiEkisu, Japan
BioZate	Whey	Antihypertension	Peptides	Davisco Foods, US
VERISOL®	Collagen	Antiaging	Peptides	GELITA Inc., US
Remake Cholesterol Block	Soy protein	Hypo-cholesterolemic	Peptide (CSPHP)	Kyowa Hakko, Japan

peptides. Currently, there are hundreds of peptide sequences that have been associated with bioactivity [25]. Typically, these have been identified through an empirical approach involving endopeptidase hydrolysis of the selected protein source to produce a hydrolysate that is then fractionated and subjected to a bioactivity assay. Commercial endopeptidases that are used to access such bioactive peptides tend to be those with a relatively tight specificity for cleaving specific peptide bonds, thus maximizing the abundance of the intact peptide of interest. Examples of such enzymes include trypsin, chymotrypsin, and pepsin. Bioinformatics based approaches have also been used extensively to search for new sources of bioactive sequences. Known bioactive peptide sequences are used to mine for similar sequences in protein and DNA databases and *in silico* analysis methods applied to determine their propensity for bioactivity [25]. One limitation to accessing specific peptide fragments is the lack of commercial availability of highly specific proteases. Those proteases mentioned earlier (porcine trypsin, chymotrypsin, and pepsin) are from animal origin, which provides ethical and regulatory barriers for some applications. The opportunity to make available microbially derived highly specific proteases is discussed below.

2.2.6 Enzymes to Meet Future Needs

The increasing demand for existing and new types of food proteins will correlate with a demand for existing and new enzymes, and enzyme-based processes. As certain food proteins, e.g. some plant proteins, move through the "protein value chain," enzymes will be selected based on the features required to bring about the required effect. It is expected that for simple protein modification (Value Level 3), for most protein types the current commercial proteases will be suitable. In certain cases, where hydrolysis of relatively resistant proteins is required, specific combinations of different types of proteases (different biological sources and different specificities) will be invaluable. Where hydrolysis processes dictate the need for specific conditions of pH or temperature, there may be a future need to identify protease enzymes that can operate (in an economically viable manner) under such conditions. Previously, we have seen the demand for thermostable asparaginase (see Acrylaway HighT). Perhaps there will be a need for new acidic proteases with the desirable specificity for more efficient and effective hydrolysis of acid whey protein. One can foresee a need for more thermolabile proteases that can be deactivated under relatively mild temperatures in hydrolysates that are particularly temperature sensitive.

Revenue from bioactive peptides is expected to grow more than twofold by 2026 as compared to 2016 [26]. In order to access very specific bioactive peptides, there is a potential future need for new highly specific endopeptidases. There are several bioinformatics resources available in the public domain for those scientists wanting to identify pre-commercial enzymes. EBI-EMBL offers access to a range of bioinformatic tools. One such tool, the Merrops peptidase database, contains a vast amount of information, e.g. specificity, inhibitors, optimal conditions, and literature references, for thousands of protease enzymes [27].

In addition to certain plant proteins gaining increasing interest (and developing through the protein value chain), there are numerous novel sources of food proteins emerging. The academic literature is rife with descriptions of the potential benefits and challenges of proteins from insects, algae, new mycoprotein sources, numerous proteinaceous waste streams, and proteins produced by "synthetic biology" [28]. Where there is a need and/or opportunity to isolate proteins from these sources, there will be a potential requirement for non-protease enzymes to assist in their isolation through breakdown of macromolecules of lipid and polysaccharide origin. Furthermore, each novel protein will possess distinct biochemical and structural properties, presenting a potential need for new proteases or protein modifying enzymes to add value. It is important to state that while there is significant demand to identify novel food proteins (driven by population increase and innovation), there will be some market resistance. The Western consumer is generally very conservative and resistant to major change in food types. Furthermore, due to religious, cultural, and historical reasons, worldwide tastes are incredibly diverse. It will be interesting to see the fate of, for example, insect and algal proteins over the next decade.

Allergenicity of both existing and new food proteins is a further challenge. Allergenicity is associated with the immune response to certain protein and peptide types [29]. Tolerase G has been developed to help reduce the levels of trace allergenic peptides from gluten. Perhaps there will be future opportunity to develop new enzymes to hydrolyze known, and hitherto unidentified, allergenic food proteins; non-hydrolytic protein modifications may also be of benefit. As proteins themselves, it will be increasingly important to demonstrate the absence of allergenicity of any new enzymes used in food processing applications.

The glutaminase enzyme generates the amino acid glutamate from glutamine. Food stuffs and food ingredients with certain levels of glutamate have enhanced "umami" (often described as "meaty" or "savory") flavor [30]. Typically, glutaminase is used after the use of endo- and exo-peptidase to maximize levels of the glutamine substrate for conversion to glutamate. Glutaminase can be used in the manufacture of yeast extract (for flavor applications) and concentrated dairy flavors such as enzyme modified cheese. Looking ahead, the food industry will continue to strive for innovative new flavors and there will be continued demand for new enzymes to help access these flavors. There is a continuing interest in identifying new flavor compounds or variants of existing flavors that are associated with peptide fragments or amino acids. Such peptides and amino acids maybe precursor compounds to the production of new flavors, e.g. producing specific Maillard compounds through release of specific amino acids. Novel enzymes have the potential to assist in accessing said flavors.

The cross-linking enzyme transglutaminase currently finds extensive use for binding proteins in the production of processed or restructured meat and fish products. Under the optimum conditions, the enzyme can efficiently form isopeptide linkages, leading to its common description "meat-glue." This aggressive cross-linking is not suited to certain applications such as binding of proteins in solutions, suspensions, or emulsions such as egg white, processed cheese, and yoghurt. Hence, there is potential future need for cross-linking enzymes with a more subtle mode of action. Enzymes such as peroxidase, laccase, tyrosinase, and glutathione oxidase have been described in the academic literature, but their industrial exploitation in protein cross-linking is yet to be fully realized [31].

Many of the future opportunities for developing innovative protein ingredients will require novel enzymes. Discovery and development of industrial novel enzymes was once the preserve of the large enzyme manufacturing companies e.g. Novozymes A/S, AB Enzymes GmbH, and Dupont Biosciences. Over the past 10 years, the explosive development in molecular biology, bioinformatics, and biotechnological processing has paved the way for smaller specialty enzyme companies to develop their own capabilities for enzyme discovery and manufacture. Chapter 1.2 described these technological advances. These specialty enzyme companies, generally speaking, provide enzyme discovery and development to meet the needs of individual customers or smaller market opportunities (<$6 million per novel enzyme per year). Hence, more specialist (and lower volume) needs for novel enzyme can be realized. Each specialist enzyme company has their individual portfolio of capabilities including metagenomic-based enzyme discovery, protein engineering, activity screening, and bioinformatic (*in silico*) screening. Some specialist enzyme companies operate primarily in the enzyme discovery space while others can additionally manufacture the enzymes they develop. BRAIN GmbH, c-LEcta GmbH, and Biocatalysts Ltd are each notable examples of such specialist enzyme companies.

Acknowledgments

Thanks to Denise Gallagher, Danielle Lloyd, Karen Lancashire and Stuart West (Biocatalysts Ltd) for assistance in the preparation of this chapter. Thanks also to Oliver May for provision of reference material for Tolerase G.

References

1 Tanford, C. and Reynolds, J. (eds.) (2002). *Natures Robots*. Oxford: Oxford University Press.
2 US Department of Agriculture, Agricultural Research Service, Nutrient Data Laboratory. USDA National Nutrient Database for Standard Reference, Legacy. Version Current: April 2018. https://www.ars.usda.gov/nea/bhnrc/ndl
3 Friedman, M. (1996). Nutritional value of proteins from different food sources. A review. *J. Agric. Food Chem.* 44 (1): 6–29.
4 OECD/FAO (2015). OECD-FAO Agricultural Outlook. OECD Agriculture Statistics (database). http://dx.doi.org/10.1787/agr-outl-data-en.
5 Research and Markets (2017). Protein Ingredients Market – Forecast to 2022.
6 Lam, R.S.H. and Nickerson, R.T. (2013). Food proteins: a review on their emulsifying properties using a structure-function approach. *Food Chem.* 141 (2): 975–984.

7 Aryee, A.N.A., Agyei, D., and Udenigwe, C.C. (2018). Impact of processing on the chemistry and functionality of food proteins. In: *Proteins in Food Processing*, Woodhead Publishing Series in Food Science, Technology and Nutrition, 2e (ed. R.Y. Yada), 27–45. Woodhead Publishing.
8 Kilcawley, K.N., Wilkinson, M.G., and Fox, P.F. (2002). Determination of key enzyme activities in commercial peptidase and lipase preparations from microbial or animal sources. *Enzyme Microb. Technol.* 31 (3): 310–320.
9 Amri, E. and Mambaya, F. (2012). Papain, a plant enzyme of biological importance: a review. *Am. J. Biochem. Biotechnol.* 8 (2): 99–104.
10 Yongjing, W. (2017). A method for producing a yeast extract. WO Patent 2017114403, Application 7 June 2017.
11 Maynard, F., Salvatore, D., and Thevenier, A. (2017). Milk based protein hydrolysates and compositions thereof. WO Patent 2016156077, Application 6 October 2017.
12 Ahmad, S. and Ahmed, S.M. (2014). Application of membrane technology in food processing. In: *Food Processing: Strategies for Quality Assessment*, Food Engineering Series (ed. A. Malik, Z. Erginkaya, S. Ahmad and H. Erten), 379–394. New York: Springer.
13 Kilara, A. and Vaghela, M.N. (2018). Whey proteins. In: *Proteins in Food Processing*, Woodhead Publishing Series in Food Science, Technology and Nutrition, 2e (ed. R.Y. Yada), 93–126. Woodhead Publishing.
14 Ryan, M.D. and Walsh, G. (2016). The biotechnological potential of whey. *Rev. Environ. Sci. Biotechnol.* 15 (8): 479–498.
15 Gottschall, G.Y. and Kies, M.W. (1942). Digestion of beef by papain. *Food Res.* 7: 373–381.
16 Victor, C. (1938). Process of preparing soft curd milk. US Patent 2,115,505, filed 1 April 1937 and issued 26 April 1938.
17 Naik, L., Mann, B., Bajaj, R. et al. (2013). Process optimization for the production of bio-functional whey protein hydrolysates: adopting response surface methodology. *Int. J. Pept. Res. Ther.* 19 (3): 231–237.
18 Mutilangi, W.A.M., Punyam, D., and Kilara, A. (2006). Functional properties of hydrolysates from proteolysis of heat-denatured whey protein isolate. *J. Food Sci.* 61 (2): 270–275.
19 Health & Nutrition Application Sheet (2016). Novozymes Formea™ TL 1200 BG for partially hydrolyzed infant formula.
20 Malav, O.M., Talukder, S., Palanisamy, G., and Chand, S. (2013). Meat analogue: a review. *Crit. Rev. Food Sci. Nutr.* 55 (9): 1241–1245.
21 dos Santos Aguilar, J.G. and Sato, H.H. (2018). Microbial proteases: production and application in obtaining protein hydrolysates. *Food Res. Int.* 105: 252–262.
22 Janssen, G., Christis, C., Kooy-Winkelaar, Y. et al. (2015). Ineffective degradation of immunogenic gluten epitopes by currently available digestive enzyme supplements. *PLoS ONE* 10 (6): e0128065.
23 Salden, B.N., Monserrat, V., Troost, F.J. et al. (2015). Randomised clinical study: *Aspergillus niger*-derived enzyme digests gluten in the stomach of healthy volunteers. *Aliment. Pharmocol. Ther.* 42 (3): 273–285.

24 Greiner-Stoeffele, T. and Struhalla, M. (2017). Heat-stable asparaginase for reducing acrylamide in food stuffs or stimulants. US Patent 9,723,854, granted 8 August 2017.

25 Li-Chan, E.C.-Y. (2015). Bioactive peptides and protein hydrolysates: research trends and challenges for application as nutraceuticals and functional food ingredients. *Curr. Opin. Food Sci.* 1: 28–37.

26 Persistence Market Research (2016). Global Market Study on Bio Active Protein and Peptides.

27 Rawlings, N.D., Barrett, A.J., and Finn, R.D. (2016). Twenty years of the MEROPS database of proteolytic enzymes, their substrates and inhibitors. *Nucleic Acids Res.* 44: 343–350.

28 Jan, S. (2018). Advancing aquaculture to meet growing demand. *Agric. Res.* 66 (1): 1–6.

29 EFSA Panel on Dietetic Products, Nutrition and Allergies (NDA) (2014). Scientific opinion on the evaluation of allergenic foods and food ingredients for labelling purposes. *EFSA J.* 12 (11): 3894.

30 Zhao, C.J., Schieber, A., and Ganzle, M.G. (2016). Formation of taste-active amino acids, amino acid derivatives and peptides in food fermentations – a review. *Food Res. Int.* 89: 39–47.

31 Buchert, J., Selinheimo, E., Kruus, K. et al. (2007). Using crosslinking enzymes to improve textural and other properties of food. In: *Novel Enzyme Technology for Food Applications*, Woodhead Publishing Series in Food Science, Technology and Nutrition (ed. R. Rastall), 101–126. Woodhead Publishing.

2.3

Dairy Enzymes

Peter Dekker

DSM Food Specialties, Alexander Fleminglaan 1, 2613 AX Delft, the Netherlands

2.3.1 Introduction

The dairy industry is an important part of the food chain in large parts of the world. Numerous different food products are produced from the milk of different mammals, and every year new innovative products based on milk ingredients appear on the market. These products may be in the savory, sweet, and sour segments in the food chain, and dairy products can be used at every meal with its astonishing variety of cheeses, fermented milk products, desserts, spreads, beverages, powders, and products made from these in many more applications. Since the number and variety of products made from milk is so high, the processing methods and manufacturing procedures are also widely different. Also, the composition of the milk may vary. Although almost 90% of the world dairy market is based on bovine milk, a substantial amount of milk comes from buffalo (mainly in India; the world's largest dairy market), goat, sheep, and some other species [1]. Obviously, the milk composition differs between these different sources. Also, the milk composition can vary from early after calving until later in the season, it can vary from one cow to another, and vary depending on the feed of the animals. Many dairies will try to avoid variation in their production process by standardization of the milk, and in many countries the seasonal differences are diminished by calving throughout the year [2].

The predictability of the effects of any specific enzyme in diverse dairy products will obviously benefit from standardization of the milk, and this may stimulate the use of enzymes in this industry. However, it must be kept in mind that the development of a new enzyme for the food industry is a time-consuming and costly process. This is not only because of the cumbersome R&D trajectory to screen, isolate, and develop a new enzyme product, but especially because of the extensive application testing and the elaborate registration process that precedes the market introduction of a new enzyme. Owing to the timing and costs of developing new enzymes, relatively few enzymes that are being investigated in the R&D phase will make it to a commercial product. Commercial success will also depend on the market size of the possible application, which is small for fragmented markets, and also depends on the willingness of the dairy producers

Industrial Enzyme Applications, First Edition. Edited by Andreas Vogel and Oliver May.

to adopt a new enzyme and/or process. Despite these limitations, the dairy industry has been blessed with a number of new market introductions from different enzyme producers in recent years.

This chapter will discuss different commercial enzymes that are currently being used in the dairy industry. Based on the reasons discussed above, many new enzymes that are being investigated in academic or company laboratories will not make it into a commercial product. Therefore, the focus here will solely be on commercially available enzyme products and their developments described in public (patent) literature. Main producers of these enzyme products for the dairy industry are Chr. Hansen, Dupont, Novozymes, DSM Food Specialties, Amano, Meito, Shin Nihon, and Godo, but also many distributors sell enzymes for the dairy industry.

Enzymes that are currently being used in the dairy industry may originate from animal, plant, or microbial material. Both the so-called "classical enzymes" and enzymes produced using recombinant DNA technology are currently being used in the dairy industry, and will be discussed later in this chapter. Most of these enzymes act on the main ingredients present in milk: protein, fat, and lactose, but there are notable exceptions.

One important limitation to the use of enzymes in the dairy industry is the presence of contaminating side activities. Milk and dairy products are very sensitive to spoilage due to unwanted oxidation, lipolysis, or proteolysis that may lead to severe off-flavors during either heating steps or during storage of the final product. The enzymes used in dairy applications have to have strict specifications on side activities even more than in other food applications. Off-flavors in dairy can occur in unexpected ways and are not always based on enzymatic conversion of the main constituents, protein, fat, or lactose. For example, with lactases it was found that the presence of arylsulfatase side activity in lactase preparations can lead to the release of *p*-cresol in milk during storage, leading to a nasty medicinal off-flavor [3]. Also, invertase side activity, the enzyme that splits sucrose, could lead to an altered sweetness perception when lactase was used in the production of sweetened yoghurt (see below). Therefore, one should be aware of side activities that may not only affect the dairy matrix itself but may also affect added ingredients that are used in the production of the final dairy product. Also, enzyme (side) activities may end up in side streams of the main dairy product, such as whey. Such contaminants (e.g. amylase activity in rennet) may be harmless in the main application but can have a detrimental effect on the applicability of side streams, and should therefore be avoided.

Four main groups of enzymes used in the dairy industry can be distinguished, which will be discussed separately here:

- The first group is the coagulants, which is among the dairy enzymes and also commercially the most important one. Coagulants are used to produce many different cheeses and therefore have a relatively broad market penetration.
- Enzymes used for cheese ripening and generation of cheese flavors are used for both cheese and enzyme modified cheese (EMC). Since each enzyme generates a specific flavor profile the potential market of individual enzymes is, per definition, limited.

- Lactases are used for the production of lactose-reduced or lactose-free dairy products for lactose-intolerant people, but can also be applied for sugar reduction, prevention of crystallization, galacto-oligosaccharide (GOS) production, and other technical applications. The market for lactases is growing steadily.
- The last group of enzymes that will be discussed is being used for miscellaneous applications that are different from the other three groups. Many innovative new developments belong to this group of enzymes, the full market potential of which is not entirely clear yet.

Application of (proteolytic) enzymes in the production of hydrolysates of (dairy) proteins, mainly for use in infant nutrition, will not be discussed here but has been referred to in Chapter 2.2.

2.3.2 Coagulants

Coagulants, or rennets, are proteases used in the cheese making process to specifically cut κ-casein and thereby initiate the coagulation of the casein micelles leading to curd formation (rennets are reviewed in [4, 5]). There are numerous variations on the cheese making process leading to thousands of different cheeses, but most processes use a coagulant for initiating the curd–whey separation. The coagulants used in the dairy industry are aspartic endoproteases (EC 3.4.23.X), which, under the conditions of the cheese-making process, are highly specific for the Phe–Met bond between residues 105 and 106 of bovine κ-casein. This split leads to the release of the carboxy terminal part of κ-casein (the casein-glycomacropeptide: cGMP) into the milk serum. The amino terminal part (the para-κ-casein) stays attached to the casein micelle. The rate of the splitting of κ-casein cannot be described by Michaelis–Menten kinetics. During commercial renneting, one enzyme molecule is available for roughly 30 casein micelles for which the splitting rate was found to be dependent on the diffusion of the enzyme from one micelle to the next [6]. The splitting of κ-casein changes the surface of the casein micelles from hydrophilic and negatively charged into hydrophobic and neutral. After ~70% of the κ-casein is split, flocculation of the milk can be noted by eye, initiated by the coagulation of the casein micelles. This flocculation is dependent on the Ca^{2+} activity (accessible concentration) in the milk, as well as the temperature, and is accelerated by the splitting of the residual κ-casein. After most κ-casein is cut and flocculation is complete, gel formation starts and the coalesced casein micelles bind together with the milk fat globules in a loose gel. Syneresis, the separation of whey liquid from the gel, and compacting of the gel is enhanced by acidification, heating, and mechanical cutting. The liquid whey is separated from the curd, which is pressed and salted (or the reverse, dependent on the cheese process), sometimes molded, and ripened. The renneting process is schematically depicted in Figure 2.3.1.

Coagulation activity in all rennets on the market is measured using a certified method: the IMCU (International Milk Clotting Unit). The method has been described in the Food Chemical Codex, and is available as International Dairy Federation (IDF) Standard [7]. This method is based on the measurement of the

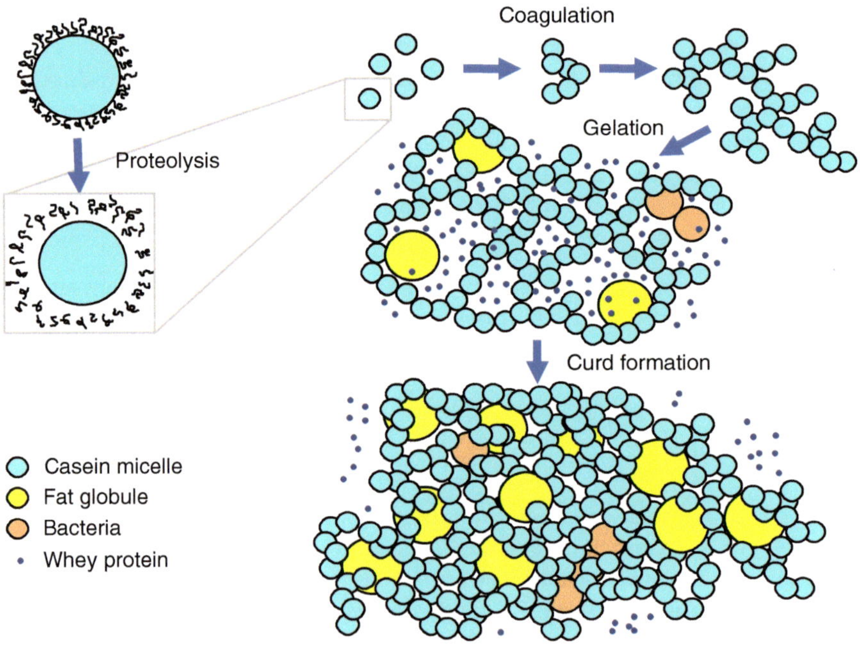

Figure 2.3.1 Schematic representation of the cheese making process. Rennet will remove the hairy outer layer of casein micelles, after which coagulation, gelation, and syneresis will lead to curd formation. Whey proteins are expelled from the curd in this process.

time of appearance of flocks of a renneted solution of skim milk powder on the wall of a rotating test tube. Similar methods have been developed for microbial rennet and ovine/caprine rennet. The coagulation time in the test solution is compared to the coagulation time with a standardized rennet solution. For the different commercial coagulants different standards are used, which are available via the different producers of rennet (e.g. Chr. Hansen and DSM Food Specialties), and these greatly reduce variation in the outcome of the assay. However, as mentioned above, clotting of the milk is a secondary reaction of the splitting of the κ-casein and therefore it is important that the exact conditions of the test are followed to avoid mistakes in the interpretation. Despite all the efforts to standardize the activity assay of the different rennet solutions, the relation of the IMCU dosage with the gel-cutting time in commercial cheese making is still dissimilar for the different rennet types. Most suppliers of rennet address this by providing products with different enzyme concentrations so the volume dosage can stay the same when switching from one rennet to another. An extensive discussion on the analytical methods used for measuring rennet activity is described in the literature [4].

Besides the essential contribution to the coagulation process, all rennets have an effect on the ripening of the cheese by slowly digesting the α- and β-caseins. This activity affects the cheese texture and generates peptides that can be used by the lactic acid bacteria in the cheese to generate flavor. An important aspect in this is the coagulation versus proteolysis (*C*/*P*) activity of the specific rennet, indicating the activity of a rennet on κ-casein versus the proteolysis of other caseins. As mentioned above, the coagulation activity is measured using the standardized IMCU assay, but this is not the case for the proteolytic activity. Proteolytic assays are most often based on the release of aromatic amino acids from a casein substrate, but pH, temperature, time, and ionic conditions may vary from one protocol to another. Since every supplier of rennet uses a different method to measure the proteolytic activity, it is difficult to compare rennets from different suppliers solely based on the reported C/P values. Even when the values of the C/P assay are normalized and the C/P of standard fermentation-produced chymosin (FPC) is set at 1, different conditions in the assay may still lead to a different outcome of the C/P when another rennet is tested.

2.3.2.1 Traditional Rennets

Coagulants are probably one of the first enzymes used by humans in a food-making process (see also Chapter 1.1). There is evidence that ancient Greeks and Romans were aware of the existence of cheese and valued it, and the ancient Egyptians were thought to be the first to discover rennet 4000–5000 years ago. It is not unlikely that the invention of cheese making would have happened by serendipity [8]. Milk may have been stored at ambient temperature in bags made from (calf) stomachs. Remnants of aspartic proteases present in such stomachs, plus the growth and acidification by the natural milk flora, may have led to coagulation. Movement of the bags and acidification would have led to syneresis, leading to curd formation. The awareness that the valuable nutrients of milk could be preserved in this process will have spurred the perfection of the technology and led to the numerous different cheese types we are used to nowadays. Also the near absence of lactose in cheese, in contrast to milk, will have increased the popularity of cheese as food stock since a large part of the world population is lactose intolerant (see also Section 2.3.4) [9].

Traditional rennet is a salt extract from the fourth stomach (the abomasum) of suckling calves. It is therefore a side product of veal production. Calf rennet is not a pure enzyme preparation but contains several aspartic proteases, and sometimes even lipases (in case of rennet paste – see also the discussion below on cheese ripening enzymes). The aspartic proteases that are normally present in calf rennet are chymosin (EC 3.4.23.4), pepsin (EC 3.4.23.1), and the minor component gastricsin (EC 3.4.23.3) [10]. Different qualities of traditional calf rennet are on the market with different amounts of pepsin compared to the chymosin content. Chymosin is the "renneting enzyme" in calf rennet, but pepsin can have a role in flavor formation during cheese ripening. Chymosin is very specific for the Phe105–Met106 bond in κ-casein, while pepsin is less specific. In general, a rennet with higher chymosin to pepsin content is regarded as superior, but the ideal composition of the rennet will depend very much on the cheese type and process.

Chymosin is produced as a prepro-chymosin molecule [11]. The pre-peptide targets the enzyme to the extracellular space and is removed in this process. The pro-sequence keeps the protease inactive and can be removed in two steps by an autocatalytic process. First, at low pH, part of the pro-peptide is removed, forming a pseudo-chymosin [12]. Second, the pseudo-chymosin is split into the mature enzyme at higher pH. This maturation can also occur directly in one step at ~pH 5, but is much faster via the two-step process. Chymosin can be present in two isoforms in calf rennet, which are called chymosin A and B, depending on the genetic makeup of the calf from which it was extracted. Both forms may have some ragging at the amino terminal end [13]. This indicates that the splitting site between the pro-part and the active chymosin can vary to some extent. Although both chymosin forms vary only by one amino acid in their primary amino acid sequence (Asp244Gly), the biochemical properties of the two chymosin isoforms are not exactly the same. Chymosin A is unstable and it will easily degrade, while chymosin B, although more stable than chymosin A, has considerably less enzymatic activity than chymosin A.

There has been some confusion about a third form of chymosin, chymosin C. This form was first thought to be a degradation product of chymosin A, but later it was isolated, characterized, and identified as a genuine isoform [14]. Apparently, chymosin C does not occur very often in calf rennet.

Nowadays, traditional calf rennet is still frequently used in cheese making, but its use is mainly limited to traditional cheese making countries in Europe. This is because many traditional European cheeses require the use of animal rennet to carry a specific brand name or a geographical indication. Therefore, in traditional cheese producing countries such as the Netherlands, France, and Italy, calf rennet is still the main coagulant used for cheese production. Calf rennet is available in different strengths and both as liquid and powder, from many different suppliers.

Rennets extracted from other mammals, such as goat, sheep, or camel, are also commercially available, but they are mainly used in the production of traditional cheeses made from the milk of these species. Calf rennet is often not very active on the κ-casein present in these milks, explaining the preferred use of these alternative mammalian sources.

Some traditional Iberian sheep cheeses such as Serra da Estrela or Serpa in Portugal and Manchego and La Serena in Spain are made with plant-based rennets, extracted from, e.g. thistle. The use of these rennets gives a distinct flavor development that characterizes these cheeses. Interestingly, recently a cardoon rennet produced by recombinant DNA technology has been on offer commercially by the Canadian firm Fytozimus Biotech under the trade name Cyprozime™. This development will allow the production of a vegetable rennet product of constant quality and availability, and may lead to a more widespread use of vegetable rennet in the dairy industry.

2.3.2.2 Microbial Rennets

With the growing worldwide demand for cheese in the 1950s and 1960s of the previous century, a shortage of calf rennet occurred leading to a sharp increase in calf stomach prices. This price effect led to the search for alternative sources of rennet enzymes. An extensive search for microbial sources of aspartic proteases that could potentially replace calf rennet in the cheese making process led to the selection of enzymes of specific fungi that showed the right properties. Although in these early days more sources of fungal aspartic proteases had been commercialized, and until today many alternative sources have been suggested in the academic literature [5], the only microbial rennets that are currently available commercially for the cheese maker are produced with either *Rhizomucor miehei* or *Cryphonectria parasitica*.

The rennet of *C. parasitica* is for sale by a limited number of suppliers and is also only used in specific cheeses. In the scientific literature the enzyme is called endothiapepsin (EC 3.4.23.22). The enzyme is relatively proteolytic and thermolabile and therefore it is mainly used in cheeses where its activity can be inactivated during the cooking step at high temperature that is part of the process of cheeses such as Emmentaler and Gruyere. The enzyme is available as Thermolase® from Chr. Hansen and Suparen® from DSM Food Specialties. The specialty is that this enzyme has a different specificity on κ-casein, and cleaves between Ser104 and Phe105, compared to other rennets that cleave after Phe105.

By far the most widely used microbial rennet is produced with *R. miehei* (previously *Mucor miehei*). In the past, also rennet from *Rhizomucor pusillus* has been marketed but this has now been largely replaced by the *R. miehei* variant. In the scientific literature, the aspartic protease responsible for curd formation is called mucorpepsin (EC 3.4.23.23; [15]). Different variants, grades, and formulations of this enzyme are available from numerous different suppliers, e.g. Fromase® from DSM Food Specialties, Hannilase® from Chr. Hansen, Marzyme® from Dupont, or Milase® from CSK, but many other suppliers are available to the cheese maker.

Microbial rennet is produced by industrial fermentation on sugar and other non-animal derived ingredients, and can therefore be used for the production of vegetarian cheese. This property gives it an advantage in the market. Besides this, most *R. miehei* based rennets are also kosher and halal, allowing it to be used for the production of such cheeses. The fermentative production of microbial rennet also allows a stable supply, independent of the calf stomach price, and a constant quality. Furthermore, the production of microbial rennet is more sustainable than calf rennet, due to the plant origin of fermentation ingredients, which excludes any possibilities of carrying over of animal pathogens and viruses. Mucorpepsin is also very versatile and can be used for most cheese types.

While nowadays award-winning cheeses have been produced using microbial rennet, and a substantial amount of the total world cheese production is based on microbial rennet, in the older literature two problems were described when using these enzymes for cheese production:

1. First of all, the original enzyme produced by *R. miehei* (and especially the one from *R. pusillus*) is thermostable. A large proportion of the rennet ends up in the whey after whey/curd separation, and can survive the subsequent pasteurization step. Currently, for many cheese makers, whey quality is at least as important as the cheese itself and therefore a contamination with active rennet must be avoided.
2. The original microbial enzymes were not very pure, and unwanted side activities could lead to off-flavor formation during ripening. This is why in older literature microbial rennets are only recommended for young cheeses.

The first problem with heat-stable microbial rennet has been solved with the development of specific grades of the enzyme that have increased lability by using a chemical oxidation process. As part of the production process the enzyme is treated with an oxidizing agent such as peroxides, peroxyacids, or hypochlorite, which will lead to oxidation of specific methionine residues in the enzyme molecule [16, 17]. As a result, the thermostability is decreased while the activity stays largely intact (Figure 2.3.2). Such grades are termed TL (thermolabile) or XL (extra labile) and are available from most suppliers. The different treatments can lead to slightly different behavior in cheese making and ripening, besides the thermolabilization. The whey obtained from cheese made with the best variants of such rennets is very clean and no remnants of microbial rennet activity remain.

The second problem that was noticed in the past was mainly due to insufficient purity of the microbial enzymes. Most manufacturers of microbial rennet have improved the product quality by strain improvement and productivity increase, reducing the ratio of contaminant to coagulant, inactivation of side activities by, e.g. pH treatment, and/or purification of mucorpepsin using, e.g. chromatography. The enzyme products obtained this way can have a very high purity with only a single band/peak present when analyzed by gel electrophoresis or size exclusion chromatography. Consequently, the purified grades of microbial rennets are without doubt among the purest classical enzymes available to the food industry. Examples of such high-purity microbial rennet are Fromase® XLG from DSM Food Specialties, Milase® Premium from CSK, or Microlant® Supreme from Chr. Hansen.

Besides their use as coagulants, rennet enzymes also have an activity on caseins in the cheese matrix to generate peptides that can be used by the proteases and peptidases of the lactic acid bacteria present in the cheese matrix, and thereby initiate cheese ripening. In this function, mucorpepsin has a slightly different specificity than chymosin. While the main targets of chymosin in the cheese are αs1-casein and the C-terminus of β-casein, mucorpepsin has a higher preference for β-casein. For this reason, the texture and flavor development during

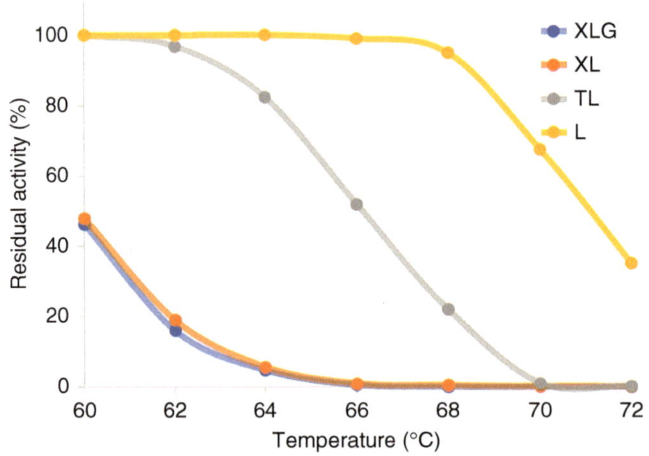

Figure 2.3.2 Residual activity (%) of different Fromase® grades after incubation in whey for 20 seconds at the indicated temperatures (°C). Fromase® L is untreated, while the other grades have received treatment with various different oxidizing agents.

cheese ripening may be slightly different, also depending on the cheese starter culture, the cheese recipe, composition of the cheese, and the storage conditions. Excellent cheeses can however be made with microbial rennet. More information about aspartic proteases used in cheese making can be found in recent reviews on this subject [5].

2.3.2.3 Fermentation Produced Chymosin

With the development of recombinant DNA technology in the 1980s of the previous century, soon the idea appeared to clone and overexpress the bovine chymosin gene in a microorganism that could be cultivated under food-grade conditions. The idea was that using this technology a reliable source of chymosin could be obtained with exactly the same properties as classical calf rennet, but with the additional benefit of not being animal derived, so having a reliable and stable supply, quality, and pricing [4]. The early experiments on overexpression of chymosin in *Escherichia coli* and *Saccharomyces cerevisiae*, reported in 1983, can be regarded as hallmarks of modern biotechnology [18, 19]. Chymosin was actually the first registered food enzyme that was produced on commercial scale using recombinant DNA technology. First trials in this respect were performed by Pfizer, using *E. coli* as production host, but soon Genencor developed a filamentous fungus (*Aspergillus niger*), and Gist-brocades a dairy yeast (*Kluyveromyces lactis*) as host. Although the *E. coli* product is not on the market anymore, the latter two products are still produced to date respectively as Chymax® by Chr. Hansen and Maxiren® by DSM Food Specialties. Development of suitable production strains for chymosin took for both microorganisms a long time, and many rounds of strain improvement were required to obtain a commercially interesting productivity. This indicates that chymosin is a difficult enzyme to produce using recombinant DNA technology. Consequently, to date these two products make up the majority of the world market for FPC.

All commercially available FPCs are based on bovine chymosin B (except for Chymax M – see below). The reason to choose this isoform for the overexpression is the good stability of the final product, an important criterion for industrial scale production of enzymes. FPC has been registered and is being used in many countries in the world, and is widely regarded as the most superior rennet for cheese making. The performance of FPC in the cheese making process is similar to that of classical high-quality calf rennet, so it is often very easy for the cheese maker to switch coagulants. Additional benefit of FPC is its use for vegetarian cheese, and its kosher and halal status. Also, since FPC is made by recombinant DNA technology, the product is more pure, of constant quality, and has stable pricing. That is, FPC does not contain any pepsin activity; it is 100% chymosin B. This fact improves cheese yield and ensures controlled flavor development during cheese ripening [4].

The latest development in FPC is the appearance of variants of the chymosin enzyme, mainly focused on quality improvements for specific cheese types. This new development was initiated by Chr. Hansen with the introduction of Chymax® M in 2008. This FPC is based on the recombinant production of a camel chymosin [20]. Camel chymosin is active on bovine milk and has higher coagulation versus proteolysis (*C*/*P*) activity compared to other coagulants, both due to a higher specific activity on κ-casein (C), and a lower generic proteolytic activity. Especially the lower proteolytic activity of this enzyme is an interesting aspect since it may lead to an increased shelf-life of, mainly, young cheeses. Time for slicing and shredding of, e.g. pizza cheese is limited by the softening of the cheese during storage. This softening is thought to be due to the proteolytic activity of the rennet, and was shown to decrease for low moisture part-skim Mozzarella, when Chymax® M was used for the coagulation instead of regular FPC [21].

This development has led to the introduction of a new bovine FPC variant with similar properties as the camel enzyme; Maxiren® XDS, by DSM Food Specialties. This FPC has an even higher *C*/*P* and also shows increased shelf life of young cheeses. Additionally, cheese produced using this coagulant can be sliced/shredded earlier due to efficient knitting of the cheese matrix during the first few days after production. This property is thought to be due to an efficient first cut in αs1-casein, in contrast to the situation with the camel enzyme, which is slow in this cut. After this cut, the product αs1-1 casein is not further degraded when Maxiren® XDS has been used in the cheese making process. Since the first cut in αs1-casein is important for early texture development in cheese, and further degradation of caseins for later softening, the observed properties of the cheeses are in line with the improved selectivity of the enzyme [22].

These and other recent developments point to a direction that rennets will become more specialized, and different rennets will be offered for different cheese types. Seeing the huge number of different cheese types with different textures and flavors and the vast number of different starter cultures that are available to the cheese maker, it is not more than logical that also rennets will be given a more prominent role. Noticing the important role that rennets play in texture and flavor formation in cheese besides its renneting activity, a trend toward specialization of the enzymes is desirable for cheese makers and is therefore to be expected in the coming years.

2.3.3 Ripening Enzymes

Besides the coagulants, some other enzymes are also available to the cheese maker to steer the texture and flavor formation in cheese. Cheese ripening enzymes may be added during regular cheese making either in the cheese milk or during salting (for e.g. Cheddar), but most of these enzymes are currently used to make cheese flavors via EMC or in the production of processed cheese. When ripening enzymes are added to the cheese milk before curd–whey separation, the remaining enzymes in the whey may lead to flavor defects when the enzymes are not sufficiently inactivated at pasteurization. As already mentioned above, whey quality is an important property for the economic operation of a cheese plant. Therefore, in the selection of ripening enzymes for use in cheese, this aspect should always be given sufficient attention.

The cheese ripening enzymes that are currently on the market can be classified into two groups, being either proteolytic or lipolytic. There are a number of different companies that sell cheese ripening enzymes, and most companies have different enzymes available, being animal, plant, or microbe derived. Only a few cheese ripening enzymes are produced using recombinant DNA technology. Companies that are active in selling proteolytic and lipolytic enzymes for cheese and EMC are, e.g. Biocatalysts, Amano, DSM Food Specialties, and Dupont.

EMC is not cheese but a process flavor. Typically, commercial production of EMC involves the simultaneous hydrolysis of cheese curd by proteolytic and lipolytic enzymes [23]. Different enzymes and blends are available to generate a specific flavor in the EMC. Emulsifying salts are often added in EMC production to solubilize the protein by chelating Ca^{2+}. The solubilized protein is an effective emulsifier of the fat. The most commonly used emulsifying salts are those that bind calcium weakly, giving a good degree of emulsification and producing a soft, easily melted cheese slurry. Such an emulsified slurry is a good substrate for both proteases and lipases. In contrast to flavor formation in cheese, starter cultures are not often added in EMC production, but may be present from the cheese curd that is used to make the EMC slurry.

2.3.3.1 Proteases/Peptidases

The use of proteases and peptidases for the production of dairy protein hydrolysates for infant, clinical or sports nutrition is important to reduce possible allergenic reactions and improve the digestibility of dairy protein. This is an important application of these enzymes and has been discussed in Chapter 2.2. The focus here will be on their effect on cheese ripening.

For this review proteases are defined as endo-acting hydrolases that split the peptide bonds in proteins. Peptidases are defined as exo-acting hydrolases that release amino acids from either the amino terminal (aminopeptidases) or the carboxy terminal (carboxypeptidases) end of proteins. Therefore, proteases will split up the substrate and generate larger peptides and few amino acids, thereby leading to changes in the structure of the protein substrate. Peptidases will yield free amino acids while there is only little change in the structure of the protein substrate. The generated peptides and amino acids can be used by the cheese cultures to generate flavor effects. Obviously, the type of flavor that is generated in cheese will depend both on the specific amino acids that are released and the ability of the cheese culture to use these amino acids for flavor formation. Therefore, the effect of added proteases and peptidases can never be seen separately from the activity of the culture and a composition that has been optimized to maximize the potential in cheese ripening [24].

From this argumentation it also becomes clear that pure exopeptidases will be more suitable for flavor generation during cheese ripening, where no textural defects are allowed, while endoproteases will be more suitable for the EMC application that requires a high degree of hydrolysis, to maximize flavor release. In EMC, exopeptidases may also be added to convert peptides into free amino acids, thereby reducing bitterness. Therefore, a mixture of proteases and peptidases with broad specificity is generally desirable in the production of EMC [23].

Both endoproteases and exopeptidases and combinations thereof are commercially available but, as indicated above, their use should be fine-tuned in each application. These enzymes are available under the brand names Accelase®, Accelerzyme®, Promod®, Flavorpro®, Flavorzyme®, and others, from different suppliers.

2.3.3.2 Lipases/Esterases

Lipases appear to have a bigger effect in flavor production in EMC than in natural cheese, which can be related directly to differences in the emulsified status of the lipids in the EMC substrate. Consequently, also lipases as cheese ripening enzymes are mainly used in the production of EMC. The original application of lipases in generating cheese flavor comes from the use of rennet paste in Italian cheeses such as Romano and Provolone. Rennet pastes were extracted from the stomach including pregastric tissue (esophagus and tongue) of calves, kid goats, or lambs [25]. The cheeses made with this rennet paste had a specific "piccante" taste typical for these Italian cheeses. Nowadays, the pregastric esterase extract of all three animals is produced separately from the rennet and can therefore be better dosed for optimal flavor formation.

Pregastric esterases exhibit a unique specificity for the release of small fatty acids of 4–10 carbon chain length and it hydrolyzes specifically the triglycerides at the sn-1,3 position of the milk fat, which gives the distinct "piccante" flavor [25]. Hardly any long-chain fatty acids, which often give a soapy taste, are released by the enzyme during ripening. All three enzyme forms from calve, kid goat, and lamb are commercially available from different companies, some even in kosher variants, and give small differences in the final flavor palate in the cheese.

Similar to the situation with animal rennet, microbial variants of pregastric lipase have been sought to have a stable supply of pure, kosher, halal, and vegetarian lipases for cheese flavor development. It has been extremely difficult to find microbial lipases that can generate the same flavor profile as the animal pregastric esterases. Most microbial lipases release more long-chain fatty acids from milk fat, and the flavor profile becomes therefore more reminiscent of fungal ripened cheeses. One microbial lipase that comes most close to the specificity of pregastric esterase is the lipase of *R. miehei*, the same organism that provided us with microbial rennet (see above). This microbial lipase is available as Piccantase® A from DSM Food Specialties, or SpiceIt® MR from Chr. Hansen.

An interesting development in this respect is the availability of Piccantase® KLC. This enzyme is the calf derived pregastric esterase produced by recombinant DNA technology with the dairy yeast *K. lactis* [26]. Obviously, the fatty acid profile in cheese and the developed flavor produced by this enzyme resemble that of the animal pregastric esterases. Differences in flavor may still be due to the higher purity of this microbial esterase compared to the animal variants.

2.3.4 Lactases

Besides protein and lipid, the third major metabolite in milk is lactose. Lactose is synonymous to β-D-galactopyranosyl-(1,4)-D-glucose and is a disaccharide unique to mammalian milk. Special β-galactosidases are capable of hydrolyzing lactose into glucose and galactose, and these lactases are nowadays frequently used in the dairy industry. Main application of lactases is the production of lactose-reduced or lactose-free dairy products. A substantial part of the world population is not able to digest, or has difficulty in digesting, lactose in the small intestine. This is due to the reduction of the expression of the human lactose/phlorizin hydrolase, since the enzyme is not required anymore after weaning. Although this is a natural condition that humans share with all other mammals, the condition is often called lactose intolerance. When lactose-containing dairy food is eaten by a lactose intolerant person, lactose is digested by the bacteria in the large intestine, which may lead to complaints such as bloating and diarrhea (Figure 2.3.3), depending on the severity of the condition and the amount of lactose ingested. See recent reviews for an elaborate discussion and references on lactose intolerance [27, 28].

In tribes that traditionally live with cattle it became an evolutionary advantage to be able to digest lactose and thereby obtain the ability to make use of the nutrients that came with the milk obtained from their cattle. In several parts of the world different mutations in the lactose/phlorizin hydrolase gene have been spread that lead to the preservation of the expression of the gene in

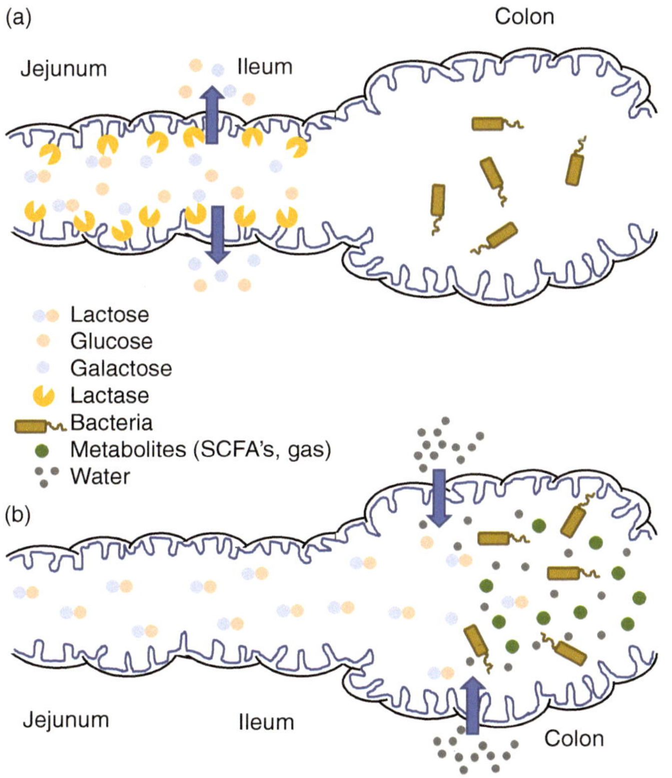

Figure 2.3.3 Lactose digestion at lactase sufficiency (a) or lactase deficiency (b). Source: Corgneau et al. [27]. Adapted with permission of Taylor & Francis.

adulthood [9]. Main areas with a lactose-tolerant population are Northern Europe, and descendants in Northern America and Australia, parts of Northern India, Pakistan, and Arabia, and selected tribes in Africa such as Tutsi, Tuareg, Fulani, and Maasai. However, lactose intolerance in adulthood must be regarded as the natural condition in most humans and other mammals.

2.3.4.1 Neutral Lactase

Neutral lactases are one of the most important enzymes used in the dairy industry nowadays, and their use is increasing. The main application of these enzymes is for the production of lactose-reduced and lactose-free milk for people with lactose intolerance. In most milk products the lactose content is reduced to either <0.5% or <0.1%, although the general trend is to reduce the lactose content even further to <0.01%. This low amount of lactose gives some debate on the suitability of common analytical methods to measure lactose in milk, stated to be insufficient to make a distinction between lactose and the minor GOSs present in lactase-treated milk [29].

In industrial practice, there are two common ways to produce lactose-free milk [30]. In the pre-hydrolysis (or batch) method the neutral lactase is added to cold non-sterile milk and incubated for ~24 hours before pasteurization or ultra-heat treatment (UHT) and packaging. Pre-hydrolysis requires a relatively large amount of enzyme, but no residual lactase can be found in the final milk anymore due to the heat inactivation. In the post-hydrolysis method the enzyme is added after milk sterilization, just before packaging. Obviously, with this method, the enzyme has to be sterile when added to the milk, and this can be achieved by either buying a pre-sterilized enzyme, or filter-sterilize the enzyme in the factory just before addition to the milk line. For both methods specialized equipment and consumables are required and these are available from, e.g. Tetrapak. Hydrolysis of lactose will take place during the quarantine period that is required for UHT milk, and the enzyme is not heat inactivated anymore.

Besides making milk suitable for drinking by lactose-intolerant people there are several other applications in the dairy industry that benefit from the hydrolysis of lactose. Since the reaction products glucose and galactose are sweeter than lactose, neutral lactases are also used to reduce the addition of sugar to dairy products such as yoghurts and other desserts. Additionally, hydrolysis of lactose may prevent lactose crystallization (sandiness) in products such as ice cream and sweetened concentrated milk. In Dulce de Leche, lactase treatment increases the production of caramel flavor by enhancing the Maillard reaction due to the higher monosaccharide content. Dairy products ranging from cheeses to crème fraiche, and many others, are nowadays also offered as lactose-free versions.

Most commonly used neutral lactase in the dairy industry is the β-galactosidase from *K. lactis* [31]. This enzyme can be obtained from many different suppliers, but there are actually only a few producers: Godo Shusei Company, Amano Enzymes, and Nagase, all producing the enzyme in Japan, and DSM Food Specialties with production sites in Europe. The Japanese producers use resellers to market their enzymes and these can be obtained under the brand names Lactozyme®, Godo YNL-2®, Halactase®, Prozyn®, and many others. DSM

Food Specialties mainly sell their Maxilact® brand themselves, which has been available to the dairy industry already for 50 years. The main differences between the different enzymes are their formulation, with different composition and strength, and the purity, with different side activities present or absent.

A continuous point of attention for lactases is the possible presence of enzymatic side activities. While in the past the main focus of the producers has been on the purification and the removal of proteases from the enzyme preparations, which could lead to off-flavor formation and textural changes in milk during processing or storage [32], this problem has largely been solved and recent attention has been on the presence of other side activities. Two developments that have improved the quality and shelf-life of low lactose dairy products are discussed here:

- The first example for this is the removal of arylsulfatase activity from lactase preparations. It was found that arylsulfatase can convert sulfonated-cresol, which is naturally present in many types of milk, into *para*-cresol. This in turn can lead to medicinal or animalic off-flavor during storage of especially UHT milk [3]. The first arylsulfatase-free lactase was developed almost 10 years ago and has since been on the market under the brand name Maxilact® LGX, or LAGX, with great success. Major improvements in flavor stability during storage of UHT milk have been found by many producers of lactose-free UHT milk.
- A second improvement has been the removal of invertase from lactase preparations. Invertase converts sucrose into glucose and fructose, and as such may impact the flavor and texture of sugared dairy products. With the increased use of lactase as possible solution for sugar reduction in, e.g. yoghurt, also the presence of invertase in lactase preparations became a problem for the industry. Currently, an invertase-free lactase is available under the brand name Maxilact® LGi.

Some of the current commercial lactases are of such high quality that hardly, or no problems occur anymore due to side activities although this is not true for all lactases on the market [33, 34]. At present, the main shelf life limitation for lactose-free UHT milk seems to be due to the development of Maillard products during storage at ambient temperature. By splitting lactose into glucose and galactose the reactivity of lactose-free milk in the Maillard reaction is much higher than that of lactose in regular milk. This may lead to increased browning and off-flavor development during storage, especially if the storage temperature is high. Consequently, the shelf life of lactose-free UHT milk is only three months or even less, depending on the storage conditions, while regular UHT milk may have a shelf life of —six to nine months.

A recent development in the field of neutral lactases is the production of lactases using recombinant DNA technology. The main advantages of this technology are obviously the possibility to supply new enzymes at commercial scale by expressing the gene for a preferred lactase in another organism, which can often be done at high efficiency leading to relatively pure and cost-effective production. The disadvantage of this technology is the reluctance from customers, regulatory authorities, and consumers in some countries to use these enzymes. A nice new example of a *Bifidobacterium* lactase produced in the bacterial enzyme expression host *Bacillus licheniformis* is the product Saphera® of Novozymes. This product can now also be obtained from other suppliers under different names. The *Bifidobacterium* lactase has some different properties than the *Kluyveromyces* enzyme discussed above, such as lower pH optimum and higher temperature stability. Therefore, the application of this enzyme in different applications should be fine-tuned to obtain the best results.

This new development has led to the introduction of another novel lactase. DSM Food Specialties has very recently launched a variant *Kluyveromyces* lactase, Maxilact® Smart, with improved properties such as higher specific activity allowing a dosage reduction in most applications. It will be interesting to see how the recombinant and self-cloned lactases will be adopted by the dairy producers in the future, and which new developments will still be in stock.

2.3.4.2 Acid Lactase

Besides neutral lactase, also β-galactosidases with an acid pH optimum have been commercially available for a long time. Several companies supply these enzymes and there are both classical and self-cloned versions available. Acid lactases are produced by filamentous fungi and the most common enzyme that is now commercially available is the lactase of *Aspergillus oryzae*. Despite their availability, acid lactases have found little application in the dairy industry. One of its applications is in hydrolysis of lactose in acid whey, e.g. from draining the curd in Greek yoghurt production, in order to make it more sweet and palatable. Acid lactase is also sometimes used to remove residual lactose in acidified dairy products such as quark or cream cheese. Another use in dairy is the production of GOSs using this enzyme (see below).

By far the most common use of acid lactases is however as nutritional supplement. The enzyme is sold as small pills to consumers, which are taken together or before a lactose containing meal. The enzyme is stable and active in the stomach, will split the lactose that is ingested, and thereby prevent gastrointestinal problems. This is a convenient fail-safe solution for people with lactose intolerance, e.g. when dining out. The European Food Safety Authority (EFSA) gave a positive opinion on the health benefit claim in 2009, which was included in the EU register as "Lactase enzyme improves lactose digestion in individuals who have difficulty digesting lactose" [28]. This nutritional use of enzymes is outside the scope of this chapter on enzymes used in the dairy industry and will be discussed in Chapter 3.1.

2.3.4.3 GOS Production

A popular application of lactases is their use in the production of GOSs and other lactose-derived prebiotics [35]. The use of GOS in especially infant nutrition has increased dramatically in recent years, and therefore a lot of research and attention has been given to this in public and patent literature. GOS is thought to have beneficial effects on the development of a healthy intestinal flora, especially having a Bifidogenic stimulation, analogous to what is thought to be the effect of human milk oligosaccharides (HMOs). Therefore, GOSs have been added to infant formula together with other so-called prebiotics [36].

GOS is a polymer of galactose residues added to the non-reducing end of a lactose molecule and can come in different forms. There can be a difference in length of the polymer measured in DP (degree of polymerization), indicating the number of monosaccharides that make up the chain. Enzymatic polymerization normally leads to relatively short chains having a DP between 2 and 8, with most GOS products having a DP of 3–4. DP depends both on enzyme and process conditions, and can therefore be modified if required, e.g. by fractionation of the product.

Another aspect that can differ from one GOS preparation to another is the bond formation. While lactose contains a β1-4 bond, in GOS also β1-2, β1-3, and β1-6 bonds can be found in different positions in the molecule. The bond formation is very dependent on the type of enzyme that is used in the production, and a different enzyme will normally lead to another composition of the GOS product [37].

Taking both DP and bond formation into account it must be clear that none of the GOS products that are commercially available are pure chemical substances. They are rather blends of many different molecules that may all have a specific role in the stimulation of the intestinal flora, or other beneficial effects.

Many companies have developed technology to produce GOSs at industrial scale with different enzymes, processes, and end product characteristics. The production processes have in common that they start with a solution of lactose at a high concentration to which an enzyme is added, either as free enzyme in batch or sometimes used in immobilized forms. High lactose concentration is essential for the efficient production of GOS: the lactose competes with water as acceptor for the galactose that is bound to the enzyme after splitting of the lactose (see Figure 2.3.4). Surplus of water will lead to both hydrolysis of lactose and hydrolysis of the just produced GOS, reducing the final yield. Preferably the

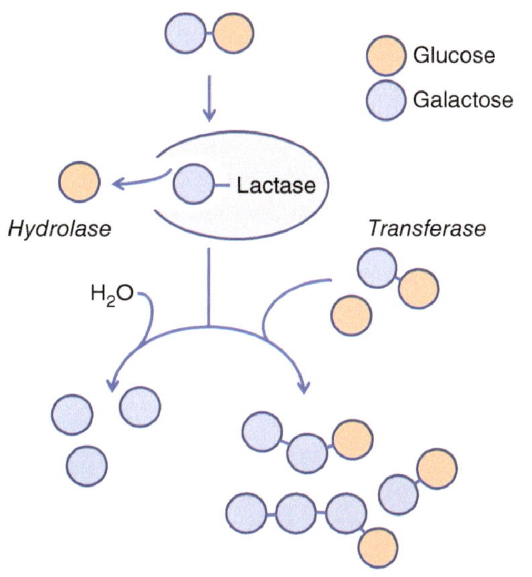

Figure 2.3.4 Lactase binds to the galactose moiety of lactose while expelling glucose. The enzyme-bound galactose can either be released by hydrolysis with water (left part) or by a carbohydrate (right part). In the transferase reaction GOS will be formed in different size and bond formation, depending on the acceptor and the enzyme that is used.

process is performed at high temperature, since the solubility of lactose is higher, leading to a higher yield and also preventing microbial contamination. For the same reason, the conditions are preferably acidic.

These constraints to the process conditions will also affect the choice of a suitable enzyme. For example, the lactase of *K. lactis*, which is most commonly used for the production of lactose-free dairy products, is not very suitable for making GOS since it is inactivated at temperatures >45 °C and a pH lower than 5.5. Therefore, several other lactases with more preferred properties have been developed for the production of GOS. Some examples (but not comprehensive) of GOS products are as follows:

Vivinal® GOS by Royal FrieslandCampina is produced using Biolacta, a lactase isolated from *Bacillus circulans*. This is one of the most popular GOS products available, supported with a lot of nutritional data [36]. The Biolacta enzyme is obtained from Amano, and interestingly, the form of the enzyme that is responsible for the GOS formation is a degradation product (or isoform) of the full-length enzyme [38]. This property has also been found in lactases from other bacteria, where the smaller forms showed more synthesis activity and the larger forms of the enzyme were more hydrolytic [39]. Main GOS forms present in Vivinal® GOS have a DP of 2–5 with most glycosidic bonds in a β1-4 configuration.

The company Clasado uses a lactase from *Bifidobacterium bifidum* for the production of their GOS product Bimuno®. Consequently, the composition of the Bimuno GOS is different from Vivinal GOS. Bimuno® GOS has a relatively high DP of 1–3, possibly since the monosaccharides and lactose are not completely separated from the GOS produced [40]. The main bonds in Bimuno® GOS are β1-6 and β1-3, making it distinct from Vivinal.

Another interesting GOS product is Oligomate® of Yakult that is produced using the lactase of *Sporobolomyces singularis* [41]. The ability of this organism to produce GOS has been noticed a long time ago and a peculiarity is that the enzyme applied in this process seems to have very little hydrolytic activity. The DP of the GOS produced by this enzyme is rather small, and the main species is DP3 with few larger oligosaccharides. The main bond formed is β1-4, but also other GOS forms such as β1-3 and β1-2 are found.

Many other GOS products are currently on the market, and some of these are produced using the acid lactase of *A. oryzae*, described above. This enzyme mainly forms GOS with a DP of 2–6 and β1-6 bonds. *K. lactis* lactase is commercially only used to remove residual lactose after GOS synthesis. One has to be careful with this approach since this lactase can also digest many different GOS species, thereby reducing the yield.

The use of β-galactosidases for GOS production is an active researched field, so new developments are expected to take place in the future, including the development of new enzymes. The holy grail in the infant nutrition industry will however be the addition of HMOs, and it is at present unclear if enzyme technology will be the prevailing technology for the production of this broad class of oligosaccharides.

2.3.5 Miscellaneous Enzymes

Besides the enzymes discussed above, several other enzymes are available for miscellaneous functions in the dairy industry. Most of these enzymes are targeted to a specific market segment, and are not widely used in the industry. However, since they are available and used at commercial scale it is interesting to mention them. By no means this chapter intends to be a complete representation of all commercial enzymes used in the dairy industry; it just gives a compilation of the interesting variety of enzymes and possible application areas one can think of in developing new enzymes.

2.3.5.1 Oxidases/Peroxidases

A number of different oxidases and peroxidases are commercially available. Obviously, the action of these enzymes is limited to applications where either oxygen or peroxide is present or added/generated. Because of this, the application possibilities are limited, but nevertheless the introduction of these enzymes for a number of creative application possibilities has been successful.

Glucose oxidase (EC 1.1.3.4) converts glucose and oxygen into hydrogen peroxide and glucono-δ-lactone, which is chemically transformed into gluconic acid [42]. The enzyme finds its use in the preservation of packaged shredded cheese. The enzyme reduces the oxygen in the package and forms some peroxide, both of which have a preservative effect. The enzyme is produced by *A. niger*, and has been available from DSM Food Specialties under the brand name Delvozyme GO.

Hexose oxidase (EC 1.1.3.5) has a similar effect as glucose oxidase, but has broader substrate specificity [42]. This enzyme is mainly marketed for the reduction of browning of pizza cheese during baking. Pasta filata type of cheeses still contain a significant amount of carbohydrates, and especially galactose leads to burning of cheese blisters during baking due to the Maillard reaction. Conversion of galactose in galacturonic acid can reduce such burning. The enzyme originates from the red alga *Chondrus crispus*, produced with *Hansenula polymorpha*, and available from Dupont.

Lactose oxidase is again similar to the oxidases mentioned above but has a preference for lactose, thereby producing lactobionic acid. LactoYIELD® is a lactose oxidase provided by Chr. Hansen and the enzyme originates from *Microdochium nivale* and expressed in *Fusarium venenatum*. The main application that has been mentioned for the dairy industry is the separate production of lactobionic acid from whey using the enzyme, which can then be added to cheese curd in order to increase cheese yield.

Peroxidase has been in use in the dairy industry for the bleaching of whey. Especially Cheddar whey is strongly colored due to the presence of annatto, and this reduces the whey value. Maxibright®, developed by DSM Food Specialties is a peroxidase from *Marasmius scorodonius* and expressed in *A. niger* and can be used for bleaching whey with favorable properties and flavor while avoiding benzoate residue, which can be a problem for regular benzoylperoxide bleaching [43]. The idea is that the enzyme splits the norbixin in annatto to a colorless product, with only a small amount of peroxide required.

2.3.5.2 Phopholipases

Phospholipids are part of the milk fat globular membrane (MFGM) and play an important role in stabilizing the fat globules. Some dairy products, such as butter milk powder, are enriched in phospholipids. Phospholipases may either remove one of the acyl chains (PL-A1 or PL-A2 dependent on which acyl-chain is split), or the polar side group including (PL-C) or excluding the phosphate group (PL-D). Obviously different phospholipases will generate different products that can have a different application effect as described below.

Phospholipase A1 (EC 3.1.1.32) is being sold for the dairy industry by Chr. Hansen under the trade name Yieldmax®. The enzyme is mentioned to increase cheese yield in pasta filata and pizza cheese achieved through increased fat retention at whey/curd separation [44]. Also, the enzyme is mentioned to have an effect on moisture retention and flavor/texture development. Phospholipase A1 splits the acyl chain at position sn-1 of phospholipids, thereby generating a lysophospholipid with increased emulsification properties. Although the claim for increased yield can be debated, since most cheese manufacturers will isolate the fat from the whey and re-use it in their process, this enzyme seems to have found its place in the dairy industry.

Acyltransferase has been developed by Dupont under the name FoodPro® Cleanline. This is a unique enzyme that transfers an acyl chain from a phospholipid to cholesterol. The final products are a lysophospholipid and a cholesterol ester. In literature the enzyme is also known as glycerophospholipid-cholesterol acyltransferase (GCAT – EC 2.3.1.43). It is mentioned that the enzyme can delay the fouling of the heat exchanger during heat treatment in the production of UHT milk, when the milk is pretreated with the enzyme [45]. The idea is that the formed lysophospholipids can bind more efficiently to denatured protein, thereby preventing it from aggregating with other protein and prevent binding to the surface of the UHT equipment.

2.3.5.3 Cross-linking Enzymes

Protein cross-linking would be a useful property in the dairy industry to provide texture and structure without using additives. Many different enzymes have been claimed to work in the cross-linking of proteins. Many of these are oxidases or peroxidases that form oxygen radicals that are responsible for this effect. In the dairy matrix such oxidative enzymes are often not useful since the presence of oxygen radicals can also lead to off-flavor formation and other defects.

Transglutaminase (TG – EC 2.3.2.13) has been commercially available for a long time and has found its main use in the meat industry to cross-link separator meat. The enzyme cross-links glutamate and lysine side chains in proteins without the requirement for oxygen. Although there are many publications on the use of TG for food, the commercial use in dairy has only been recent and was delayed due to the selectivity of the enzyme for casein, the presence of a TG inhibitor in milk, and especially the price of the enzyme. Some manufacturers have solved these issues and first yoghurt and cheese products applying TG have found their way to the market. A microbial *Streptoverticillium* TG is produced by Ajinomoto and sold to the dairy industry under the brand name Activa® YG. More information can be found in recent reviews [46].

2.3.5.4 Preservation

Enzymes can also be used to extend the shelf life of dairy products and/or prevent microbial growth. Although the number of enzymes used for this purpose is limited, the use of glucose oxidase for shredded cheese preservation is a good example (see above).

Lysozyme (EC 3.2.1.17), most commonly isolated from hen egg white, has been offered for a long time for the prevention of late blowing in cheese production caused by butyric fermentation of germinated *Clostridium tyrobutyricum* spores. However, the commercial use has always been limited since a cost-effective solution against these cheese ripening defects was the addition of nitrate salts in the cheese. Prohibition in many countries of the use of this preservation method may open up new possibilities for the use of this enzyme in cheese production.

2.3.6 New Developments

Many new enzymes have been described in public and patent literature for having an effect in various dairy applications, but have not reached commercialization yet. Interesting data for use in dairy applications can be found for the effect of PNGase, phospholipase C, protein-glutaminase, sialidase, laccase, lactoperoxidase, sulfhydryl oxidase, lysyl oxidase, and numerous other enzymes with a broad range of specific dairy application benefits [47–53]. With the current trend in the food industry on "natural," "sustainable," and "clean-label," we will undoubtedly see more new enzymes being commercialized and used for the development of exciting new dairy products.

References

1 Gerosa, S. and Skoet, J. (2012). Milk availability: trends in production and demand and medium-term outlook. ESA Working paper No. 12-01. FAO/UN.

2 Jelen, P. (2011). Standardization of fat and protein content. In: *Encyclopedia of Dairy Sciences*, 2e (ed. J.W. Fuquay, P.F. Fox and P.L.H. McSweeney), 545–549. Academic Press.

3 De Swaaf, M.P.M., van Dijk, A.A., Edens, L., and Dekker, P.J.T. (2006). Enzyme preparation yielding a clean taste. EP1954808 B1.

4 Harboe, M., Broe, M.L., and Qvist, K.B. (2010). The production, action and application of rennet and coagulants. In: *Technology of Cheesemaking*, Chapter 3 (ed. B.A. Law and A.Y. Tamime). Blackwell Publishing Ltd.

5 Yegin, S. and Dekker, P. (2013). Progress in the field of aspartic proteinases in cheese manufacturing: structures, functions, catalytic mechanism, inhibition and engineering. *Dairy Sci. Technol.* 93: 565–594.

6 Walstra, P., Geurts, T.J., Noomen, A. et al. (1999). Principles of cheese making. In: *Dairy Technology: Principles of Milk Properties and Processes*, Chapter 21, 541–553. Marcel Dekker Inc.

7 Technical Committee ISO/TC 34/SC5 (2007). Milk-determination of total milk-clotting activity of bovine rennets. ISO 11815|IDF 157.

8 The Nibble (2005). The history of cheese: cheese history from an ancient Nomad's Horseback to today's luxury cheese cart. thenibble.com/REVIEWS/main/cheese/cheese2/history.asp (accessed 07 March 2019).

9 Curry, A. (2013). The milk revolution. *Nature* 500: 20–22.

10 Andrén, A. (1992). Production of prochymosin, pepsinogen and progastricsin, and their cellular and intracellular localisation in bovine abomasal mucosa. *Scand. J. Clin. Lab. Invest.* 52: 59–64.

11 Foltmann, B. (1999). General and molecular aspects of rennets. In: *Cheese: Chemistry, Physics and Microbiology* (ed. P.F. Fox). Gaithersburg: Aspen Publication Inc.

12 Pedersen, V.B., Christensen, K.A., and Foltmann, B. (1979). Investigations on the activation of bovine prochymosin. *Eur. J. Biochem.* 94: 573–580.

13 Lilla, S., Caira, S., Ferranti, P., and Addeo, F. (2001). Mass spectrometric characterisation of proteins in rennet and in chymosin-based milk-clotting preparations. *Rapid Commun. Mass Spectrom.* 15: 1101–1112.

14 Rampilli, M., Larsen, R., and Harboe, M. (2005). Natural heterogeneity of chymosin and pepsin in extracts of bovine stomachs. *Int. Dairy J.* 15: 1130–1137.

15 Yegin, S., Fernandez-Lahore, M., Salgado, A.J.G. et al. (2011). Aspartic proteinases from *Mucor spp.* in cheese manufacturing. *Appl. Microbiol. Biotechnol.* 89: 949–960.

16 Colein, B.A.V. and Delecourt, R.L.F. (1981). Thermally sensitive microbial rennet, a process for its preparation and its application for cheese preparation. EP0027834 (A1).

17 Cornelius, D. (1982). Process for decreasing the thermal stability of microbial rennet. US4348482 (A).

18 Emtage, J.S., Angal, S., Doel, M.T. et al. (1983). Synthesis of calf prochymosin (prorennin) in *Escherichia coli*. *Proc. Natl. Acad. Sci. U.S.A.* 80: 3671–3675.

19 Mellor, J., Dobson, M.J., Roberts, N.A. et al. (1983). Efficient synthesis of enzymatically active calf chymosin in *Saccharomyces cerevisiae*. *Gene* 24: 1–14.

20 Kappeler, S.R., van den Brink, J.M., Rahbek-Nielsen, H. et al. (2006). Characterization of recombinant camel chymosin reveals superior properties for the coagulation of bovine and camel milk. *Biochem. Biophys. Res. Commun.* 342: 647–654.

21 Moynihan, A.C., Govindasamy-Lucey, S., Jaeggi, J.J. et al. (2014). Effect of camel chymosin on the texture, functionality, and sensory properties of low-moisture, part-skim Mozzarella cheese. *J. Dairy Sci.* 97: 85–96.

22 Dekker, P.J.T., de Jong, R.M., Tabeling, M.D., and Muijlwijk, C.M. (2013). Improved enzyme variants. WO2013164481 (A1).

23 Kilcawley, K.N., Wilkinson, M.G., and Fox, P.F. (1998). Enzyme-modified cheese. *Int. Dairy J.* 8: 1–10.

24 Kilcawley, K.N., Nongonierma, A.B., Hannon, J.A. et al. (2012). Evaluation of commercial enzyme systems to accelerate Cheddar cheese ripening. *Int. Dairy J.* 26: 50–57.

25 Nelson, J.H., Jensen, R.G., and Pitas, R.E. (1977). Pregastric esterase and other oral lipases—a review. *J. Dairy Sci.* 60: 327–362.

26 Parenicova, L., Busink, R., and van der Laan, J.M. (2010). Pregastric esterase and derivatives thereof. EP2406372 (B1).

27 Corgneau, M., Scher, J., Ritie-Pertusa, L. et al. (2017). Recent advances on lactose intolerance: tolerance thresholds and currently available answers. *Crit. Rev. Food Sci. Nutr.* 57: 3344–3356.

28 Kies, A.K. (2014). Authorised EU health claims related to the management of lactose intolerance: reduced lactose content, dietary lactase supplements and live yoghurt cultures. In: *Foods, Nutrients and Food Ingredients with Authorised EU Health Claims*, Chapter 9 (ed. M.J. Sadler), 177–211. Woodhead Publishing.

29 van Scheppingen, W.B., van Hilten, P.H., Vijverberg, M.P., and Duchateau, A.L.L. (2017). Selective and sensitive determination of lactose in low-lactose dairy products with HPAEC-PAD. *J. Chromatogr. B* 1060: 395–399.

30 Harju, M. (2004). Chromatographic and enzymatic removal of lactose from milk. *Bull. IDF* 389: 4–8.

31 Dekker, P.J.T. (2015). Enzymes exogenous to milk in dairy technology: β-D-galactosidase. In: *Reference Module in Food Sciences* (ed. G.W. Smithers, L. Day, P. Ferranti, et al.), 1–8. Elsevier.

32 Mittal, S.B., Newell, G., Hourigan, J.A., and Zadow, J.G. (1991). The effect of protease contamination in lactase on the flavour of lactose-hydrolyzed milks. *Aust. J. Dairy Technol.* 46: 46–47.

33 Jensen, S., Jansson, T., Eggers, N. et al. (2015). Storage-induced changes in the sensory characteristics and volatiles of conventional and lactose-hydrolyzed UHT processed milk. *Eur. Food Res. Technol.* 240: 1247–1257.

34 Troise, A.D., Bandini, E., De Donno, R. et al. (2016). The quality of low lactose milk is affected by the side proteolytic activity of the lactase used in the production process. *Food Res. Int.* 89: 514–525.

35 Illanes, A., Guerrero, C., Vera, C. et al. (2016). *Lactose-Derived Prebiotics – A Process Perspective*. London: Academic Press. ISBN: 978-0-12-802724-0.

36 Van Leusen, E., Torringa, E., Groenink, P. et al. (2014). Industrial applications of galactooligosaccharides. In: *Food Oligosaccharides. Production, Analysis and Bioactivity* (ed. F.J. Moreno and M.L. Sanz). UK. ISBN: : IFT Press, Wiley Blackwell. ISBN: 978-1-118-42649-4.

37 Torres, D.P., Gonçalves, M., Teixeira, J.A., and Rodrigues, L.R. (2010). Galacto-oligosaccharides: production, properties, applications, and significance as prebiotics. *Compr. Rev. Food Sci. Food Saf.* 9: 438–454.

38 Song, J., Abe, K., Imanaka, H. et al. (2011). Causes of the production of multiple forms of β-galactosidase by *Bacillus circulans*. *Biosci. Biotechnol. Biochem.* 75: 268–278.

39 Jørgensen, F., Hansen, O.C., and Stougaard, P. (2001). High-efficiency synthesis of oligosaccharides with a truncated β-galactosidase from *Bifidobacterium bifidum*. *Appl. Microbiol. Biotechnol.* 57: 647–652.

40 Clasado (2013). GRAS Exemption Claim for Galacto-Oligosaccharides. GRAS Notice (GRN) No. 484.

41 Yakult (2010). Generally Recognized as Safe (GRAS) Determination for the Use of Galacto-Oligosaccharides (GOS) in Foods and Infant Formulas. GRAS Notice (GRN) No. 334.

42 McSweeney, P. (2016). Enzymes exogenous to milk in dairy technology: catalase, glucose oxidase, glucose isomerase and hexose oxidase. In: *Reference Module in Food Science* (ed. G.W. Smithers, L. Day, P. Ferranti, et al.). Elsevier.

43 Campbell, R.E. and Drake, M.A. (2014). Enzymatic bleaching in commercial colored Cheddar whey retentates. *Int. Dairy J.* 38: 148–153.

44 Lilbæk, H.M., Broe, M.L., Høier, E. et al. (2006). Improving the yield of Mozzarella cheese by phospholipase treatment of milk. *J. Dairy Sci.* 89: 4114–4125.

45 Larsen, N.E. and Søe, J.B. (2010). Method for producing ultra-heat treatment milk. US2010215803 (A1).

46 Romeih, E. and Walker, G. (2017). Recent advances on microbial transglutaminase and dairy application. *Trends Food Sci. Technol.* 62: 133–140.

47 Dekker, P.J.T., Metselaar, R., and Sein, A. (2011). Use of sialidase in dairy technology. WO2011095477A1.

48 Ercili Cura, D., Lantto, R., Lille, M. et al. (2009). Laccase-aided protein modification: effects on the structural properties of acidified sodium caseinate gels. *Int. Dairy J.* 19: 737–745.

49 Jacobsen, J., Wind, S.L., Rasholt, E.L., and van den Brink, J.M. (2014). N-glycosidase F improves gel firmness in fermented milk products. *Int. Dairy J.* 38: 169–173.

50 Miwa, N., Yokoyama, K., Wakabayashi, H., and Nio, N. (2010). Effect of deamidation by protein-glutaminase on physicochemical and functional properties of skim milk. *Int. Dairy J.* 20: 393–399.

51 Sein, A. and Uijen, H.M.W.J.C. (2016). Use of phospholipase C. WO2016166149A1.

52 Swaisgood, H.E. (1980). Sulphydryl oxidase: properties and applications. *Enzyme Microb. Technol.* 2: 265–272.

53 Van Dijk, A.A., Dekker, P.J.T., Westerlaken, I., and Bakhuis, J.G. (2008). Novel lysyl oxidases. WO2008113799.

2.4
Enzymatic Process for the Synthesis of Cellobiose

Birgit Brucher[1] and Thomas Häßler[2]

[1] c-LEcta GmbH, Perlickstr. 5, Leipzig, Germany
[2] Pfeifer & Langen GmbH & Co. KG, Dürener Str. 40, Elsdorf, Germany

2.4.1 Enzymatic Synthesis of Cellobiose

Cellobiose, a naturally occurring disaccharide consisting of two β1-4 linked glucose monomers, can be produced in an enzymatic cascade reaction using two phosphorylases (Figure 2.4.1). The first enzyme in the enzymatic synthesis of cellobiose is sucrose phosphorylase (SP). This enzyme belongs to the GH-13 family (family assignment can be found at the public Carbohydrate-active enZYme (CAZy) database) and can be classified as a retaining phosphorylase. Interestingly, sucrose phosphorylases can also act as glycosyl transferases and have gained considerable attention in recent years for the synthesis of other carbohydrates, such as kojibiose, nigerose [1, 2], and α- D-glucosides [3]. Cellobiose phosphorylases (CP) catalyze the formation of cellobiose from glucose and α-glucose 1-phosphate in the second reaction step. CPs are classified to the GH-94 family. The crystal structure of the cellobiose phosphorylase from *Cellulomonas uda* has been elucidated [4]. Previous studies involving enzyme engineering of CP enzymes include expanding the substrate spectrum of the enzyme to the phosphorolysis of lactose by enzyme engineering [5].

The production of cellobiose using SP and CP can be performed as a one-pot reaction. Reaction processes using several enzymes in a one-pot reaction for multistep synthesis offer significant advantages compared to sequential reactions such as reduced production costs and no necessity for intermediate product recovery [6]. However, multienzyme reactions are often more challenging concerning process design due to more complex reaction kinetics or the use of enzymes with different optima for reaction parameters such as temperature or pH-value. For this reason, despite significant advances in the development of multienzyme reactions in recent years by enzyme engineering and reaction modeling [7], few processes have found application in industry.

Industrial Enzyme Applications, First Edition. Edited by Andreas Vogel and Oliver May.

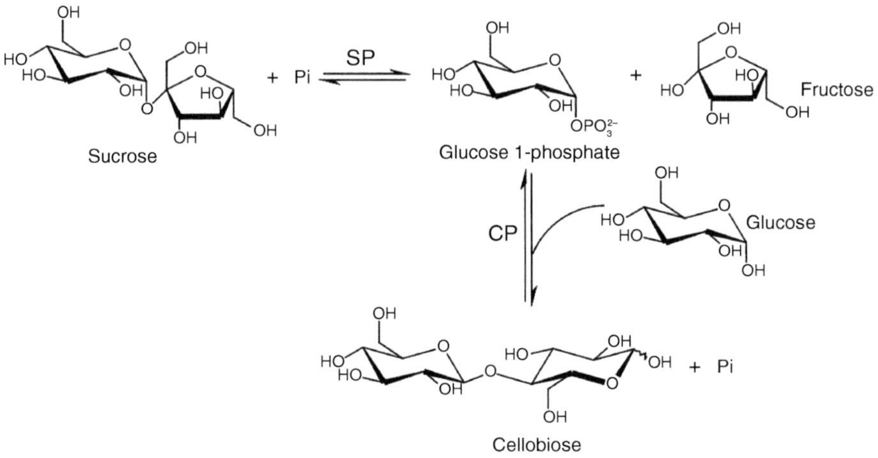

Figure 2.4.1 Enzymatic synthesis of cellobiose from sucrose (SP, sucrose phosphorylase; CP, cellobiose phosphorylase).

2.4.2 Cellobiose – Properties and Applications

Cellobiose is the basic repeat structural unit of cellulose and thus the main component of plant cell walls. Cellobiose is a reducing sugar and compared to sucrose it has a lower water activity and a lower sweetness, being rated as 0.2 times relative to sucrose.

The properties of cellobiose are very different from sucrose and are much more similar to those of lactose. Hence, cellobiose can be used in similar applications as lactose. Pfeifer & Langen GmbH & Co. KG in cooperation with potential customers from industry investigated a broad range of applications. Selected applications can be found in the field of beverages, meat and sausage products, comprimates (pills), and encapsulation technologies.

As a reducing sugar cellobiose can undergo Malliard reactions leading to a browning effect similar to sucrose, which is mostly wanted in baked products. Thus, cellobiose has extensive potential to be used in baking.

Furthermore, cellobiose can be used for more efficient encapsulation of sensitive compounds in foodstuff such as fats and oils with high amounts of polyunsaturated fatty acids (PUFAs) or flavors in order to prevent oxidation reactions. PUFAs undergo a strong exogenous and endogenous destruction resulting in off-flavor, rancid odor, reduction of nutritionally value, and formation of free radicals [8–10]. The challenge in the production of functional foodstuff is maintaining the general properties and especially the sensory profiles that the consumers expect [11]. Linoleic acid, a polyunsaturated omega-6 fatty acid, has a relatively low shelf life and short oxidative induction time [9, 12] and was therefore chosen as a reference substance for assessing the effect of cellobiose addition to the encapsulation process. It could be shown that the encapsulation efficiency of linoleic acid increased with increasing cellobiose content in the matrix with maltodextrin and starch. Additionally, the peroxide value, which is a measure of the primary oxidation of oils, decreased with increasing amounts of cellobiose in the matrix, e.g. the peroxide value was reduced by 60% when linoleic acid was encapsulated with 20% cellobiose and stored at 35 °C for eight weeks compared to encapsulation without cellobiose. After three weeks of storage, a sensory panel confirmed a significantly less rancid odor of samples encapsulated with cellobiose compared to samples encapsulated without cellobiose. Furthermore, a similar encapsulation procedure for volatile compounds such as flavors, e.g. limonene, resulted in the deceleration of oxidative destruction and occurrence of off-flavors after storage [13].

Another application of cellobiose is in beverages. In general, low-sugar or sugar-free soft drinks have a negative sensory profile, a reduced mouth feeling, frequently perceived as unpleasant by-products and/or aftertaste, and an inferior flavor profile compared to sucrose sweetened beverages. Cellobiose has a positive impact on the sensory profile of carbohydrate reduced and other carbohydrate drinks, i.e. drinks, which in addition to cellobiose contain other carbohydrates or a reduced amount of other carbohydrates. It acts as a texturizing agent and thus gives carbohydrate reduced or carbohydrate drinks a better mouth feeling. Moreover, cellobiose masks the off-flavor of sweeteners and thus enables low sugar compositions with an improved flavor profile. Cellobiose was also tested as an ingredient in jams and baked products. Owing to its low sweetness it was particularly suitable to improve mouth feeling of alternative sweetened or sucrose reduced products with surprisingly improved overall sensory profile in the applications. Cellobiose was also suitable to improve the whitening effect when used as fondant and applications thereof [13, 14].

Carbohydrates are traditionally used in meat, sausages, and offal products. In such food preparations carbohydrates have specific functions. These include a sweet taste, support, and control of fermentation processes as a nutrient for the used microbiological cultures, e.g. in raw sausage and ham, intensification of browning and browning taste due to Maillard reactions and effect on water activity (a_w-value). However, the carbohydrates commonly used in food preparations have several drawbacks. Glucose, fructose, and sucrose are comparatively quickly metabolized by microorganisms, which is often undesirable because long fermentation times are preferred to allow the complex flavors to fully develop and mature. Other carbohydrates such as maltodextrin ferment too slowly or not at all. For many food preparations the conventionally used carbohydrates glucose, fructose, or sucrose are too sweet whereas lactose, which has much lower sweetness, has only limited applicability due to the occurrence of lactose intolerance in an estimated 65% of the global population. In products such as Bratwurst, meat loaf, cooked sausage, grilled sausage, fried and grilled minced meat products, and other meat products, cellobiose leads to an accelerated browning reaction (browning intensity and tan color), which is at least as advantageous as the use of lactose or even better. In addition, cellobiose can be used to lower the a_w-value of food preparations. The a_w-value has a direct impact on the growth of microorganisms. Thus, a decrease in water activity increases the shelf life of food preparation. The low sweetness of cellobiose allows the use of higher amounts in the product without impairing the sensory profile. Most importantly, cellobiose gives a pleasant taste profile to processed or fermented meat and sausage products, which was significantly favored in sensory panel trails [15].

Cellobiose has also been assigned to be prebiotic, by supporting the growth of probiotic microorganism. During digestion of cellobiose the production of health promoting short-chained fatty acids has been observed [16–25]. Feeding studies with pigs by Otsuka et al. [26] further confirmed the prebiotic effect. The weight gain of pigs fed with a diet containing cellobiose was significant higher than the control group fed without cellobiose. The authors postulated that pigs that were fed with cellobiose are healthier and develop more consistently [26]. Because humans are not able to degrade cellobiose since the β1-4 glycosidic bond is resistant toward enzymatic and chemical degradation in the gastrointestinal tract, cellobiose in food preparations does not affect the blood glucose and insulin level. Internal studies at Pfeifer & Langen GmbH & Co. KG have shown that cellobiose was resistant to gastric acidity, to hydrolysis by intestinal cells, and to gastrointestinal absorption. Additionally, cellobiose was fermented by the intestinal microflora resulting in the induction of beneficial volatile fatty acids and stimulated selectively the growth and/or activity of intestinal bacteria that contribute to health and well-being, e.g. *Enterococcus faecium* or *Bacteroides vulgatus*.

Another type of application for cellobiose is as an excipient. Pills and effervescent tablets often comprise lactose, sucrose, and sugar alcohols, which have an effect on the glycemic index. When cellobiose was uses as a tableting filler agent it fulfilled the desired properties in terms of good compressibility, resulting in hardness, good abrasion resistance, shelf life, or thermal stability of the pills and no necessity for disintegrants for rapid tablet dissolution [13]. Owing to its properties and low hygroscopicity cellobiose is perfectly suitable as a tableting filler agent. Cellobiose also has the potential for clinical usage. Using mice as a model, it was shown that cellobiose has a preventive effect on patients with inflammatory bowel diseases [27].

2.4.3 Existing Routes for Cellobiose Synthesis

Chemical synthesis of cellobiose is challenging and suffers from low yield and efficiency. Several enzymatic routes for the synthesis of cellobiose have been explored. For example, cellobiose was produced by enzymatic hydrolysis of sugar beet pulp with a yield of 35 % [28]. Alternatively, cellobiose can be synthesized enzymatically using phosphorylases. Suzuki et al. [29] described the synthesis of cellobiose from starch with a glucan-phosphorylase and a cellobiose phosphorylase in a two-step reaction. The individual enzymatic reactions require different reaction conditions. While the first step (starch phosphorolysis) requires high phosphate concentration, the second reaction is severely inhibited by high phosphate concentrations. In order to reduce the phosphate concentration between both steps, electrodialysis was applied. In total, the raw material efficiency and overall yield were relatively low with 29% from starch.

The synthesis of cellobiose from sucrose using phosphorylases was first described by Nippon Petrochem and the National Food Research Institute [30]. The authors demonstrated cellobiose synthesis in a one-pot reaction from 200 mM sucrose with 70–73% yield. In the first reaction step, sucrose is cleaved to α-glucose 1-phosphate and fructose by the action of a sucrose phosphorylase (Figure 2.4.1). In the second step, cellobiose phosphorylase catalyzes the formation of cellobiose from α-glucose 1-phosphate and glucose. Sucrose is a well-defined substrate as opposed to starch or cellulose. The free energy of the glycosidic bond of sucrose is very high compared to other di- and oligosaccharides. The net conversion of sucrose in cellobiose is therefore thermodynamically clearly favored. Moreover, sucrose is ubiquitously available at high purity level. The process has the advantage of using a cheap, clean, and readily available substrate, a high raw material efficiency, and high yield. Additionally, both reaction steps can be performed simultaneously in one reaction vessel, which has a positive impact on production costs. We therefore decided to use this route to develop a production process for cellobiose.

2.4.4 Enzyme Development

In a first step, we tested different wild-type sucrose phosphorylase (SP) and cellobiose phosphorylase (CP) enzymes and tested various parameters that were relevant for the process. This included activity, stability, recombinant expression level, pH–activity profile, and applicability at high substrate concentration.

For both enzymes, six potential candidates were identified by literature and database research and cloned as codon-optimized genes for recombinant expression in *Escherichia coli*. Additionally, the c-LEcta *in-house* genomic and metagenomic libraries were screened using both degenerate primers based on known SP and CP enzyme sequences, and an activity based screening assay. As expected, the primer based screening yielded enzymes with high similarity to already known sequences, whereas the activity based screening gave access to new sequences. The most interesting candidates were further characterized (Table 2.4.1). Based

Table 2.4.1 Characterization of different SP and CP enzymes.

Enzyme	Source	Activity yield	Temperature optimum (°C)	pH optimum
CP	*Cellulomonas uda*	++++	40–50	6.5
	Cellvibrio gilvus	+		
	Thermotoga neapolitana	++	80	6.0
	Ruminococcus flavefaciens	++		
	Paenibacillus polymyxa	+		
	Bifidobacterium dentium	++++	40–50	5.5–6.0
	Environmental library	+		
SP	*Bifidobacterium adolescentis*	++++	40	6.5
	Clostridium saccharolyticum	+		
	Streptococcus mutans	++++	40	6.5
	Environmental library	++++	40	6.5

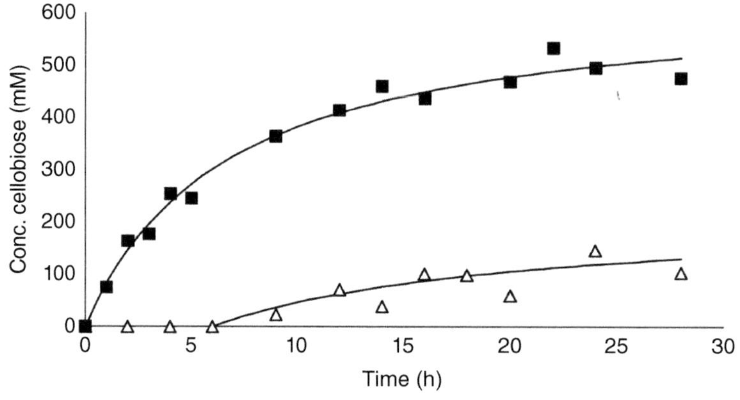

Figure 2.4.2 Cellobiose production in a one-pot reaction using SP wild type and CP wild type or CPCu17-13 (sucrose, glucose: 750 mM; NaMES-buffer: 50 mM; sodium/potassium phosphate buffer: 100 mM, pH 6.3; SP wild-type: 8 PU/ml; CP: 15.8 SU/ml) Δ CPCuda wild type, ■ CPCu17-13.

on the recombinant activity yield in *E. coli* and their pH and temperature optima, the SP from *Bifidobacterium adolescentis* and the CP from *Cellolumonas uda* were selected.

The application of these two enzymes in the one-pot synthesis of cellobiose using high substrate concentrations revealed that the CP wild-type enzyme constitutes a major bottle neck (Figure 2.4.2). This is probably due to its inhibition by glucose [31]. However, for industrial applications higher product concentrations are necessary in order to achieve a cost-efficient production process. It was therefore decided to improve the efficiency of the cellobiose phosphorylase in the cellobiose production process by enzyme engineering.

This can be achieved in a number of ways, for example, by improving the specific activity of the enzyme, by decreasing inhibition toward any of the compounds present in the enzymatic reaction, or by improving the K_m-value of the enzyme for a substrate that is not present in saturating concentrations, in this case α-glucose 1-phosphate. In order to find variants that may possess any of these desirable features, libraries were screened in parallel at moderate and increased substrate concentrations for high cellobiose yield. Thus, libraries can be screened directly for variants with advantageous qualities for the desired application even without further information concerning the mechanism of the reaction limitations. Beneficial mutations were combined in several subsequent combinatorial libraries. Target residues for mutations were selected based on a semi-rational approach using c-LEcta's MDM approach (see Chapter 1.2) and included the analysis of the enzyme's crystal structure. The selected residues were not limited to the active site of the enzyme but expanded to residues from the putative substrate tunnel and the surface of the enzyme. In order to avoid overly large combinatorial libraries, several smaller libraries were created instead, each containing only some of the beneficial mutations identified in the previous round of engineering. Overall, four rounds of engineering were performed containing less than 500 variants in total (Table 2.4.2).

Table 2.4.2 Overview of the cellobiose phosphorylase libraries screened.

Round	Library type	No. of libraries	No. of variants
1	Single mutants	1	77
2	Combinatorial	2	56
3	Single mutants	3	175
4	Combinatorial	13	170

Table 2.4.3 SU/PU ratio compared to the wild-type enzyme at different substrate concentrations.

Variant	Round	Added mutations to previous round	Fold improvement in SU/PU ratio at substrate concentration		
			250 mM	500 mM	750 mM
CPCuda wild-type	—		1	1	n.d.[a)]
CPCu5-2	2	Q161M/R188K/D196N/ A220L/L705T	0.8	1.1	n.d.
CPCu11-16	4	Y164F /K283A/A512V	2.4	2.9	>2.5
CPCu17-13	4	F164Y/S169V /T788V	15.9	28.2	>29.6

1SU, the amount of enzyme that generates 1 μmol cellobiose per minute from the indicated concentrations of glucose and α-glucose 1-phosphate in 50 mM MES-buffer pH 6.5 at 30 °C. 1 PU, the amount of enzyme that generates 1 μmol of α-glucose 1-phosphate in one minute from 10 mM cellobiose in 75 mM potassium phosphate buffer pH 7 at 30 °C.

a) n.d., not detected. The activity was below the limit of detection of the assay used for the determination of synthesis activity. However, some cellobiose formation by the wild-type enzyme could be observed after longer reaction times (Figure 2.4.2).

Improved variants from each round were further characterized with regard to the initial activity for the synthesis of cellobiose from varying substrate concentrations of 250, 500, and 750 mM glucose and α-glucose 1-phosphate (synthesis units, SUs) as well as the initial activity for cellobiose phosphorolysis (phosphorolysis units, PUs) (Table 2.4.3). The final variant, CPCu17-13 with 11 mutations, showed greatly improved activity for the synthesis of cellobiose compared to the wild-type enzyme not only at moderate substrate concentration but also at higher substrate concentrations of up to 750 mM α-glucose 1-phosphate and glucose (Table 2.4.3), which means that the variant possesses less substrate and/or product inhibition than the wild-type enzyme. This in turn allowed an efficient synthesis of cellobiose in a one-pot synthesis reaction using 750 mM sucrose, which was not possible with the wild type or the best variant from round 2 (Figure 2.4.2).

2.4.5 Process Development

Process development started in parallel to the development of the enzymes and food applications tests.

From the beginning it became obvious that from an economic point of view that neither substrate components nor enzymes can be wasted and need to be recycled. Additionally, the co-product fructose has to be purified in order to commercialize this as an additional valuable product. Therefore, process development necessitated the development of an efficient cellobiose synthesis process as well as a purification process including the recycling of enzymes, valuable co-products and water.

2.4.5.1 Synthesis of Cellobiose

The conversion of sucrose to cellobiose is a two-step enzymatic cascade (Figure 2.4.1). The preferred process of an enzymatic cascade is a one-pot reaction. This leads to reduced investment and operating costs. The process complexity in general is simpler, but the process control can be more complex at the same time. As the CP enzyme shows inhibition by phosphate, another advantage of a one-pot reaction for the synthesis of cellobiose is that low phosphate concentration can be used. This allows the enzyme to work more efficiently.

Reaction conditions should be kept at moderate temperatures in order to minimize the formation of color generated due to the presence of reducing sugars and phosphates.

Through enzyme engineering of the CP a simple and efficient one-pot reaction could be developed, which allowed the synthesis of cellobiose at industrially relevant substrate concentrations with approximately 70% yield (Figure 2.4.2).

2.4.5.2 Purification of Cellobiose

Customers desire a product of high purity and processability. In general, reducing sugars tend to undergo Maillard reactions at high temperatures, an effect which is even further accelerated by the presence of phosphate [32]. Thus, any purification step should be performed at moderate temperatures in order to avoid undesirable browning of the product. Several purification technologies alone and in combination were studied to achieve a white crystalline product.

As the first step in the purification of cellobiose, an ultrafiltration step was developed for the separation and recycling of the enzymes. Next, different technologies were tested for the separation of the saccharides and phosphate. The process liquor contains a challenging composition of components with similar molecular weight (fructose and glucose or sucrose and cellobiose) and ionic compounds (phosphate, α-glucose-1-phosphate, sodium, or potassium). For the separation of saccharides membrane technologies and chromatography were investigated. Defined process solutions were prepared to identify suitable purification technologies. Size exclusion chromatography for saccharides and ion exchange chromatography for charged compounds are well-established technologies. However, in order to purify cellobiose and fructose, two separate chromatographic steps would be necessary. Different resins were tested and shown to be suitable for the separation of saccharides. As a rather expensive technology, however, chromatography should be avoided if possible. As an alternative, filtration technologies, such as nanofiltration, osmosis, and electrodialysis were studied. Screening of different membranes for filtration technologies proved this method to be suitable for the separation of saccharides. However, as with chromatography, this technology is not economically viable. Electrodialysis, on the other hand, could be shown to be an efficient technique for the separation and recovery of phosphate and α-glucose 1-phosphate. Thus, electrodialysis was introduced as the second step in the purification of cellobiose. Both phosphate and α-glucose-1-phosphate are valuable compounds that can be used for subsequent cellobiose synthesis reaction. This further enhances raw material- and cost-efficiency of the process.

Spray drying, crystallization, and ethanol precipitation are options to achieve a solid product. In a proof of concept study, a chromatographically purified cellobiose solution was submitted to all three process options. All three technologies were suitable; however, crystallization was the most promising one. The solubility curves indicate the potential of separating cellobiose from other saccharides by crystallization, because it shows a very different solubility profile compared to the other saccharides (Figure 2.4.3). The presence of fructose, glucose, and sucrose constitute a challenge for the development of a crystallization process for cellobiose with no or only minimal color formation. Therefore, vacuum crystallization and cooling crystallization were investigated further, because they allow for concentration of the process solution under gentle conditions. The final crystallization process of cellobiose separates up to 80% of the cellobiose content from a matrix containing cellobiose, fructose, glucose, sucrose, and water.

The solution is then submitted to a chromatographic step separating the second product, fructose, from the remaining saccharides.

In the end, a simple and easily scalable process was developed (Figure 2.4.4) [35]. For the highest cost-efficiency, the enzymes are retained in the synthesis reaction by ultrafiltration and unreacted substrates are reintroduced into the cellobiose synthesis reaction.

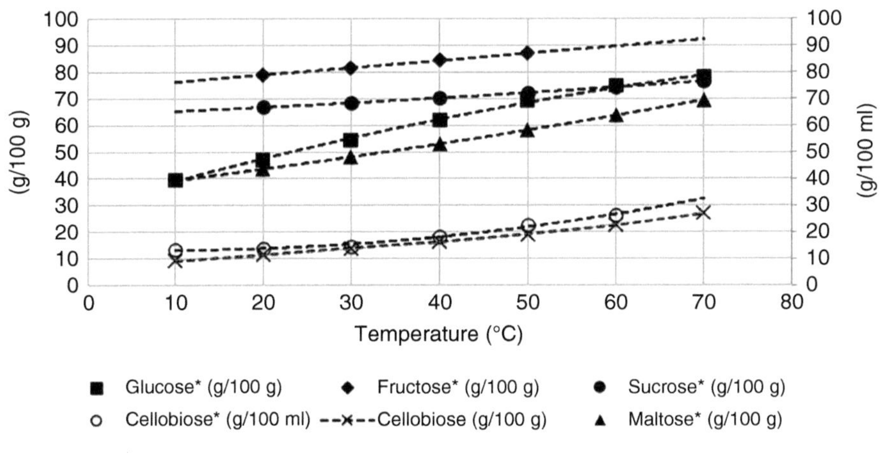

Figure 2.4.3 Solubility curve of saccharides; *Data from Refs. [33, 34].

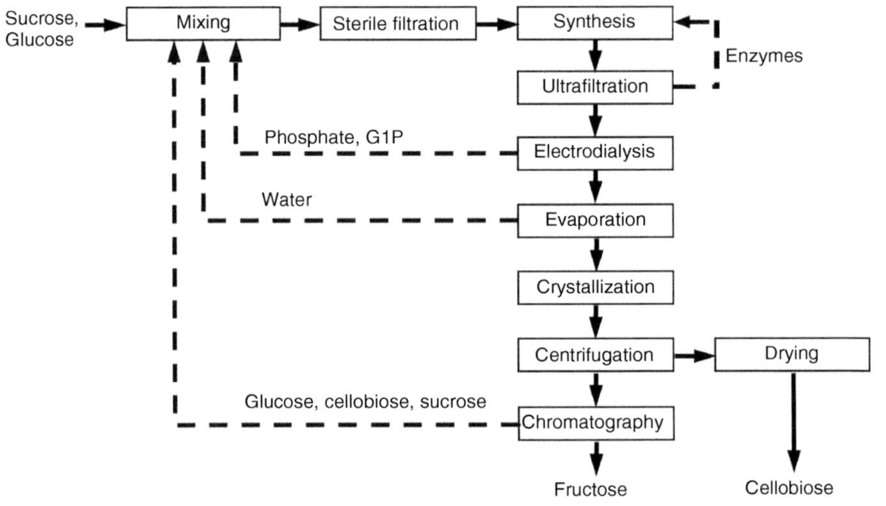

Figure 2.4.4 Process scheme of the enzymatic production of cellobiose from sucrose.

2.4.6 Summary and Future Perspective

Through enzyme engineering the drawbacks of the wild-type CP enzyme could be overcome to allow for cellobiose synthesis at industrially relevant substrate concentrations. Additionally, through process development involving the recycling of the substrates and enzymes and an efficient purification process for cellobiose and the co-product fructose, an overall efficient multienzyme process for the production of cellobiose could be realized.

Currently, cellobiose production has been upscaled at Savanna Ingredients GmbH into pilot scale. The capacity is around 100 t/a of cellobiose as a crystalline powder. Cellobiose has already been evaluated as a feed material in Germany by the "Normenkommission für Einzelfuttermittel im Zentralausschuss der Deutschen Landwirtschaft." Even though cellobiose is one of the most abundant building blocks in nature as a constituent of cellulose, it is considered as a new ingredient and will require a Novel Food Approval for application in food by the European Commission.

References

1 Beerens, K., De Winter, K., Van de Walle, D. et al. (2017). Biocatalytic synthesis of the rare sugar kojibiose: process scale-up and application testing. *J. Agric. Food Chem.* 65 (29): 6030–6041.

2 Kraus, M., Görl, J., Timm, M., and Seibel, J. (2016). Synthesis of the rare disaccharide nigerose by structure-based design of a phosphorylase mutant with altered regioselectivity. *Chem. Commun. (Camb.)* 52 (25): 4625–4627.

3 Goedl, C., Sawanghan, T., Wildberger, P., and Nidetzky, B. (2010). Sucrose phosphorylase: a powerful transglucosylation catalyst for synthesis of

α-D-glucosides as industrial fine chemicals. *Biocatal. Biotransform.* 28 (1): 10–21.

4 Van Hoorebeke, A., Stout, J., Kyndt, J., and De Groeve, M. (2010). Crystallization and X-ray diffraction studies of cellobiose phosphorylase from *Cellulomonas uda. Acta Crystallogr., Sect. F* 66: 346–351.

5 De Groeve, M., De Baere, M., Hoflack, L. et al. (2009). Creating lactose phosphorylase enzymes by directed evolution of cellobiose phosphorylase. *Protein Eng.* 22 (7): 393–399.

6 Ricca, E., Brucher, B., and Schrittwieser, J.H. (2011). Multi-enzymatic cascade reactions: overview and perspectives. *Adv. Synth. Catal.* 353 (13): 2239–2262.

7 Honda, K. (2017). Industrial applications of multistep enzyme reactions. In: *Biotechnology of Microbial Enzymes* (ed. G. Brahmachari), 433–450. Elsevier Inc.

8 Jacobsen, C. (1999). Sensory impact of lipid oxidation in complex food systems. *Lipid/Fett* 101 (12): 484–492.

9 Krist, S., Buchbauer, G., and Klausberger, C. (2008). *Lexikon der pflanzlichen Fette und Öle*. Wien: Springer-Verlag.

10 Richter, R. (2010). *Handbuch für Lebensmittelchemiker*. Berlin, Heidelberg: Springer-Verlag.

11 Lache, A. (2000). Stabilisierung von ungesättigten Glyceriden gegen Autoxidation durch molekulare Verkapselung mit Stärkederivaten. Dissertation. Düsseldorf: Heinrich-Heine University.

12 Matthäus, B. (2011). Welches Fett und Öl zu welchem Zweck? Merkmale und Spezifikationen von Ölen und Fetten, Deutsche Gesellschaft für Fettwissenschaft e.V. (DGF).

13 Koch, T.J., Kaufmann, B., Ludovici, K. et al. (2016). Cellobiose in Zusammensetzungen zum Verzehr oder zur Einnahme. DE202016008304.

14 Koch, T.J., Kaufmann, B., Ludovici, K., and König, S. (2016). Cellobiose-haltiges Getränk. WO2016116620.

15 Koch, T.J., Ludovici, K., Kaufmann, B. et al. (2016). Cellobiose in fermented meat and sausage products. WO2016116627.

16 Basholli-Salihu, M., Mueller, M., Unger, F.M., and Viernstein, H. (2013). The use of cellobiose and fructooligosaccharide on growth and stability of bifidobacterium infantis in fermented milk. *Food Nutr. Sci.* 4 (12): 1301–1306.

17 Heinritz, S.N., Weiss, E., Seifert, J. et al. (2017). Effect of cellobiose supplementation on in vitro fermentation activity and bacterial numbers of porcine inocula. *J. Anim. Physiol. Anim. Nutr.* 102 (2): 474–482.

18 Nakamura, S., Oku, T., and Ichinose, M. (2004). Bioavailability of cellobiose by tolerance test and breath hydrogen excretion in humans. *Nutrition* 20 (11): 979–983.

19 Ruiz-Matute, A.I., Brokl, M., Sanz, M.L. et al. (2011). Effect of dextransucrase cellobiose acceptor products on the growth of human gut bacteria. *J. Agric. Food Chem.* 59: 3693–3700.

20 Sanz, M.L., Gibson, G.R., and Rastall, R.A. (2005). Influence of disaccharide structure on prebiotic selectivity in vitro. *J. Agric. Food Chem.* 53 (13): 5192–5199.

21 Umeki, M., Oue, K., Mochizuki, S. et al. (2004). Effect of Lactobacillus rhamnosus KY-3 and cellobiose as synbiotics on lipid metabolism in rats. *J. Nutr. Sci. Vitaminol.* 50 (5): 330–334.

22 Umeki, M., Oue, K., Mori, M. et al. (2005). Fluorescent in situ hybridization analysis of cecal microflora in rats simultaneously administrated *Lactobacillus rhamnosus* KY-3 and cellobiose. *Food Sci. Technol. Res.* 11 (2): 168–170.

23 van Zanten, G.C., Krych, L., Roeytioe, H. et al. (2014). Synbiotic *Lactobacillus acidophilus* NCFM and cellobiose does not affect human gut bacterial diversity but increases abundance of lactobacilli, bifidobacteria and branched-chain fatty acids: a randomized, double-blinded cross-over trial. *FEMS Microbiol. Ecol.* 90: 225–236.

24 van Zanten, G.C., Sparding, N., Majumder, A. et al. (2015). The differential proteome of the probiotic *Lactobacillus acidophilus* NCFM grown on the potential prebiotic cellobiose shows upregulation of two β-glycoside hydrolases. *BioMed Res. Int.* 1–10.

25 Yacine, A., Aicha, T.M., Boumediene, M. et al. (2012). The in vitro effect of a symbiotic combination between cellobiose and two probiotic strains toward two pathogenic bacteria. *J. Life Sci. (El Monte, CA, USA)* 6 (6): 615–623.

26 Otsuka, M., Ishida, A., Nakayama, Y. et al. (2004). Dietary supplementation with cellooligosaccharide improves growth performance in weanling pigs. *Anim. Sci. J.* 75 (3): 225–229.

27 Nishimura, T., Andoh, A., Hashimoto, T. et al. (2010). Cellobiose prevents the development of dextran sulfate sodium (DSS)-induced experimental colitis. *J. Clin. Biochem. Nutr.* 46 (2): 105–110.

28 Kohl, A., Brück, T., Gerlach, J. et al. (2012). Cellobiose production from biomass. WO2012001102.

29 Suzuki, M., Kaneda, K., Nakai, Y. et al. (2009). Synthesis of cellobiose from starch by the successive actions of two phosphorylases. *New Biotechnol.* 26 (3–4): 137–142.

30 Taniguchi, H., Sasaki, T., and Kitaoka, M. (1998). Method for producing cellobiose. JP2815023B2.

31 Nidetzky, B., Griessler, R., Schwarz, A., and Splecht, B. (2004). Cellobiose phosphorylase from *Cellulomonas uda*: gene cloning and expression in *Escherichia coli*, and application of the recombinant enzyme in a 'glycosynthase-type' reaction. *J. Mol. Catal. B: Enzym.* 29: 241–248.

32 Rizzi, G.P. (2004). Role of phosphate and carboxylate ions in maillard browning. *J. Agric. Food Chem.* 52 (4): 953–957.

33 Bubnik, Z., Kadlec, P., and Urban, D. (1995). *Sugar Technologists Manual: Chemical and Physical Data for Sugar Manufacturers and Users*. Berlin: Bartens.

34 Shimada, K., Ichihara, T., Tsumiya, T. et al. (2009). Purification method and production method for cellobiose. WO2006112316.

35 Koch, T.J., Häßler, T., and Kipping, F. (2016). Process for the enzymatic preparation of a product glucoside and of a co-product from an educt glucoside. WO2016038142.

2.5

Emerging Field – Synthesis of Complex Carbohydrates. Case Study on HMOs

Dora Molnar-Gabor[1], Markus J. Hederos[1], Sebastian Bartsch[2], and Andreas Vogel[2]

[1] *Glycom A/S, Kogle Alle 4, 2970 Hørsholm, Denmark*
[2] *c-LEcta GmbH, Perlickstr. 5, 04103 Leipzig, Germany*

2.5.1 Introduction to Human Milk Oligosaccharides (HMOs)

2.5.1.1 Discovery and Function of HMOs

Numerous research studies and decades of collected experience claim breastfeeding to be associated with multiple health benefits for infants including lower susceptibility to infections, lower risk to develop obesity, asthma and/or allergic complications, and improved cognitive outcomes [1–7]. In addition, babies who get mother's milk exclusively for the first six months in life without any formula provided show lower score in terms of ear infections, respiratory illnesses, and/or diarrhea [8]. Already at the end of the nineteenth century, awareness of the importance of nutrition was rapidly growing and was identified to have a key role in the health of infants. This led to increased research activities in associated fields. In the early twentieth century it was observed that breastfed infants had approximately seven times higher survival rate than the non-breastfed infants. These findings stimulated the scientific community to focus on the identification of the natural components in human milk that contribute to these beneficial effects. As a consequence, in 1930s the first human milk carbohydrate fraction was identified – the real starting point of the human milk oligosaccharide (HMO) research field [9]. In the following years the first HMOs were isolated, structurally characterized, and their function investigated, such as influence on gut microbial composition and effects on the immune system [10]. Furthermore, their structural characterization provided the missing link to elucidate the enzymatic basis for the human histo-blood group and Lewis epitopes, which are structurally and biochemically related to HMOs [11, 12].

HMOs are a complex and structurally diverse family of glycans present in mother's milk, typically in concentrations of 5–20 g/l [13]. They build up the third most abundant component of human milk after lactose (60–80 g/l) and lipids (30–40 g/l) [14]. The structure and composition of the HMOs found

Industrial Enzyme Applications, First Edition. Edited by Andreas Vogel and Oliver May.

in human milk are different from the oligosaccharide profile found in other mammalians. For example, the oligosaccharides found in cow milk are structurally less complex and are present in significantly lower concentration [15]. As a consequence, cow milk is not an adequate source to compensate for the complex oligosaccharides contained in human milk. Accordingly, there is a need for an industrial scale procedure that enables the application of HMOs as food ingredient, with the main demand from infant formula producers.

Meanwhile, over 200 different HMOs have been identified with more than 140 structures fully elucidated [16, 17]. The large diversity and quantity of HMOs has formed predominately during the recent hominid evolution, most likely as consequence of host–pathogen and host–symbiont trade-off [18, 19], which suggests that various biological functions and roles are correlating to the different individual structures. It has been scientifically established that HMOs are fibers passing undigested through the upper gastrointestinal tract and act as highly selective substrates for advantageous microbes in the colon (i.e. prebiotic effect). In addition, HMOs can reduce pathogens in the infant intestinal microbiota firstly by acting as soluble decoy receptors, and secondly by selectively allowing the natural and healthy HMO-utilizing microbiota to outcompete them [20, 21]. Furthermore, there is emerging evidence that HMOs might play a role in the cognitive development of an infant's brain [22, 23]. Moreover, recent studies are proposing HMOs to be attributed to antimicrobial and antibiofilm properties [24–26]. Undoubtedly, HMOs offer various health benefits to infants, and may moreover be beneficial for older children and adults, as summarized in recent reviews [27, 28]. Thus, it is not surprising that this class of compounds is of highest interest for the dairy, food and nutritional industry worldwide [29].

2.5.1.2 Structure of HMOs

HMOs are built from five monosaccharide building blocks: D-glucose (Glc), D-galactose (Gal), *N*-acetylglucosamine (GlcNAc), L-fucose (Fuc), and sialic acid (*N*-acetylneuraminic acid; Neu5Ac). They all contain the disaccharide lactose (Gal(β1-4)Glc) in the reducing end. Lactose is the simplest core structure of HMOs, which can be decorated with fucose and/or sialic acid residues. Further elongation of lactose in a linear or branched manner, occurring by two types of disaccharides, lacto-*N*-biose (Gal(β1-3)GlcNAc) (Type I) or *N*-acetyllactosamine (Gal(β1-4)GlcNAc) (Type II), leads to a diverse but defined set of HMO core structures. Fucosylation and/or sialylation of the core backbones results in a broad array of HMOs; in Figure 2.5.1 the positions for fucosylation and sialylation are depicted [17].

The relative abundance and composition of HMOs varies between mothers and normally decreases during the course of lactation [30–33]. The most abundant neutral HMOs are 2′-fucosyllactose (2′-FL), 3-fucosyllactose (3-FL), 2′,3-difucosyllactose (DFL or LDFT), lacto-*N*-tetraose (LNT), lacto-*N*-*neo*-tetraose (LNnT), lacto-*N*-fucopentaose I (LNFP-I), lacto-*N*-fucopentaose II (LNFP-II), lacto-*N*-fucopentaose III (LNFP-III), lacto-*N*-fucopentaose V (LNFP-V), lacto-*N*-difucohexaose I (LNDFH-I), lacto-*N*-difucohexaose II (LNDFH-II), etc. The most representative acidic HMOs are

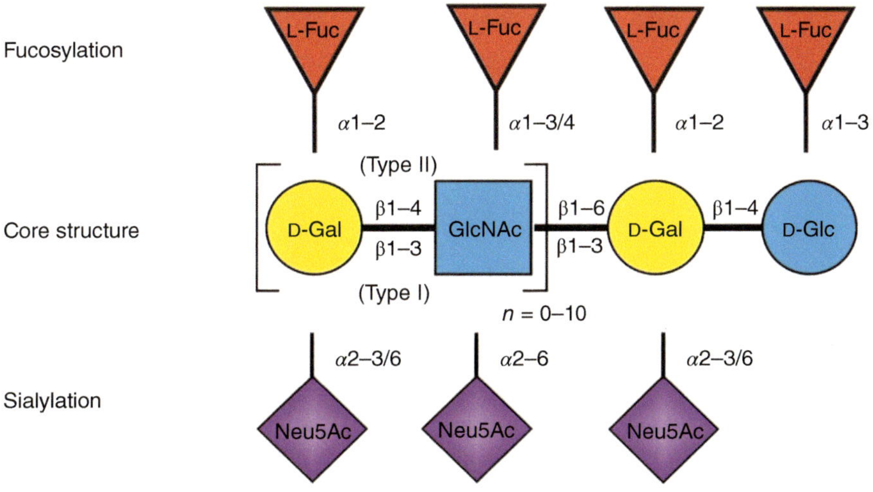

Figure 2.5.1 General structure of HMOs. The schematic representation of monosaccharides according to IUPAC and IUBMB recommendations. General structure and linkage types present in HMOs are indicated.

3′-sialyllactose (3′-SL), 6′-sialyllactose (6′-SL), sialyl-lacto-*N*-tetraose a (LST-a), sialyl-lacto-*N*-tetraose b (LST-b), sialyl-lacto-*N*-*neo*-tetraose c (LST-c), etc. [33–36] (Figure 2.5.2).

Despite all the known benefits of HMOs, a widespread use of them has not been possible due to the lack of large-scale production processes that allow overcoming the difficulties connected to the high structural and compositional diversity. Instead, structurally much simpler synthetic prebiotic carbohydrate polymers such as galacto-oligosaccharides (GOS) and inulin have been added to infant formula for many years as an alternative and as an attempt to mimic HMOs [37, 38]. However, these simple mimics are not able to reproduce the range of HMO functions and are also not naturally present in breast milk at any significant degree, and therefore development of industrial feasible technologies for large-scale HMO production has been of high interest for a long time with recent breakthrough technology having been developed by a number of biotechnology companies such as Glycom A/S, Jennewein Biotechnology GmbH, Glycosyn LLC, and InBiose, to name a few [39].

2.5.1.3 HMO Production, Regulatory Authorizations, and Commercial Launch – Historical Overview

Historically, HMOs have been isolated from human milk for research purposes such as structural identification and early biological studies. For obvious reasons, this approach is unsuitable for commercialization. Chemical synthesis has been thoroughly investigated as one alternative for large-scale HMO production. During 2005–2012 Glycom A/S produced 2′-FL and LNnT in ton scale by chemical synthesis as a strategy to bring forward preclinical and clinical programs. However, from a commercial perspective the chemical synthesis suffers

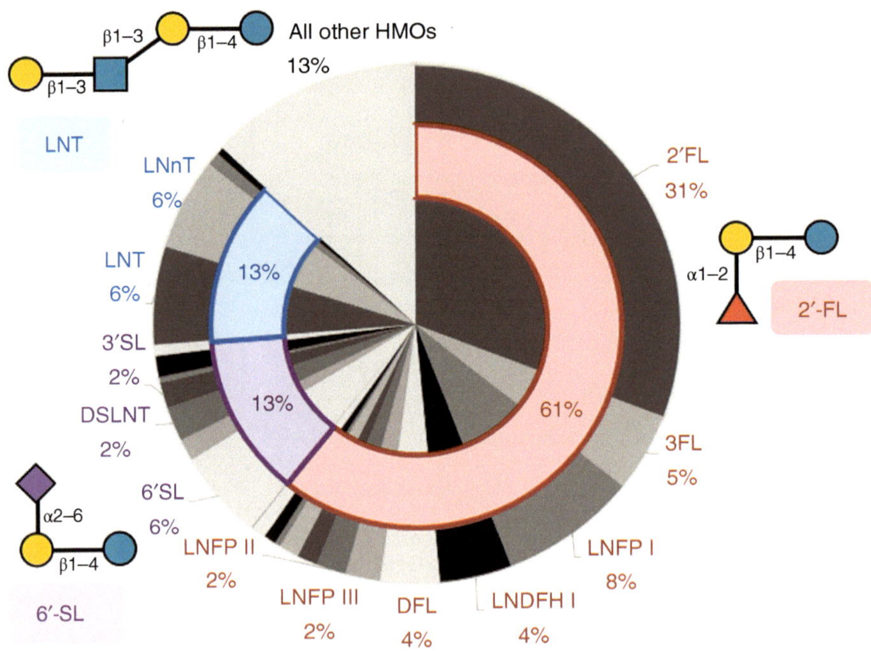

Figure 2.5.2 Relative abundance of HMOs in human milk. Three of the more abundant HMO structures are depicted as examples for each of the three classes of sialylated (purple), fucosylated (red), and neutral core HMOs (blue). Source: Graph reproduced from [35] with permission from Elsevier.

from limitations to make HMOs a vital long-term business. The complexity, total number of reaction steps, waste management, environmental impact, limited availability, and cost of raw materials make the end-product prohibitively expensive for dairy and nutrition applications. Still, the quality and purity of the HMOs accessed by chemical synthesis was extremely high with a fully controlled impurity profile [40–43]. The chemically synthesized 2′FL and LNnT were used in clinical trials in 2013/2014, sponsored by Nestec Ltd., demonstrating that the infant formula supplemented with these two human-identical milk oligosaccharides (HiMOs) is safe, well-tolerated, and supports age-appropriate growth [44]. Moreover, the study showed a correlation between HMO supplemented formula and lower parent-reported morbidity (particularly bronchitis) and medication use (antipyretics and antibiotics). The positive outcome of the clinical trials stimulated future technology development at Glycom A/S, and a shift of the production platform from chemical synthesis toward biotechnology was initiated. Compared to the chemical synthesis, which involves multistep processes using carefully designed protecting group strategy and relatively expensive raw materials and chemicals, the biotechnological approach using highly specific enzymes enclosed in cellular fermentation factories appeared to be more feasible for large-scale HMO production, also providing an advantageous cost picture.

In 2014 Glycom A/S filed for novel food approval of 2′-FL and LNnT derived from chemical synthesis, and obtained EFSA opinions for them in

July 2015 [45, 46] and marketing authorizations in March 2016 [47, 48]. These approvals constitute the very first novel food approvals for any HMO in the EU. Between 2012 and 2016 Glycom A/S developed a robust and scalable fermentation technology for 2′-FL and LNnT and confirmed and notified both compounds as substantial equivalent novel food ingredients for use in the EU. In the same time period Jennewein Biotechnologie GmbH (2015/2017), DuPont Nutrition & Health in collaboration with InBiose (2017), and FrieslandCampina in collaboration with Glycosyn LLC (2018) developed their own technologies for 2′-FL production based on microbial fermentation, and have received regulatory approval for the use of their products as novel food ingredient.

In the United States Glycom notified generally recognized as safe (GRAS) of 2′-FL and LNnT from chemical synthesis to the FDA in October 2014 and received FDA *"no further question"* response letters in September and October 2015, respectively. Jennewein notified GRAS of 2′-FL from microbial fermentation with *Escherichia coli* BL21 in March 2015 and obtained the FDA response letter in November 2016, while Glycom notified 2′-FL and LNnT from microbial fermentation with *E. coli* K-12 in May and August 2016, respectively, and obtained both response letters in November 2016. Glycosyn LLC and FrieslandCampina co-notified their 2′-FL in November 2017 and received the response letter in April 2018, while DuPont notified around January 2018 and also received a response letter in April 2018.

A new player and additional HMO appeared in March 2018, when GRAS notification of chemoenzymatically synthesized 3′-SL produced by GeneChem Inc. received a positive response letter from FDA.

With the regulatory approvals for the HMO ingredients infant nutrition manufacturers have included these into novel and improved infant formulas. Abbott Nutrition and Nestlé Nutrition launched their first formulas containing 2′-fucosyllactose (plus lacto-*N*-neotetraose in case of Nestlé) almost simultaneously in October 2016 in the United States and Spain, respectively. In October 2016 Abbot Nutrition launched Similac® Pro-Advance and Pro-Sensitive on the US market and followed up in Mexico in October 2017, in Hong Kong in November 2017, and in the Middle East in March 2018. In 2017 Nestlé expanded their initial launch to their bigger brand NAN® Optipro Supreme 2 HM-O^2 in Spain, followed by launches of NAN Optipro HM-O^1 with 2′-FL in the Near East (KSA, UAE, QA, BHR, KW) and Good Start® Optipro Supreme HM-O^1 in Mexico in January 2018; Wyeth Nutrition launched the follow-on formula Wyeth Iluma® with 2′-FL in Hong Kong. This short summary reflects the events around the initial launches (up to 2018), but with both global infant nutrition players progressing with their launch plans, expanding into more diverse brands (including mainstream brands), and other players entering the arena soon, HMOs will soon become available in most markets of the world.

The new infant formula launched by Nestlé in the EU contains 2′-FL and LNnT produced with the technology developed by Glycom A/S. This is the very first time a combination of HMOs has been added to infant formula and made available to the consumer.

In addition to the obvious application in infant nutrition, HMOs have also become of high interest for non-infant applications and being launched are dietary supplements and medical foods (called "food for special medical purpose," [FSMP], in the EU). At the beginning of 2018 in the United States, Metagenics has launched a food supplement for children called MetaKids™ Nutrition Powder and a medical food called UltraGI Replenish®. In late 2018 Glycom launched a dietary supplement and a medical food for the clinical dietary management of IBS in the United States and EU under the brand names Holigos™ Digestive Health and Holigos IBS.

2.5.2 Glycom A/S Technologies Toward Commercial HMO Production

Current technologies at Glycom A/S are intended to provide a broad palette of structurally diverse HMOs. For this purpose two different carbohydrate transfer catalyzing enzyme classes are used – Leloir glycosyltransferases (GTs) and transglycosylases (TGs). The Leloir GTs are the principal enzymes that catalyze glycoside synthesis, utilizing activated mono- or diphosphonucleotide sugar donors. These enzymes are supposed to be involved in the natural HMO biosynthesis even though the fundamental knowledge in this field is rather limited [49, 50]. TGs and Leloir GTs both catalyze glycosyl transfer (transfer of a glycosyl residue from a donor substrate to an acceptor) but act differently on a mechanistic level. TGs are structurally more related to glycoside hydrolases (GHs) and catalyze the intra- or intermolecular substitution of the anomeric position of glycosides.

For industrial utilization the Leloir GTs are the first and primary choice by employing engineered microorganisms suitable to express and deliver a continuous supply with adequate nucleotide-activated sugar donor substrate and matched GTs to produce the targeted HMO. Using GTs and recombinant host, Glycom A/S has developed an *in vivo* fermentation technology enabling individual HMO production toward 2′-FL, LNnT, 3-fucosyllactose (3-FL), difucosyllactose (DFL), lacto-*N*-tetraose (LNT), 3′-sialyllactose (3′-SL), 6′-sialyllactose (6′-SL), and lacto-*N*-fucopentaose I (LNFP-I).

When the larger and more complex HMO structures are targeted – especially those with increased degree of fucosylation and/or sialylation such as LST-c, LST-a, and LNDFH-I – the *in vivo* technology can cause problems in terms of formation of carbohydrates that are not found in mother's milk, i.e. non-HMOs. This is mainly a consequence of a general increased complexity with the increased number of GTs to be co-expressed and sugar-nucleotide levels to be controlled. Naturally, the removal of the non-HMO fraction from these fermentations will have a high impact on the requirements of the downstream and purification processes. Therefore, to expand the toolbox for production of the more complex HMOs, Glycom A/S has developed in collaboration with c-LEcta GmbH an *in vitro* technology based on TGs where the key feature is the generation of HMO mixtures containing only structurally known HMOs without the non-HMO carbohydrates. This allows mimicking more closely the natural diversity found in human milk and opens up also the opportunity to design personalized formulas as supplements to human milk and significantly widening the structural space of commercially available HMOs.

2.5.2.1 Whole Cell Microbial Fermentation to HMOs (*In Vivo* Process)

Microbial fermentation is an established and well-recognized manufacturing process for food and pharmaceutical ingredients. Amino acids, antibiotics, and vitamins are all examples of compounds produced by fermentation using engineered microorganisms [51, 52]. Prokaryotic microorganisms are not known to be equipped with a natural function to synthesize HMOs. However, as their biochemistry provides the necessary building blocks (e.g. sugar nucleotides), the remaining metabolic gap can be closed by selective recombinant DNA technology and metabolic engineering of microbial host organisms and has already been demonstrated to be a feasible and economically viable technology to access large quantities of HMOs [53–55].

Glycom's *in vivo* technology is based on a bacterial host organism that has been carefully engineered with essential features and knock-outs to be able to produce various HMOs [56, 57]. In these processes lactose is added during the fermentation process and is taken up into the cell by the transporter lactose permease (lacY). The natural β-galactosidase (lacZ) activity is deleted by genetic manipulation to avoid lactose degradation. For HMO production from lactose, expression of the adequate nucleotide activated sugar donor is required together with the expression of matching glycosyl transferase enzyme (GT) suitable to transfer the sugar moiety from the activated sugar donor into a certain position of the lactose molecule as acceptor. Glycom's 2′-FL first-generation production strain schematically presented in Figure 2.5.3 is a two plasmid strain aiming to increase the intracellular GDP-L-fucose (guanosine 5′-diphospho-β-L-fucose) concentration in combination with expression of an efficient α1-2-fucosyltransferase enzyme. The plasmids also carry the genes of the sucrose operon, which enable the strain

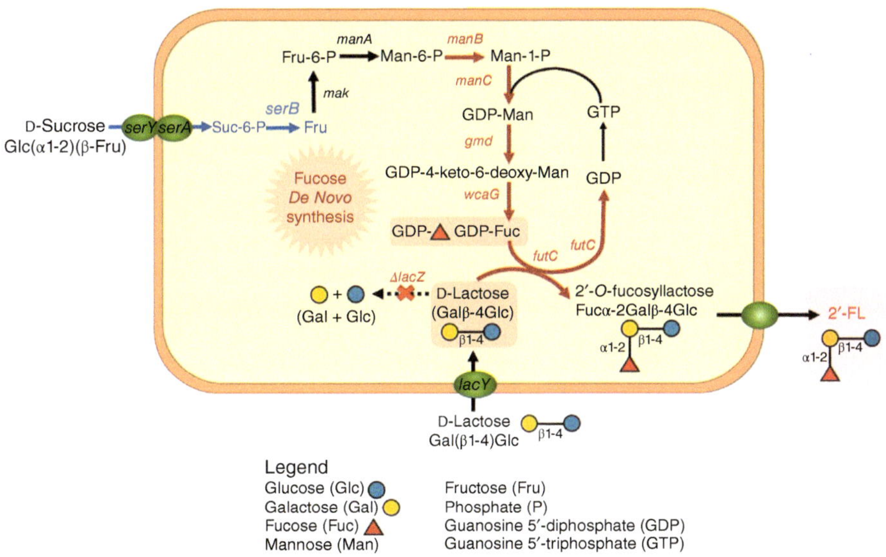

Figure 2.5.3 Schematic overview of engineered bacteria for 2′-FL production. Source: Graph reproduced from [35] with permission from Elsevier.

to grow on sucrose as the sole, relatively cheap carbon source. By paring the sucrose genes adequately onto two plasmids and growing the cells on sucrose medium, the use of antibiotics is eliminated and plasmid loss during fermentation is minimized. The resulting strain is highly stable, reliable, and produces extraordinarily high titers of 2′-FL. The high production efficiency of the strain is beneficial for product purification (i.e. increasing the ratio of product to potential impurities) at the start of the downstream purification process.

There are many ways to genetically modify bacterial host cells for the production of HMOs. Plasmid-born expression systems, such as the Glycom 2′-FL production strain described above, are typically characterized by high plasmid copy numbers that impose a high metabolic burden on the cells, thereby limiting their resources for (recombinant) protein production. At Glycom we work on finding alternatives to plasmid-based protein expression, i.e. genome-based expression that will ensure stable and selection marker-free expression of the genes of interest. Also, we work on modifying the ability of the bacterial host cells to take up lactose, to export the HMO produced, and on modifying metabolic pathways that support overexpression of the required nucleotide activated sugar donor.

One of the most appealing features of microbial fermentation is the scale-up potential of the process. In a fermentation upscaling program crucial parameters that require optimization include oxygen supply, media composition, and feed strategy – factors all important to keep the strain in non-stressed conditions and allowing it to do what it is made for, a high performing HMO producing cell factory. A well-controlled fermentation process is not worth much without an efficient downstream process (DSP), which includes purification and isolation of the HMO from the fermentation broth. The high-quality requirement on the end product requires special consideration in the DSP development. Especially important in the commercial production of HMOs by fermentation is the requirement of complete removal of the microbial production strain, absence of expressed proteins, lipopolysaccharides, DNA, and trace metals. Glycom has developed an efficient HMO purification process suitable for industrial scale, which is composed of an initial biomass removal by ultrafiltration followed by concentration of the broth using nanofiltration. Ion exchange chromatography ensures removal of inorganic salts and charged organic biomolecules, and activated charcoal treatment removes color bodies. Finally, the target carbohydrate components are polished with a crystallization step. Overall, this HMO downstream and purification process provides high-purity HMOs, safe to be used as ingredient in infant formula and other nutritional applications.

Evidently, *in vivo* fermentation technology is so far the most powerful method to produce HMOs in large quantities. The technological improvements in HMO production including isolation and purification have created an exciting opportunity to study the benefits of HMOs in detail. Access to the pure single HMOs and mixtures thereof in unprecedented quantities now allows the assessment of their metabolic fate and efficacy and their structure–function relationship using the entire spectrum from *in vitro* and *in vivo* models to human clinical studies.

2.5.2.2 The Glycom *In Vitro* Concept to Diversify HMO Blends

The *in vitro* diversification technology relies on the use of simple HMOs acting as donor/acceptor substrates that are treated with highly specific TGs that are rarely accessible directly from nature. Using the HMOs as substrates provides clear benefits since they offer remarkable properties in term of costs, safety, stability, and solubility compared to the usually described activated donors, i.e. the GT donors CMP-sialic acid (cytidine-5′-monophospho-N-acetylneuraminic acid) and GDP-fucose, and TG donors *para*-nitrophenyl-sialic acid (pNP-sialic acid), pNP-fucose, and fluoro-fucose. The reaction system provides a specific product mixture, called "HMO profile," containing the HMO donor and acceptor, the new HMO product plus liberated lactose. This offers a straightforward use of the HMO profile after simple DSP (mainly enzyme removal) since no component of the profile by itself would be considered as a concern to be left in final product – all are well-known compounds present in mother's milk. This HMO diversification approach is advantageous compared to the traditional oligosaccharide *in vitro* synthesis, where chemically activated donors are used liberating unnatural leaving groups (e.g. pNP), which are difficult to remove to acceptable levels for nutritional applications. In addition, the HMO profile is an advantageous starting point for purification and isolation of the formed single HMO product compared to *in vivo* fermentation since it provides a more defined product composition, especially in cases where the degree of fucosylation and/or sialylation increases [58].

2.5.2.3 Validation of the HMO Diversification Concept with Non-optimized Enzymes

In order to validate the HMO diversification concept, potential TGs were selected that could recognize simple lactose-derived HMOs as donors (3-FL, 3′-SL, and 6′-SL) and transfer a fucosyl- (Fuc) or sialic acid (SA) residue to HMO acceptors, thus leading to the formation of newly decorated HMO products (Figure 2.5.4). Based on literature search, the following three candidates were investigated in more detail: α1-3/4 transfucosidase (α1-3/4 TF) from *Bifidobacterium longum* subsp. *infantis* (BiTF L321P, GH29) [59, 60], α2-3 transsialidase (α2-3 TS) from *Trypanosoma cruzi* (TcTS, GH33) [61], and the α2-6 transsialidase from *Photobacterium damselae* JT0160 primary known as α2-6 sialyltransferase (Pd2,6ST, GT80) [62]. These enzymes were produced in 1–10 l scale fermentation processes and purified via his-tag affinity chromatography to afford enough biocatalyst amounts for the validation process.

Since the first test of HMO profile creation gave promising results, we aimed to demonstrate a wide applicability of the technology toward HMO formation. Thus, each enzyme was studied for their ability to decorate different types of HMO molecules. Acceptors and donors were selected in such a manner that the stipulated HMO products would be already isolated and structurally identified. The two tetrasaccharides LNT and LNnT were chosen as acceptors since they represent major core structures (lactose excluded) from which seven additional fucosylated and/or sialylated HMOs can be derived. Finally, the approach

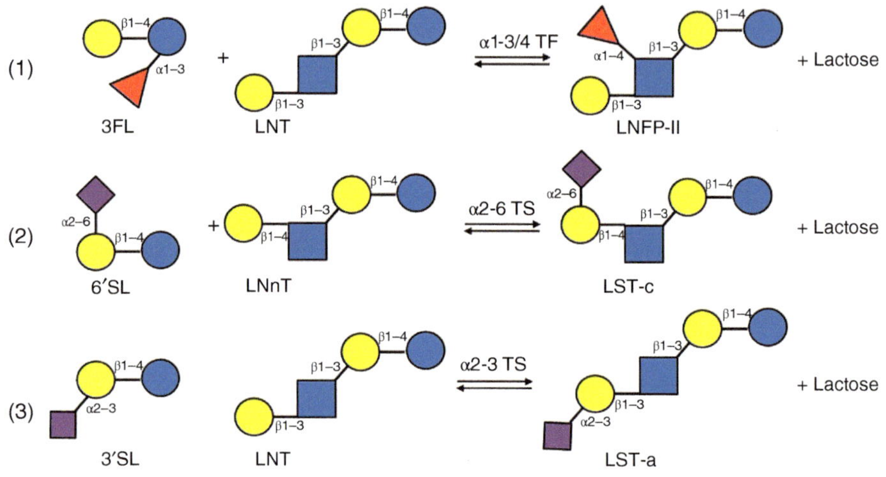

Figure 2.5.4 Reaction scheme for the transfucosidase and transsialidase as a proof-of-concept for HMO diversification.

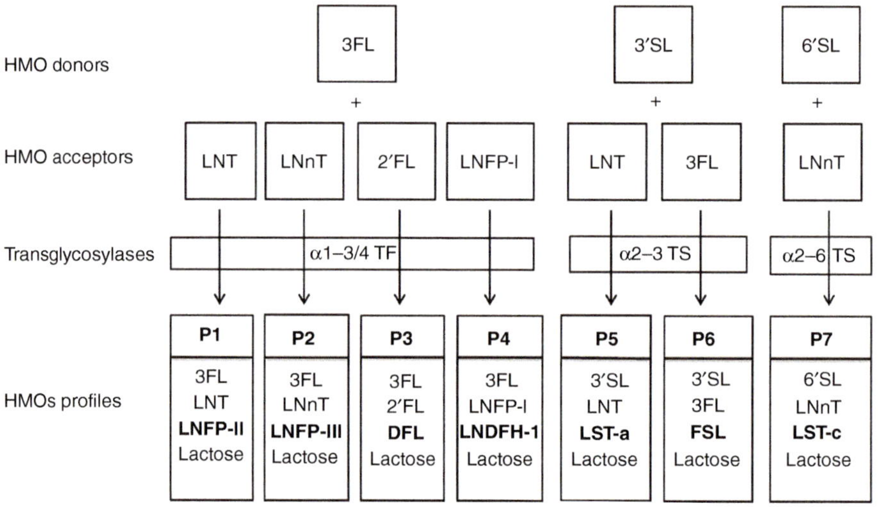

Figure 2.5.5 Diversification of HMOs into seven HMO profiles using different combinations of HMO donors and acceptors with different transglycosylases.

was successfully applied to already fucosylated acceptors (2′-FL, LNFP-I, and 3-FL), increasing their degree of decoration by generating di-fucosylated or monofucosylated/monosialylated structures. In total, seven different reactions leading to the HMO profiles P1–P7 (Figure 2.5.5) were studied in 1 ml volume to confirm product formation, optimized by varying different reaction parameters, and scaled up to 10–20 ml for product isolation and subsequent structural identification. Seven new HMOs were purified by preparative HPLC (quantity range: 10–600 mg) and characterized by NMR and/or MS/MS. Interestingly, the seven HMO profiles generated provide and sum up to totally 14 different

HMOs representing a majority of the most abundant HMOs in typical samples of human breast milk [63–66].

In profiles P1–P4 the α1-3/4 TF displayed very high specificity, preventing by-product formation. The main concern was the moderate conversion yield and the competing hydrolysis activity that makes the reaction difficult to control. On the contrary, the α2-6 TS involved in P7 showed a low level of hydrolysis but generated high levels of by-products (non-HMOs). For example, the low enzyme specificity resulted in the internal sialylation of LNnT forming two additional sialylated oligosaccharides. Finally, in profiles P5–P6 the α2-3 transsialidase (α2-3 TS) produced minor amounts of side products and showed low level of hydrolytic activity. Based on these results the α2-3 TS wild type (WT) was considered for direct scale-up development.

The synthetic properties of the α1-3/4TF and the specificity of the α2-6TS toward LST-c formation were not good enough to be considered for a scale-up development. Therefore, an enzyme engineering program was initiated between Glycom A/S and c-LEcta GmbH to remove the unwanted hydrolytic activity of the α1-3/4TF and the side product formation of α2-6TS. Additionally, both enzymes were optimized for industrial processes by improving the thermo- and process stability (enzyme production, storage, and application), and high efficiency (productivity of >100 g product per hour per gram of biocatalyst).

2.5.3 Enzyme Development

2.5.3.1 Optimization of the α1-3/4 Transfucosidase

The template enzyme for engineering the transfucosylation reaction was *B. longum* subsp. *infantis* (BiTF), which has been shown to have little synthetic activity, and initial enzyme engineering studies were undertaken [58]. Profile 2 was initially chosen as the reaction for optimization. The objective for the engineering was primarily the suppression of the hydrolytic activity toward LNFP-III and 3-FL to improve the yield of the transferase product LNFP-III (Figure 2.5.6).

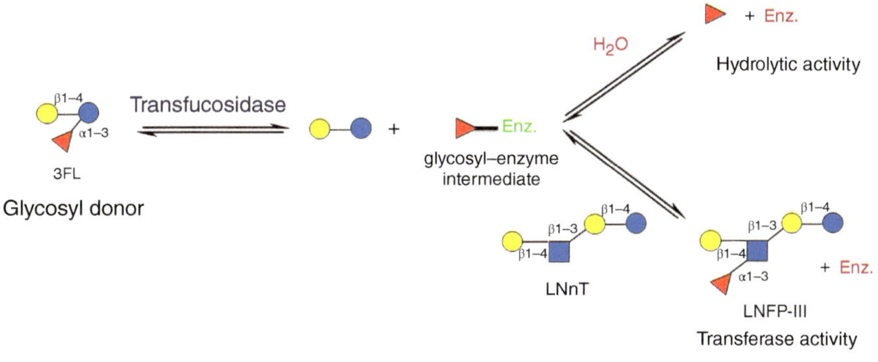

Figure 2.5.6 Reaction mechanism of transfucosylase and illustration of the two options from the glycosyl–enzyme intermediate leading to hydrolysis or transfucosylation.

Transfucosidases use a double-replacement mechanism comprising several steps. First the fucosyl-donor substrate, in this case 3-FL, enters the active site of the enzyme. The fucosyl residue is bound by several hydrogen bonds and is correctly oriented in the active site toward the two catalytic residues: aspartate 172 (D172) acts as a nucleophile and attacks the anomeric carbon atom of the fucosyl residue. A glutamate residue (E217) on the opposite position in the active site serves as a general acid/base, protonates the glycosidic oxygen, and promotes the leaving of lactose. This leads to the formation of a covalent glycosyl–enzyme intermediate. In the second step the breakdown of the intermediate usually proceeds by the attack of water at the anomeric center catalyzed by the general acid/base residue. Fucose is released as the product from the hydrolysis from the active site with overall retention of anomeric configuration (Figure 2.5.6) [67]. Alternatively, an alternative acceptor molecule can replace water leading to the transferase product, which is the desired reaction.

Transfucosidases belong to the enzyme class of glycoside hydrolases (GH29) and their predominant activity is hydrolysis. The transferase reaction is only a minor side activity. Under process-like conditions (100 g/l 3-FL, 142 g/l LNnT, 2 g/l enzyme) the BiTF wild-type enzyme shows over 30-fold higher hydrolytic activity compared to the transferase activity (Table 2.5.1). Previous enzyme engineering resulted in a single point mutation (L321P) showing some improvements in the transferase activity over hydrolysis [68]. Still, the hydrolytic activity of the variant was 4.5-fold higher compared to the transferase activity.

Therefore, the enzyme had to be engineered to avoid the activation of water and to promote the binding and activation of an acceptor molecule.

The binding of 3-FL in the active side was analyzed by automated docking of the product into the crystal structure of BiTF (pdb: 3UET) [60]. After molecular dynamics simulations in water for energy minimization of the *in silico* model, all residues responsible for the binding of 3-FL in the active site were identified. The fucosyl residue was found to be very specifically bound by several hydrogen bonds in the active site for a precise orientation of the substrate. The lactose moiety was coordinated by only very few hydrogen bonds. The binding of the desired product was studied as well in detail using *in silico* substrate docking and molecular dynamics simulations. Based on the generated models several residues could be identified, which directly or indirectly influence the binding 3-FL and the acceptor molecule LNnT to result in the desired LNFP-III product.

Table 2.5.1 Reaction kinetics of BiTF and its variants for transferase activity and hydrolysis.

Entry	Mutation	Synthesis of LNFP-III (U/mg)	Hydrolysis of LNFP-III (U/mg)	Hydrolysis of 3-FL (U/mg)	Synthetic performance S/H ratio
Wild type	None	6.5	208	132	0.03
L321P[a)]	L321P	1.5	7	25	0.2
BiTF-111	A174H	0.3	0.01	0.08	31

a) Mutation described by Osanjo et al. [68] for fucosidase from *Thermotoga maritima*.

Consequently, the first mutant library focused on the binding side of LNFP-III, selecting all residues that directly or indirectly influence the binding of the substrate and acceptor molecules. Residues that were essential for a correct orientation of the substrates were excluded from mutagenesis. Based on the suggested reaction mechanism [67], the hydrolytic activity will be reduced if no water molecules can be activated for a nucleophilic attack by the catalytic residues D172 and E217. We applied c-LEcta's proprietary MDM technology platform (Multi Dimensional Mutagenesis), a bioinformatic platform that includes an automated routine for analyzing the sequence for structural, functional, and homology features and guides the choice of sites and amino acid substitutions (see also Chapter 1.2). This resulted in the selection of 94 amino acid residues to be mutated and the deliberated selection of individual amino acid substitutions on each site. Each variant, containing a single mutation, was physically created and tested in biotransformation reactions for LNFP-III production using 200 mM of LNnT as acceptor and 3-FL as donor.

The single point mutation A174H (BiTF-111) was identified, which reduced the hydrolytic activity by a factor of more than 10 000 and improved the synthesis/hydrolysis ratio 2000-fold compared to the WT (Table 2.5.1).

With the BiTF-111 variant, the enzyme was transformed from a GH to a real transglycosidase, with transfucosylation being the predominant reaction (Figure 2.5.7).

This leads to a significant change in the reaction process. Owing to the little hydrolytic activity, much higher yields of up to 60% LNFP-III could be reached (Figure 2.5.7). The hydrolytic side reaction has been reduced to such an extent that even after prolonged incubation of the reaction mixture with the enzyme no significant hydrolysis could be detected. Instead, a thermodynamic equilibrium is reached. However, the molecular mechanism behind this dramatic change of the enzymatic properties has not been fully investigated. *In silico* automated docking studies with different substrates and molecular dynamics simulations lead to the assumption that the sterically more demanding residue at position

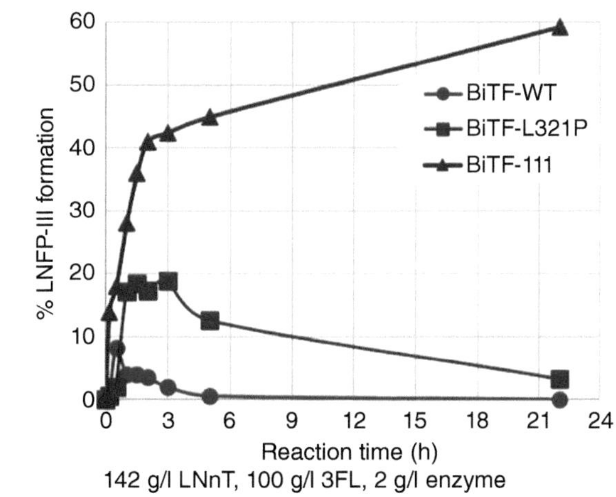

Figure 2.5.7 Effect of the single mutations on transfucosylated product formation. Biotransformation of 3-FL and LNnT to LNFP-III using process-like conditions (142 g/l LNnT, 100 g/l 3-FL, and 2 g/l crude enzyme extract).

174 prevents water from being correctly oriented for the hydrolytic attack at the glycosyl–enzyme intermediate.

A downside of this mutation is that BiTF-111 showed ~20-fold less activity for LNFP-III synthesis and no activity for the synthesis of the other target compounds. Accordingly, further optimization of the enzyme was pushed forward.

Further libraries were generated and screened to improve the activity and improve the thermostability, which is important for process and storage stability of the enzyme. Additional substrates were included into the screening to develop a final variant showing high process performance with a wide range of different donor and acceptor molecules. In total, 15 mutant libraries were designed using c-LEcta's MDM technology platform and thousands of variants subjected to a multiparameter screen under process-like conditions.

Various intermediate enzyme variants were obtained (BiTF-236, -357) and characterized, finally yielding variant BiTF-641, which contains only four mutations (W135F, A174N, N274A, E413R). Variant BiTF-641 shows nearly no hydrolytic side activity while the performance for LNFP-III synthesis could be further improved sixfold compared to BiTF-111 (Table 2.5.2). BiTF-641 can also be used for the production of di-fucosylated products and is applicable for the synthesis of all other target compounds. The thermostability, measured as the melting temperature T_m, was improved by 11 °C. The hydrolytic activity is negligible and not relevant in the synthesis reaction for the complex HMOs.

2.5.3.2 Optimization of the α2-6 Transsialidase

An α2-6 transsialidase was required for the transfer of a sialic acid moiety from 6′-sialyllactose (6′-SL) via an α2-6 linkage to lacto-*N*-neo-tetraose (LNnT) in order to form 6′-*O*-sialyllacto-*N*-neotetraose (LST-c). Initially, the α2-6 transsialidase from *P. damselae* (Pd2,6ST) [62] was investigated and showed the desired activity. However, the enzyme also showed significant side activity toward the formation of undesired products due to unspecific transsialidation of the internal galactose residue of LNnT (Figure 2.5.8). Only 20% of the desired LST-c was formed while 23% of 6SLNnT1 and 11% of double sialylated 6DSLNnT were formed. 6SLNnT1 and 6DSLNnT are not found in human milk and should therefore be avoided.

Table 2.5.2 Characterization and comparison of different generations of engineered BiTF variants and the starting wild-type (WT) enzyme.

	Transferase activity				Hydrolysis of	
Entry	LNFP-III (U/mg Lyo)	LNFP-II (U/mg Lyo)	LNDFH-I (U/mg Lyo)	DFL (U/mg Lyo)	LNFP-III (U/mg Lyo)	T_m (°C)
Wild type	6.5	n.d.	n.d.	n.d.	208	52
BiTF-111	0.3	n.d.	0.02	n.d.	0.01	58
BiTF-236	1.2	n.d.	0.02	n.d.	0.01	62
BiTF-357	0.97	n.d.	0.45	n.d.	0.02	51
BiTF-641	1.71	1.65	1.08	1.19	0.06	63

n.d., no data.

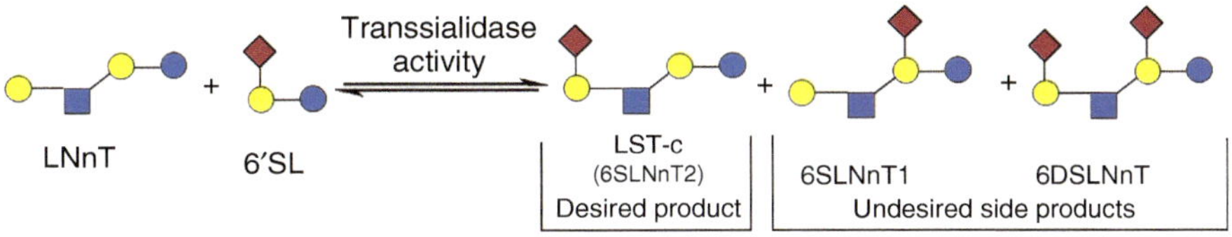

Figure 2.5.8 Reaction catalyzed by Pd2,6ST in the target reaction.

The enzyme belongs to the GT-80 family of glycosyltransferases, which consists of several members with α2-3 and α2-6 specificity. A striking feature of several members within this family is multifunctionality. They display sialyltransferase (ST) activity, catalyzing the α2-3/2-6-sialyltransfer from CMP-Neu5Ac to galactoside acceptors, but at the same time they display α2-3 and α2-6-transsialidase (TS) activity, accounting for the transfer of sialic acid from an donor substrate to an acceptor substrate [69]. It was proposed that the mechanism of the transsialidase activity is similar to the two-step double-displacement mechanism of well-characterized transsialidase from *T. cruzi* (member of the glycoside hydrolase family GH33), where the sialic acid is first transferred from the donor substrate to form a covalent sialyl–enzyme intermediate, followed by transfer to an alternate acceptor substrate [70]. However, more recently it was shown that the activity depends on the presence of CMP in the active site of the enzyme. It could also be demonstrated that CMP-Neu5Ac is generated as an intermediate from a reversed sialyation reaction and can then be hydrolyzed or transferred to an alternate acceptor substrate [69].

Another group of potential starting enzymes for the desired reaction are the α2-3 transsialidases from *Trypanosoma* species showing very low hydrolytic side activity. The rational to use these enzymes for building an α2-6-linkage is based on reports for the *Trypanosoma congolense* transsialidase [71] and *Trypanosoma brucei* transsialidase [72] demonstrating their acceptance of 6′-SL as well.

We started with the test of further candidates toward selectivity and activity in order to identify the best starting enzyme for a subsequent enzyme optimization. In this respect, eight different wild-type enzymes were analyzed for their suitability to perform α2-6 sialidase transfer (Table 2.5.3). The sialyltransferase from *Photobacterium leiognathi* JT-SHIZ-119 (PlST, Accession: BAI49484) [75] was selected from this selection because it showed the highest selectivity and good soluble expression level in *E. coli*. The enzyme was not stable over a period of 24 hours at 30 °C (35% residual activity), which is predicative for process stability, and thus needed to be stabilized for the application. A hydrolytic side reaction was, however, not an issue for PlST.

Based on structural investigations using high-quality homology models, c-LEcta used its proprietary MDM technology to predict several mutations at 192 positions in proximity to the substrate binding site, which were supposed to promote the selectivity of the enzyme for LST-c production as well as the overall activity. The library was screened by HPLC to identify active and, in terms of selectivity, improved variants. Twelve positions were identified to increase the selectivity of the enzyme for LST-c production. A complex recombination library was designed combining the four most promising positions (A218, N222, G156, G349). The best variant from this library screening, PlST-078

Table 2.5.3 Characterization of different candidates toward activity and selectivity. Only those enzymes expressed in *E. coli* were characterized.

Organism	Soluble expression in *E. coli*	Activity with 6′SL/LNnT	Ratio desired product/side products[a]	References
Bifunctional sialyltransferases				
Photobacterium damselae JT-ISH-0160	+	+	49%	[70]
Photobacterium sp. JT-ISH-224	++	+	78%	[73]
Photobacterium leiognathi JT-SHIZ-145	++	+	44%	[74]
Photobacterium leiognathi JT-SHIZ-119	+++	+	82%	[75]
Trypanosoma transsialidases (mainly α2-3*)*				
Trypanosoma brucei	−	−	n.d.	[76]
Trypanosoma congolense	−	−	n.d.	[77]
Trypanosoma congolense	−	−	n.d.	[77]
Trypanosoma cruzi	−	−	n.d.	[78]

a) Selectivity for LST-c production calculated by the amount of LST-c divided by the sum of observed side products.
n.d., no data.

(A218Y-N222D-G349C), showed a significantly improved regioselectivity. The product purity is dependent on the rate of conversion: after a certain amount of the desired product has been reached, the formation of side products becomes predominant. With the mutant, high product purities of 99% up to a conversion of 47% was achieved (Figure 2.5.9).

Further engineering rounds focused on improving the process stability, since the first-generation variant showed instabilities when used in a 48 hours reaction

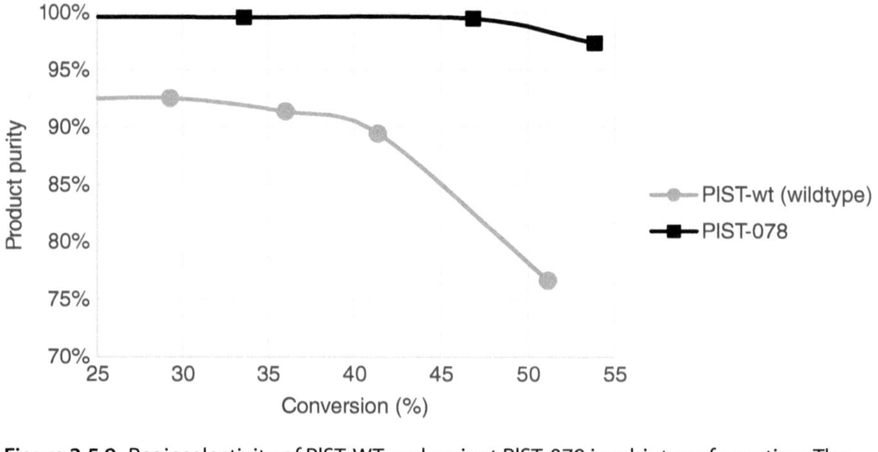

Figure 2.5.9 Regioselectivity of PIST-WT and variant PIST-078 in a biotransformation. The product purity depends on the conversion.

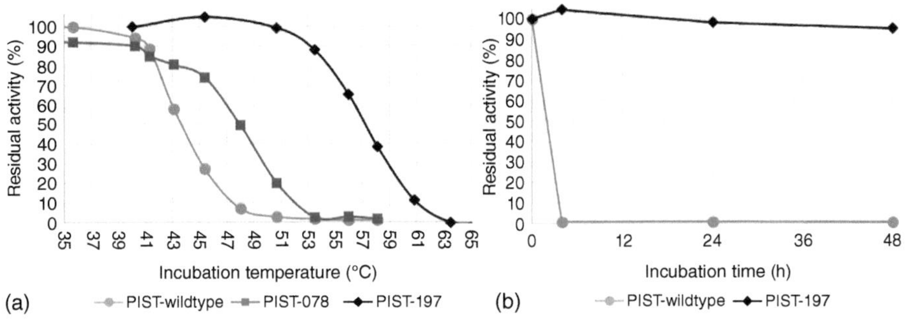

Figure 2.5.10 Thermostability improvements of PlST variants in comparison to WT. (a) Residual activity after 15 minutes of incubation at different temperatures. (b) Incubation under process conditions at 40 °C over 48 hours.

at 30 °C and very low stability at 40 °C, which was defined as a desired reaction temperature at that time in the development. This shortcoming was overcome by the final variant PlST-197, which incorporated five mutations (A218Y, N222R, G349S, S412P, D451K) responsible for an increase in the thermostability by 13 °C and, more importantly, perfect stability under process conditions over 48 hours at 40 °C (Figure 2.5.10). The latter turned out to be an important feature as it allowed to set the reaction temperature to even 55 °C in the final process (Table 2.5.4).

2.5.4 Applications of the Optimized Enzymes for the HMO Profiles

With the optimized enzymes in hand and with access of HMO donors and acceptors purified after single product *in vivo* fermentation, the concept was ready to be tested and valorized in preparative scale (>100 mmol scale) mimicking conditions and using unit operations suitable for industrial *in vitro* synthesis of the complex HMO products. The two optimized enzymes, α1-3/4 transfucosidase (BiTF-641) and α2-6 transsialidase (PlST-197), were used in development laboratories to provide large quantities of the HMO profiles P1–P4 and P7, respectively. In addition, the use of the non-engineered α2-3 transsialidase (TcTS, GH33) was elaborated to make LST-a (P5) and 3-fucosyl-3′-sialyllactose (FSL) (P6) in preparative scale. Before scale-up, reaction conditions were screened and tested out. In the end, suitable conditions and parameters were conducted in 250–500 g scale to demonstrate the feasibility of the processes for large-scale HMO profile production, targeting product quality suitable for dietary and nutritional applications.

2.5.4.1 Scale-Up of the Lacto-*N*-fucopentaose III (LNFP-III), Sialyl Lacto-*N*-neotetraose (LST-c), and Sialyl Lacto-*N*-tetraose (LST-a) HMO Profiles

Downscaling of a potential HMO profile production line resulted in well-known and non-complicated unit operations. In summary, the enzymatic reaction is executed in a conventional chemical reactor where pH, temperature, and agitation

Table 2.5.4 Screened conditions for generation of the HMO profiles P2, P5, and P7.

			Parameter							
			Temperature (°C)		pH[b)]		Donor/acceptor ratio[c)]		Time (h)[d)]	
Profile	Product	Enzyme[a)]	Test range	Selected	Test range[b)]	Selected	Test range	Selected	Test range	Selected
P2	LNFP-III	BiTF-641	20–70	55	5.0–8.0	~6.5	3 : 1–1 : 3	2 : 1	1–24	4
P5	LST-a	TcTS, GH33	20–55	55	5.0–8.0	~6.5	2 : 1–1 : 3	2 : 1	1–24	4
P7	LST-c	PlST-197	20–55	55	5.0–9.0	~6.5	2 : 1–1 : 3	2 : 1	1–24	4

a) Enzyme load 0.1 mg/ml reaction solution.
b) Various buffers are tolerated by the enzymes.
c) 150 mM concentration of donor and acceptor in the 1 : 1 ratio reaction mixture. The ratio HMO donor/acceptor is flexible and normally chosen depending on the targeted ratio of the different HMOs in the HMO profile of interest. Thus, by adjusting this ratio the HMO profile composition can be tailored given the application and/or needs.
d) The reaction time was chosen to achieve maximum substrate conversion in the shortest time, which is beneficial for productivity, and minimize the risk of microbial contamination, which could be a high risk for aqueous Carbohydrate-containing solutions.

are well controlled. When the enzymatic reaction is finished the enzyme is effectively adsorbed on low amounts of activated charcoal after a short heat cycle where the protein is denaturated. This is followed by filtration to remove the charcoal adsorbent holding the enzymes. Concentration by nanofiltration is the next operation, which is an excellent method to pre-concentrate an aqueous solution of HMOs without product loss – which then provides a concentrated solution of the HMO profile ready to be spray dried to finally deliver the HMO profile mix as a final powder.

In parallel to testing the DSP unit operations, the engineered and optimized enzymes were studied focusing on the parameters temperature, pH, reaction time, and ratio of HMO donor and acceptor (Table 2.5.4). For the pH control different standard buffer systems can be used, e.g. phosphate, citrate, and acetate. Given the selected parameters, the typical product formation was found to be approximately 50%, in good agreement with the small-scale data generated during enzyme optimization.

Based on the results obtained the scale-up program from 10–20 ml to 1–2 l was successful since the performance of the optimized enzymes in the HMO diversification was confirmed. In addition, considering the chosen experimental conditions and unit operations, the concept was concluded to be scalable and ready for the next step in an industrialization program offering high-quality HMO mixes well suitable for dietary and nutritional applications.

2.5.5 Conclusion and Perspective

The development described here is exemplary for the maturity of industrial enzyme development. The HMO synthesis was initially done by pure chemistry, which was important to get access to the compounds for further studies. However, simplification of the synthetic routes, strict control of regio- and chemoselectivity, scalable production processes, and simple HMO purification could be achieved only by enzymatic processes. The development was conducted in a very open-minded manner, with chemical, *in vivo,* and *in vitro* processes being investigated in parallel. In the end, a set of *in vivo* and *in vitro* biotechnological processes are now in place to access the different single HMOs and HMO profiles.

Enzymes are the catalysts of choice for the synthesis of complex oligosaccharides. It is crucial to gain expertise on the specific requirements, limitations, and prospects on the different enzyme classes that can be utilized. In this project, we used GTs for *in vivo* processes and TGs for *in vitro* processes. A third enzyme class, phosphorylases, is outlined in Chapter 2.4. The key for the application of TGs was the optimization by enzyme engineering. Various features such as improving S/H ratio, activity, selectivity, and thermostability were optimized and reached values exceeding our initial expectations. It turned out that enzyme engineering is particularly well suited for these enzymes to turn an initial synthetic activity into a commercially valuable synthesis process.

Glycom has made the necessary scientific breakthroughs for commercial production of HMOs, and they are now available for the first time. These oligosaccharides are structurally identical to those present in mother's milk and are of excellent purity, and safe for use. This a big step forward to close the nutritional gap between breast milk and infant formula. Now with a broad palette of HMOs available, a new chapter in human health is ready to be written, where the HMOs can offer lifetime benefits to individuals.

References

1 Oddy, W.H. (2002). *Breastfeed. Rev.* 10: 5–18.
2 Quigley, M.A., Carson, C., Sacker, A., and Kelly, Y. (2016). *Eur. J. Clin. Nutr.* 70: 1420–1427.
3 Yan, J., Liu, L., Zhu, Y. et al. (2014). *BMC Public Health* 14: 1267.
4 Oddy, W.H. (2017). *Ann. Nutr. Metab.* 70 (Suppl. 2): 26–36.
5 Munblit, D., Peroni, D.G., Boix-Amoros, A. et al. (2017). *Nutrients* 9: 894.
6 Horta, B.L., de Mola, C.L., and Victora, C.G. (2015). *Acta Paediatr.* https://doi.org/10.1111/apa.13139.
7 Victora, C.G., Bahl, R., Barros, A.J. et al. (2016). *Lancet* 387: 475–490.
8 Moore, M.L. (2001). *J. Perinat. Educ.* 10: 38–41.
9 Kunz, C. (2012). *Adv. Nutr.* 3: 430S–439S.
10 Bode, L. (2012). *Glycobiology* 22: 1147–1162.
11 Grollman, E.F., Kobata, A., and Ginsburg, V. (1969). *J. Clin. Invest.* 48: 1489–1494.
12 Hod, E.A., Spitalnik, P.F., and Spitalnik, S.L. (2009). Carbohydrate blood groups. In: *Rossi's Principles of Transfusion Medicine*, 4e (ed. T.L. Simon, E.L. Snyder, B.G. Solheim, et al.), 89–108. Blackwell Publishing Ltd.
13 Xu, G., Davis, J.C., Goonatilleke, E. et al. (2017). *J. Nutr.* 147: 117–124.
14 Ballard, O. and Morrow, A.L. (2013). *Pediatr. Clin. North Am.* 60: 49–74.
15 Urashima, T., Taufik, E., Fukuda, K., and Asakuma, S. (2013). *Biosci. Biotechnol. Biochem.* 77: 455–466.
16 Urashima, T., Fukuda, K., Kitaoka, M. et al. (2011). *Milk Oligosaccharides*, 1e. New York: Nova Biomedical Books. ISBN-13: 978-1611228311.
17 Chen, X. (2015). *Adv. Carbohydr. Chem. Biochem.* 72: 113–190.
18 Bishop, J.R. and Gagneux, P. (2007). *Glycobiology* 17: 23R–34R.
19 Springer, S.A. and Gagneux, P. (2016). *J. Proteomics* 135: 90–100.
20 Kitaoka, M. (2012). *Adv. Nutr.* 3: 422S–429S.
21 De Leoz, M.L.A., Kalanetra, K.M., Bokulich, N.A. et al. (2015). *J. Proteome Res.* 14: 491–502.
22 Oliveros, E., Ramirez, M., Vazquez, E. et al. (2016). *J. Nutr. Biochem.* 31: 20–27.
23 Wang, B. (2009). *Annu. Rev. Nutr.* 29: 177–222.
24 Ackerman, D.L., Craft, K.M., Doster, R.S. et al. (2018). *ACS Infect. Dis.* 4: 315–324.
25 Ackerman, D.L., Doster, R.S., Weitkamp, J.H. et al. (2017). *ACS Infect. Dis.* 3: 595–605.
26 Lin, A.E., Autran, C.A., Szyszka, A. et al. (2017). *J. Biol. Chem.* 292: 11243–11249.
27 Bode, L. and Jantscher-Krenn, E. (2012). *Adv. Nutr.* 3: 383S–391S.

28 Elison, E., Vigsnaes, L.K., Rindom Krogsgaard, L. et al. (2016). *Br. J. Nutr.* 116: 1356–1368.

29 Chichlowski, M., German, J.B., Lebrilla, C.B., and Mills, D.A. (2011). *Annu. Rev. Food Sci. Technol.* 2: 331–351.

30 Smilowitz, J.T., Lebrilla, C.B., Mills, D.A. et al. (2014). *Annu. Rev. Nutr.* 34: 143–169.

31 Sprenger, N., Lee, L.Y., De Castro, C.A. et al. (2017). *PLoS ONE* 12: e0171814. https://doi.org/10.1371/journal.pone.0171814.

32 Kunz, C., Meyer, C., Collado, M.C. et al. (2017). *J. Pediatr. Gastroenterol. Nutr.* 64: 789–798.

33 McGuire, M.K., Meehan, C.L., McGuire, M.A. et al. (2017). *Am. J. Clin. Nutr.* 105: 1086–1100.

34 Thurl, S., Munzert, M., Boehm, G. et al. (2017). *Nutr. Rev.* 75: 920–933.

35 Bych, K., Mikš, M.H., Johanson, T. et al. (2019). *Curr. Opin. Biotechnol.* 56: 130–137.

36 Austin, S., De Castro, C.A., Benet, T. et al. (2016). *Nutrients* 8: 346.

37 Donovan, S.M., Gibson, G.R., and Newburg, D.S. (2009). *Prebiotics in Infant Nutrition*, Mead Johnson Nutrition Monographs. https://www.meadjohnson.com/pediatrics/us-en/sites/hcp-usa/files/LB2329-Prebiotics.pdf.

38 Ackerman, D.L., Craft, K.M., and Townsend, S.D. (2017). *Carbohydr. Res.* 437: 16–27.

39 Bode, L., Contractor, N., Barile, D. et al. (2016). *Nutr. Rev.* 74: 635–644.

40 Agoston, K., Bajza, I., Dekany, G. et al. (2011). Glycom A/S, WO 2011/150939, Priority date 1 June 2010, WO2011150939A1.

41 Dekany, G., Bajza, I., Boutet, J. et al. (2010). Glycom A/S. WO2010115935A1.

42 Bajza, I., Dekany, G., Agoston, K. et al. (2011). Glycom A/S, WO 2011/100980, Priority date 19 February 2010, WO2011100980A1.

43 Kovacs, I., Bajza, I., Hederos, M. et al. (2013). Glycom A/S, WO2013044928A1.

44 Puccio, G., Alliet, P., Cajozzo, C. et al. (2017). *J. Pediatr. Gastroenterol. Nutr.* https://doi.org/10.1097/MPG.0000000000001520.

45 EFSA-NDA-Panel (2015). *EFSA J.* 13: 4184, 4132. https://doi.org/4110.2903/j.efsa.2015.4184. www.efsa.europa.eu/en/efsajournal/doc/4184.pdf. Scientific opinion on the safety of 2′-O-fucosyllactose as a novel food ingredient pursuant to Regulation (EC) No 258/97.

46 EFSA-NDA-Panel (2015). *EFSA J.* 13: 4183, 4132. https://doi.org/4110.2903/j.efsa.2015.4183. www.efsa.europa.eu/en/efsajournal/doc/4183.pdf. Scientific opinion on the safety of lacto-N-neotetraose as a novel food ingredient pursuant to Regulation (EC) No 258/97.

47 EU (2016). *Off. J. Eur. Union* 59 (L70): 27–31. http://eur-lex.europa.eu/legal-content/EN/ALL/?uri=CELEX:32016D30376. Commission Implementing Decision (EU) 2016/376 of 11 March 2016 authorising the placing on the market of 2′-O-fucosyllactose as a novel food ingredient under Regulation (EC) No 258/97 of the European Parliament and of the Council (notified under document C(2016) 1423).

48 EU (2016). *Off. J. Eur. Union* 59 (L70): 22–26. http://eur-lex.europa.eu/legal-content/EN/ALL/?uri=uriserv:OJ.L_.2016.2070.2001.0022.2001.ENG. Commission Implementing Decision (EU) 2016/375 of 11 March 2016 authorising the placing on the market of lacto-N-neotetraose as a novel food ingredient under Regulation (EC) No 258/97 of the European Parliament and of the Council (notified under document C(2016) 1419).
49 Bode, L. (2013). *Human Health Handbooks*, vol. 5 (ed. S. Zibadi, R.R. Watson and V.R. Preedy), 515–531. Wageningen: Wageningen Academic Publishers.
50 Urashima, T., Hirabayashi, J., Sato, S., and Kobata, A. (2018). *Trends Glycosci. Glycotechnol.* 30: SE51–SE65.
51 Theisen, M. and Liao, J.C. (2017). *Industrial Biotechnology*, 149–181. Wiley-VCH.
52 Chen, X., Zhou, L., Tian, K. et al. (2013). *Biotechnol. Adv.* 31: 1200–1223.
53 Fierfort, N. and Samain, E. (2008). *J. Biotechnol.* 134: 261–265.
54 Dumon, C., Bosso, C., Utille, J.P. et al. (2006). *ChemBioChem* 7: 359–365.
55 Drouillard, S., Driguez, H., and Samain, E. (2006). *Angew. Chem.* 45: 1778–1780.
56 Priem, B., Gilbert, M., Wakarchuk, W.W. et al. (2002). *Glycobiology* 12: 235–240.
57 Merighi, M., McCoy, J.M., and Heidtman, M.I. (2012). Glycosyn LLC, USA. US20120208181A1.
58 Saumonneau, A., Champion, E., Peltier-Pain, P. et al. (2016). *Glycobiology* 26: 261–269.
59 Ashida, H., Miyake, A., Kiyohara, M. et al. (2009). *Glycobiology* 19: 1010–1017.
60 Sakurama, H., Fushinobu, S., Hidaka, M. et al. (2012). *J. Biol. Chem.* 287: 16709–16719.
61 Frasch, A.C. (2000). *Parasitol. Today* 16: 282–286.
62 Huynh, N., Li, Y., Yu, H. et al. (2014). *FEBS Lett.* 588: 4720–4729.
63 Champion, E., Vogel, A., Bartsch, S., and Dekany, G. (2016). Glycom A/S, Den. WO2016063261A1.
64 Champion, E., McConnell, B., and Dekany, G. (2016). Glycom A/S, Den. WO2016063262A1.
65 Vogel, A., Schmiedel, R., Champion, E., and Dekany, G. (2016). Glycom A/S, Den. WO2016199069A1.
66 Champion, E., McConnell, B., and Dekany, G. (2016). Glycom A/S, Den. WO2016199071A1.
67 Liu, S.-W. and Li, Y.-K. (2009). *J. Chin. Chem. Soc.* 56: 850–858.
68 Osanjo, G., Dion, M., Drone, J. et al. (2007). *Biochemistry* 46: 1022–1033.
69 Mehr, K. and Withers, S.G. (2016). *Glycobiology* 26: 353–359.
70 Cheng, J., Huang, S., Yu, H. et al. (2010). *Glycobiology* 20: 260–268.
71 Tiralongo, E., Schrader, S., Lange, H. et al. (2003). *J. Biol. Chem.* 278: 23301–23310.
72 Montagna, G., Cremona, M.L., Paris, G. et al. (2002). *Eur. J. Biochem.* 269: 2941–2950.
73 Tsukamoto, H., Takakura, Y., Mine, T., and Yamamoto, T. (2008). *J. Biochem.* 143: 187–197.

74 Yamamoto, T., Hamada, Y., Ichikawa, M. et al. (2007). *Glycobiology* 17: 1167–1174.

75 Mine, T., Katayama, S., Kajiwara, H. et al. (2010). *Glycobiology* 20: 158–165.

76 Nakatani, F., Morita, Y.S., Ashida, H. et al. (2011). *Biochem. Biophys. Res. Commun.* 415: 421–425.

77 Koliwer-Brandl, H., Gbem, T.T., Waespy, M. et al. (2011). *BMC Biochem.* 12: 39.

78 Cremona, M.L., Sánchez, D.O., Frasch, A.C.C., and Campetella, O. (1995). *Gene* 160: 123–128.

Part III

Enzyme Applications for Human and Animal Nutrition

3.1

Enzymes for Human Nutrition and Health

Yoshihiko Hirose

Enzyme Techno, 733 Nagamatsu, Ogaki, Gifu 503-0997, Japan

3.1.1 Introduction

Industrial enzymes are extracted from animals and plants or produced by fermentation of microorganisms. People sometimes take enzymes to improve their digestion and prevent uncomfortable symptoms. Proteases such as papain, bromelain, and ficin that help in the digestion of proteins can be found in some fruits, for example, whereas amylases such as diastase that aid the digestion of starch and dextrin can be present in vegetables. Enzymes can be useful for human nutrition and healthcare. This enzyme application is dealt with in this chapter.

3.1.2 Current Problems of Enzymes in Healthcare Business

There are many enzymes provided for healthcare in the market. Lactase, α-galactosidase (ADG), dextranase, and laccase are some examples of typical commercial digestive enzymes. However, there are two issues regarding enzymes in the healthcare market: one is the country-specific regulations for (the status of) enzymes and the other is the status of genetically modified enzymes (GMO enzymes).

In 1971, the Ministry of Health, Labor, and Welfare (MHLW) in Japan issued a guideline stating that enzymes such as amylase, lipase, protease, lactase, and maltase are recognized common drugs.[1] Therefore, these digestive enzymes are not sold as supplements in Japan. The regulations in some EU and Asian countries are similar to those in Japan and enzyme drugs are sold at drug stores as common drugs. In the United States, on the other hand, these enzyme products are sold at the supermarket as dietary or nutritive supplements. Enzyme products are generally produced under Good Manufacturing Practice (GMP) guidelines and sold as prescription drugs or over-the-counter (OTC) drugs at drug stores (Table 3.1.1).

1 Notice of pharmaceutical practice chief of the bureau issued on 1 June 1971.

Industrial Enzyme Applications, First Edition. Edited by Andreas Vogel and Oliver May.

Table 3.1.1 Basic status of enzymes in each country (area).

	Common enzymes (amylase, protease, lipase, cellulase, hemicellulase, lactase, etc.)	Plant enzymes (bromelain, papain, etc.)
Japan	Drug	Food additive
USA	Food	Food
EU	Drug	Food additive
Asian	Drug	Food additive

In GMO enzymes, some amino acids in the variants are replaced by protein engineering techniques to improve their stability and specificity. The definition of GMO enzymes is ambiguous from time to time. When enzymes are produced in their original expression system and the amino acid sequence of enzymes is the same as the original enzymes, enzymes are recognized as non-GMO enzymes. The technique is used to increase the productivity and the case is a natural occurrence. But if the same enzymes are produced in a different expression system to increase the productivity the enzymes are considered GMO enzymes. Some GMO enzymes are already accepted for food processing; however, GMO enzymes are basically not allowed as dietary supplement. Enzyme manufacturers are required to provide the correct information.

3.1.3 Enzymes in Existing Healthcare Products

3.1.3.1 Digestive Enzymes

Digestive enzymes discussed in this chapter belong to the enzyme classes of amylases, proteases, and lipases, which support digestion of macronutrients such as carbohydrates, proteins, and lipids in general foods. Cellulase and hemicellulase, or pancreatin from porcine pancreas are sometimes added to digestive enzymes. The first commercial microbial enzyme was TAKA-diastase. Its production patent was filed in September, 1894, in the United States by Dr. Takamine [1, 2]. TAKA-diastase was launched by the company Park Davis in the United States in 1895 and by Sankyo in Japan in 1899. Many digestive enzyme products have been provided from pharmaceutical companies. Digestive enzymes are sold as a prescription or OTC drugs in Japan, with the respective sales estimated at \$8M and \$40M. They are applied to give relief in case of, for example, stomach discomfort, overeating, heartburn, and nausea. For future applications, the use of these enzymes is investigated to improve nutrition for the aged people and good quality of life (QOL) in patients with pancreatic insufficiency.

3.1.3.1.1 Digestive Enzymes in United States

Digestive enzymes are sold as dietary supplement at drug stores or supermarkets in the United States. The components of the product are similar to those of enzyme drugs in Japan and contain multienzymes with vitamins and prebiotics.

In 2011, an American company provided immobilized lipase cartridge (http://www.alcresta.com/relizorb%E2%84%A2-immobilized-lipase), which is used to hydrolyze esters of long-chain polyunsaturated fatty acids (LCPUFAs), such as docosahexaenoic acid, eicosapentaenoic acid, and arachidonic acid. These omega-3 and omega-6 fatty acids are essential to maintain good health and are easily absorbed as monoglycerides or free fatty acids.

3.1.3.1.2 Therapeutic Digestive Enzymes

Highly concentrated digestive enzymes extracted from porcine pancreas are sold as enzyme replacement therapy to treat maldigestion, malabsorption, and malnutrition as a result of exocrine pancreatic insufficiency associated with cystic fibrosis or chronic pancreatitis. They are produced under c-GMP compliance and controlled production methods of the same mixture of digestive enzymes. As these enzymes are not stable under the acidic condition found in the stomach, they are enteric-coated preparations that keep the enzymes active in the stomach. They are solely sold as prescription drugs, with the worldwide sales estimated around $800M (https://www.cff.org/).[2] The other therapeutic enzyme products are glucocerebrosidase, a kind of β-glucosidase, for Gaucher's disease, α-galactosidase for Fabry disease, and α-glucosidase for Pompe disease. These lysosomal storage diseases are caused by deficiency of single enzymes that metabolize lipids, glycoproteins, and oligosaccharides. The total market of enzyme replacement therapy is estimated at more than $3B.

3.1.3.2 Acid Lactase

People who are lactose intolerant are not able to properly digest lactose because of a lack of lactase [3–5]. Majority of the people stop producing lactase between two and five years of age, and as a consequence cannot digest lactose anymore. About 80% of people from Asian, African, Native American, or Mediterranean descent and approximately 15–20% of American adults are lactose intolerant to some extent [6–8]. In this case, acid lactase can be used for enzyme replacement therapy [9]. The sales of lactase in the US market were estimated to be about $650M in 2016 (Table 3.1.2).

Lactase catalyzes the hydrolysis of lactose to galactose and glucose. (Figure 3.1.1) There are three types of lactase, i.e. fungal acid lactase, bacterial lactase, and yeast neutral lactase [10, 11], which differ in pH stability. Fungal

Table 3.1.2 Population ratio of lactose intolerance.

Caucasian	10%
African American	75%
Chinese	93%
Tai	98%
Australians/Aborigine	85%

2 World Enzymes (Freedonia), 2015.

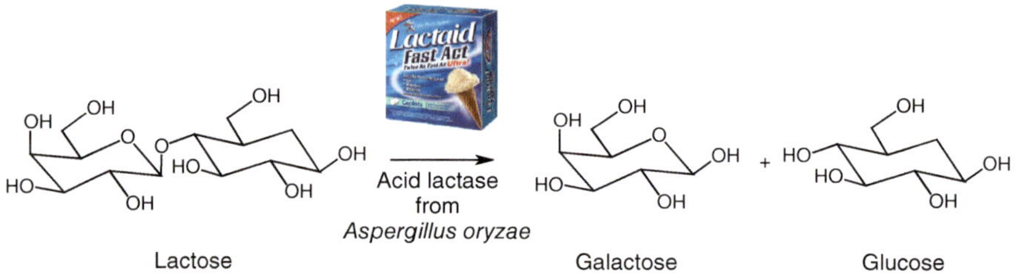

Figure 3.1.1 Enzymatic hydrolysis of lactose by lactase in stomach. Source: https://www.lactaid.com/products/lactaid-fast-act-caplets.

lactase is stable between pH 4 and 7, which is a very desirable property suitable for conditions in stomach after meal, making it the preferred lactase used as dietary supplement. Fungal acid lactase can be either from *Aspergillus oryzae* or from *Penicillium multicolor*, with the former being a very popular healthcare supplement in the United States and some European countries. As described above, lactase products are sold as only a prescription drug to the infants in Japan, Asian countries, and many European countries. Yeast lactase is stable at pH 7–9 and shows very high activity. This lactase is mainly used for lactose-free dairy products to remove lactose in milk products.

3.1.3.3 α-Galactosidase (ADG)

Some people suffer from discomfort, bloating, diarrhea, and gases in the body after eating beans, grains, nuts, broccoli, or many other vegetables. These high-fiber products contain oligosaccharides such as stachyose and raffinose that are indigestible by human intestinal enzymes and are fermented in the lower intestine and colon to produce the gases that cause discomfort. ADG derived from *Aspergillus niger* catalyzes the hydrolysis of these oligosaccharides to galactoses and sucrose, and as such reduces the physical discomfort [12]. (Figure 3.1.2) This enzyme is sold as dietary supplement in supermarkets or through internet except in Japan, with sales estimated around $40M [13, 14].

3.1.3.4 Dextranase

Dextran is a polymer of glucose with mainly α-1,6-glycoside and partially α-1,3-glycoside linkage. The use of the dextranase enzyme from *Chaetomium gracile* [15] is an efficient method for hydrolyzing dextrans in sugar production. This results in reduced viscosity of sugar syrup and improved process efficiency and recovery. Dextran can accumulate on the teeth and, hence, causes dental stone and plaque. A Japanese healthcare company has been selling a toothpaste containing dextranase named Clinica to remove scale or plaque and prevent dental decay since 1981 [16, 17] (Figure 3.1.3).

3.1.3.5 Glucose Oxidase

Glucose oxidase derived from *A. niger* is commercially available for diagnostic purposes to measure glucose level in the blood. It is also applied for oral care

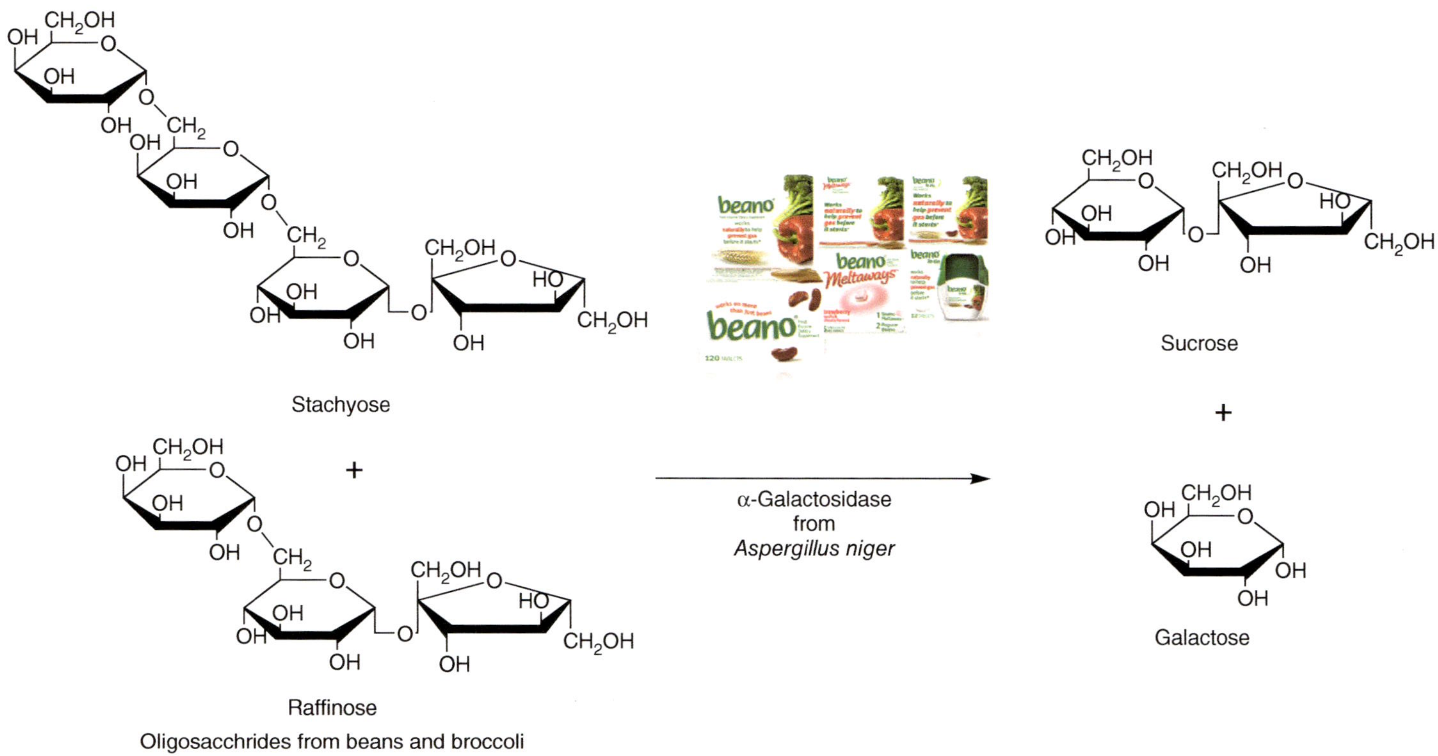

Figure 3.1.2 Enzymatic hydrolysis of oligosaccharides by α-galactosidase in stomach. Source: https://www.beanogas.com/anti-gas-pills/.

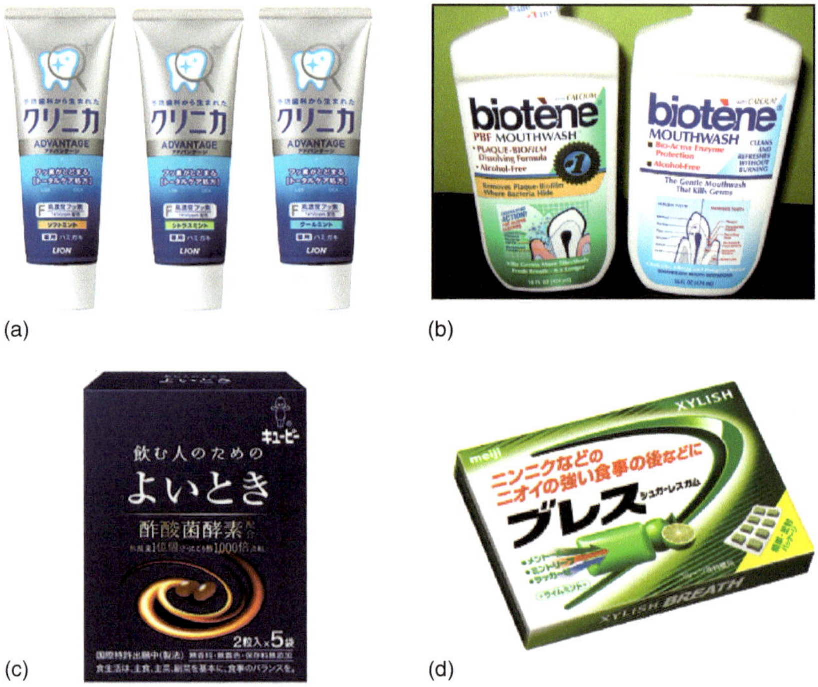

Figure 3.1.3 Healthcare products containing other enzymes. (a) Dextranase in toothpaste; (b) glucose oxidase in mouthwash; (c) alcohol dehydrogenase and aldehyde dehydrogenase in Acetobacter extract; and (d) laccase (polyphenol oxidase) in chewing gum.

in mouthwash products in the United States. This enzyme oxidizes glucose to gluconolactone and produces hydrogen peroxide as by-product. This hydrogen peroxide is able to sterilize bacteria and as such prevent bad breath in the mouth (Figure 3.1.3).

3.1.3.6 Acetobacter Enzymes

Gluconacetobacter sp. produces two enzymes, acetaldehyde dehydrogenase and alcohol dehydrogenase, which can convert ethanol to acetic acid [18, 19].

A Japanese company started to provide the supplement in Japanese market as an extract in 2016. These enzymes may protect and keep the function of liver from overdrinking and minimize hangover (Figure 3.1.3).

3.1.3.7 Laccase (Polyphenol Oxidase)

A Japanese company sells chewing gum containing laccase derived from *Trametes hirsute*. This laccase is a food additive capable of changing the taste and odor of chewing gum for bad breath prevention. Chewing gum contains micro-encapsulated laccase with some mediator. When chewing, laccase oxidizes phenolic compounds, which results in changed taste and odor of the gum [20] (Figure 3.1.3).

3.1.4 New Enzyme Developments in Healthcare Products

3.1.4.1 Transglucosidase

Transglucosidase (TG) is used in the food industry to prepare oligosaccharides from maltose and maltotriose. For more than 10 years, a Japanese enzyme company has investigated applying this enzyme as dietary supplement to prevent the increase of postprandial blood glucose level [21–23]. The working mechanism of TG is shown in Figure 3.1.4. It was shown that 450 kU(150 mg) and 900 kU(300 mg) of TG resulted in the formation of oligosaccharides (both *in vitro* and *in vivo*), preventing glucose levels from rising when tested in healthy volunteers [24–28].

TG derived from *A. niger* is a very common enzyme and shows two different activities: transglucosidase activity and α-glucosidase activity. The former rearranges the sugars and catalyzes the formation of panose, isomaltose, and isomaltotriose, which are oligosaccharides difficult to digest. The latter activity leads to an undesired reaction in which panose and isomaltose are gradually hydrolyzed (Figure 3.1.5).

TG shows good activity and effective stability between pH 4 and 6, which is similar to the condition encountered in the stomach after a meal. TG is a safe food additive registered in Japan and France.

There are some approaches to reduce and inhibit the hydrolysis activity of TG. The most popular and effective method is to prepare the best mutant by protein engineering technique. However, as the purpose is to apply this enzyme inhibitor as dietary supplement, only non-GMO enzymes are allowed. Therefore, the best method would be to use the existing enzyme in combination with a natural material showing α-glucosidase inhibitor activity. There are diabetic drugs with strong α-glucosidase inhibiting activity. Besides these, a new material of *Salacia reticulata* containing salacinol and kotalanol apart from a mild α-glucosidase inhibiting activity was recently found (Figure 3.1.6). When combining TG and Salacia, hydrolysis activity of TG was reduced to 1/3 of its original activity, without affecting the formation of panose from maltose (Y. Hirose, unpublished data).

3.1.4.2 Laccase

Many elderly people try to use commercial hair dyeing products to look young. However, there are allergy risks associated with to some dyeing reagents and chemicals. Therefore, the use of a safer dyeing product based on enzymes instead of risky chemicals was investigated.

The basic formulation of commercial hair dyeing products was established in 1883 in France and is still being used. The formulation typically consists of *p*-phenylene diamine (PPD), hydrogen peroxide, and ammonia solution. These three reagents are harmful for human hairs or skins, and – in the case of PPD – can cause serious allergy. The surface of hairs is formed with cuticle protein, which mainly consists of acidic amino acids. When the pH of the

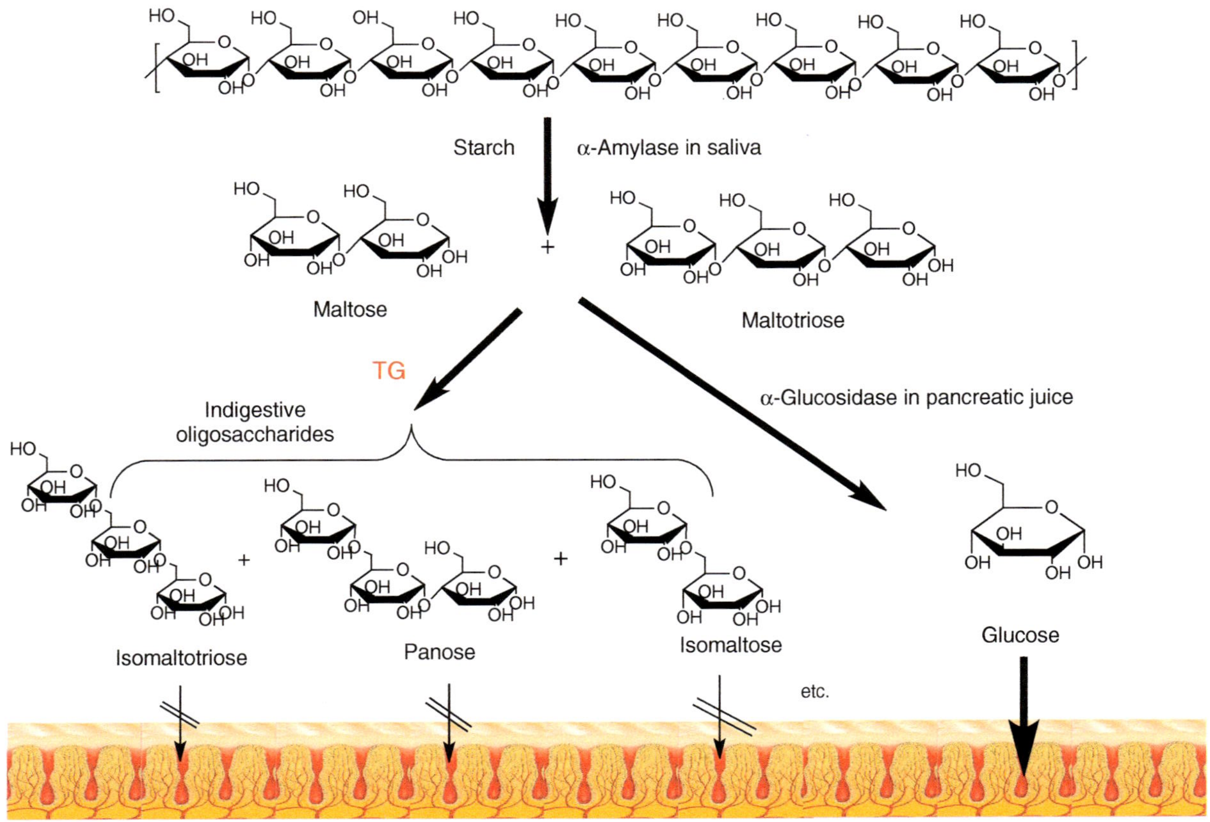

Figure 3.1.4 Prevention of postprandial glucose absorption level by TG.

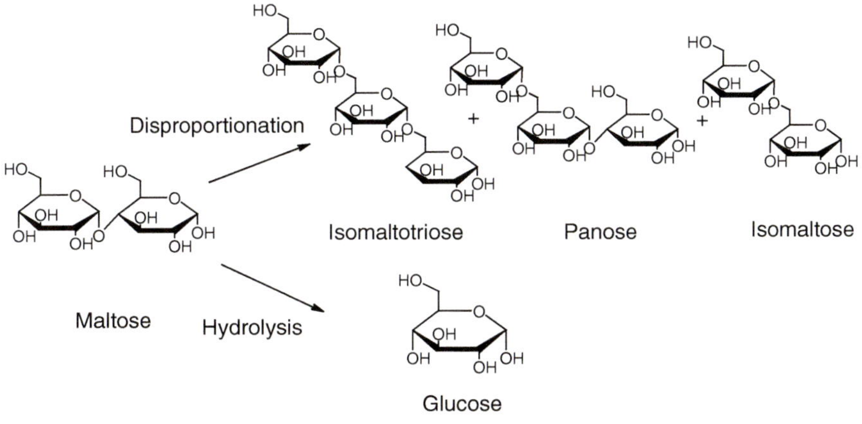

Figure 3.1.5 Enzymatic reaction of TG.

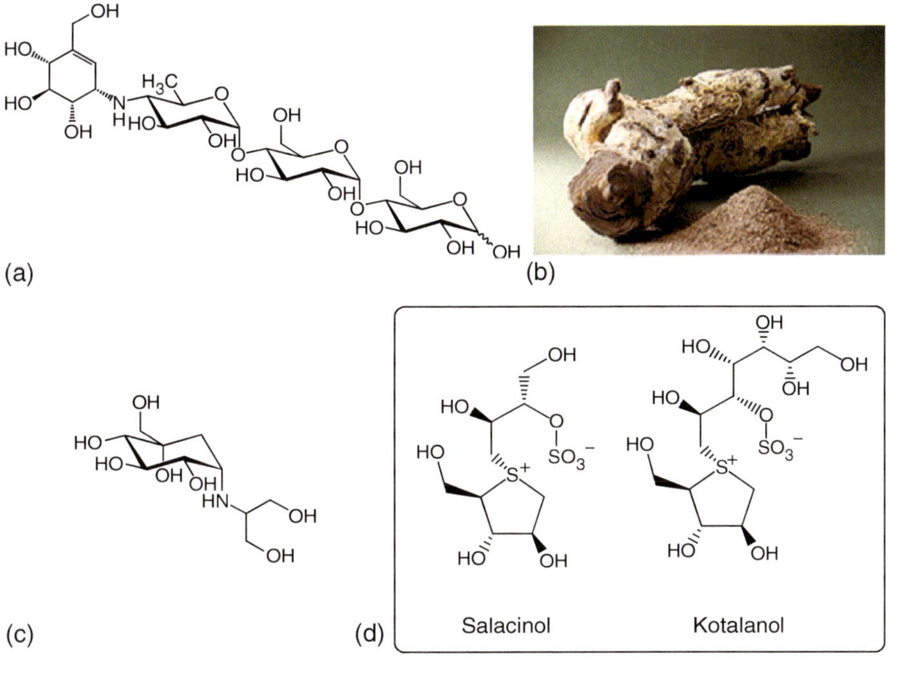

Figure 3.1.6 α-Glucosidase inhibitors (α-GIs). (a) Glucobay (acarbose) Bayer 100 mg/meal, 3 times/day; (b) *Salacia reticulata* (ayurvedic medicine); (c) Basen (voglibose) Takeda 0.2 mg/meal, 3 times/day; and (d) α-GIs in *Salacia oblonga*, *Salacia reticulata*, and *Salacia chinensis*.

hair surface is high, the cuticle is loosened, and dyeing reagents and hydrogen peroxide can permeate into the cuticle, allowing hair dyeing to go well.

Melanin's biosynthesis pathway is shown in Figure 3.1.7. 5,6-Dihydroxyindole (DHI) is a melanin precursor and is known as one of the safer reagents for hair dyes [29]. The application of DHI for hair dyeing has been reported since the

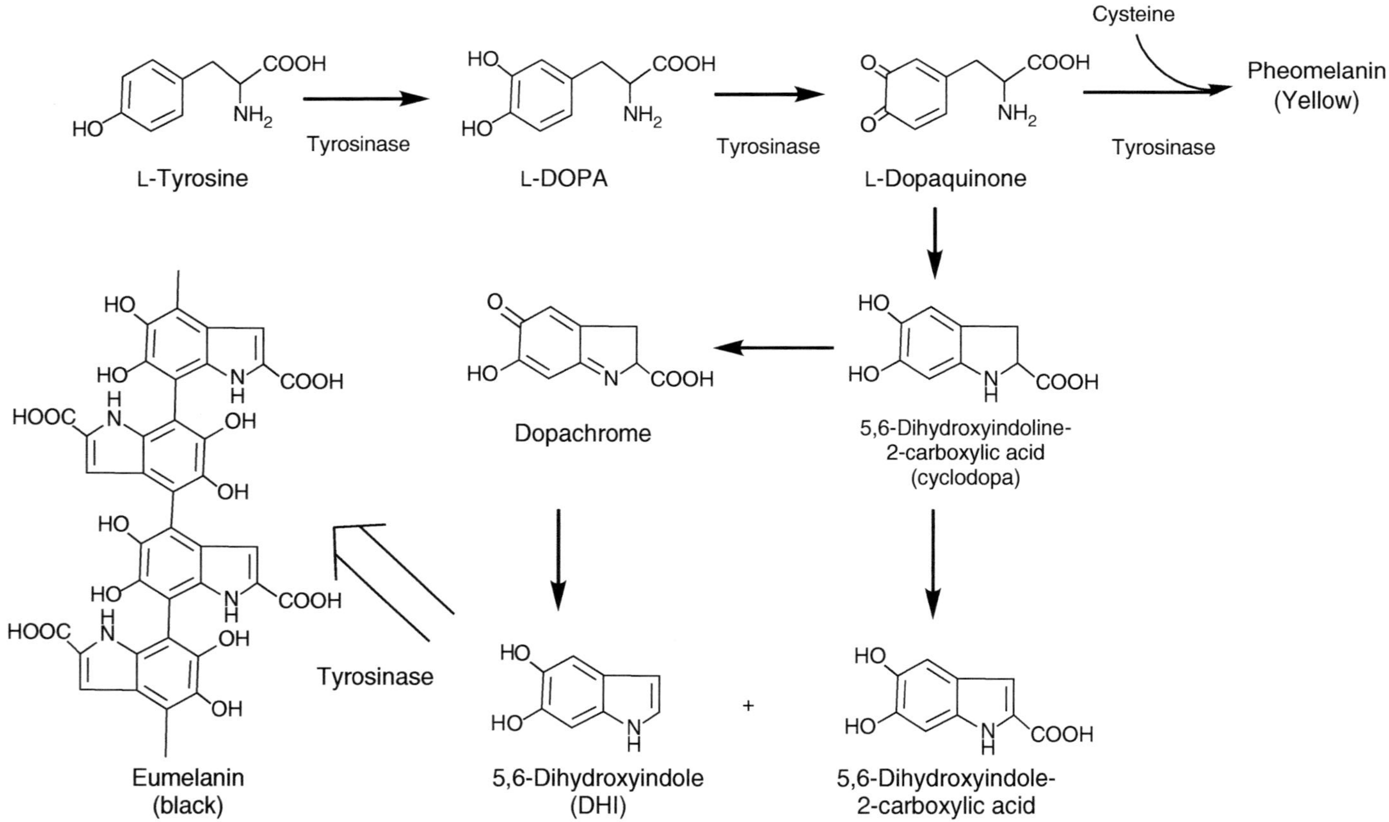

Figure 3.1.7 Melanin's biosynthesis pathway.

1970s after which many healthcare companies tried to commercialize the product [30]. One drawback of this product is that it requires oxidation, which can be slow or unreliable using natural air oxidation. To improve this process commercial laccase [31, 32] derived from *Myrothecium verrucaria* is a more recent and promising development area [33–36].

References

1 Takamine, J. (1894). Process of making diastatic enzyme. USP525823.
2 Advertisement of Parke, Davis & Co. (1895). *Am. J. Pharm.* 67: 12.
3 Sibley, E. (2004). Genetic variation and lactose intolerance: detection methods and clinical implications. *Am. J. Pharm.* 4: 239–245.
4 Mattherws, S.B., Waud, J.P., Roberts, A.G., and Campbell, A.K. (2005). Systemic lactose intolerance: a new perspective on an old problem. *Postgrad. Med. J.* 81: 167–173.
5 Lomer, M.C., Parkes, G.C., and Sanderson, J.D. (2008). Review article: lactose intolerance in clinical practice--myths and realities. *Aliment. Pharmacol. Ther.* 27: 93–103.
6 Jackson, K.A. and Savaiano, D.A. (2001). Lactose maldigestion, calcium intake and osteoporosis in African-, Asian-, and Hispanic-Americans. *J. Am. Coll. Nutr.* 20: 198S.
7 Kanai, H. and Akino, T. (2003). Function of lactase and its application: food processing technology with enzyme utilization (Japanese). *Food Chem.* 9: 76–84.
8 Swagerty, D.L. Jr., Walling, A.D., and Klein, R.M. (2002). Lactose intolerance. *Am. Fam. Phys.* 65: 1845–1850.
9 DiPalma, J.A. and Collins, M.S. (1989). Enzyme replacement for lactose malabsorption using a beta-D-galactosidase. *J. Clin. Gastroenterol.* 11: 290–293.
10 Goto, M. (2012). Fungal lactase (Japanese). *Milk Sci.* 61: 239–245.
11 Catani, J. S. and Yocum, F. K. (2008). Lactase-containing comestibles and related methods. WO/2008/030828A1.
12 Kligerman, E. A. (2002). Compositions and method for reducing gastro-intestinal distress due to alpha-D-galactosidase-linked/containing sugars. USP6344196B1.
13 Di Stefano, M., Miceli, E., Gotti, S. et al. (2007). The effect of oral alpha-galactosidase on intestinal gas production and gas-related symptoms. *Dig. Dis. Sci.* 52: 78–83.
14 Adya, S. and Elbein, A.D. (1977). Glycoprotein enzymes secreted by *Aspergillus niger*: purification and properties of alpha-galactosidase. *J. Bacteriol.* 129: 850–856.
15 Horita, M., Suguro, T., Obayashi, K., and Akahori, T. (1989). Studies on dextranase production. Part I. Dextranase production in "fed-batch" culture (Japanese). *Nippon Nogeikagaaku Kaishi* 63: 837–844.
16 Amimoto, A., Okamoto, H., Tamazawa, O. et al. (1978). In vitro and clinical effects of *chaetomium gracile* dextranase on the release of reducing sugar and on the reduction of dental plaque. *J. Dent. Health* 28: 89–98.

17 Ishibashi, K., Hattori, A., Ikegami, Y., and Kato, S. (1981). Preparation method of tablets including dextranase. JP37969.
18 Matsushita, K., Matsutani, M., and Yakushi, T. (2010). Adaptive evolution of acetic acid bacteria and application of the adaptive ability to development of high temperature fermentation system (Japanese). *J. Brew. Soc. Jpn* 105: 730–737.
19 Narita, K., Seitou, Y., Matsushita, K., et al. (2015). Screening of acetic acid bacteria having high activity of aldehyde dehydrogenase. The Society for Bioscience and Bioengineering, Japan (Abstract), 114.
20 Shimizu, S. and Nakatani, T. (1999). Laccase and the production method, JP2984552.
21 Kariya, K., Ogawa, T., and Jou, K. (2001). Effect on gut flora of oligosaccharide synthesizing enzymes (Japanese). *Digestion & Absorption* 23: 107–109.
22 Kimura, S., Ogawa, T., Kariya, K and Yanase, H. (2000). Enzyme composition and use thereof. JP2000-325045, EP0968719A2, US6042823.
23 Ishihara, S. (2016). Modified alpha-glucosidase and applications of same. WO/2012124520, JP5992902B2.
24 Mizoshita, T., Jou, T., Sasaki, M. et al. (2006). Effect on gut flora and blood glucose of synthesizing oligosaccharide enzymes (Japanese). *Digestion & Absorption* 28: 79–81.
25 Sako, T., Mori, A., Lee, P. et al. (2010). Supplementing transglucosidase with a high-fiber diet for prevention of postprandial hyperglycemia in streptozotocin-induced diabetic dogs. *Vet. Res. Commun.* 34: 161–722.
26 Sasaki, M., Joh, T., Koikeda, S. et al. (2007). A novel strategy in production of oligosaccharides in digestive tract: prevention of postprandial hyperglycemia and hyperinsulinemia. *J. Clin. Biochem. Nutr.* 41: 191–196.
27 Sasaki, M., Imaeda, K., Okayama, N. et al. (2012). Effects of transglucosidase on diabetes, cardiovascular risk factors and hepatic biomarkers in patients with type 2 diabetes: a 12-week, randomized, double-blind, placebo-controlled trial. *Diabetes Obes. Metab.* 14: 379–382.
28 Sasaki, M., Ogasawara, N., Funaki, Y. et al. (2013). Transglucosidase improves the gut microbiota profile of type 2 diabetes mellitus patients: a randomized double-blind, placebo-controlled study. *BMC Gastroenterol.* 13: 81–87.
29 Obata, H., Ishida, H., Hata, Y. et al. (2004). Cloning of a novel tyrosinase-encoding gene (melB) from *Aspergillus oryzae* and its overexpression in solid-state culture (Rice Koji). *J. Biosci. Bioeng.* 97: 400–405.
30 Lan, J. and Kotolle, J. (2002). Coloring oxidized components including laccase of kelatin and coloring method using their components, JP2002-509086.
31 Murao, S. and Tanaka, N. (1981). A new enzyme "Bilirubin Oxidase" produced by *Myrothecium verrucaria* MT-1. *Agric. Biol. Chem.* 45: 2383–2384.
32 Tanaka, N. and Murao, S. (1985). Reaction of bilirubin oxidase produced by *Myrothecium verrucaria* MT-1. *Agric. Biol. Chem.* 49: 843–844.
33 Mizutani, K., Toyoda, M., Sagara, K. et al. (2010). X-ray analysis of bilirubin oxidase from *Myrothecium verrucaria* at 2.3 A resolution using a twinned crystal. *Acta Crystallogr. Sect. F Struct. Biol. Cryst. Commun.* 66: 765–770.
34 Maeda, H., Sasaki, I., Hirose, Y. et al. (1992). Toxicity of Bilirubin and detoxification by PEG-bilirubin oxidase conjugate: a new tactic for treatment of

jaundice. In: *Poly (ethylene glycol) Chemistry: Biotechnical and Biomedical Applications* (ed. J.M. Harris), 153–169. New York: Plenum Press.

35 Hirose, Y. and Sato, Y. (2012). Dyeing agent and use for same. WO2012/153791A1.

36 Sato, Y. and Hirose, Y. (2013). Dyeing keratinous fibers the with indole analogues. WO2013/099296 A1.

3.2

Enzyme Technology for Detoxification of Mycotoxins in Animal Feed

Dieter Moll

BIOMIN Research Center, Technopark 1, 3430 Tulln, Austria

3.2.1 Introduction to Mycotoxins

The secondary metabolism of filamentous fungi is capable of producing great chemical diversity. Many molecules of this chemical diversity have little or no known biological activity and seem to just collectively form the molecular variety pool from which fungi can evolve highly potent small molecule ligands for biological macromolecules [1]. Ligands with sufficiently high and specific binding affinity can interfere with essential molecular processes of cells, are therefore toxic, and help fungi occupy or defend their biological niches. In that way, such ligands give fungi an advantage in natural selection. Several fungal molecules with high binding affinity to important biomolecular targets were discovered and developed for use as antibiotics, antifungals, cholesterol-lowering substances, and immunosuppressants and for other applications [2]. However, some fungal molecules were classified as mycotoxins because of their toxicity for animals and humans. Mycotoxins comprise toxins, which are produced by filamentous fungi and to which animals or humans are inadvertently exposed [3]. In most cases, this inadvertent exposure is through consumption of mycotoxin-contaminated feed or food, but exposure through inhalation or skin contact is also possible. Mushroom poisons are not comprised within the definition of mycotoxins, because fruiting body forming fungi do not normally grow on crop plants or harvested crop, and while these poisons may accidentally be consumed by mushroom hunters, they are not normally encountered as contaminants of regular animal or human diets.

Mycotoxins are produced by many different filamentous fungi, and the most important genera are *Aspergillus, Penicillium,* and *Fusarium*. Some of these fungi thrive on stored grain or other feed or food if humidity is sufficiently high, and are called postharvest mycotoxin producers. There are also preharvest mycotoxin producers, which include powerful plant pathogens such as *Fusarium* species. These fungi infect plants on the field, and while postharvest mycotoxin formation can or at least could typically be prevented by sound storage conditions, preharvest mycotoxin formation is notoriously difficult to control.

Industrial Enzyme Applications, First Edition. Edited by Andreas Vogel and Oliver May.

Mycotoxins probably played a role in animal and human nutrition and health throughout history, and were associated with events described in the Old Testament of the Bible or historic records [4]. Apparently, even Neanderthals were aware of bioactive ingredients in molds, and may have deliberately used antibiotic-producing *Penicillium* for medication [5]. The mycotoxicosis caused by ergot alkaloids of *Claviceps purpurea*, a pathogen of rye, was endemic in central Europe in the previous centuries [6], and vasoconstriction and other symptoms caused by ergot alkaloids, historically called "St. Anthony's Fire," continue to be a concern in animal nutrition [7].

Modern mycotoxin research began after the discovery that Brazilian groundnut meal caused acute and fatal hepatotoxicity in turkeys [8, 9]. Strains of *Aspergillus flavus* were identified, and the active principle, aflatoxin, was isolated with the help of a bioassay with day-old ducklings [10]. The molecular structure of aflatoxin was determined [11], and aflatoxin was found not only to be toxic for all tested animal species but also carcinogenic [12]. Subsequently, other important mycotoxins were discovered, including ochratoxin A [13], zearalenone [14], deoxynivalenol [15], and the fumonisins [16]. Today, more than 300 fungal metabolites are classified as mycotoxins. Analytical methods to detect and quantify their concentration in feed or food are available and continue to be developed [17]. These methods have allowed to conduct global surveys of the prevalence of mycotoxins in grains, feed, and food, and showed that the presence of mycotoxins is the rule rather than the exception [18]. However, not all mycotoxin contaminations that can be detected with sensitive analytical methods are also of toxicological concern. In Europe, the United States, and many other countries, legislation for maximum acceptable concentrations in food or feed is available for several mycotoxins. In these countries, mycotoxins can be considered a minor and sporadic health concern for humans, which is dwarfed by infectious diseases and cultural diseases including obesity, alcoholism, and smoking. However, the frequency of occurrence of certain types of cancer or birth defect is associated with mycotoxin exposure [19, 20], and in countries with less rigid or no safety regulations, mycotoxins continue to be a major human health issue [21].

Farm animals all over the world are habitually exposed to mycotoxins. Although acute cases of mycotoxin poisoning may be rare, reduction of animal performance caused by low levels of mycotoxin contamination is common and widespread [22]. Typical effects of mycotoxins in animal feed are reduced weight gain [23] and immunomodulation, which can make animals more susceptible to infectious diseases [24, 25]. Not only mycotoxin-caused health effects but also economic effects are difficult to quantify. If losses could be quantified more accurately, the drive for implementation of mycotoxin mitigation strategies would be even stronger. However, calculations to reach estimates were made [26] and indicate that mycotoxins cause substantial annual losses in global animal husbandry.

3.2.2 Mycotoxin Mitigation Strategies

The mycotoxin problem would ideally be solved by prevention of mycotoxin contamination rather than remediation. Preharvest mycotoxin contamination could be prevented if crop cultivars were resistant to fungal infection. Fungal resistance, for instance, *Fusarium* resistance of wheat or maize, is determined by many phenotypic traits. The lack of highly resistant genotypes or single, major resistance genes makes breeding of resistant cultivars difficult. Furthermore, plant breeders have to consider also agronomic traits, yield, resistance to other pests, and local adaptation of their cultivars. Cultivars with moderate *Fusarium* resistance have been bred [27], and quantitative trait loci for *Fusarium* resistance in the wheat genome were identified and are used in breeding programs [28]. Plant biotechnology has also been employed, and host-induced gene silencing, which is based on production of small interfering RNA molecules in transgenic plants to silence genes of a fungal pathogen, has been reported to have potential for fungal infection and mycotoxin control [29]. Genetic modification of plants to enzymatically biotransform and detoxify mycotoxins in transgenic plants has also been attempted [30]. The use of transgenic insect-resistant maize (*Bt* maize) may also be considered a mycotoxin control measure, because reduction of insect damage also reduces fungal infection [30].

Several agricultural practices were described to reduce fungal infection of crop plants and mycotoxin formation on the field [31]. Fungicides can help reduce *Fusarium* head blight in wheat, but their application is not straightforward and the efficacy is influenced by several factors [32]. Crop rotation and tillage can also be used with mycotoxin management in mind. Predictive modeling, based on geographic location and weather data, is being established to support farmers with *Fusarium* management decisions such as time of fungicide application or time of harvest [33]. Biological approaches to mycotoxin management have also been pursued. The aflatoxin-producing ability of *A. flavus* isolates is known to range from highly aflatoxigenic to atoxigenic, and field application of atoxigenic strains together with understanding of population dynamics has helped to reduce aflatoxin contamination [34]. Atoxigenic strains of *Fusarium* species also exist [35], but such atoxigenic relatives are better suited as biocontrol agents for the vegetable and fruit pathogen *Fusarium oxysporum* [36] than for the major cereal pathogens including *Fusarium verticillioides* and *Fusarium graminearum*. Fumonisin production [35] and deoxynivalenol production [37] are pathogenicity factors for *F. verticillioides* and *F. graminearum*, respectively, and when atoxigenic strains are released for biocontrol, they tend to be outcompeted quickly. Biocontrol of these cereal pathogens has also been pursued with bacterial isolates, including *Bacillus* species [38], or fungal strains [39].

Postharvest mycotoxin formation can, contrary to preharvest mycotoxin formation, be prevented by monitoring and management of stored grain [40]. However, postharvest mycotoxins still remain a concern in animal and human nutrition in less industrialized parts of the world [41].

The appropriate and primary approach of agriculture and the food and the feed industry is to prevent fungal infection and mycotoxin contamination. Nevertheless, grain that is produced and stored even with current best practice methods is usually still contaminated. Therefore, several decontamination approaches were explored to remove or destroy mycotoxins [31, 42, 43]. Physical removal is possible, because the distribution of mycotoxins in a batch of grain is very inhomogeneous. Broken kernels and dust are typically most highly contaminated, and grain cleaning will typically reduce the mycotoxin contamination. Some mills use optical sorting technology in the grain cleaning process to remove ergot sclerotia or highly contaminated, misshaped, or discolored kernels. However, such technology is not usually utilized in animal feed mills. The mycotoxin concentration is highest on and near the surface of kernels. Therefore, milling fractions for human consumption can be depleted of mycotoxins, but such fractionation would not be economically viable for animal feed production. Mycotoxins near the surface can also be destroyed by exposure of kernels to nonthermal atmospheric pressure plasma, but the technology has not proceeded beyond an experimental stage [44].

Mycotoxins are temperature stable, and therefore thermal treatment of grain or animal feed would not be an option for reduction of the relevant mycotoxins. Chemical treatment with ozone or ammonia has been explored, mostly for aflatoxin detoxification, but not routinely implemented [45, 46]. Such chemical treatments would inevitably reduce nutritional value. Storage and processing of grain for animal feed is dry, and the feed is typically provided in dry form, which limits the possibilities for chemical degradation of mycotoxins. For deoxynivalenol and other trichothecenes, the most promising chemical detoxification approach is treatment with sodium metabisulfite, but wet preservation of feed seems to be required [47, 48].

Since mycotoxins cannot be sufficiently prevented or depleted during grain production, storage, and processing for animal feed, there is a demand for animal feed additives for mitigation of mycotoxin toxicity. Such feed additives require water for activity, and therefore work in the gastrointestinal tract of animals. Typically, the additives contain an adsorbent, to which mycotoxins, dissolved in the chymus, are supposed to bind to prevent their resorption from the intestinal tract. A high affinity binding interaction between aflatoxin and certain types of bentonite clay was found [49], and special clays were and still are used as feed additives [50, 51]. Although aflatoxin binding in the gastrointestinal tract is not as complete as *in vitro*, analysis of aflatoxin biomarkers has shown that aflatoxin exposure of animals [52], and also of humans [53], can be reduced. The same adsorption approach to amelioration of mycotoxins in animal nutrition has also been tried for other mycotoxins. However, aflatoxin is unique with its π-electron system of a planar molecular structure (Figure 3.2.1), and no adsorption agents with similarly high binding affinity and selectivity have been found for other mycotoxins such as the *Fusarium* toxins. Nevertheless, such adsorption agents are being used in animal nutrition, even when evidence for activity is scarce [54, 55]. A frequently used kind of adsorption agent is composed of yeast cell wall fractions [56, 57]. Some of the commercially available feed additives also contain chemoprotectants or immunostimulants, because a common effect of mycotoxins is to compromise the immune system.

Among contaminants in animal feed, mycotoxins have a distinct and unifying feature: They are biological molecules, and as such, they should be amenable to biodegradation. Indeed, biodegradation of mycotoxins has been studied since the early days of mycotoxin research [58] with a prospect for application for mycotoxin detoxification [59, 60]. However, only one feed additive with verified mycotoxin biodegradation activity, Mycofix® from BIOMIN, has so far become commercially available [61]. Biodegradation and biological reactions in general

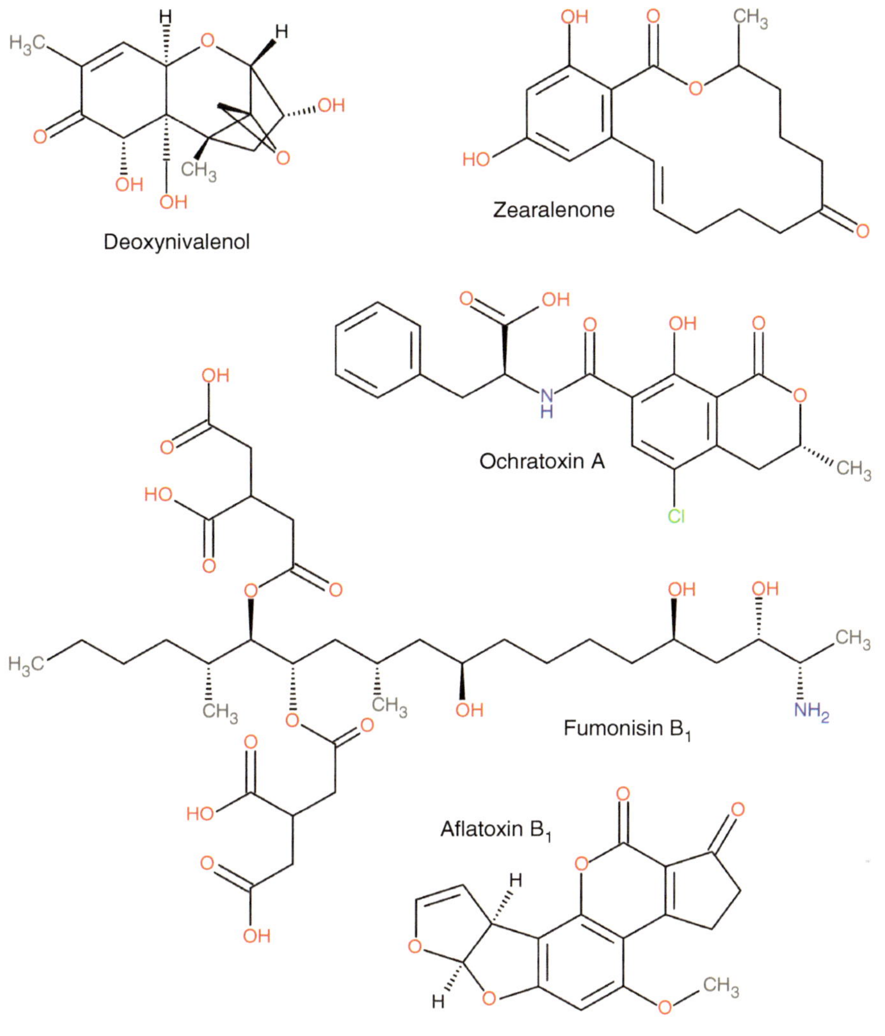

Figure 3.2.1 The "Big Five" mycotoxins.

require water, and therefore activity is again constrained to the gastrointestinal tract. Technological applications for mycotoxin biotransformation have first been realized with the bacterial isolate BBSH797 and with the yeast *Trichosporon mycotoxinivorans*. Bacterium BBSH797 is a member of the gram-positive *Coriobacteriaceae* isolated from bovine rumen, and it converts deoxynivalenol to the nontoxic metabolite DOM-1 by oxidoreductive biotransformation [62, 63]. The basidiomycete yeast *T. mycotoxinivorans* has been isolated from the intestinal tract of termites, and studied for its hydrolysis of ochratoxin A and cleavage of zearalenone [64, 65]. Both microorganisms are cultivated by large-scale fermentation, harvested from fermentation broth, and lyophilized for dry formulation of a feed additive [66]. Recently, biological mycotoxin detoxification technology has evolved to include a recombinant enzyme, FUM*zyme*®, for hydrolytic degradation of fumonisins [67, 68].

3.2.3 Enzyme Applications

The biological process of digestion of feed in the gastrointestinal tract of an animal begins with mechanical crushing, addition of water, and mashing in the mouth. Teeth and salivary glands enable this initial processing of feed and generation of chymus. The digestive tract of poultry is different, and although accomplished by the crop, proventriculus, and gizzard, normally with the help of gastroliths, the essential process of turning dry feed into a mashed pulp is the same. Enzymes secreted into the mouth and stomach, or crop, proventriculus, and gizzard, and further down the digestive tract cleave biological macromolecules to small, soluble unit molecules. Only soluble molecules can be resorbed from the gastrointestinal tract and become available as nutrients for an animal's metabolism.

Mycotoxins, which are typically present in feed because the best agricultural practice fails to prevent their formation and no practical and effective detoxification methods are available, will also be released from the feed matrix and become soluble in the course of digestion. In the normal course of events, mycotoxins will be resorbed and taken up into the bloodstream, perhaps to be modified by liver enzymes, and to exert their respective toxic mechanism. The intestinal tract itself, where the mycotoxin concentration is typically highest, is also a target for mycotoxin toxicity [69]. Conventional mycotoxin remediation feed additives contain mycotoxin binding agents with the aim to reduce the concentration of soluble, bioavailable mycotoxin. However, the digestive tract may also be regarded as an elaborate, continuous enzyme reactor. Digestion is an enzymatic process. The set of endogenous enzymes, secreted into the gastrointestinal lumen by the animal, is complemented by microbial enzymes, most importantly in the rumen of ruminants or the cecum or colon of monogastric animals. This natural set of enzymes is often amended by feed enzymes for animals in agricultural production to enhance nutrient bioavailability and utilization. Phytase is probably the most widely used feed enzyme in animal nutrition, and because it helps to reduce phosphate emission and to save scarce phosphate minerals that are conventionally added to feed, it makes an important contribution to sustainability of animal husbandry [70] (see also Chapter 3.3). However, other feed enzymes to enhance digestion of non-starch polysaccharides and proteases are also available [71]. For all these reasons, detoxification of mycotoxins seems like a feasible and attractive opportunity for development and application of an enzyme technology, to fill a technology gap, solve an existing problem, and to support animal health and agricultural sustainability. The technological challenges are conspicuous. The concentration of mycotoxins in feed is low, typically in the microgram per kilogram (ppb) range, which requires high substrate affinity of an enzyme. Chymus is a highly concentrated and complex matrix, which requires high selectivity of an enzyme. The detoxification reaction is in competition with resorption, which requires an enzyme to be fast. The price that farmers can afford to pay for feed additives is low and the commercial margin is small, which requires an enzyme to have a simple, cheap production process and high specific activity. The amount of feed to be treated is high, which requires enzyme production in a bulk, large-scale bioprocess. The reaction conditions in the gastrointestinal tract are relatively constant as far as temperature, pH, salt, and water concentration are concerned, but the conditions cannot be changed, as might be possible for other enzyme technologies, for instance, in the chemical industry. Furthermore, the digestive tract contains proteases and will also digest feed enzymes. For these reasons, an enzyme for mycotoxin detoxification may need to be adapted to the reaction conditions by enzyme engineering. At the same time, an enzyme can also be optimized for tolerance of feed processing, and especially feed pelleting, where feed is conditioned with steam and pressed to pellets at temperatures of 85 °C or even higher.

So far, only one enzyme has passed the scrutiny of regulatory authorities and become commercially available: The fumonisin esterase FUM*zyme*® developed by BIOMIN. The research and development pathway for FUM*zyme*® has been long indeed: It included the microbiology and molecular biology of discovering an enzyme; analytical methods and assay development; the biochemistry of enzyme purification and characterization; enzyme engineering; recombinant gene expression and production host strain development; animal feeding studies and analyses of biomarkers to determine efficacy, the required dose, and suitable formulations; the development of a production bioprocess including fermentation, downstream processing, drying and formulation; studies for feed pelleting and storage stability; and all efficacy and safety studies required for regulatory approval. Any next-generation mycotoxin detoxification feed enzymes for other mycotoxins will need to pass these steps in a similar way in the course of research and development. For this reason, it will be instructive to regard FUM*zyme*® as a role model and look in more detail how a new enzyme technology has become available in this case.

3.2.4 FUM*zyme*®

3.2.4.1 The Substrate: Fumonisins

Fumonisins B_1 and B_2 (FB_1 and FB_2, Figure 3.2.1) are the most prevalent and important members of the most recently discovered class of the major mycotoxins, the fumonisins [72]. The two molecules are identical except for one hydroxyl group, which FB_1 has at carbon atom 10 and which FB_2 is lacking. Already for years before their first isolation [16] and structure determination [73] in South Africa, *F. verticillioides* (*Fusarium moniliforme*), a pathogen of maize, was known to be toxic and carcinogenic [74]. The fungal pathogen and its toxicity were also already known to be of global relevance, and a research group in the United States was working on identification and evaluation of the mycotoxin at the same time [75]. The fumonisins were confirmed to elicit equine leukoencephalomalacia, an often fatal malady of horses known from field outbreaks caused by moldy maize [76]. Horses succumb to necrotic brain lesions. Fumonisins are carcinogenic for rodents [77], and the International Agency for Research on Cancer (IARC) classified fumonisins in group 2B, "possibly carcinogenic to humans" (IARC monographs on the evaluation of carcinogenic risks to humans, volume 82, 2002). Such human carcinogenicity has been suspected based on correlations between fumonisin exposure and esophageal cancer [78]. Fumonisins are also teratogenic and cause neural tube defect in mouse embryos, and they are suspected to affect embryonic neural tube development similarly in humans [20, 79]. In pigs, fumonisins cause pulmonary edema [80] and are hepatotoxic [81]. However, their effect on the immune response and the resulting decreased natural resistance against infectious disease [82–84], which has been observed for low dietary fumonisin concentrations, may have a bigger impact on the health and productivity of farmed pigs.

The molecular mechanism of fumonisin toxicity and carcinogenicity is to block the active site of ceramide synthase by a tight binding interaction, which inhibits sphingolipid biosynthesis and causes accumulation of intermediates including sphinganine, sphingosine, and their respective 1-phosphates, and depletion of complex sphingolipids such as gangliosides. This molecular mechanism was initially suspected based on the molecular similarity between fumonisins and sphingoid bases, and subsequently experimentally verified [85–87]. Sphingoid bases are involved as messenger molecules in cell signaling in several cellular processes including cell differentiation and apoptosis [88]. Complex sphingolipids also have vital biological functions including a role in regulation of membrane fluidity and formation of lipid domains [89], and as extracellular receptors involved in the immune response, the nervous system, control of metabolism, and cancer [90]. The molecular diversity of sphingolipids is even greater than their known biological functions [91]. For these reasons, the single molecular mechanism of ceramide synthase inhibition leads to such a range of toxic and carcinogenic effects of fumonisins in various animals and tissues [92]. The vital importance of sphingolipid homeostasis for eukaryotic cells is further emphasized by the fact that fungi have evolved several other toxins, in addition to fumonisins, to interfere with sphingolipid metabolism. These toxins include myriocin, an inhibitor of serine palmitoyltransferase, which catalyzes the first step of sphingolipid biosynthesis [93].

Fumonisin-induced changes in sphingolipid concentrations can be measured in blood [94] and organs. These concentration changes correlate with fumonisin exposure, and can be measured for low fumonisin contamination in feed, which does not produce clinical signs [95]. Therefore, determination of sphingolipid concentrations can serve as a biomarker of fumonisin effects on animals, and has been explored as a biomarker also for humans [96]. Sphingolipid concentrations can also be measured in urine, which enables noninvasive sampling [97, 98].

Although *F. verticillioides* and contaminated maize may be responsible for most of the animal or human nutritional exposure to fumonisins, several other fungi, including *Aspergillus niger* [99] and *Tolypocladium* sp. [100], are also known to produce fumonisins.

Regulatory authorities of many countries have set recommendations for maximum concentrations of fumonisins in feed and food. In the European Union, for instance, the guidance value for fumonisins B_1 and B_2 in pig feed is 5 mg/kg (2006/576/EC). Such guidance values are set "as low as reasonably achievable (ALARA)" and not necessarily as "no effect" concentrations. For instance, 5 mg/kg fumonisins in feed are known to affect sphingolipid concentrations in pigs [95, 101].

3.2.4.2 Enzyme Discovery

BIOMIN was founded as a company for animal feed additives in rural Lower Austria in 1983. The company's mission has been to promote natural animal health and performance through wholesome nutrition, as illustrated by the company's slogans "...the natural way" or "naturally ahead" (www.biomin.net). BIOMIN has been providing natural feed additives as substitutes for antibiotic growth promoters to help maintain animal productivity without antibiotics and the associated problem of driving antibiotic resistance gene spread in bacteria including possible human pathogens. The approach to mycotoxin contamination of animal feed has been driven by the same desire to harness natural genetic resources. A central working hypothesis is that mycotoxins are biologically formed, natural toxins, and as such, they are biodegradable. The mycotoxin mission at the BIOMIN Research Center is to find natural, microbial mycotoxin biodegradation, and to harness and develop it for use as a feed additive for gastrointestinal mycotoxin degradation and detoxification in animals. When the fumonisins became a new target for research on biodegradation, BIOMIN's Mycofix® already contained lyophilized biomass of two microbes for degradation reactions for other mycotoxins: Bacterium BBSH797 for deoxynivalenol and related trichothecenes [62], and *T. mycotoxinivorans* for zearalenone and ochratoxin A [64, 65]. The first goal of BIOMIN's fumonisin project was also to isolate microbial strains that can metabolize fumonisins.

The project started with a PhD thesis in collaboration between BIOMIN and the Institute of Environmental Biotechnology at the University for Natural Resources and Life Sciences Vienna ("BOKU") Department of Agrobiotechnology IFA-Tulln in Lower Austria. The PhD student Martin Täubel cultivated microbes from several different habitats as mixed microbial cultures. These habitats, from which he took samples, included intestinal contents from the small and large intestine of slaughtered pigs, freshly collected from the butcher's, and rumen fluid from cattle. Marc Lemmens at the Institute of Biotechnology in Plant Production at IFA-Tulln was investigating the fumonisin production potential of various *F. verticillioides* strains on maize lines cultivated in Austria, and kindly provided fumonisin-contaminated maize cobs for use as a source of microbes. Soil samples were taken from various spots and soil compositions. The screening program for microbial fumonisin degradation also included hundreds of isolates from the BIOMIN strain collection. These isolates comprised bacterial strains, yeast strains, and fungal strains, some from public strain collections such as DSMZ, ATCC, and IFO, and some proprietary isolates from previous projects. At the time, microbial fumonisin degradation was already described by Jonathan Duvick and coworkers at Pioneer Hi-Bred in the literature and in patent applications. Catabolic pathways for fumonisins were known for the yeasts *Exophiala spinifera* and *Rhinocladiella atrovirens*, and for a bacterium with the isolate number 2412.1, tentatively assigned to be a *Sphingomonas* or *Xanthomonas* species [102, 103]. Such microbial strains were also included in BIOMIN's screening program, and the fumonisin degrading activity of *E. spinifera* was confirmed. The screening program depended on analytical methods for quantification of fumonisin B_1 and intermediates of catabolic pathways. Although a commercially available ELISA kit for fumonisin quantification (Romer Labs® AgraQuant) was used for a part of the screening, the HPLC and LC–MS analytics contributed by Elsa Vekiru at the Center of Analytical Chemistry at IFA-Tulln proved to be essential.

The microbial cultivations included many different media, aerobic and anaerobic incubations, minimal media and mineral solutions, and a range of added FB_1 concentrations for enrichment of active strains. Anaerobic cultures derived from the gastrointestinal tract of animals showed partial FB_1 degradation, and hydrolysis of fumonisins in the gastrointestinal tract has also been described for several animal species [97, 104, 105]. However, aerobic cultures derived from soil samples showed more complete, robust, and reproducible degradation of FB_1. Individual pure strains were isolated from such cultures, and some of them produced hydrolyzed FB_1 (HFB_1, Figure 3.2.2) as intermediate of catabolism. The strains were taxonomically assigned by 16S rDNA sequencing, their kinetics of catabolism of FB_1 and HFB_1 in various media was compared, and a patent application (WO2006053357) was filed. The strain with the fastest fumonisin degradation kinetics, *Sphingopyxis macrogoltabida* MTA144, produced HFB_1 as the first intermediate of the catabolic pathway. This first intermediate was also produced by clarified cell lysate, and the activity was inducible with fumonisins. *S. macrogoltabida* MTA144 catalyzed fumonisin degradation also in rich and complex matrices that were prepared to simulate chymus, but it required aerobic conditions to thrive. Together with BIOMIN management, R&D director Gerd Schatzmayr decided to attempt development of a recombinant enzyme as feed additive for gastrointestinal fumonisin detoxification, rather than development of a native microbial strain for that purpose, as has been done in previous projects. A collaboration with BOKU's Institute of Applied Microbiology was initiated, and the author was hired to strengthen the now required molecular biology expertise at BIOMIN.

Our colleagues at the Institute of Applied Microbiology followed two parallel approaches to attempt cloning of the fumonisin detoxification genes. One approach was to purify the first enzyme of the catabolic pathway, an esterase that releases the two tricarballylic acid (TCA) side chains by hydrolytic cleavage to convert FB_1 to HFB_1 (Figure 3.2.2). Fractions with enriched enzymatic activity were obtained by chromatography. The other approach was focused on amplification of the esterase gene by PCR with degenerate primers. Consensus sequence elements of carboxylesterases were chosen and back-translated to "wobble" primers. Primer design was also facilitated by the sequence for the fumonisin esterase FccA of bacterium 2412.1, which had been published in a patent application (US6538177). A PCR fragment was obtained and sequenced, the flanking regions were obtained by inverse PCR, and eventually the biochemical approach of native enzyme purification became obsolete. Several more rounds of inverse PCR and Y-linker PCR were performed until over 20 kb of

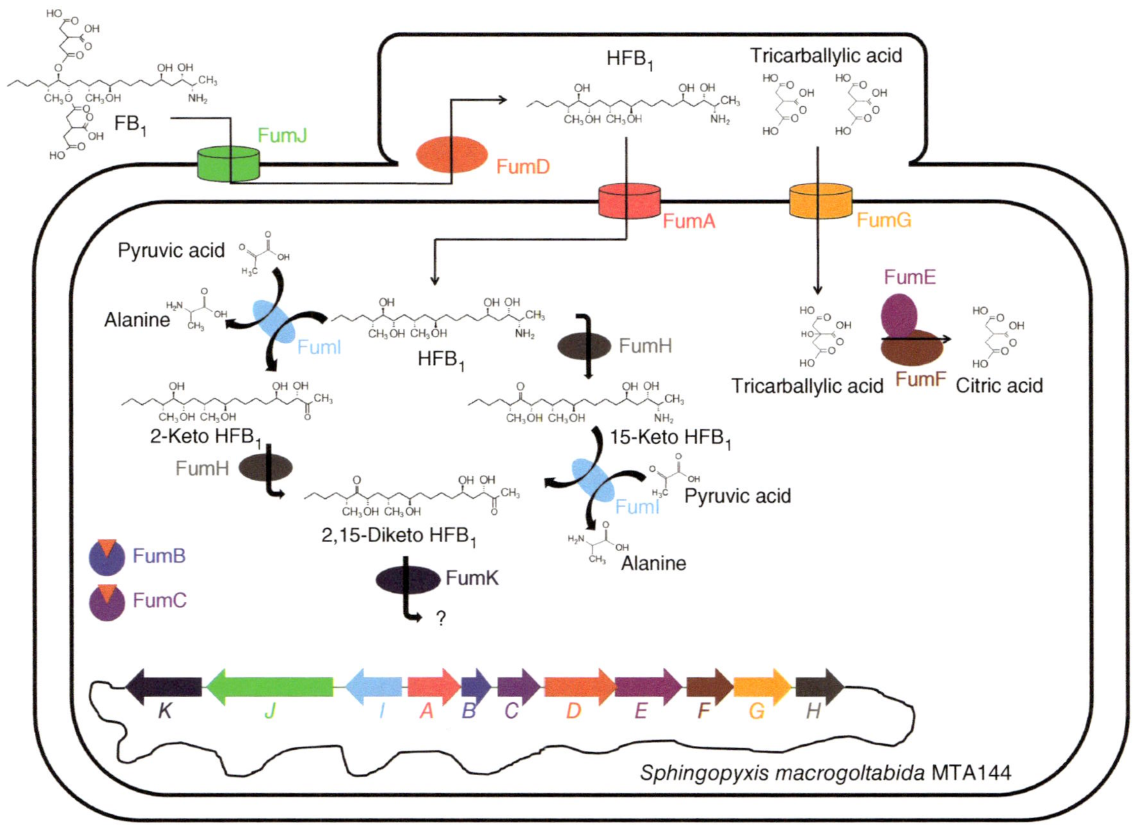

Figure 3.2.2 The pathway of fumonisin catabolism in *Sphingopyxis macrogoltabida* MTA144.

genomic DNA sequence, comprising the genes for fumonisin catabolism, were available [106]. Whole genome sequencing soon became affordable enough to confirm the DNA sequence. Following standard nomenclature for bacterial genes, we named the genes for fumonisin catabolism the *fum* genes, starting with *fumA* as the gene proximal to a predicted promoter. We cloned and expressed the *fumD*, *fumI*, *fumH* and *fumK* genes and used the recombinant enzymes to produce, and verify by NMR, the fumonisin catabolism intermediates shown in Figure 3.2.2. Sequence comparison with the fumonisin catabolism gene cluster of bacterium 2412.1 revealed amino acid sequence identities between 50% and 82%, and the genes were in different order and orientations. The amino acid sequence identity between the fumonisin esterases FumD and FccA was 54%. We found that bacterial fumonisin catabolism involved a transaminase, rather than an aminooxidase, as had been found for *E. spinifera* [102] and postulated for the bacterial pathway, and cloned a homologous transaminase of bacterium 2412.1 [107].

The presence of a subset of three genes of the cluster, *fumE*, *fumF*, and *fumG*, suggested that MTA144, just like 2412.1, could utilize not only the core chain of fumonisins but also the TCA side chains, import them, and convert them to citric acid [108]. We also found that fumonisin catabolism was inducible. When biomass was grown with or without FB_1 in the medium, harvested by centrifugation, lysed by passing through a French press, and clarified by centrifugation, only the clear lysate from biomass that had been induced with FB_1 in the growth medium was able to hydrolyze added FB_1 quickly. Furthermore, we recorded growth curves and found that the increase of optical density of MTA144 cultures correlated with the concentration of FB_1 added to minimal medium, which was evidence that the strain could utilize FB_1 as carbon and energy source. Comparison of the fumonisin catabolism gene clusters of the bacterial strains MTA144 and 2412.1 further suggested that the two gene clusters shared a common ancestor way back in evolutionary time, and the ability to utilize fumonisins as a niche substrate must have provided a selective advantage. All these findings suggested that the enzymes for fumonisin degradation were shaped and optimized by natural evolution, and not merely side activities of promiscuous enzymes. However, the question remained how to develop an animal feed detoxification technology and which of the enzymes to utilize.

3.2.4.3 Enzyme Selection

Fumonisins have an amino group at carbon atom 2 (Figure 3.2.1). This amino group is equivalent to the 2-amino group of sphinganine (Figure 3.2.3), which gets acylated by ceramide synthase as a key step of sphingolipid biosynthesis in eukaryotic cells from yeast to humans. The 2-amino group of fumonisins seems to be important for high-affinity binding and blocking of ceramide synthase, and acetylation of this amino group [109, 110], reaction with a reducing sugar [97, 111], or other modifications [112] reduce toxicity. Hydrolyzed fumonisins function as substrate, rather than inhibitor, of ceramide synthase, and their 2-amino group can get acylated [113]. Formation of such *N*-acyl-HFB_1 has been found to occur not only in cultured cells but also *in vivo* in rats after

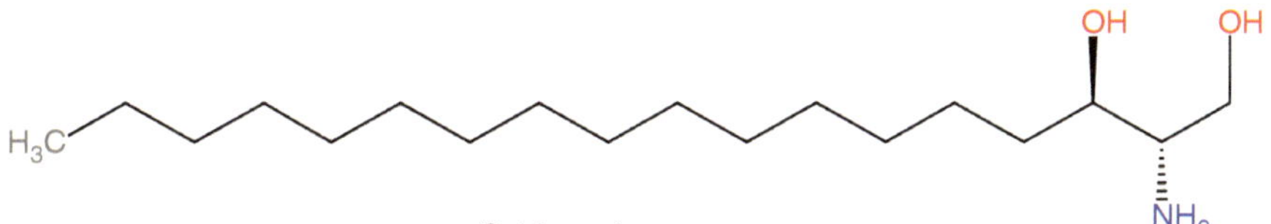

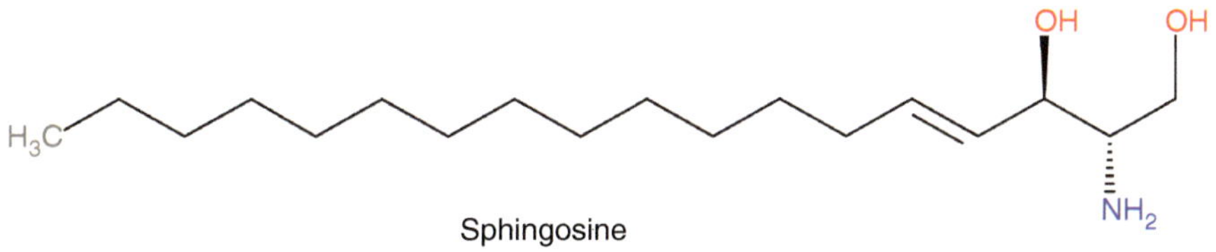

Figure 3.2.3 Sphinganine and sphingosine are key intermediates of sphingolipid metabolism. The ratio of their concentrations has been used as biomarker of fumonisin toxicity [96–98].

intraperitoneal injection of HFB_1 [114]. Congeners of *N*-acyl-HFB_1 with fatty acids of various chain lengths are potent inhibitors of ceramide synthase, and are therefore toxic. For these reasons, we thought the enzyme activity of HFB_1 transaminase FumI would be crucial for our detoxification technology.

We cloned the *fumI* gene, but gene expression in *Escherichia coli* produced mostly inclusion bodies, and we could not get useful yield of soluble, active FumI with *Pichia pastoris* as gene expression host either. We probed and tweaked *E. coli* expression to obtain sufficient yield of FumI for characterization of the HFB_1 transaminase activity [115]. The enzyme needed not only pyridoxal phosphate as prosthetic group but also an α-keto acid as co-substrate and amino group receptor for activity [116]. A technological application as animal feed additive, where high enzyme activities must be provided cheaply and in large quantities, seemed difficult. Contrary to FumI, the fumonisin esterase FumD could be produced as secreted protein in *P. pastoris* with good initial yields [106]. The enzyme was robust, active in clear cell lysates of MTA144 where FumI and FumH were inactive, and its hydrolytic mechanism requires no cofactors. We considered a detoxification technology based on fumonisin esterase FumD only, and scrutinized the literature about toxicity of the reaction products, hydrolyzed fumonisins, again [117].

Residual toxicity of hydrolyzed fumonisins, although lower than without hydrolysis, was reported for feeding trials with rats with large amounts of *F. verticillioides* culture material treated by alkaline cooking [118–120]. Alkaline cooking is a traditional food processing method for maize meal in Latin America. It is called nixtamalization and causes hydrolysis or partial hydrolysis of fumonisins [121]. For the experimentally nixtamalized fungal culture materials, concentrations of partially hydrolyzed fumonisins were not reported, and hydrolysis may have been incomplete. Later studies reported that nixtamalization, at least when carried out with contaminated corn rather than fungal culture material [122], abolished fumonisin toxicity [123, 124]. Feeding trials with pure HFB_1 showed that HFB_1 was not toxic and not teratogenic [109, 125–127]. Studies of toxicity of HFB_1 in *in vitro* systems, where various cultured cells, embryos, or plant tissues were exposed to sometimes high concentrations of HFB_1, showed variable results. Some studies reported toxicity of HFB_1 [109, 128, 129], while others reported reduced or no toxicity of HFB_1 [110, 130–138].

A reconciled picture emerged for us from the extant literature: Hydrolyzed fumonisins have little binding affinity to ceramide synthase and are, as such, nontoxic. When a high concentration is provided, ceramide synthase can convert enough hydrolyzed fumonisins to ceramide synthase inhibiting *N*-acyl HFB to elicit toxicity through interference with normal sphingolipid metabolism. Hydrolyzed fumonisins are poorly resorbed from the gastrointestinal tract [139], and therefore, after oral uptake of doses that are representative for practically relevant fumonisin exposure, HFB_1 concentrations in the body are too low for sufficient *N*-acyl HFB to accumulate to cause ceramide synthase inhibition. These mechanisms are also important for human nutrition, because nixtamalization causes partial hydrolysis of fumonisins in this traditional Latin American maize meal alkaline cooking process. Nixtamalization helps reduce fumonisin toxicity, because hydrolyzed fumonisins, at least in the concentrations that can emerge from naturally contaminated maize, have no oral toxicity. We decided to investigate oral toxicity of hydrolyzed fumonisins in our own experiment with piglets. An extract of *F. verticillioides* culture material with 700 mg/l fumonisins was hydrolyzed with fumonisin esterase FumD, or mock-treated, and given to piglets by gavage. The piglets received 2.8 μmol FB_1 or HFB_1/kg body weight/d, which is a fumonisin exposure equivalent to feed contamination of about 40 mg/kg. Only the piglets exposed to intact fumonisin, but not the piglets exposed to hydrolyzed fumonisins, showed elevation of the sphingolipid biomarker and evidence for toxicity [140]. We proceeded to develop fumonisin detoxification technology based on fumonisin esterase FumD.

Bioinformatics predicted that the fumonisin esterase FumD had an N-terminal signal peptide and was secreted to the periplasm of *S. macrogoltabida* MTA144. For recombinant gene expression in *E. coli* and *P. pastoris*, we cloned the part of the gene encoding the predicted mature enzyme, and confirmed activity. Subsequently, we purified the native enzyme with the help of a rabbit anti-FumD antiserum raised against the recombinant enzyme, and confirmed the N-terminus of the mature, processed fumonisin esterase by mass spectrometric analysis of tryptic fragments.

3.2.4.4 Enzyme Activity Assays

Analytical standards for fumonisins and intermediates of catabolism were bought or prepared for the early project phases of enzyme discovery and initial characterization. Initially, we measured fumonisin esterase activity by taking time point samples from reaction courses with FB_1 as substrate and by determining concentrations of residual FB_1 and generated HFB_1 by LC–MS/MS [106]. A fully isotope labeled standard, ^{13}C-FB_1, was commercially available from Romer Labs, added to each sample, and used to compensate for fluctuations in recorded peak areas. However, generation of HFB_1 from FB_1 requires two catalytic cycles. The two versions of the reaction intermediate, partially hydrolyzed FB_1 ($pHFB_1$), were detectable in reaction time courses. They were unstable and interconvert, and therefore we could not produce reliable standards for accurate quantification. We still preferred to quantify reaction product formation, rather than less accurate substrate consumption. Each reaction cycle releases one of two identical residues of TCA, and we decided to use TCA as the analyte to quantify. We purchased TCA in analytical standard quality, developed an accurate but fast LC–MS/MS method, and wrote an SOP for fumonisin esterase activity measurement. Time courses of TCA formation, when generated with the right, sufficiently low enzyme concentration, were linear and the slope could be used for accurate determination of enzyme activity. This activity assay was also the basis for the fumonisin esterase activity unit definition: One unit is the enzymatic activity that releases 1 μmol TCA per minute from 100 μM FB_1 in 20 mM Tris-Cl buffer pH 8.0 with 0.1 mg/ml BSA at 30 °C.

Later on in the project, in the course of enzyme engineering, a collaboration partner explored the possibility of a photometric activity assay. They tested a range of acyl esters of *para*-nitrophenol as colorimetric substrates for fumonisin esterase. 4-Nitrophenyl 2-(trimethylsilyl)ethyl carbonate (pNSi, CAS 80149-80-0) was the best of the tested substrates. Activity was at least 100-fold lower than with FB_1 as substrate and assays were sensitive to foreign esterase activity in the sample, and therefore not ideally suitable for activity determination of crude enzyme preparations. However, the pNSi activity assay was invaluable for an enzyme engineering project to improve temperature stability of fumonisin esterase FumD.

For fumonisin esterase quantification, we generated polyclonal antisera. First, we generated mouse anti-FumD antiserum for Western Blotting [106] and a competitive ELISA. Then, a commercial provider immunized rabbits with a FumD preparation from us, and we developed a double-antibody sandwich ELISA, where the rabbit anti-FumD serum was used for coating. Subsequently, we had to change the detection antiserum from mouse to guinea pig, because the amount of the mouse antiserum, although of high titer and specificity, was not enough. In the double-antibody sandwich ELISA for our thermostable, engineered version of fumonisin esterase, we used a goat antiserum for plate coating and a rabbit antiserum for detection.

3.2.4.5 Enzyme Characterization and Evaluation

Expression of the *fumD* gene in *E. coli* allowed to experimentally confirm fumonisin esterase activity and verify that the cloned gene was functional. However, the recombinant enzyme accumulated mostly in the form of inclusion bodies. Our efforts to refold active enzyme after isolation from inclusion bodies remained without success. It would have been difficult to make a preparation of pure enzyme for characterization, and we needed to explore high yield production for possible development of a feed enzyme anyway. We tried *Saccharomyces cerevisiae* as expression host, but failed to obtain detectable enzyme activity. However, expression in *P. pastoris* worked well, and good yields of secreted, recombinant FumD were obtained. There was no apparent difference in enzyme activity between FumD with a C-terminal 6xHis-tag and tag-free FumD.

We fermented *P. pastoris* GS115, with *fumD* cloned into the vector pPICZalphaA for secreted production of C-terminally 6-His-tagged FumD under control of the methanol-inducible AOX1 promoter, in a 5 l fermenter. Culture supernatant was cleared from biomass by centrifugation, and applied to immobilized metal affinity chromatography (IMAC) resin for affinity purification of FumD. The purity of the enzyme preparation was verified by SDS-PAGE, and the concentration was determined with Bradford reagent. Aliquots were made to test various storage conditions, and this enzyme preparation was used for the initial enzyme characterization. FumD was active between pH 4.0 and pH 10.0 with a maximum at pH 8.0. The temperature optimum was 30 °C. Different buffers were tolerated, but high salt concentration reduced activity. FumD regained full activity when neutralized after incubation at pH 2.0. Kinetic parameters were measured for optimal incubation conditions for FumD based on initial FB_1 consumption and HFB_1 generation rates for various FB_1 starting concentrations. Later, kinetic parameters were determined with better accuracy and confidence based on initial TCA generation rates.

We tried to evaluate the potential of FumD for application as a feed enzyme by comparing fumonisin hydrolysis rates in buffer with rates in various complex matrices including a duodenum simulation buffer. Subsequently, we collected pig intestines fresh from the local butcher's, tied off and cut about 20 cm sections from the duodenum and jejunum, injected FB_1 and FumD from opposite ends, mixed, incubated, and measured FB_1 hydrolysis. The measured generation of HFB_1 was considered an encouragement to continue to evaluate and develop FumD as fumonisin detoxification feed enzyme.

3.2.4.6 Enzyme Feeding Trials and Biomarker Analysis

FumD was a very rewarding enzyme to work with in the early stages of R&D. It could be produced easily in good quantities, it had high specific activity also at the low fumonisin concentrations that are typical for animal feed, it showed no special requirements for specific buffer conditions, and it was stable and robust to handle. At the time, we had an ongoing collaboration with Isabelle Oswald at the INRA in Toulouse, France, to study effects of mycotoxin co-contamination of feed with piglets [141]. The opportunity presented itself to investigate FumD activity *in vivo* in a feeding trial with piglets. We produced the enzyme by fermentation of our recombinant *P. pastoris* strain in a 20 l fermenter at the BIOMIN Research Center, harvested fermentation broth and removed biomass by centrifugation, freeze-dried the supernatant, measured enzyme activity, and shipped the preparation to Toulouse. Bertrand Grenier, the PhD student in Toulouse, spread feed for the experiment, and artificially contaminated feed with lyophilized *F. verticillioides* culture material to contain fumonisins at a concentration of 6 mg/kg, on plastic sheets on the floor. He dissolved the enzyme preparation in water and used a plant spray bottle to spray it evenly on the feed. We received samples of this and the control feed mixes in Tulln, and verified the concentrations of fumonisins, enzyme activity of 40 U/kg, and that fumonisins had not been hydrolyzed in the course of spray application except to a proportion of less than 1%. Groups of six piglets per group were fed the experimental diets *ad libitum* for 35 days [101]. Blood samples were taken weekly and sent to Tulln for sphingolipid analysis. An increase of the Sa/So ratio in blood of each individual piglet of the fumonisin group was found over the course of the experiment, but this increase was prevented by FumD (Figure 3.2.4). These results were our first evidence that fumonisin esterase was active in the gastrointestinal tract and catalyzed hydrolytic detoxification of fumonisins in exposed piglets. A whole set of additional analyses including hematology and blood biochemical parameters, response to a vaccine, cytokine gene expression, and histopathology of hepatic, pulmonary, and intestinal tissues confirmed fumonisin detoxification [101].

The results of this first feeding experiment with FumD at the INRA Toulouse provided proof of technical feasibility for enzymatic gastrointestinal fumonisin detoxification. Our next goal was to find out what minimum concentration of FumD in feed was required. We continued to focus on pigs, which are known to be highly sensitive to fumonisins. The "Biological Ingredient Formulation Team" at the BIOMIN Research Center in Tulln used different drying and processing technologies to produce different formulations of FumD. These formulations included wax-coated FumD on a cellulose carrier, shellac-coated FumD on sugar beads, FumD compacted in an inulin matrix, and freeze-dried and spray dried FumD with carrier. The goal was to compare these formulations to see if any of them could enhance delivery of FumD to the gastrointestinal tract and enable better fumonisin hydrolysis with fewer enzyme units per kilogram feed. The feeding experiment took place at BIOMIN's Center of Animal Nutrition. Four different FumD concentrations covering 3 orders of magnitude were tested. With so many formulations and concentrations to test, there were only two piglets per experimental group. The feed was artificially contaminated with *F. verticillioides* culture material to a final fumonisin concentration of 30 mg/kg, which allowed a clear sphingolipid biomarker response to be measured after two weeks. Concentrations of fumonisins and hydrolysis products in feces were also measured. No difference between the formulations was observed. FumD at the highest tested concentration caused complete detoxification. One tenth of this concentration in feed caused partial detoxification, and concentrations that were lower by another 1 or 2 orders of magnitude caused no detoxification.

The next feeding trial, also at BIOMIN's Center for Animal Nutrition, allowed us to investigate fumonisin detoxification for naturally, rather than artificially, contaminated feed to narrow down on the required dose of FumD, and to get statistics from more piglets per group. We bought one ton of maize with a fumonisin contamination of 52 mg/kg FB_1 and 18 mg/kg FB_2, and mixed feed with a final fumonisin concentration of 2.5 mg/kg. The maize was also contaminated with aflatoxin and therefore bentonite for aflatoxin binding was included in the feed mix, as would be typical for pig farming. Groups of six piglets were fed the experimental diets for 42 days. Sphingolipid concentrations in blood and fumonisin and (partially) hydrolyzed fumonisin concentrations in feces were measured for all piglets, and sphingolipid concentrations in lung, liver, and kidney were measured for two or three piglets per group [68]. The lowest tested FumD concentration, 15 U/kg, enabled complete detoxification, and higher concentrations provided no additional benefit.

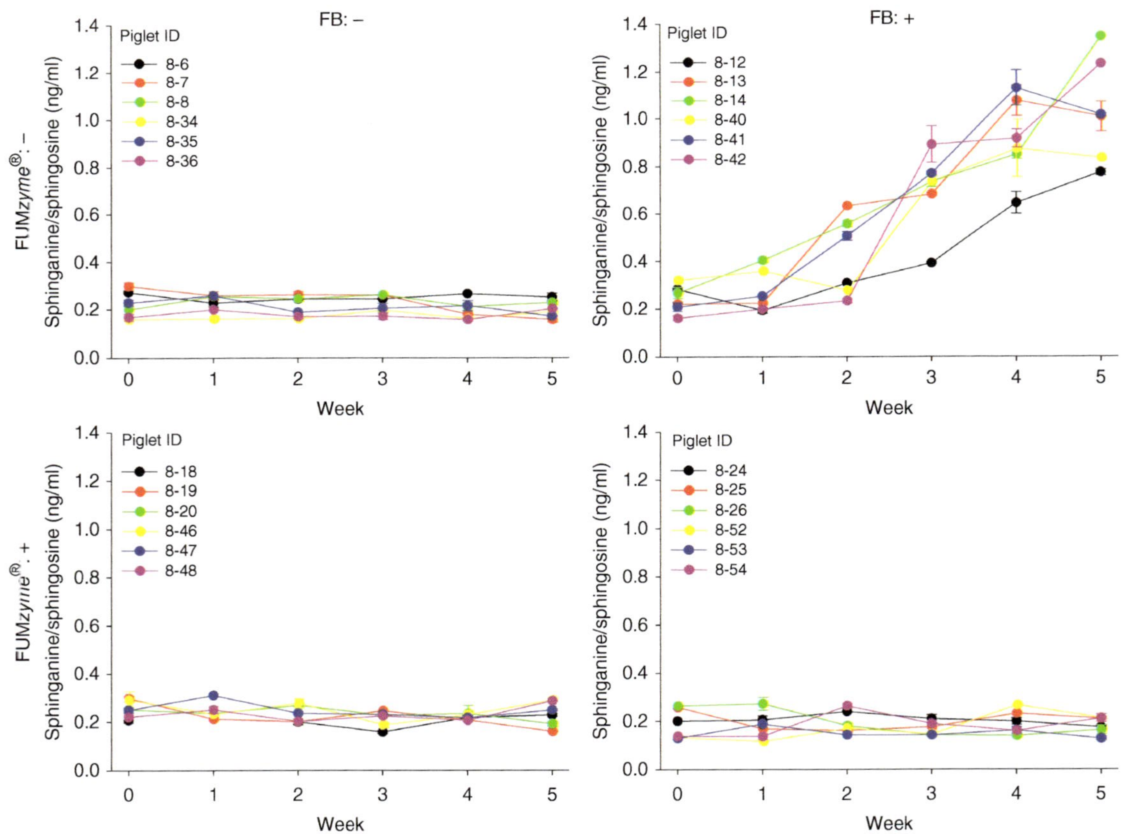

Figure 3.2.4 The ratio of sphinganine to sphingosine (Sa/So ratio) in blood samples of individual pigs (six per group), measured over the six weeks time course of a feeding trial [101] with fumonisin contaminated feed (6 mg/kg, groups in the right column) and FUM*zyme*® (groups in the bottom row).

Following the encouraging results of these feeding trials, the decision to begin with safety and efficacy studies required for assembly of a dossier to file with regulatory authorities was made, and the name FUM*zyme*® was coined for the feed enzyme to be. More feeding trials were performed to confirm FUM*zyme*® activity and the recommended dose [67]. In a collaboration with Purdue University, sphingolipid biomarker analysis was applied for measuring fumonisin effect in chickens [142], and a FUM*zyme*® application for chickens was developed [143].

3.2.4.7 Enzyme Engineering

We used the thermofluor assay with Sypro Orange to determine the thermostability of FumD and found that the thermal denaturation temperature of FumD was 48 °C. This result matched previously measured residual enzyme activities after heat treatment. Animal feed mixtures are frequently processed by pelleting, which involves exposure to heat. Animals tend to prefer feed pellets over mash feed of the same composition, waste less feed, and show better feed conversion ratios with pellets. Mash feed is conditioned with steam to gain a set temperature, the softened mash is pressed through the holes of a die, and blades cut the nascent pellets to the desired length, determined by the blade rotation speed. The newly formed pellets are cooled by ventilation. Animal feed enzymes are typically engineered for thermostability [70], so that they can be added in the course of normal mash feed mixing and tolerate the pelleting process. We tested FumD in a pilot-scale pellet press and found that after conditioning and pelleting at 75 °C, we could recover no more than 8% of the initial fumonisin esterase activity, and the tested higher temperatures caused complete inactivation. Feed enzymes can be encapsulated and coated to improve their tolerance to temperature exposure and pelleting, and we tested and used such coatings successfully. However, such enzyme formulation adds to the production cost. Intrinsic thermostability would be preferred for a feed enzyme, and we embarked on an enzyme engineering project with the goal of making a thermostable variant of FumD for an intrinsically pelleting stable version of FUM*zyme*®.

With little previous experience in enzyme engineering, structural computational biology or high-throughput screening, we contacted companies that could offer the required technology in a collaboration. Our partner of choice set up a high-throughput screening assay for increased thermostability, with pNSi as colorimetric substrate. Every enzyme variant with increased thermostability and unreduced activity for hydrolysis of pNSi had to be tested for activity with FB_1 as substrate with our LC–MS/MS based analytical method. The project started with random mutagenesis and combination of verified, beneficial mutations. Rounds of random mutagenesis and combinations of mutations were repeated, and improved enzyme variants were generated. Sequence regions, where beneficial mutations tended to accumulate, emerged and were subjected to focused mutagenesis. Selected amino acid positions and combinations of amino acid positions were subjected to saturation mutagenesis. The choice of positions for such mutagenesis was supported by molecular models of FumD generated by homology modeling and energy minimization refinement. At the same time, we entered into a collaboration with the "Center of Optimized Structural Studies"

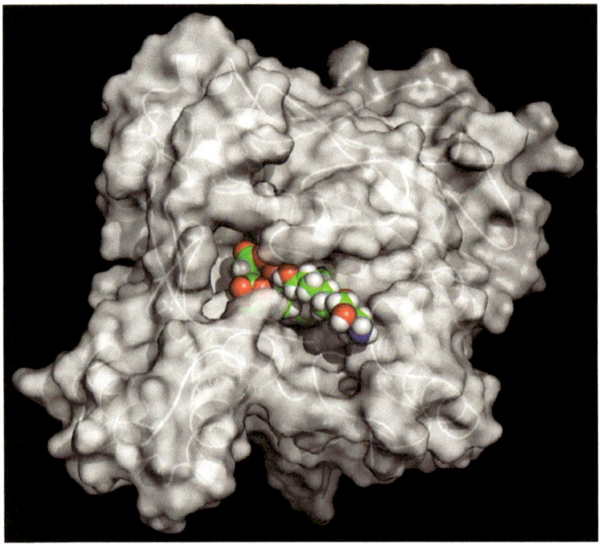

Figure 3.2.5 X-ray crystal structure of fumonisin esterase FumD with bound substrate FB_1.

(COSS) at the Max F. Perutz Laboratories of the University of Vienna. The team of Kristina Djinovic-Carugo crystallized FumD and solved the X-ray crystal structure. Subsequently, they also crystallized thermostable variants of FumD, and co-crystallized FB_1 with an inactive variant of FumD, and provided the atom coordinates to support further, structure-guided rational enzyme engineering (Figure 3.2.5). Several thermostable versions of FumD were produced and compared in pelleting experiments and feeding trials. An enzyme variant called FumD_comb11 was chosen, and development of a high-yielding production strain and bioprocess started, and studies required for regulatory approval were initiated. FumD_comb11 differs from the wild-type amino acid sequence of FumD in 23 positions. Its thermal denaturation temperature, determined by the thermofluor assay, was increased by about 30 °C, and it tolerated feed pelleting with good recovery of activity.

3.2.4.8 Enzyme Production

Pichia pastoris gave good yield of secreted FumD in our early efforts to produce recombinant fumonisin esterase [106]. Encouraging results from *in vitro* enzyme characterization meant that we needed more recombinant enzyme for feeding experiments and feed pelleting experiments, and that we needed to explore the possibility of future large-scale production of FumD. *P. pastoris* seemed like an encouraging recombinant gene expression system to pursue further.

In our initial *P. pastoris* strain for FumD production, the *fumD* gene was under control of the methanol inducible *AOX1* promoter and fused to the sequence encoding the N-terminal secretion signal of the *S. cerevisiae* α-factor in a commercially available vector and *P. pastoris* host strain. We had a codon-optimized version of the *fumD* gene synthesized and cloned it, with the same secretion signal, under control of the *GAP* promoter for constitutive expression without methanol induction. The yield was equally good, and to further optimize recombinant FumD production with *P. pastoris*, we joined the "Austrian Center of Industrial Biotechnology" (ACIB). In this research collaboration, we were involved in studies of recombinant protein folding and secretion [144] and engineering of *P. pastoris* for enhanced secreted FumD production. We also entered into collaboration with a specialized technology company that helped us with large-scale clone screening, comparison of host strains with altered expression levels of genes involved in folding and secretion, and promoter variants. They also helped us with development and upscaling of a fermentation protocol, optimized for the selected, best production clone. In parallel, we also investigated FumD production by recombinant filamentous fungi. The product titers reached with *Trichoderma reesei* were lower. The product titers reached with *Myceliophthora thermophila* were equal to initial titers reached with *P. pastoris*, but fermentation supernatants also had high concentrations of secreted host proteins, and the stability of FumD was compromised by proteolytic degradation. In comparison, *P. pastoris* allowed FumD production with shorter fermentation times, and fermentation supernatants were virtually free of host proteins and suitable for downstream microfiltration, ultrafiltration, and drying. The production bioprocess was also scalable and robust for implementation at several manufacturing sites, where FUM*zyme*® is now produced by fermentation at multi cubic meter scale. The long history of safe use, availability of the whole genome sequence [145, 146], and complete documentation of the strain history in molecular detail enabled regulatory authorities to approve of *P. pastoris* as production host.

3.2.4.9 Enzyme Registration

FUM*zyme*® was the first enzyme of its kind, and there were no precedents to guide a path to approval by regulatory authorities. In fact, when the FUM*zyme*® project was in the early phases of research and development, there was not even a class of feed additives in which FUM*zyme*® could have been registered in Europe [147]. However, among the goals of the European Commission is also the authorization of feed additives, after scientific evaluation, for "improving the quality of feed and the quality of food from animal origin, or to improve the animals' performance and health, e.g. providing enhanced digestibility of the feed materials." The European Food Safety Authority (EFSA) recommended the introduction of a new functional group for detoxifiers [148]. The European Commission appreciated that "new feed additives have been developed which suppress or reduce the absorption, promote the excretion of mycotoxins or modify their mode of action and thereby mitigate possible adverse effects of mycotoxins on animal health." Commission Regulation No. 386/2009 of 12 May 2009 added the new point "m" to the list of functional groups of feed additives of Regulation No. 1831/2003: "substances for reduction of the contamination of feed by mycotoxins: substances that can suppress or reduce the absorption, promote the excretion of mycotoxins or modify their mode of action." The Commission stressed that future availability of feed additives authorized for mycotoxin detoxification would have no effect on existing maximum or guidance concentrations for mycotoxins. Only feed that is lawfully on the market and below the set limits could be treated to improve the quality of the feed. One consequence of this reasonable clause was that we had to be able to provide evidence for activity of FUM*zyme*® for low fumonisin contaminations. Analysis of biomarkers for fumonisin, and in particular the sphingolipid analysis, were not only required for technology development but also for registration.

FUM*zyme*® was the first, and is to date (spring 2019) still the only recombinant enzyme authorized in this group. The evaluation was done by EFSA and required more than 1000 pages of documents about studies and experiments performed to investigate the safety and efficacy of FUM*zyme*®. These documents included a description of the manufacturing process, technological properties, methods of analyses, information regarding user or worker safety, toxicological studies including acute toxicity, genotoxicity, subchronic toxicity, reproduction toxicity, toxicological studies with HFB_1 and TCA, systemic toxicity, effects on eyes and skin, effects on the respiratory system, allergenicity, exposure assessment and methods to control exposure, documentation for the recombinant production strain, and verification that the product is free of the genetically modified host or recombinant DNA. Evidence for efficacy was provided in the form of several feeding trials and biomarker analysis. Considering the volume of documents and the size of the task, the 16 months, including the time required to provide additional documents and answer questions, between submission of a dossier and publication of an opinion by EFSA does not seem long.

Registration might sometimes be regarded as something that stands between R&D and a product, and in the way of being able to get the product on the market and start recovering the R&D expenses. However, registration can also be regarded as a process that helps ensure that a product is safe and that it works properly. In this respect, regulatory requirements can be support for those responsible for R&D of a product in the business environment of a company. A business drive to start product sales and recovery of R&D expenses might cause a temptation to expedite late stages of development. Therefore, the safety and efficacy studies required for a dossier can actually be in favor of R&D staff and their intrinsic interest to make sure that the technology they develop works robustly and is safe. One example is the minimum required dose, which needs to be defined for a dossier. On one hand, a higher minimum dose will provide a safety margin and ensure that a technology works also under unfavorable conditions. On the other hand, a lower minimum dose will give a better cost margin and allow selling at a more competitive price. In this trade-off, the efficacy studies required for a dossier can work in favor of technology robustness.

3.2.5 Future Mycotoxinases

We are attempting to repeat the success of FUM*zyme*® for other classes of mycotoxins. Some mycotoxins are more difficult targets than others. Zearalenone and ochratoxin A have bonds that can be hydrolyzed, similar to the ester bond of fumonisins, and hydrolytic reaction mechanisms are suitable for feed enzymes.

Zearalenone is a mycoestrogen produced by *F. graminearum* and typically co-occurs with deoxynivalenol in wheat and maize from temperate climate zones [149]. Zearalenone is a lactone (Figure 3.2.1), and the ester bond can be hydrolyzed. A mycoparasitic fungus called *Gliocladium roseum* was found to be capable of hydrolyzing the lactone ester of zearalenone [150], and the gene for the lactonase was cloned [151]. Kinetic parameters for the lactonase reaction were determined [152]. At the low mycotoxin concentrations typically encountered in animal feed, the zearalenone lactonase is much less active and has much lower specific activity than FUM*zyme*®. Nevertheless, we attempted a feeding trial with recombinant *G. roseum* zearalenone lactonase and zearalenone-contaminated feed, but found no evidence for enzyme-catalyzed hydrolysis of zearalenone. Another microbial conversion of zearalenone to a non-estrogenic derivative is catalyzed by *T. mycotoxinivorans* [153]. The conversion involves a redox mechanism and requires more than one enzyme, which makes it difficult to apply for recombinant enzyme-based animal feed detoxification. Therefore, we looked for new enzymes. We enriched and screened mixed microbial cultures, isolated from the environment, and isolated bacterial strains with zearalenone hydrolysis activity. We cloned and expressed the gene for a zearalenone lactonase, and characterized the encoded enzyme and related enzymes that were mined from databases as belonging to the same sequence family. The specific activity and activity at low zearalenone concentration were higher than for the zearalenone lactonase from *G. roseum*, and feeding trials with recombinant enzyme showed gastrointestinal zearalenone hydrolysis. However, engineering of an enzyme variant with sufficient activity and thermostability for commercial application is still a challenge.

Ochratoxin A [154] is another mycotoxin that is amenable to detoxification by a hydrolytic mechanism. The amide bond (Figure 3.2.1) can be cleaved by some peptidases and amidases [155, 156], and ochratoxin-α, which is the remaining derivative after phenylalanine is hydrolyzed off, is much less toxic [157, 158]. *Aspergillus niger* has an ochratoxin A hydrolyzing amidase [159], and the gene was cloned and the recombinant enzyme characterized [160]. The reported enzyme activity at low concentration of ochratoxin A may not be sufficient for an application for detoxification of animal feed.

Aflatoxin B_1 has long been known to be amenable to microbial biotransformation and detoxification reactions [58]. However, the lactone ester bond of aflatoxin B_1 is contained in a planar ring structure with a conjugated electron system and it is therefore not susceptible to enzymatic hydrolysis. Reported biotransformation reactions of aflatoxin B_1 are based on redox mechanisms. The specificity and activity of laccases is too low for technological application [161]. The best characterized enzymes, $F_{420}H_2$ dependent reductases [162, 163], are also difficult to harness for a technological application because of their requirement for a specific cofactor and redox-regeneration of the cofactor. For the time being, adsorption to certain bentonite clay minerals remains the best available method for amelioration of aflatoxin contamination in animal feed [51].

Although deoxynivalenol and related trichothecenes contain an epoxide ring (Figure 3.2.1), enzymatic hydrolysis of this epoxide ring has never been reported.

Steric hindrance may be responsible for making this particular epoxide ring inaccessible for enzymatic hydrolysis. Reduction of the epoxide ring is catalyzed by bacterium BBSH797 [62], which is used as a feed additive for deoxynivalenol detoxification, and several other microbes [164–166], but catalysis of this reaction with recombinant enzymes does not seem to be feasible. Recently, a two-step redox reaction mechanism for inversion of the stereoisomeric conformation at carbon atom 3 of deoxynivalenol has been described [167–169]. The reaction product, 3-epi-DON, is nontoxic [63, 170]. It remains to be seen if a combination of two redox-coupled enzymes can be employed for a deoxynivalenol detoxification technology.

Ergopeptines, the most important class of ergot alkaloids [171], are also amenable to biotransformation by microbial enzymes [172, 173]. The reaction product lysergic acid has much lower vasoconstrictive activity than the parent ergopeptines [171]. However, several circumstances have prevented the development of a detoxification technology based on recombinant feed enzymes, including the likely requirement of more than one enzyme, possible psychoactivity of lysergic acid, and the difficulty of providing feed additives for pasture grazing animals that are exposed to ergot alkaloid poisoning [174].

3.2.6 Conclusions

Biocatalysis based on recombinant enzyme technology has entered modern civilized life and brought innovations and new or improved technological processes, often largely unnoticed, in many ways. Enzyme-catalyzed detoxification of mycotoxins in animal feed is yet another example. FUM*zyme*® is making a contribution to more sustainable and also more profitable animal husbandry by preventing the compromise of natural animal vigor and disease resistance caused by fumonisin mycotoxins. Enzymes for other classes of mycotoxins will follow. However, challenges remain. A technological challenge is to provide enzymes with high specific activity for low substrate concentration, so that a small enzyme dose is sufficient for fast mycotoxin detoxification before resorption, and in the set reaction conditions of the gastrointestinal tract. Moreover, enzymes have to be sufficiently thermostable to tolerate feed pelleting. A commercial challenge is that feed enzymes have small economic margins and must be provided cheap and in large quantities, which requires high yield production strains and bioprocesses, and, again, high specific activity. Although much can be learned from the example of FUM*zyme*®, each enzyme is different and requires its own characterization and engineering, and each mycotoxin is different and requires its own set of analytical methods and biomarkers. Furthermore, nature does not seem to have evolved enzymes which are as amenable to engineering and feed enzyme technology development as fumonisin esterase for all mycotoxins.

The future may also bring applications for enzymatic mycotoxin detoxification outside of the gastrointestinal tract of animals. FUM*zyme*® has already been successfully tested in the maize bioethanol process, where it enables the side product distillers dried grains and solubles (DDGS), which is sold for use as animal feed, to be produced free of contaminating fumonisins. Applications for human nutrition also seem feasible, and FUM*zyme*® has already been tried successfully in an experimental maize milling process. However, the food for the population that is most at risk from mycotoxin contamination is typically derived from subsistence farming and not industrially processed. FUM*zyme*® has been considered and tested for household-scale maize washing for African rural communities and may have potential for such noncommercial applications [175, 176].

Mycotoxin detoxification may be peculiar among recombinant enzyme applications, because the long-term hope and vision would be that the technology becomes obsolete. One day, plant breeding and agricultural practice may enable production of grain without mycotoxin contamination, and enzymatic degradation and detoxification will no longer be required. However, until then, mycotoxin detoxification is a perfect example for how recombinant enzyme technology can contribute to animal welfare, and thus to human welfare and more sustainable nutrition and existence of the human population on Earth.

References

1 Firn, R.D. and Jones, C.G. (2000). MicroOpinion: the evolution of secondary metabolism – a unifying model. *Mol. Microbiol.* 37: 989–994.
2 Hoffmeister, D. and Keller, N.P. (2007). Natural products of filamentous fungi: enzymes, genes, and their regulation. *Nat. Prod. Rep.* 24: 393–416.
3 Bennett, J.W. and Klich, M. (2003). Mycotoxins. *Clin. Microbiol. Rev.* 16: 497–516.
4 Schoental, R. (1995). Climatic changes, mycotoxins, plagues, and genius. *J. R. Soc. Med.* 88: 560–561.
5 Weyrich, L.S., Duchene, S., Soubrier, J. et al. (2017). Neanderthal behaviour, diet, and disease inferred from ancient DNA in dental calculus. *Nature* 544: 357–361.
6 Schiff, P.L. (2006). Ergot and its alkaloids. *Am. J. Pharm. Educ.* 70: 98.
7 Strickland, J.R., Looper, M.L., Matthews, J.C. et al. (2011). Board-invited review: St. Anthony's Fire in livestock: causes, mechanisms, and potential solutions. *J. Anim. Sci.* 89: 1603–1626.
8 Blount, W.P. (1961). Turkey "X" disease. *Turkeys (J. Brit. Turkey Fed.)* 9: 55–58.
9 Wogan, G.N. (1966). Chemical nature and biological effects of the aflatoxins. *Bacteriol. Rev.* 30: 460–470.
10 Sargeant, K., Sheridan, A., O'Kelly, J., and Carnaghan, R.B.A. (1961). Toxicity associated with certain samples of groundnuts. *Nature* 192: 1096–1097.
11 Asao, T., Büchi, G., Abdel-Kader, M.M. et al. (1963). Aflatoxins B and G. *J. Am. Chem. Soc.* 85: 1706–1707.
12 Lancaster, M.C., Jenkins, F.P., and Philp, J.M. (1961). Toxicity associated with certain samples of groundnuts. *Nature* 192: 1095–1096.
13 van der Merwe, K.J., Steyn, P.S., Fourie, L. et al. (1965). Ochratoxin A, a toxic metabolite produced by *Aspergillus ochraceus* Wilh. *Nature* 205: 1112–1113.

14 Urry, W.H., Wehrmeister, H.L., Hodge, E.B., and Hidy, P.H. (1966). The structure of zearalenone. *Tetrahedron Lett.* 7: 3109–3114.

15 Yoshizawa, T. and Morooka, N. (1973). Deoxynivalenol and its monoacetate: new mycotoxins from *Fusarium roseum* and Moldy Barley. *Agric. Biol. Chem.* 37: 2933–2934.

16 Gelderblom, W.C., Jaskiewicz, K., Marasas, W.F. et al. (1988). Fumonisins – novel mycotoxins with cancer-promoting activity produced by *Fusarium moniliforme*. *Appl. Environ. Microbiol.* 54 (7): 1806–1811.

17 Berthiller, F., Brera, C., Iha, M.H. et al. (2017). Developments in mycotoxin analysis: an update for 2015–2016. *World Mycotoxin J.* 10: 5–29.

18 Streit, E., Naehrer, K., Rodrigues, I., and Schatzmayr, G. (2013). Mycotoxin occurrence in feed and feed raw materials worldwide: long-term analysis with special focus on Europe and Asia. *J. Sci. Food Agric.* 93: 2892–2899.

19 Liu, Y. and Wu, F. (2010). Global burden of aflatoxin-induced hepatocellular carcinoma: a risk assessment. *Environ. Health Perspect.* 118: 818–824.

20 Marasas, W.F.O., Riley, R.T., Hendricks, K.A. et al. (2004). Fumonisins disrupt sphingolipid metabolism, folate transport, and neural tube development in embryo culture and *in vivo*: a potential risk factor for human neural tube defects among populations consuming fumonisin-contaminated maize. *Am. Soc. Nutr. Sci.* 134: 711–716.

21 Wu, F., Groopman, J.D., and Pestka, J.J. (2014). Public health impacts of foodborne mycotoxins. *Annu. Rev. Food Sci. Technol.* 5: 351–372.

22 Bryden, W.L. (2012). Mycotoxin contamination of the feed supply chain: implications for animal productivity and feed security. *Anim. Feed Sci. Technol.* 173: 134–158.

23 Dersjant-Li, Y., Verstegen, M.W.A., and Gerrits, W.J.J. (2003). The impact of low concentrations of aflatoxin, deoxynivalenol or fumonisin in diets on growing pigs and poultry. *Nutr. Res. Rev.* 16: 223.

24 Oswald, I.P., Marin, D.E., Bouhet, S. et al. (2005). Immunotoxicological risk of mycotoxins for domestic animals. *Food Addit. Contam.* 22: 354–360.

25 Sharma, R.P. (1993). Immunotoxicity of mycotoxins. *J. Dairy Sci.* 76: 892–897.

26 Mitchell, N.J., Bowers, E., Hurburgh, C., and Wu, F. (2016). Potential economic losses to the USA corn industry from aflatoxin contamination. *Food Addit. Contam. Part A* https://doi.org/10.1080/19440049.2016.1138545.

27 Miedaner, T. (1997). Breeding wheat and rye for resistance to *Fusarium* diseases. *Plant Breed.* 116: 201–220.

28 Buerstmayr, H., Ban, T., and Anderson, J.A. (2009). QTL mapping and marker assisted selection for *Fusarium* head blight resistance in wheat: a review. *Plant Breed.* 128: 1–26.

29 Machado, A.K., Brown, N.A., Urban, M. et al. (2017). RNAi as an emerging approach to control *Fusarium* head blight disease and mycotoxin contamination in cereals. *Pest Manag. Sci.* 74: 790–799.

30 Duvick, J. (2001). Prospects for reducing fumonisin contamination of maize through genetic modification. *Environ. Health Perspect.* 109: 337–342.

31 Kabak, B., Dobson, A.D.W., and Var, I. (2006). Strategies to prevent mycotoxin contamination of food and animal feed: a review. *Crit. Rev. Food Sci. Nutr.* 46: 593–619.
32 Mesterházy, Á., Tóth, B., Varga, M. et al. (2011). Role of fungicides, application of nozzle types, and the resistance level of wheat varieties in the control of *Fusarium* head blight and deoxynivalenol. *Toxins (Basel)* 3: 1453–1483.
33 van der Fels-Klerx, H.J. and Booij, C.J.H. (2010). Perspectives for geographically oriented management of fusarium mycotoxins in the cereal supply chain. *J. Food Prot.* 73: 1153–1159.
34 Mehl, H.L., Jaime, R., Callicott, K.A. et al. (2012). *Aspergillus flavus* diversity on crops and in the environment can be exploited to reduce aflatoxin exposure and improve health. *Ann. N.Y. Acad. Sci.* 1273: 7–17.
35 Glenn, A.E., Zitomer, N.C., Zimeri, A.M. et al. (2008). Transformation-mediated complementation of a FUM gene cluster deletion in *Fusarium verticillioides* restores both fumonisin production and pathogenicity on maize seedlings. *Mol. Plant-Microbe Interact.* 21: 87–97.
36 Fravel, D., Olivain, C., and Alabouvette, C. (2003). *Fusarium oxysporum* and its biocontrol. *New Phytol.* 157: 493–502.
37 Bai, G.H., Desjardins, A.E., and Plattner, R.D. (2002). Deoxynivalenol-nonproducing *Fusarium graminearum* causes initial infection, but does not cause disease spread in wheat spikes. *Mycopathologia* 153: 91–98.
38 Khan, N., Maymon, M., and Hirsch, A. (2017). Combating fusarium infection using *Bacillus*-based antimicrobials. *Microorganisms* 5: 75.
39 Saravanakumar, K., Li, Y., Yu, C. et al. (2017). Effect of *Trichoderma harzianum* on maize rhizosphere microbiome and biocontrol of *Fusarium* Stalk rot. *Sci. Rep.* 7: 1–13.
40 Magan, N. and Aldred, D. (2007). Post-harvest control strategies: minimizing mycotoxins in the food chain. *Int. J. Food Microbiol.* 119: 131–139.
41 IARC (2015). *Mycotoxin Control in Low- and Middle- Income Countries.* Lyon: International Agency for Research on Cancer.
42 Karlovsky, P., Suman, M., Berthiller, F. et al. (2016). Impact of food processing and detoxification treatments on mycotoxin contamination. *Mycotoxin Res.* 32: 179–205.
43 Park, D.L. (1993). Perspectives on mycotoxin decontamination procedures. *Food Addit. Contam.* 10: 49–60.
44 ten Bosch, L., Pfohl, K., Avramidis, G. et al. (2017). Plasma-based degradation of mycotoxins produced by *Fusarium*, *Aspergillus* and *Alternaria* species. *Toxins (Basel)* 9: 97.
45 Udomkun, P., Wiredu, A.N., Nagle, M. et al. (2017). Innovative technologies to manage aflatoxins in foods and feeds and the profitability of application – a review. *Food Control* 76: 127–138.
46 Womack, E.D., Brown, A.E., and Sparks, D.L. (2014). A recent review of non-biological remediation of aflatoxin-contaminated crops. *J. Sci. Food Agric.* 94: 1706–1714.
47 Tran, A.T., Kluess, J., Berk, A. et al. (2018). Detoxification of *Fusarium*-contaminated maize with sodium sulphite – *in vivo* efficacy with special

emphasis on mycotoxin residues and piglet health. *Arch. Anim. Nutr.* 72: 58–75.

48 Dänicke, S., Kersten, S., Valenta, H., and Breves, G. (2012). Inactivation of deoxynivalenol-contaminated cereal grains with sodium metabisulfite: a review of procedures and toxicological aspects. *Mycotoxin Res.* 28: 199–218.

49 Masimango, N., Remacle, J., and Ramaut, J.L. (1978). The role of adsorption in the elimination of aflatoxin B_1 from contaminated media. *Eur. J. Appl. Microbiol. Biotechnol.* 6: 101–105.

50 Phillips, T.D., Kubena, L.F., Harvey, R.B. et al. (1988). Hydrated sodium calcium aluminosilicate: a high affinity sorbent for aflatoxin. *Poult. Sci.* 67: 243–247.

51 Phillips, T.D., Sarr, A.B., and Grant, P.G. (1995). Selective chemisorption and detoxification of aflatoxins by phyllosilicate clay. *Nat. Toxins* 3: 204–213.

52 Di Gregorio, M.C., Jager, A.V., Souto, P.C.M.C. et al. (2017). Determination of serum aflatoxin B_1-lysine to evaluate the efficacy of an aflatoxin-adsorbing feed additive in pigs fed an aflatoxin B_1-contaminated diet. *Mycotoxin Res.* 33 (2): 93–102.

53 Mitchell, N.J., Kumi, J., Johnson, N.M. et al. (2013). Reduction in the urinary aflatoxin M1 biomarker as an early indicator of the efficacy of dietary interventions to reduce exposure to aflatoxins. *Biomarkers* 18: 391–398.

54 Kong, C., Shin, S.Y., and Kim, B.G. (2014). Evaluation of mycotoxin sequestering agents for aflatoxin and deoxynivalenol: an *in vitro* approach. *SpringerPlus* 3: 346.

55 De Mil, T., Devreese, M., De Baere, S. et al. (2015). Characterization of 27 mycotoxin binders and the relation with *in vitro* zearalenone adsorption at a single concentration. *Toxins (Basel)* 7: 21–33.

56 Yiannikouris, A., François, J., Poughon, L. et al. (2004). Alkali extraction of β-D-glucans from *Saccharomyces cerevisiae* cell wall and study of their adsorptive properties toward zearalenone. *J. Agric. Food Chem.* 52: 3666–3673.

57 Yiannikouris, A., André, G., Poughon, L. et al. (2006). Chemical and conformational study of the interactions involved in mycotoxin complexation with β- D-glucans. *Biomacromolecules* 7: 1147–1155.

58 Ciegler, A., Lillehoj, E.B., Peterson, R.E., and Hall, H.H. (1966). Microbial detoxification of aflatoxin. *Appl. Microbiol.* 14: 934–939.

59 Karlovsky, P. (1999). Biological detoxification of fungal toxins and its use in plant breeding, feed and food production. *Nat. Toxins* 7: 1–23.

60 McCormick, S.P. (2013). Microbial detoxification of mycotoxins. *J. Chem. Ecol.* 39: 907–918.

61 Hahn, I., Kunz-Vekiru, E., Twarużek, M. et al. (2015). Aerobic and anaerobic *in vitro* testing of feed additives claiming to detoxify deoxynivalenol and zearalenone. *Food Addit. Contam. – Part A Chem. Anal. Control. Expo. Risk Assess.* 32: 922–933.

62 Fuchs, E., Binder, E.M., Heidler, D., and Krska, R. (2002). Structural characterization of metabolites after the microbial degradation of type A trichothecenes by the bacterial strain BBSH 797. *Food Addit. Contam.* 19: 379–386.

63 Pierron, A., Mimoun, S., Murate, L.S. et al. (2016). Microbial biotransformation of DON: molecular basis for reduced toxicity. *Sci. Rep.* 6: 29105.

64 Molnar, O., Schatzmayr, G., Fuchs, E., and Prillinger, H. (2004). *Trichosporon mycotoxinivorans* sp. nov., a new yeast species useful in biological detoxification of various mycotoxins. *Syst. Appl. Microbiol.* 27: 661–671.

65 Politis, I., Fegeros, K., Nitsch, S. et al. (2005). Use of *Trichosporon mycotoxinivorans* to suppress the effects of ochratoxicosis on the immune system of broiler chicks. *Br. Poult. Sci.* 46: 58–65.

66 Schatzmayr, G., Zehner, F., Täubel, M. et al. (2006). Microbiologicals for deactivating mycotoxins. *Mol. Nutr. Food Res.* 50: 543–551.

67 Masching, S., Naehrer, K., Schwartz-Zimmermann, H.E. et al. (2016). Gastrointestinal degradation of fumonisin B_1 by carboxylesterase FumD prevents fumonisin induced alteration of sphingolipid metabolism in Turkey and swine. *Toxins (Basel)* 8 (3): 84.

68 Schwartz-Zimmermann, H.E., Hartinger, D., Doupovec, B. et al. (2018). Application of biomarker methods to investigate FUM*zyme* mediated gastrointestinal hydrolysis of fumonisins in pigs. *World Mycotoxin J.* 11: 201–214.

69 Pierron, A., Alassane-Kpembi, I., and Oswald, I.P. (2016). Impact of two mycotoxins deoxynivalenol and fumonisin on pig intestinal health. *Porc. Heal. Manag.* 2: 21.

70 Lei, X.G. (2003). Phytase enzymology, applications, and biotechnology. *Biotechnol. Lett.* 25: 1787–1794.

71 Ravindran, V. and Son, J.-H. (2011). Feed enzyme technology: present status and future developments. Recent patents food. *Nutr. Agric.* 3: 102–109.

72 Scott, P.M. (2012). Recent research on fumonisins: a review. *Food Addit. Contam. Part A* 29: 242–248.

73 Bezuidenhout, S.C., Gelderblom, W.C.A., Gorst-Allman, C.P. et al. (1988). Structure elucidation of the fumonisins, mycotoxins from *Fusarium moniliforme. J. Chem. Soc., Chem. Commun.* 1730: 743.

74 Marasas, W.F.O. (2001). Discovery and occurrence of the fumonisins: a historical perspective. *Environ. Health Perspect.* 109: 239–243.

75 Voss, K.A., Plattner, R.D., Bacon, C.W., and Norred, W.P. (1990). Comparative studies of hepatotoxicity and fumonisin B_1 and B_2 content of water and chloroform/methanol extracts of *Fusarium moniliforme* strain MRC 826 culture material. *Mycopathologia* 112: 81–92.

76 Kellerman, T.S., Marasas, W.F., Thiel, P.G. et al. (1990). Leukoencephalomalacia in two horses induced by oral dosing of fumonisin B_1. *Onderstepoort J. Vet. Res.* 57: 269–275.

77 Gelderblom, W.C.A., Abel, S., Smuts, C.M. et al. (2001). Fumonisin-induced hepatocarcinogenesis: mechanisms related to cancer initiation and promotion. *Environ. Health Perspect.* 109: 291–300.

78 Sydenham, E.W., Thiel, P.G., Marasas, W.F.O. et al. (1990). Natural occurrence of some *Fusarium* mycotoxins in corn from low and high esophageal cancer prevalence areas of the Transkei, Southern Africa. *J. Agric. Food Chem.* 38: 1900–1903.

79 Gelineau-van Waes, J., Voss, K.A., Stevens, V.L. et al. (2009). Maternal fumonisin exposure as a risk factor for neural tube defects. *Advances in Food and Nutrition Research*. Elsevier Inc, vol. 56, pp. 145–181 (Chapter 5).
80 Harrison, L.R., Colvin, B.M., Greene, J.T. et al. (1990). Pulmonary edema and hydrothorax in swine produced by fumonisin B_1, a toxic metabolite of *Fusarium moniliforme. J. Vet. Diagn. Invest.* 2: 217–221.
81 Haschek, W.M., Motelin, G., Ness, D.K. et al. (1992). Characterization of fumonisin toxicity in orally and intravenously dosed swine. *Mycopathologia* 117: 83–96.
82 Devriendt, B., Gallois, M., Verdonck, F. et al. (2009). The food contaminant fumonisin B_1 reduces the maturation of porcine CD11R1$^+$ intestinal antigen presenting cells and antigen-specific immune responses, leading to a prolonged intestinal ETEC infection. *Vet. Res.* 40: 1–14.
83 Oswald, I.P., Desautels, C., Laffitte, J. et al. (2003). Mycotoxin fumonisin B_1 increases intestinal colonization by pathogenic *Escherichia coli* in pigs. *Appl. Environ. Microbiol.* 69: 5870–5874.
84 Taranu, I., Marin, D.E., Bouhet, S. et al. (2005). Mycotoxin fumonisin B_1 alters the cytokine profile and decreases the vaccinal antibody titer in pigs. *Toxicol. Sci.* 84: 301–307.
85 Merrill, A.H., Sullards, M.C., Wang, E. et al. (2001). Sphingolipid metabolism: roles in signal transduction and disruption by fumonisins. *Environ. Health Perspect.* 109: 283–289.
86 Riley, R.T., Enongene, E., Voss, K.A. et al. (2001). Sphingolipid perturbations as mechanisms for fumonisin carcinogenesis. *Environ. Health Perspect.* 109: 301–308.
87 Wang, E., Norred, W.P., Bacon, C.W. et al. (1991). Inhibition of sphingolipid biosynthesis by fumonisins. Implications for diseases associated with *Fusarium moniliforme. J. Biol. Chem.* 266: 14486–14490.
88 Bartke, N. and Hannun, Y.A. (2009). Bioactive sphingolipids: metabolism and function. *J. Lipid Res.* 50: S91–S96.
89 Sonnino, S., Prinetti, A., Mauri, L. et al. (2006). Dynamic and structural properties of sphingolipids as driving forces for the formation of membrane domains. *Chem. Rev.* 106: 2111–2125.
90 Lopez, P.H. and Schnaar, R.L. (2009). Gangliosides in cell recognition and membrane protein regulation. *Curr. Opin. Struct. Biol.* 19: 549–557.
91 Merrill, A.H. (2011). Sphingolipid and glycosphingolipid metabolic pathways in the era of sphingolipidomics. *Chem. Rev.* 111: 6387–6422.
92 Desai, K., Sullards, M.C., Allegood, J. et al. (2002). Fumonisins and fumonisin analogs as inhibitors of ceramide synthase and inducers of apoptosis. *Biochim. Biophys. Acta, Mol. Cell. Biol. Lipids* 1585: 188–192.
93 Wadsworth, J.M., Clarke, D.J., Mcmahon, S.A. et al. (2013). The chemical basis of serine palmitoyltransferase inhibition by myriocin. *J. Am. Chem. Soc.* 135: 14276–14285.
94 Wang, E., Ross, P.F., Wilson, T.M. et al. (1992). Increases in serum sphingosine and sphinganine and decreases in complex sphingolipids in ponies given feed containing fumonisins, mycotoxins produced by *Fusarium moniliforme. J. Nutr.* 122: 1706–1716.

95 Riley, R.T., An, N.H., Showker, J.L. et al. (1993). Alteration of tissue and serum sphinganine to sphingosine ratio: an early biomarker of exposure to fumonisin-containing feeds in pigs. *Toxicol. Appl. Pharmacol.* 118: 105–112.

96 Shephard, G.S., Van Der Westhuizen, L., and Sewram, V. (2007). Biomarkers of exposure to fumonisin mycotoxins: a review. *Food Addit. Contam.* 24: 1196–1201.

97 Hahn, I., Nagl, V., Schwartz-Zimmermann, H.E. et al. (2015). Effects of orally administered fumonisin B_1 (FB_1), partially hydrolysed FB_1, hydrolysed FB_1 and *N*-(1-deoxy-D-fructos-1-yl) FB_1 on the sphingolipid metabolism in rats. *Food Chem. Toxicol.* 76: 11–18.

98 Riley, R.T., Hinton, D.M., Chamberlain, W.J. et al. (1994). Dietary fumonisin B_1 induces disruption of sphingolipid metabolism in sprague-dawley rats: a new mechanism of nephrotoxicity. *J. Nutr.* 124: 594–603.

99 Frisvad, J.C., Smedsgaard, J., Samson, R.A. et al. (2007). Fumonisin B_2 production by *Aspergillus niger*. *J. Agric. Food Chem.* 55: 9727–9732.

100 Mogensen, J.M., Møller, K.A., Von Freiesleben, P. et al. (2011). Production of fumonisins B_2 and B_4 in *Tolypocladium* species. *J. Ind. Microbiol. Biotechnol.* 38: 1329–1335.

101 Grenier, B., Bracarense, A.P.F.L., Schwartz, H.E. et al. (2013). Biotransformation approaches to alleviate the effects induced by *Fusarium* Mycotoxins in swine. *J. Agric. Food Chem.* 61: 6711–6719.

102 Blackwell, B.A., Gilliam, J.T., Savard, M.E. et al. (1999). Oxidative deamination of hydrolyzed fumonisin B_1 (AP1) by cultures of *Exophiala spinifera*. *Nat. Toxins* 7: 31–38.

103 Duvick, J., Rood, T., Maddox, J., and Gilliam, J. (1998). Detoxification of mycotoxins in planta as a strategy for improving grain quality and disease resistance. In: *Developments in Plant Pathology: Molecular Genetics of Host-Specific Toxins in Plant Disease* (ed. K. Kohmoto and O. Yoder), 369–381. the Netherlands: Kluwer Academic Publishers.

104 Fodor, J., Balogh, K., Weber, M. et al. (2008). Absorption, distribution and elimination of fumonisin B1 metabolites in weaned piglets. *Food Addit. Contam. Part A* 25: 88–96.

105 Shephard, G.S., Thiel, P.G., Sydenham, E.W. et al. (1994). Determination of the mycotoxin fumonisin B_1 and identification of its partially hydrolysed metabolites in the faeces of non-human primates. *Food Chem. Toxicol.* 32: 23–29.

106 Heinl, S., Hartinger, D., Thamhesl, M. et al. (2010). Degradation of fumonisin B_1 by the consecutive action of two bacterial enzymes. *J. Biotechnol.* 145: 120–129.

107 Heinl, S., Hartinger, D., Thamhesl, M. et al. (2011). An aminotransferase from bacterium ATCC 55552 deaminates hydrolyzed fumonisin B_1. *Biodegradation* 22: 25–30.

108 Lewis, J.A., Horswill, A.R., Schwem, B.E., and Escalante-Semerena, J.C. (2004). The tricarballylate utilization (tcuRABC) genes of *Salmonella enterica* serovar Typhimurium LT2. *J. Bacteriol.* 186: 1629–1637.

109 Gelderblom, W.C.A., Cawood, M.E., Snyman, S.D. et al. (1993). Structure-activity relationships of fumonisins in short-term carcinogenesis and cytotoxicity assays. *Food Chem. Toxicol.* 31: 407–414.
110 Norred, W.P., Plattner, R.D., Dombrink-Kurtzman, M.A. et al. (1997). Mycotoxin-induced elevation of free sphingoid bases in precision-cut rat liver slices: specificity of the response and structure-activity relationships. *Toxicol. Appl. Pharmacol.* 147: 63–70.
111 Howard, P.C., Churchwell, M., Couch, L.H. et al. (1998). Formation of N-(carboxymethyl)fumonisin B_1, following the reaction of fumonisin B_1 with reducing sugars. *J. Agric. Food Chem.* 46: 3546–3557.
112 Humpf, H.U. and Voss, K.A. (2004). Effects of thermal food processing on the chemical structure and toxicity of fumonisin mycotoxins. *Mol. Nutr. Food Res.* 48: 255–269.
113 Humpf, H.-U., Schmelz, E.-M., Meredith, F.I. et al. (1998). Acylation of naturally occurring and synthetic 1-deoxysphinganines by ceramide synthase. *J. Biol. Chem.* 273: 19060–19064.
114 Seiferlein, M., Humpf, H.U., Voss, K.A. et al. (2007). Hydrolyzed fumonisins HFB_1 and HFB_2 are acylated *in vitro* and *in vivo* by ceramide synthase to form cytotoxic N-acyl-metabolites. *Mol. Nutr. Food Res.* 51: 1120–1130.
115 Hartinger, D., Heinl, S., Schwartz, H.E. et al. (2010). Enhancement of solubility in *Escherichia coli* and purification of an aminotransferase from *Sphingopyxis* sp. MTA144 for deamination of hydrolyzed fumonisin B_1. *Microb. Cell Fact.* 9: 62.
116 Hartinger, D., Schwartz, H., Hametner, C. et al. (2011). Enzyme characteristics of aminotransferase FumI of *Sphingopyxis* sp. MTA144 for deamination of hydrolyzed fumonisin B_1. *Appl. Microbiol. Biotechnol.* 91: 757–768.
117 Hartinger, D. and Moll, W. (2011). Fumonisin elimination and prospects for detoxification by enzymatic transformation. *World Mycotoxin J.* 4: 271–283.
118 Hendrich, S., Miller, K.A., Wilson, T.M., and Murphy, P.A. (1993). Toxicity of *Fusarium proliferatum*-fermented nixtamalized corn-based diets fed to rats: effect of nutritional status. *J. Agric. Food Chem.* 41: 1649–1654.
119 Voss, K.A., Bacon, C.W., Meredith, F.I., and Norred, W.P. (1996). Comparative subchronic toxicity studies of nixtamalized and water-extracted *Fusarium moniliforme* culture material. *Food Chem. Toxicol.* 34: 623–632.
120 Voss, K.A., Riley, R.T., Bacon, C.W. et al. (1998). Toxicity and sphinganine levels are correlated in rats fed fumonisin B_1 (FB_1) or hydrolyzed FB_1. *Environ. Toxicol. Pharmacol.* 5: 101–104.
121 Meredith, F.I., Torres, O.R., Saenz de Tejada, S. et al. (1999). Fumonisin B_1 and hydrolyzed fumonisin B_1 (AP_1) in tortillas and nixtamalized corn (*Zea mays* L.) from two different geographic locations in Guatemala. *J. Food Prot.* 62: 1218–1222.
122 Burns, T.D., Snook, M.E., Riley, R.T., and Voss, K.A. (2008). Fumonisin concentrations and *in vivo* toxicity of nixtamalized *Fusarium verticillioides* culture material: evidence for fumonisin-matrix interactions. *Food Chem. Toxicol.* 46: 2841–2848.

123 Voss, K., Ryu, D., Jackson, L. et al. (2017). Reduction of fumonisin toxicity by extrusion and nixtamalization (alkaline cooking). *J. Agric. Food Chem.* 65: 7088–7096.

124 Voss, K.A., Riley, R.T., Moore, N.D., and Burns, T.D. (2013). Alkaline cooking (nixtamalisation) and the reduction in the *in vivo* toxicity of fumonisin-contaminated corn in a rat feeding bioassay. *Food Addit. Contam. – Part A Chem. Anal. Control. Expo. Risk Assess.* 30: 1415–1421.

125 Collins, T.F.X., Sprando, R.L., Black, T.N. et al. (2006). Effects of aminopentol on *in utero* development in rats. *Food Chem. Toxicol.* 44: 161–169.

126 Howard, P.C., Couch, L.H., Patton, R.E. et al. (2002). Comparison of the toxicity of several fumonisin derivatives in a 28-day feeding study with female B6C3F_1 mice. *Toxicol. Appl. Pharmacol.* 185: 153–165.

127 Voss, K.A., Riley, R.T., Snook, M.E., and Gelineau-van Waes, J. (2009). Reproductive and sphingolipid metabolic effects of fumonisin B_1 and its alkaline hydrolysis product in LM/Bc mice: hydrolyzed fumonisin B_1 did not cause neural tube defects. *Toxicol. Sci.* 112: 459–467.

128 Abbas, H.K., Gelderblom, W.C., Cawood, M.E., and Shier, W.T. (1993). Biological activities of fumonisins, mycotoxins from *Fusarium moniliforme*, in jimsonweed (*Datura stramonium* L.) and mammalian cell cultures. *Toxicon* 31: 345–353.

129 Merrill, A.H., Wang, E., Gilchrist, D.G., and Riley, R.T. (1993). Fumonisins and other inhibitors of *de novo* sphingolipid biosynthesis. *Adv. Lipid Res.* 26: 215–234.

130 Abbas, H.K., Tanaka, T., and Shier, W.T. (1995). Biological activities of synthetic analogues of *Alternaria alternata* toxin (AAL-toxin) and fumonisin in plant and mammalian cell cultures. *Phytochemistry* 40: 1681–1689.

131 Flynn, T.J., Stack, M.E., Troy, A.L., and Chirtel, S.J. (1997). Assessment of the embryotoxic potential of the total hydrolysis product of fumonisin B_1 using cultured organogenesis-staged rat embryos. *Food Chem. Toxicol.* 35: 1135–1141.

132 Gilchrist, D.G., Ward, B., Moussato, V., and Mirocha, C.J. (1992). Genetic and physiological response to fumonisin and AAL-toxin by intact tissue of a higher plant. *Mycopathologia* 117: 57–64.

133 Hartl, M. and Humpf, H.U. (2000). Toxicity assessment of fumonisins using the brine shrimp (*Artemia salina*) bioassay. *Food Chem. Toxicol.* 38: 1097–1102.

134 Lamprecht, S.C., Marasas, W.F.O., Alberts, J.F. et al. (1994). Phytotoxicity of fumonisins and TA-toxin to corn and tomato. *Phyopathology* 84: 383–391.

135 Norred, W.P., Wang, E., Yoo, H. et al. (1992). *In vitro* toxicology of fumonisins and the mechanistic implications. *Mycopathologia* 117: 73–78.

136 Schmelz, E.M., Dombrink-kurtzman, M.A., Roberts, P.C. et al. (1998). Induction of apoptosis by fumonisin B1 in HT29 cells is mediated by the accumulation of endogenous free sphingoid bases. *Toxicol. Appl. Pharmacol.* 148: 252–260.

137 Tanaka, T., Abbas, H.K., and Duke, S.O. (1993). Structure-dependent phytotoxicity of fumonisins and related compounds in a duckweed bioassay. *Phytochemistry* 33: 779–785.

138 Van Der Westhuizen, L., Shephard, G.S., Snyman, S.D. et al. (1998). Inhibition of sphingolipid biosynthesis in rat primary hepatocyte cultures by fumonisin B_1 and other structurally related compounds. 36: 497–503.

139 Dantzer, W.R., Hopper, J., Mullin, K. et al. (1999). Excretion of ^{14}C-fumonisin B_1, ^{14}C-hydrolyzed fumonisin B_1, and ^{14}C-fumonisin B_1-fructose in rats. *J. Agric. Food Chem.* 47: 4291–4296.

140 Grenier, B., Bracarense, A.P.F.L., Schwartz, H.E. et al. (2012). The low intestinal and hepatic toxicity of hydrolyzed fumonisin B_1 correlates with its inability to alter the metabolism of sphingolipids. *Biochem. Pharmacol.* 83: 1465–1473.

141 Grenier, B., Loureiro-Bracarense, A.P., Lucioli, J. et al. (2011). Individual and combined effects of subclinical doses of deoxynivalenol and fumonisins in piglets. *Mol. Nutr. Food Res.* 55: 761–771.

142 Grenier, B., Schwartz-Zimmermann, H.E., Caha, S. et al. (2015). Dose-dependent effects on sphingoid bases and cytokines in chickens fed diets prepared with *Fusarium verticillioides* culture material containing fumonisins. *Toxins (Basel)* 7: 1253–1272.

143 Grenier, B., Schwartz-Zimmermann, H.E., Gruber-Dorninger, C. et al. (2017). Enzymatic hydrolysis of fumonisins in the gastrointestinal tract of broiler chickens. *Poult. Sci.* 96: 4342–4351.

144 Puxbaum, V., Mattanovich, D., and Gasser, B. (2015). Quo vadis? The challenges of recombinant protein folding and secretion in *Pichia pastoris*. *Appl. Microbiol. Biotechnol.* 99: 2925–2938.

145 Küberl, A., Schneider, J., Thallinger, G.G. et al. (2011). High-quality genome sequence of *Pichia pastoris* CBS7435. *J. Biotechnol.* 154: 312–320.

146 Sturmberger, L., Chappell, T., Geier, M. et al. (2016). Refined *Pichia pastoris* reference genome sequence. *J. Biotechnol.* 235: 121–131.

147 Kolosova, A. and Stroka, J. (2011). Substances for reduction of the contamination of feed by mycotoxins: a review. *World Mycotoxin J.* 4: 225–256.

148 Bories, G., Brantom, P., de Barberà, J.B. et al. (2008). Functional groups of additives as described in annex 1 of regulation (EC) No 1831/20031 self-task of the panel on additives and products or substances used in animal feed. *EFSA J.* 6 (12): 920.

149 Zinedine, A., Soriano, J.M., Moltó, J.C., and Mañes, J. (2007). Review on the toxicity, occurrence, metabolism, detoxification, regulations and intake of zearalenone: an oestrogenic mycotoxin. *Food Chem. Toxicol.* 45: 1–18.

150 El-Sharkawy, S. and Abul-Hajj, Y.J. (1988). Microbial cleavage of zearalenone. *Xenobiotica* 18: 365–371.

151 Takahashi-Ando, N., Kimura, M., Kakeya, H. et al. (2002). A novel lactonohydrolase responsible for the detoxification of zearalenone: enzyme purification and gene cloning. *Biochem. J.* 365: 1–6.

152 Takahashi-Ando, N., Ohsato, S., Shibata, T. et al. (2004). Metabolism of zearalenone by genetically modified organisms expressing the detoxification gene from *Clonostachys rosea*. *Appl. Environ. Microbiol.* 70: 3239–3245.

153 Vekiru, E., Hametner, C., Mitterbauer, R. et al. (2010). Cleavage of zearalenone by *Trichosporon mycotoxinivorans* to a novel nonestrogenic metabolite. *Appl. Environ. Microbiol.* 76: 2353–2359.

154 Malir, F., Ostry, V., Pfohl-Leszkowicz, A. et al. (2016). Ochratoxin A: 50 years of research. *Toxins (Basel)* 8: 191.

155 Pitout, M.J. (1969). The hydrolysis of ochratoxin A by some proteolytic enzymes. *Biochem. Pharmacol.* 18: 485–491.

156 Stander, M.A., Bornscheuer, U.T., Henke, E., and Steyn, P.S. (2000). Screening of commercial hydrolases for the degradation of ochratoxin A. *J. Agric. Food Chem.* 48: 5736–5739.

157 Wu, Q., Dohnal, V., Huang, L. et al. (2011). Metabolic pathways of ochratoxin A. *Curr. Drug Metab.* 12: 1–10.

158 Xiao, H., Madhyastha, S., Marquardt, R.R. et al. (1996). Toxicity of ochratoxin A, its opened lactone form and several of its analogs: structure–activity relationships. *Toxicol. Appl. Pharmacol.* 137: 182–192.

159 Abrunhosa, L. and Venâncio, A. (2007). Isolation and purification of an enzyme hydrolyzing ochratoxin A from *Aspergillus niger*. *Biotechnol. Lett.* 29: 1909–1914.

160 Dobritzsch, D., Wang, H., Schneider, G., and Yu, S. (2014). Structural and functional characterization of ochratoxinase, a novel mycotoxin-degrading enzyme. *Biochem. J.* 462: 441–452.

161 Alberts, J.F., Gelderblom, W.C.A., Botha, A., and van Zyl, W.H. (2009). Degradation of aflatoxin B1 by fungal laccase enzymes. *Int. J. Food Microbiol.* 135: 47–52.

162 Lapalikar, G.V., Taylor, M.C., Warden, A.C. et al. (2012). $F_{420}H_2$-dependent degradation of aflatoxin and other furanocoumarins is widespread throughout the *Actinomycetales*. *PLoS ONE* 7: e30114.

163 Taylor, M.C., Jackson, C.J., Tattersall, D.B. et al. (2010). Identification and characterization of two families of $F_{420}H_2$-dependent reductases from *Mycobacteria* that catalyse aflatoxin degradation. *Mol. Microbiol.* 78: 561–575.

164 Islam, R., Zhou, T., Young, J.C. et al. (2012). Aerobic and anaerobic de-epoxydation of mycotoxin deoxynivalenol by bacteria originating from agricultural soil. *World J. Microbiol. Biotechnol.* 28: 7–13.

165 Li, X.Z., Zhu, C., de Lange, C.F.M. et al. (2011). Efficacy of detoxification of deoxynivalenol-contaminated corn by *Bacillus* sp. LS100 in reducing the adverse effects of the mycotoxin on swine growth performance. *Food Addit. Contam. – Part A Chem. Anal. Control. Expo. Risk Assess.* 28: 894–901.

166 Yu, H., Zhou, T., Gong, J. et al. (2010). Isolation of deoxynivalenol-transforming bacteria from the chicken intestines using the approach of PCR-DGGE guided microbial selection. *BMC Microbiol.* 10: 182.

167 Carere, J., Hassan, Y.I., Lepp, D., and Zhou, T. (2018). The enzymatic detoxification of the mycotoxin deoxynivalenol: identification of DepA from the DON epimerization pathway. *Microb. Biotechnol.* 11: 1106–1111.

168 Hassan, Y.I. and Zhou, T. (2018). Addressing the mycotoxin deoxynivalenol contamination with soil-derived bacterial and enzymatic transformations targeting the C3 carbon. *World Mycotoxin J.* 11: 101–112.

169 Hassan, Y.I., He, J.W., Perilla, N. et al. (2017). The enzymatic epimerization of deoxynivalenol by *Devosia mutans* proceeds through the formation of 3-keto-DON intermediate. *Sci. Rep.* 7: 6929.
170 He, J.W., Bondy, G.S., Zhou, T. et al. (2015). Toxicology of 3-epi-deoxynivalenol, a deoxynivalenol-transformation product by *Devosia mutans* 17-2-E-8. *Food Chem. Toxicol.* 84: 250–259.
171 Klotz, J.L., Bush, L.P., Smith, D.L. et al. (2006). Assessment of vasoconstrictive potential of D-lysergic acid using an isolated bovine lateral saphenous vein bioassay. *J. Anim. Sci.* 84: 3167–3175.
172 Hahn, I., Thamhesl, M., Apfelthaler, E. et al. (2015). Characterisation and determination of metabolites formed by microbial and enzymatic degradation of ergot alkaloids. *World Mycotoxin J.* 8: 393–404.
173 Thamhesl, M., Apfelthaler, E., Schwartz-Zimmermann, H.E. et al. (2015). *Rhodococcus erythropolis* MTHt3 biotransforms ergopeptines to lysergic acid. *BMC Microbiol.* 15: 73.
174 Aiken, G.E. and Strickland, J.R. (2013). Forages and pastures symposium: managing the tall fescue-fungal endophyte symbiosis for optimum forage-animal production. *J. Anim. Sci.* 91: 2369–2378.
175 Alberts, J.F., van Zyl, W.H., and Gelderblom, W.C.A. (2016). Biologically based methods for control of fumonisin-producing *Fusarium* species and reduction of the fumonisins. *Front. Microbiol.* 7: 548.
176 Alberts, J.F., Lilly, M., Rheeder, J.P. et al. (2017). Technological and community-based methods to reduce mycotoxin exposure. *Food Control* 73: 101–109.

3.3

Phytases for Feed Applications

Nikolay Outchkourov and Spas Petkov

HUVEPHARMA EOOD, 3A, Nikolay Haytov Street, 5th floor, 1113 Sofia, Bulgaria

3.3.1 Phytase As a Feed Enzyme: Introduction and Significance

Modern animal farming practices demand efficient means of nutrient utilization with minimal environmental footprint. Enzymes used in farm animal nutrition increase the dietary value of feed ingredients by degrading indigestible components and anti-nutritional factors. Phytase in particular dominates the global feed enzyme market together with non-starch polysaccharide-degrading enzymes such as xylanase, β-glucanase, cellulase, mannase, and α-galactosidase [1–3]. Demands for high quality animal-derived protein, growing consumer awareness about health, environmental and animal nutrition benefits combined with increasing prices of feed ingredients contribute to the recent global expansion of the market of feed enzymes in general and the phytase in particular with emphasis on cost effective and environmentally friendly phosphorous utilization.

Phosphorous is an essential mineral that is required in sufficient amounts for growth and development of all animals [4]. Various biochemical pathways, physiological processes, DNA and bone integrity are directly linked to the presence of sufficient bioavailable phosphorus. However, when released to the environment in excessive quantities phosphorous can act as a pollutant [5]. Phytic acid known as inositol hexakisphosphate (IP6 or Ins(1,2,3,4,5,6)P_6 when exact numbering of the phosphate groups is required) is the principal storage form of phosphorus in plant tissues and is especially abundant in seeds and grains [6]. In its salt form phytic acid is known as phytate.

Phytase enzymes hydrolyze the stepwise dephosphorylation of IP6 to IP5 and lower inositol phosphates and inorganic phosphorous (P) [1–3]. Monogastric animals such as pigs, poultry, and fish cannot utilize phytate-bound phosphorus due to inappropriate phytase activity in their digestive tracts [7, 8]. Phytate-P content, as a percentage of total P content, is higher in cereals and in wheat by-products than in oilseed meals and legume seeds [9]. Formulated poultry and swine diets contain a high proportion (about 60–80%) of the total phosphorus bound to phytate [2, 10, 11] and thus poorly bioavailable to monogastric animals. Values for total phytate in feedingstuffs vary largely but typical phytate

Industrial Enzyme Applications, First Edition. Edited by Andreas Vogel and Oliver May.

concentrations are about 10 g/kg (range of 5–25 g/kg), which consist of 2.8 g/kg (range 2–4 g/kg) phytate-bound phosphorus [2, 12]. In addition to the reduction of the bioavailability of phosphorus, phytate also acts as an anti-nutritional factor by strong binding to essential minerals; amino acids and proteins also limit their bioavailability and thus additionally suppress growth and feed efficiency of farm animals. Supplementation of extra inorganic phosphorus to feedingstuffs can compensate for the insufficient bioavailable phosphorus but cannot counteract the anti-nutritional properties of the phytate. Subsequently, the excess of phosphorus is then released to the environment through manure and can contribute to environmental problems such as algal blooms, eutrophication of water-bodies, and fish kills [5, 13]. Inorganic phosphorus is a relatively expensive and nonrenewable mineral. On the other hand, supplementation of animal feed with phytase increases the bioavailability of phytate-bound phosphorus, diminishes the anti-nutritional effect of phytate, improves calcium, zinc, and iron utilization by animals [2, 3, 14–16], and reduces the release of phosphorus in the environment by up to 50% [1, 2, 15].

A study from Lott et al. [17] estimate that on a global scale 51 million metric tons of phytate (9.9 million metric tons of phosphorus) is sequestered annually in crop seeds and fruits each year. Phytases unlock the bioavailability of phosphorus stored in plant materials and thus generate nutritional, economic, and environmental benefits. In view of the growing human population, increasing prices of finite rock phosphate, and increasing environmental concerns [18], it is expected that use of phytases for animal nutrition will increase and play an even more prominent role for efficient phosphorus utilization in the future. Taking into consideration the ability of phytase to improve the iron and zinc bioavailability in animals [16, 19] and the substantial iron and zinc deficiency in human population prevalent in particular African and Asian countries, which partly can be attributed to low bioavailability of zinc and iron from cereals, in 2012 World Health Organization approved phytase from *Aspergillus niger* for use in human food [20]. Other new emergent phytase applications in biofuel production and food processing are expected to extend the industrial utilization of phytase-based enzymes much further [3, 21].

3.3.2 Historical Overview of the Phytase Market Development

The first description of phytate dates from two studies of Hartig from 1855 to 1856 when small round particles in various plant seeds similar in size to potato starch grains were described [22–24]. Phytase activity was described in 1907 by Suzuki et al. in rice bran [25]. Attempts to develop commercial phytase enzyme date back to 1962 [3] when International Minerals & Chemicals (IMC) foresaw the need of the market for an enzyme that would hydrolyze phytate in plant materials. A collection of more than 2000 organisms was screened for phytase activity. The expression levels in the screened strains were too low to sustain commercial manufacturing of the phytase enzyme but from this project a valuable isolate

A. niger NRRL 3135 (ATCC 66876) was identified [26], which served as a basis for commercial phytase development later in 1990. The first cloning of the *PhyA* gene was reported in 1991 by Mullaney et al. [27] and its overexpression by the team of TNO, Rijswijk, The Netherlands [28]. *PhyA* gene overexpressed in *A. niger* with the help of recombinant DNA technology served as a basis for the development of the first globally commercialized phytase Natuphos®, produced in cooperation by Gist-Brocades, The Netherlands (now DSM) and BASF, Germany. The active substance phytase is produced via fermentation by Gist-Brocades and the commercial product Natuphos was formulated by BASF. In Europe, Natuphos was given, for the first time, provisional authorization in April 1994 and was approved in the United States in November 1995. This was a 3-phytase with established efficacy in many different trials, which we would like to refer as first-generation phytase. For more details please refer to [3, 29–32].

Development of the second-generation phytase took its own parallel path and originated from the work of Dassa and coworkers. This team worked for many years on a pH 2.5 acid phosphatase from *Escherichia coli* and published peer reviewed articles since 1979 on this topic [33]. The name *appA* of the *E. coli* "acid phosphatase" and later identified as phytase was coined upon identification of a mutation on the <u>a</u>cid <u>p</u>hosphoanhydride <u>p</u>hosphohydrolase A1 (*appA1*) gene located at 22.5 minutes on the *E. coli* genetic map, strongly suggesting that *appA* is the structural gene of the acid phosphatase [34]. The *appA* gene was cloned as a pH 2.5 acid phosphatase in 1990 [35]. Later work of Greiner et al. [36] after the classical biochemical purification of phytases from *E. coli* suggested, based on partial amino acid sequence of the purified phytase, that pH 2.5 acid phosphatase appA described by Dassa et al. [35] is a 6-phytase. Much later in 1999 with the cloning of the *appA* gene and its homolog *appA2* and their overproduction in *Pichia pastoris* yeast system the first direct evidence was generated that the appA enzyme was indeed more a phytase than an acid phosphatase [37]. This work of Rodriguez et al. from Cornell University paved the way for development of the second-generation appA based 6-phytases, which had a superior animal efficacy due to better pepsin resistance and greater catalytic speed over the 3-phytase such as Natuphos® [37–40]. In addition to this, the work of Rodriguez also identified *P. pastoris* as a new expression system suitable for appA phytase overproduction. The latter served as basis of commercial development for OptiPhos® phytase commercialized initially by JBS United in United States and later acquired by Huvepharma, who introduced it in Europe and the rest of the world. Interestingly, overexpression in *P. pastoris* added a number of glycosylation sites on the surface of the OptiPhos, which are not normally present on the native protein, and affected positively the thermal stability of the recombinant appA phytase and its resistance to pepsin [37, 41] without notable effects on the catalytic rate and K_m of the enzyme.

Importantly for the market development, the appearance of OptiPhos coincided with great increase in the price of inorganic phosphorus [3], which improved the economic feasibility of using phytases for feed application and boosted the development of new phytases. Consequently, a number of products based on *appA* gene have been subsequently marketed: Phyzyme XP®

(Danisco), an *E. coli* appA phytase expressed in *Schizosaccharomyces pombe* (ATCC 5233); Finase EC® (AB Vista) based on *E. coli* appA protein overproduced in *Trichoderma reesei*; and Quantum® and Quantum Blue® (AB Vista) containing a thermostabilized appA variant [3, 42] overproduced in *P. pastoris* and *T. reesei*, accordingly. More recently, a new 6-phytase from *Citrobacter braakii* was described in the scientific literature [43]. When overproduced in *Aspergillus oryzae* this phytase was commercialized as RONOZYME HiPhos® (DSM/Novozymes). Both of the phytases from *C. braakii* and *E. coli* appA are derived from enterobacterial origin with acidic pH profiles (optimum around pH 3.5–4.5), high specific activity, and resistance to proteases [37, 43]. Interestingly, also fungal 6-phytases have been identified from *Peniophora lycii* [44] and introduced as RONOZYME P® and RONOZYME NP® from DSM/Novozymes. RONOZYME NP is a *P. lycii* variant with improved intestinal and thermal stability. The fungal PhyA and *P. lycii* phytases have pH optima around 5.5 and 4–5, respectively. See also Table 3.3.1 for additional information.

Table 3.3.1 The main commercial phytases approved for animal feed, all from the HAPhy family.

Product	Company	Protein origin	Subtype	Production host
Natuphos	BASF	*Aspergillus niger* **PhyA**	3-Phytase	*Aspergillus niger*
NatuPhos E® (in 2018)	BASF	Hybrid phytase of three bacterial sources	6-Phytase	*Aspergillus niger*
Ronozyme P	Novozymes/ DSM alliance	*Peniophora lycii* **PhyA**	6-Phytase	*Aspergillus oryzae*
Ronozyme NP	Novozymes/ DSM alliance	Improved *Peniophora lycii* intestinal and thermal stability	6-Phytase	*Aspergillus oryzae*
Ronozyme HiPhos	Novozymes/ DSM alliance	*Citrobacter braakii*	6-Phytase	*Aspergillus oryzae*
Finase EC	AB Vista	*Escherichia coli* **appA**	6-Phytase	*Trichoderma reesei*
Quantum	AB Vista	Improved *Escherichia coli* **appA** variant for thermal and gastric stability	6-Phytase	*Pichia pastoris*
Quantum Blue	AB Vista	Improved *Escherichia coli* **appA** variant for thermal and gastric stability	6-Phytase	*Trichoderma reesei*
Phyzyme®	Danisco	*Escherichia coli* **appA**	6-Phytase	*Schizosaccharomyces pombe*
Axtra Phy®	Danisco	*Buttiauxella* spp.	6-Phytase	*Trichoderma reesei*
OptiPhos	Huvepharma	*Escherichia coli* **appA**	6-Phytase	*Pichia pastoris*

3.3.3 From Phytate to Phosphorus: Step by Step Action of the Phytase

3.3.3.1 Properties of Phytate

In plants and thus in plant-based diets IP6 exists in its anionic form, phytate, as a complex salt of mainly Ca^{2+}, Mg^{2+}, and K^{+}, and in some cases it is bound to proteins and starches. This mixed salt of calcium–magnesium complexed or chelated molecule of IP6 is known as phytin. In plant seeds a number of physiological roles could be attributed to the phytate. It is used as a P reserve and energy store. Additionally, phytin immobilizes and releases a number of divalent cations needed for the control of cellular processes upon germination by the action of intrinsic plant phytases [11, 45]. The IP6 molecule contains 28.2% phosphorus per molar base and is the most abundant phosphorus constituent in swine and poultry diets. In plant-based diets the total phosphorus is present at sufficient levels to meet the nutritional requirement of the animals but most of the phosphorus is locked and nutritionally not bioavailable in the form of phytin. Secondly, phytin is most known as antinutrient because of its ability to complex mineral cations binding to amino acids or proteins in the seed and in the diets [46, 47], rendering these nutrients as well as phosphorus unavailable to the animals and thus greatly reducing the nutritional value of feedingstuffs. Phytin-bound nutrients as well as phosphorus are unavailable to the animals due to inadequate absorption in the small intestine, thereby defining the anti-nutritional properties of the phytate with adverse effects on various metabolic processes and growth parameters. In addition, phytic acid is also known to bind to proteins and starch, resulting in reduced digestibility and absorption of these nutrients as well. Proteins can bind to IP6 through electrostatic charges at low pH or through salt bridges formed at high pH [45, 48]. The binding of IP6 to proteins and starch may reduce the availability of these nutrients and their digestibility. Added to this, complexing to proteins and starches will influence the degree of ionization of the complex and the binding to minerals, potentially affecting the efficacy of different phytases.

At a pH range of 0.5–9.0, IP6 adopts a sterically stable conformation with one phosphate in the axial and five phosphates in the equatorial position (Figure 3.3.1a). At higher pH above 9.5, IP6 converts to the sterically hindered conformation with five phosphates in the axial and one phosphate in the equatorial position [50]. Figure 3.3.1b shows the mnemonic turtle-based rule for phytate numbering proposed by Bernard Agranoff with the 2-phosphate axial and pointing upwards and the carbon atoms numbered in D configuration counterclockwise around the ring, as recommended by the Nomenclature Committee of the International Union of Biochemistry [49].

Phytic acid has 12 replaceable protons acting as reactive sites: 6 are strongly acidic, with pK_a-values of 1.5–2; 2 are weakly acidic with pK_a-values of about 6; and 4 are very weakly acidic with pH of 9–11 [45, 51]. This means that due to the small size of the inositol molecule, phytic acid will carry an extremely dense negative charge at a physiological pH of 6–7 that is normally encountered in feeds and parts of the digestive tracts. This dense negative charge strongly attracts to cations such as Zn^{2+}, Cu^{2+}, Ni^{2+}, Co^{2+}, Mn^{2+}, Fe^{2+}, and Ca^{2+} in

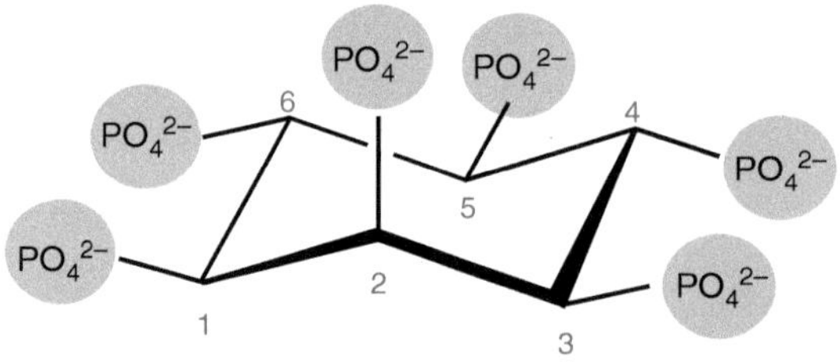

Figure 3.3.1 Rule for inositol numbering. Carbon atoms of phytate numbered anticlockwise around the inositol ring. According to the mnemonic representation proposed by Bernard Agranoff representing the inositol ring as a turtle, phosphate at carbon 2 position is pointing upwards signifying the head and the five equatorial phosphates serve as forelimbs, hind limbs, and the tail at carbon position 5. Source: Agranoff 2009 [49]. Reproduced with permission of ASBMB.

very stable complexes. IP6 phytates are known to occur in the form of a polyanion at pH 1–6 with 3–6 negative charges in the crop, proventriculus, and gizzard of poultry as well as the stomach of humans and swine [52]. The chelating ability of phytic acid varies with the metal according to the sequence: $Cu^{2+} > Zn^{2+} > Cd^{2+} > Ni^{2+} \approx Co^{2+}$ [52]. Because the physical state and solubility of phytin complex determines directly the availability of the phytate-P, understanding the interaction of IP6 with other nutrients in feeds and the digestive tract of the monogastric farm animals is of notable importance from the nutritional perspective. If the IP6 complex is precipitated then phytases will be unable to exert their enzymatic hydrolysis. Notably, acidification of the media to pH 4.0 decreased the inhibitory potency of all of the divalent cations tested. This indicates that when passing through the stomach with acidic pH, a window of opportunity opens up for phytate destruction by the phytases. Partially dephosphorylated inositol penta- (IP5), tetra- (IP4), and triphosphate (IP3) are also called phytates, but they have reduced chelating and protein-binding ability and increased solubility [53]. *In vivo* IP6 increases the excretion of endogenous minerals and amino acids in broiler chickens [54]. In addition to this, phytate is known to inhibit several digestive enzymes such as a-amylase [55], protease [56], and lipase [57]. Thus, phytases not only release phosphorus from phytic acid but also can greatly reduce via partial dephosphorylation the dense negative charge of the phytate and the anti-nutritive properties associated to this. Indeed, "extraphosphoric" effects of phytases are increasingly supported by the scientific literature [58, 59].

3.3.3.2 Phytases Structural and Functional Classification

The abundantly sequestered phosphorus in the form of phytate in plant materials such as grains and seeds stimulated the concomitant ubiquitous presence of phytate-degrading enzymes in many different organisms. Functional phytases catalyzing phosphate monoester hydrolysis of phytic acid have been reported from plants, microbes, and some animals. Phytases include structurally diverse enzymes with many different enzymatic mechanisms to cleave phosphate groups from phytate [1, 3, 60]. Today, phytases are classified into four distinct groups:

histidine acid phosphatases (HAP) [61], β-propeller phytases (BPPPhy), cysteine phytases (CPhy), and purple acid phytases (PAPhy).

3.3.3.2.1 Phytases from the Histidine Acid Phosphatases (HAP) Superfamily

HAP superfamily consists of a heterogeneous group of proteins that hydrolyze phosphate esters, optimally at low pH, and contains a conserved His residue in their active center that is transiently phosphorylated during the catalytic cycle. Other conserved residues contribute to a "phosphate pocket" and interact with the phosphoryl group of the substrate before, during, and after its transfer to the His residue. HA phosphatases that dephosphorylate phytate as a substrate are only a subgroup of the HAP superfamily, designated by Oh et al. in 2004 as histidine acid phytase (HAPhy) [62]. HAPhy are known from both eukaryotic and prokaryotic kingdoms. Members of this family HAPhy have been identified in animals, plants, and microorganisms [63, 64]. They share very little sequence homology other than the conserved site motive. The majority of the identified and all of the commercially explored phytases are from HAPhy family. Currently, all phytases registered for animal feed applications belong to the class of HAPhy and are exclusively from microbial (fungal or bacterial) origin. Table 3.3.1 lists the main commercial phytases approved for animal feed, all from HAPhy family. The HAPhy typical fold consists of a large α/β-domain and a small α-domain [1, 3, 65, 66]. A common catalytic site architecture is located at the interface of the two domains including a conserved N-terminal active site motif RHGXRXP and a C-terminal HD motif. In a two-step reaction a histidine from the RHGXRXP motif exerts nucleophilic attack on the phosphorus by forming a covalent phosphohistidine intermediate while the aspartic acid of the HD motif serves as a proton donor to the oxygen atom of the phosphomonoester bond to be hydroxylated [67–69]. Crystal structures of two of the most studied and utilized as animal feed additive HAPhy from *E. coli* pH 2.5 acid phosphatase (appA) and from *A. niger* (PhyA) are shown in Figure 3.3.2. *In vitro* studies by Wyss et al. [70] reveal two classes of HAPhy. In one class, phytases with lower substrate selectivity accept a range of different substrates but have lower specific activity for phytate, and in the second class the high specific activity for phytate is correlated with narrow substrate selectivity. Commercial phytases based on appA and PhyA are all from the second class with high specificity for phytic acid.

3.3.3.2.2 β-Propeller Phytase (BPP)

β-Propeller molecular architecture accommodates proteins with a wide range of biochemical functions and includes phytases identified in *Bacillus subtilis* [71] and in *Bacillus amyloliquefaciens* [72]. The structure of BPPPhy (also referred to as alkaline phytase) resembles a propeller with six blades. In contrast to HAPhy, the BPP class exhibits more basic pH optimum, higher thermal stability, and substrate preference for the calcium–phytate complex rather than the dissociated phytic acid [73]. Interestingly, BPP structure requires Ca^{2+} in order to create a favorable electrostatic environment for phytate binding [74]. Indeed, the crystal structure of the enzyme in complex with inorganic phosphate reveals that two phosphates and four calcium ions are tightly bound at the active site [75], suggesting that potential application of BPP in diets with high levels of Ca concentrations

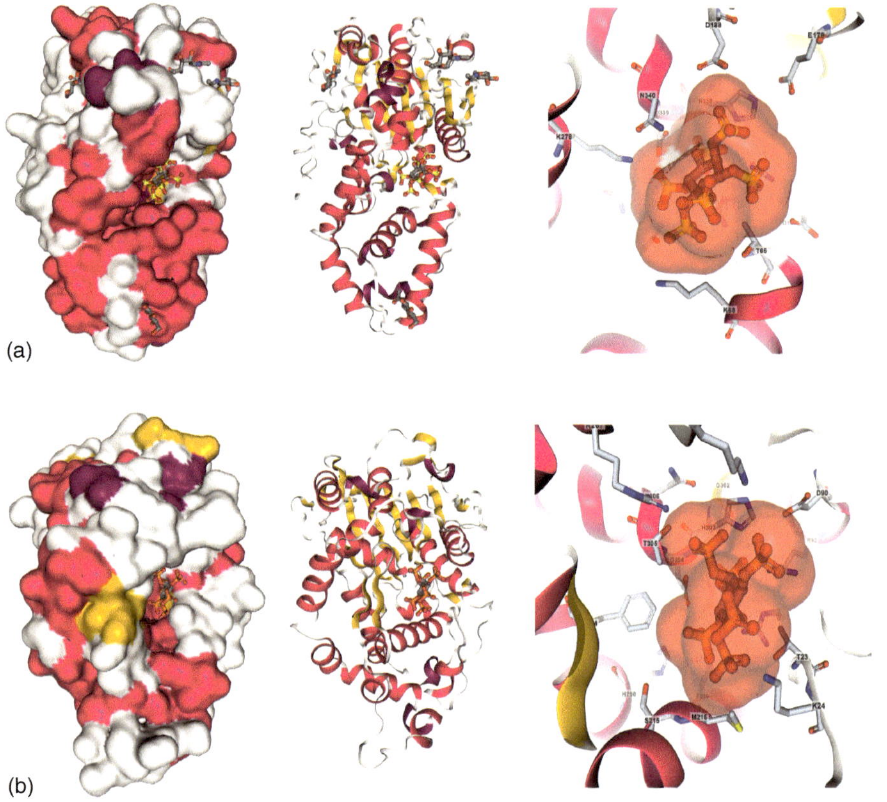

Figure 3.3.2 Crystal structures of HAPhy. Surface (left) and cartoon (middle) views of *Aspergillus niger* PhyA [65] (a) and *Escherichia coli* appA [66] (b) phytases and their interaction with phytic acid (right). α Helices are shown in magenta and β strands in yellow. Source: (a) Kostrewa et al. 1997 [65]. Reproduced with permission of Springer Nature. (b) Lim et al. (2000) [66]. Reproduced with permission of Springer Nature.

(such as laying hens diets) can be of benefit for better phosphorus utilization and phytate destruction. BPPPhy produces two different myoinositol trisphosphates as end products: Ins(2,4,6)P_3 and Ins(1,3,5)P_3 [76]. *Bacillus* phytases are also more sensitive to pepsin at low pH but more resistant to trypsin proteases at neutral pH [76]. This property proposes that BPP will be more suited to work in the small intestine of broilers and pigs. A study by Elkhalil et al. [77] suggests that the combination of HAPhy from *E. coli* acting in the gizzard with phytases of *Bacillus* acting in the intestine is a promising strategy to further improving the *in vivo* efficacy of phytases in poultry. However, so far BPP did not find its way to a commercial phytase product. Although potential applications of the Bacillus BPP phytases in animal nutrition, human health, and synthesis of lower myoinositol phosphates are discussed in the scientific literature, BPP in general has less specific activity than HAPhy and it is questionable if the sensitivity of BPP to pepsin will allow passage through the stomach for achieving their optimal pH in the intestine.

3.3.3.2.3 Cysteine Phytase (CPhy)

Another class of phytase identified from anaerobic ruminant bacterium *Selenomonas ruminantium* is CPhy (also referred as protein tyrosine phosphatase or dual-specificity phosphatase). This class of enzymes shows a phytase fold similar to the dual-specificity phosphatases consisting of one large and one small domain. The active site is located near the edge of the large domain, forming the wall of a shallow pocket to support the binding of the substrate. Two loops, the first one called the P loop and the second one called the WPD loop according to the protein tyrosine phosphatase nomenclature [78], provide support from the bottom of the active-site pocket. The P loop contains the catalytically important HCXXGXXR(T/S) sequence motif. The activity from CP from *S. ruminantium* is strongly inhibited by the presence of Fe^{2+}, Fe^{3+}, and Hg^{2+} ions and significantly retarded by Zn^{2+} ion [79]. CPhy preferentially cleaves the 5-phosphate position of IP6 and it produces IP2 via a highly unique and ordered pathway of sequential dephosphorylations: IP6, IP(1,2,3,4,6), IP(1,2,3,6), IP(1,2,3), and IP(1,2) [80]. To date, there are no commercial applications of CPhy. This is likely due to the lack of highly productive expression systems and limited heat stability [79, 80].

3.3.3.2.4 Purple Acid Phytases (PAPhy)

PAPhy sequences exhibit a high degree of similarity to purple acid phosphatases, a class of metallo-phosphoesterases. These metalloenzymes contain di-nuclear Fe^{3+}, Zn^{2+}, Mg^{2+}, or Mn^{2+} catalytic center. Initially, PAPhy were identified in germinating soybean seedlings (*Glycine max* L Merr.) [81] and later in other plant species [82]. The low phytase activity of plant PAPhy may be advantageous for the plant seed where slow and regulated phosphorus release during germination might be desirable. To date, there are no known industrial applications of PAPhy.

3.3.3.2.5 Classification of the Phytases Based on Phytate Dephosphorylation Steps

IP6 is a symmetric molecule with six dihydrogen phosphate groups on the *myo*inositol core. Phytases release P from IP6 via a stepwise pathway of site-specific dephosphorylation. Based on *in vitro* studies phytases are classified by the International Union of Pure and Applied Chemistry and the International Union of Biochemistry (IUPAC-IUB) as 3-phytases (EC 3.1.3.8, myoinositol-hexakisphosphate 3-phosphohydrolase), 4/6-phytases (EC 3.1.3.26, myoinositol-hexakisphosphate 4-phosphohydrolase), and 5-phytases (EC 3.1.3.72, myoinositol-hexakisphosphate 5-phosphohydrolase) on the basis of the carbon position on the inositol ring at which the phosphate release is initiated. While the family of 5-phytases is of limited importance with only a few reported members in lily pollen [83, 84] and *S. ruminantium* [80], most of the phytases identified to date are from the 3- or 4/6-family. Based on this classification initially it was suggested that the 4/6-phytases are from plant origin and initiate hydrolysis at the 4 or 6 position of IP6, and 3-phytases are usually of microbial origin, such as the fungal *A. niger* phytase [85–87] or the bacterial *Pseudomonas* phytase [26, 88]. However, *E. coli* appA phytase was first identified as an exception to this rule characterized as a 6-phytase [36, 37]. Interestingly, Konietzny and Greiner [89] noticed that in the pathway of stepwise dephosphorylation irrespective of the type of phytase 3- or 4/6- and their origin (fungal, bacterial, or plant), all of the acid phytase-degrading enzymes studied release five of the six phosphate groups with a final product yield of Ins(2)P. This indicates that the acidic phytase-degrading enzymes have strong preference for equatorial phosphate groups over the axial phosphate groups. Only alkaline phytase from *Bacillus* is capable of removing the phosphate group from the C-2 position of the inositol ring [76]. Figure 3.3.3 shows a model of stepwise dephosphorylation of the different types of phytases. Because the different phytases dephosphorylate different residues in the *myo*inositol backbone it could be argued that combining a fungal type 3-phytase, e.g. from *A. niger*, and a 6-phytase from *E. coli* will result in additive and/or synergistic effects on phosphorus release and animal performance. However, multiple studies [39, 77, 91, 92] on pigs and chicks do not reveal effects of combining 3- and 6-phytases on the phosphorus digestibility or growth performance. The tudy of Zimmermann et al. [93] concludes that

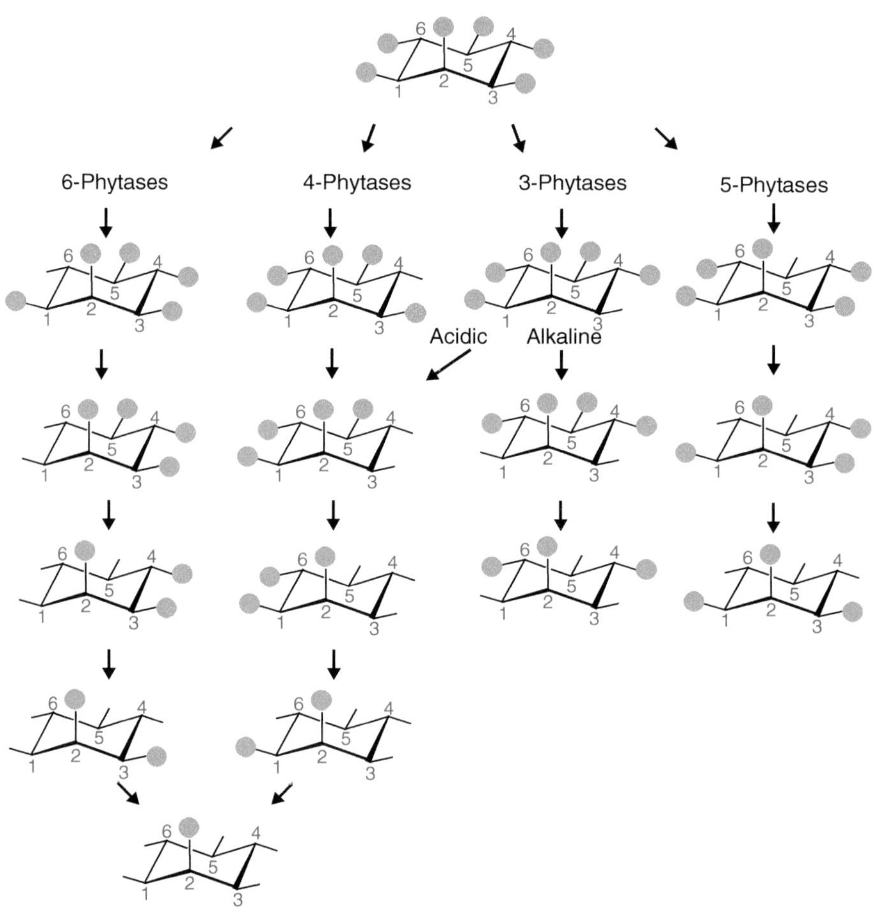

Figure 3.3.3 A model of stepwise dephosphorylation of the different phytases. Source: Reproduced from Greiner and Konietzny [1] and Hayakawa et al. [90].

the intrinsic cereal phytase from rye and wheat exhibits linear additivity in response to apparent phosphorus absorption. The lack of synergistic effects of the different phytases implies that phytases work independently of each other. Currently, it is not completely clear if reaction intermediates of 3-phytase could be a good substrate for the 6-phytase and vice versa in order to expect synergistic effects. Indeed, recent *in vitro* studies of Greiner [94] show that IP5 and IP4 intermediates are representatives of the individual phytate dephosphorylation pathways of phytase preparations of *A. niger* (3-phytase); *E. coli* (6-phytase) and rye (4-phytase) are identified as the best substrates of the phytases initiating the dephosphorylation steps. This strongly indicates that indeed the different phytases act for themselves.

Studies of Wyss et al. [70] and Greiner et al. [36, 87] reveal interesting differences between 6-phytases and 3-phytases in their stepwise kinetics of dephosphorylation of IP6. While the 3-phytases effectively manage to continue their attack on a selected IP6 molecule until it is reduced to IP1, the 6-phytases seem to first produce mostly IP4 and lower esters and then to engage into dephosphorylation of a new IP6 molecule. As a result, for the provision of similar quantities of phosphorus as determined by enzymatic activity assay, there is considerably more destruction of IP6 by a 6- compared to a 3-phytase. This will clearly influence the relative extra-phosphoric effects of the 3- versus 6-phytases, with the 6-phytases being able to destroy more molecules of IP6 per given amount of time, thereby eliminating more anti-nutritional molecules per given time.

3.3.4 Nutritional Values of Phytase in Animal Feed

As discussed above, phytate in animal feed contains sufficient phosphorus to meet the requirements and to sustain development of the target animals. Addition of dietary phytase releases phytate-bound phosphorus and increases its bioavailability so that inorganic phosphorus supplementation is unnecessary. Inorganic phosphorus equivalences of phytases have been studied extensively in different feed diets and for various species. It is generally accepted that 500 phytase units/kg releases approximately 0.8 g of digestible phosphorus equivalent to the replacement of 1.0 or 1.3 g/kg of phosphorus from mono- or dicalcium phosphate [3]. At the same time, phytase supplementation obliterates the anti-nutritional property of the phytate and thus improves calcium, zinc, and iron utilization by animals significantly [14–16, 95]. Especially consistent is the enhancement of calcium digestibility from 60% to 70% in the control diet without phytase to 70–80% in the diets with phytase supplementation. In contrast, the effect of phytase on the amino acid digestibility has been small and to some extent controversial [59, 96, 97].

3.3.5 Phytase Application As Feed Additive

In 2001, Lei and Stahl [98] proposed that an ideal phytase should possess at least three characteristics: (i) ability to effectively hydrolyze phytate phosphorus in the upper digestive tract of the animal, (ii) resilience to the 65–80 °C temperatures of feed pelleting, and (iii) cheap production costs. These characteristics are still valid today and are required in order to have a successful commercial phytase product.

3.3.6 Effective Phytate Hydrolysis in the Upper Digestive Tract of the Animal

Exogenously added phytase exerts its beneficial properties on phytate destruction in concert with a number of endogenous phytase activities normally present in the digestive tract. There are three sources of endogenous phytases existing in the digestive tract. The first phytase activity is derived from the raw plant materials used to formulate the feed [99–101]. This intrinsic feed activity comes from endogenous grain phytases and can vary significantly. Data from the Huvepharma feed analysis laboratory show variations in the range of 20–200 phytase units (ISO30024:2009 method [102]) per kg of feed. Eeckhout and De Paepe in 1994 [9] quantitatively analyzed 51 different feedingstuff raw materials used in Belgian feed mills for the presence of phytase activity, phytate phosphorus, and total phosphorus. Of the cereals analyzed, rye, triticale, wheat (and wheat by-products), and barley were rich in phytase activity. Wheat by-products, such as fine bran meal, pellets, middlings, and bran were particularly rich in phytase. However, as plant phytases are not intrinsically thermostable their activity and contribution to the total phytate hydrolysis is reduced significantly upon regular feed pelleting using high steam pelleting temperatures [11]. As phytase activity in feedingstuffs is not related to phytate phosphorus content [9] and plant phytases are not adapted to the conditions of the digestive system such as acidic pH optimum and resistance to pepsin, various authors agree [1, 2] that endogenous plant phytases have limited contribution to phytate destruction in the digestive tract. Indeed, cereal phytases appear less efficient than the microbial phytase from *A. niger*, Natuphos, in growing pigs [93] but still can provide some bioavailable phosphorus from phytate. The study of Zimmermann et al. [93] concludes that the efficacy of cereal phytases is 40% compared to that of microbial phytase.

Endogenous intestinal mucosal phytase activity is yet another source of phytase activity in the digestive tract of animals. This activity was identified in rats [103], small intestinal brush border membrane vesicles from broilers [104], and small intestine of pigs [105]. In broilers the site with the highest endogenous phytase activity was the duodenum and in pigs it was the jejunum. Interestingly, broiler chicks and mature laying hens have similar specific phytase activity but the total phytase activity in laying hens is 35% higher [104]. This suggests that the intestinal brush border phytase in chickens could contribute to -phosphorus digestibility and is probably dependent on the chicken breed or subject to regulation in response to external stimuli. Similar interplay of the effects of chicken breed and age of the animals is also documented by the study of Abudabos [106].

Lower parts of the digestive tract of monogastric animals, e.g. colon in pigs and ceca in poultry species contain normal gastrointestinal microflora, which is another source of phytase activity [107]. From the microbiome of a pig colon a very efficient *E. coli*-based phytase gene *appA2* was isolated and served as the basis for successful commercial development of OptiPhos phytase [37]. Phytase activity has also been reported for different lactic acid bacteria isolated from chicken intestine [108]. Studies on intestinal and total tract digestibility of phosphorus and phytate of Schlemmer et al. [53] and Marounek et al. [109] indicate that the microbiome in the pig colon and in caeca in broilers is a rich source of phytase activity. However, because phytate absorption occurs mainly in the small intestines it is questionable if post ileum phytate degradation leads to bioavailable phosphorus. This, as indicated by Selle et al. [2], might be of value only for coprophagic animals. Interestingly, a straightforward reverse correlation of phytate degradation in the gut and the calcium level in the diet points toward inhibitory effects of calcium on the endogenous phytase activity. This was first shown in humans by the study of Walker et al. in 1948 [110] and later in pigs by Sandberg et al. [111]. In pigs the total phytate degradation throughout the gut decreased significantly to 97%, 77%, and 45% when the dietary calcium intake increased by 4.5, 9.9, and 15 g/d, respectively. As the phytate degradation in the stomach and small intestine was almost unaffected, different supplementation of calcium carbonate predominantly affected the hydrolysis of phytate in the large intestine. Similar results were reported for rats by Pileggi et al. [112] and Wise and Gilburt [113]. Most probably, high dietary calcium content in the gastrointestinal chime affects the phytate solubility or competes with the phytase enzyme for the substrate and thereby reduces the accessibility of phytate for enzymatic hydrolysis.

It is generally accepted in the field that in order to be efficient, endogenously added phytase needs to release phosphorus from phytate in the upper digestive tract of monogastric animals. Two factors contribute to this acceptance: (i) its substrate phytate is soluble at an acidic pH giving a chance to the phytase to work in the stomach and (ii) phosphorus absorbance in the digestive tract occurs in the small intestines [89, 114]. *In vitro* studies have investigated the effects of pH on the solubility of phytate complexes and in general they agree that the higher the complex of phytic acids, proteins, and ions is the lower the pH at which it dissolves [22, 46]. Diets are composed of complex mixtures of minerals, which likely results in interactive effects in forming complexes with phytic acid. Insoluble phytate mineral precipitates and soluble mineral phytate complexes may be resistant to hydrolysis by intrinsic and supplemental phytase in the diet. Thus, phytate is soluble at the acidic pH of the stomach, giving a window of opportunity to the phytase to work in order for phosphorus absorbance in digestive tract to occur in the small intestines. Field trials using commercial phytases confirm that main site of phytate phosphorus digestibility of the commercial phytases is the stomach of pigs [114–116] and the crop of poultry [109, 117]. The study of Maenz et al. [118] analyzes the susceptibility of soluble or insoluble phytate mineral chelates to hydrolysis by microbial phytase purified from Natuphos. In solution, at pH 7.0, mineral concentrations from 0.053 mM for Zn^{2+} up to 4.87 mM for Mg^{2+} caused a 50% inhibition of phytate-P hydrolysis by microbial phytase. At neutral pH the order of mineral potency as inhibitors of phytate hydrolysis was $Zn^{2+} > Fe^{2+} > Mn^{2+} > Fe^{3+} > Ca^{2+} > Mg^{2+}$. In contrast, at low pH of 4 the inhibitory potency of all of the di- and trivalent cations tested had decreased. Protonation of the phosphate groups of the phytate molecule will greatly reduce the affinity for divalent cation binding, and therefore phytase with high affinity for phytate at low pH and high kinetic speed can use this window of opportunity more efficiently for maximum phytate destruction and/or maximum phosphate release.

Phytic acid has six phosphate ester groups that might be released at different orders and at different rates. With the purpose of understanding how phytases function in the digestive tract a kinetic description of their activity is needed with clear definition of the desired end product of the endogenously added phytase in the pathway of stepwise degradation. The last is ideally combined and supported by experimental data. The question whether a complete degradation of phytate to IP1 is needed in the early part of the digestive tract or whether degradation to intermediate IP4, IP3, or IP2 forms is sufficient for the bio-efficacy of the exogenously added phytase could be a subject of discussion. As discussed earlier, based on the activities in *in vitro* studies of Wyss et al. [70], microbial phytases can be classified into two types: (i) phytases with broad substrate specificity and (ii) phytases with high selectivity for phytic acid. The phytases with broad substrate specificity are active against a whole range of structurally diverse phosphate compounds, and dephosphorylate phytic acid to myoinositol and 2-monophosphateinositol with no major accumulation of intermediates. In contrast, phytases specific for phytic acid have much higher specific activity and during phytic acid degradation accumulate *myo*-inositol tris- and bisphosphate coupled with a progressive decrease in the rate of phosphate release. The last suggests that lower inositol phosphates are not preferred substrates for the phytases with high selectivity for phytic acid. For the provision of similar quantities of P as determined by the activity assay, there is considerably more destruction of IP6 by 6- compared to 3-phytases. Because IP6 or IP5 has much higher mineral binding strength than the IP4, IP3, and IP2 [119], the solubility of IP is greatly affected by the numbers of phosphate esters. For example, in the study of Schlemmer et al. [53] the solubility of various IP esters was accessed in the small intestine of pigs (pH 6.6); there, IP6, IP5, IP4, IP3, and IP2 are correspondingly 2%, 7%, 8%, 31%, and 75% soluble. This has also direct correlation with the aggregation ability of IP6 [120]. As reviewed in Cowieson et al. [12] this suggests that for sufficient solubility of both the minerals and IP esters at higher pH a gastric reduction to at least IP3 might be necessary. Interestingly, *in vivo* studies show that a high degree of inositol phosphorylation (IP6, IP5) strongly inhibits zinc and calcium uptake in suckling rat model and in humans [121, 122]. Efficacy of the phytase is also affected by interplay with the concentration of calcium and non-phytate phosphorus as documented by the study of Driver et al. [123]. Growth and bone quality responses to phytase were greatest at low non-phytate phosphorus levels and high calcium levels. When the Ca^{2+} level was reduced, the efficacy of the phytase also decreased. The same was observed when the non-phytate phosphorus levels were increased. Based on the endogenous phytase activity in the digestive tract of poultry and pigs toward IP4, IP3, and lower IP isoforms and the strong binding and anti-nutritional properties of the IP6 and IP5, it might appear that fast initial dephosphorylation by the feed phytases from IP6 to IP5 and IP4 is important for their biological activity. This has several implications; first, adequate destruction of the anti-nutritional effect of the phytate early in the digestive tract, and second, it leads to increased solubility of the lower inositiol IP4 and IP3 esters in the small intestine allowing access to the endogenous phytases and phosphatases. *In vivo* studies of the hydrolysis of phytate and formation of inositol phosphate isomers without or with supplementation of 3-phytase from *A. niger* (PhyA) and two 6-phytases from *E. coli* (PhyE1 and PhyE2) in different segments of the digestive tract of broilers [8] provide insights into the relative contribution of the endogenous and exogenous phytases to the stepwise phytate degradation *in vivo*. Significant phytate degradation by the endogenous phytases is observed especially in the lower digestive tract of broilers (duodenum/jejunum, ileum and ceca) whereas supplemented phytases increased IP6 hydrolysis in the crop but not in the lower ileum. Also, a typical 3-phytase and 6-phytase degradation pathway is observed for the PhyA and PhyE1/2 in the crop, confirming the previous *in vitro* studies and classification of the 3- and 6-phytases. *In vitro* studies of *E. coli* phytase show a fast progression from IP6 to IP5 and IP4 for *E. coli* phytase [36]. This result corresponds to the *in vivo* findings by Zeller et al. [8] indicating good agreement between *in vitro* enzyme properties and *in vivo* properties. In contrast to the exogenous phytase, the study of phytase activity against IP6 by Hu et al. [105] using purified jejunual pig phytase activity reveals that increasing the number of phosphate groups on IP leads to decreased intestinal phytase activity. This proposes that the intestinal phytase activity may act in concert with the exogenous microbial phytase added to the feeds. Because the phosphatases arising from the mucosa of the small intestine dephosphorylate much more efficiently partially phosphorylated *myo*inositol phosphates than IP6-phytate [124, 125], Greiner and Konietzny [1] suggest that a complete transformation of dietary phytate into myoinositol tetra- and trisphosphates in the stomach to is apparently much more important for the bio-efficacy of supplementary phytase than complete dephosphorylation of single phytate molecules. Thus, apparently an ideal phytase used as feed additives should have low pH optimum, resistance to pepsin, high affinity, and high turnover numbers for phytate and myoinositol penta- and tetra-kis inositol phosphate.

3.3.7 Kinetic Description of Ideal Phytases

While it is experimentally and theoretically not straightforward to comprehensively understand all kinetic parameters (phosphorus release and accumulation of intermediates) of how phytases interact and dephosphorylate IP6, IP5, and lower IP phosphates in the digestive tract, *in vitro* laboratory studies measuring released phosphorus per given time agree that feed phytases obey a Michaelis–Menten model kinetics [1, 89]. In general, the K_m of the enzyme for phytate is the concentration of the substrate [S] at which 1/2 of the active sites of the enzyme phytase are filled and the velocity of the reaction is half of V_{max}. K_m is an important characteristic of enzyme–substrate interactions and is independent of enzyme and substrate concentrations but it might be influenced by the ionic strength and pH. For the phytases, a high K_m at the stomach pH implies weak phytate binding and it takes more phytate to fill enzyme catalytic sites. A low K_m implies strong phytate binding; it takes less phytate to fill substrate binding sites. Rough estimates based on the content of phytate in the feed reveal that phytate concentration in the stomach of pigs and poultry ranges from about 2 to 15 mM. While this concentration is much higher than the K_m of the commercially available phytases [70] the stomach chyme of pigs and poultry is a far from ideal solution containing viscous slurry with feed particles of various sizes and thus many different microenvironments. Likely, the effective concentration of the substrate that the phytase may encounter is much lower than actually estimated. In addition to this as discussed above, phytate, by binding to proteins and by chelating minerals such as zinc, iron, calcium, and magnesium [46, 52, 89], makes various complexes that may compete with phytases for binding. When effective phytate concentrations is lower than the K_m of the enzyme at the pH and ionic strength of the digestive tract, the dephosphorylation velocity depends on the ratio of k_{cat}/K_m. Under these conditions, k_{cat}/K_m is the rate constant for the interaction of S and E and can be used as a measure of catalytic efficiency. Therefore, phytases with high k_{cat}/K_m ratio at gastric pH are likely to reach kinetic perfection in enzymatic catalysis [126], release more phosphorus, and yield superior performance in animal trials. As noticed by Greiner and Konietzny [1] the phytase members of enterobacteriaceae such as *E. coli* [38], *C. braakii* [43], and *Yersinia* spp. [127] exhibit the highest k_{cat}/K_m values reported for phytases to date. Important in accessing the k_{cat}/K_m ratio is to perform analysis at the pH of the stomach. Table 3.3.2 shows unpublished data from Outchkourov et al. as a comparison of the kinetic parameters (including k_{cat}/K_m ratio) of PhyA enzyme from *A. niger* recombinantly produced in *A. niger* compared to appA phytase enzyme from *E. coli* produced in

Table 3.3.2 Michaelis–Menten kinetic properties of recombinant appA phytase produced in *Pichia pastoris* and recombinant PhyA produced in *Aspergillus niger* in the pH range of 2–4 and temperature of 39 °C, both representative for the upper digestive tracts of pigs and poultry.

pH	2	2.5	3	3.5	4
Escherichia coli appA					
V_{max}, 10^{-6} mol/min	896	1020	2030	2800	1160
K_m, 10^{-3} M	0.374	0.460	0.901	1.22	0.522
k_{cat}, s^{-1}	692	930	1580	2170	899
k_{cat}/K_m, $(s^{-1}\ M^{-1}) \times 10^4$	186	202	175	177	172
Aspergillus niger PhyA					
V_{max}, 10^{-6} mol/min	140	412	227	153	104
K_m, 10^{-3} M	0.471	1.55	0.807	0.516	0.337
k_{cat}, s^{-1}	137	404	223	150	102
k_{cat}/K_m, $(s^{-1}\ M^{-1}) \times 10^4$	29	26	28	29	30

Source: Data from Outchkourov et al. (unpublished).

P. pastoris in the pH range of 2–4 and temperature of 39 °C, both representative of the upper digestive tracts of pigs and poultry. These data clearly illustrate the different kinetic properties of the different phytases and their dependence on the pH. Notable is the much higher V_{max} and favorable k_{cat}/K_m ratio of the appA phytase.

3.3.8 Resistance to Low pH and Proteases

In order to effectively hydrolyze phytate phosphorus in the upper digestive tract of the animal, phytases used in animal nutrition should not only be highly active, e.g. attain maximum dephosphorylation speed but also be perfectly adapted to the conditions of the stomach (or crop, proventriculus, and gizzard) such as low pH and high pepsin proteolytic activity. Resistance to proteases in general is an intrinsic property of the polypeptide chain and its three-dimensional structure. In general, phytate-degrading enzymes of microbial origin are more pH stable and thermostable than their plant counterparts [1]. Early studies on the application of *A. niger* phytase for animal feed indicate that this phytase is degraded by proteolytic digestion in the digestive tract of pigs [114]. Subsequent studies reveal that *A. niger* phytase is more pepsin resistant than plant-derived wheat phytase [128]. Rodriguez et al. [129] have investigated *in vitro* the resistance of phytases PhyA from *A. niger* expressed in the same host and appA from *E. coli* expressed in *P. pastoris* to the action of proteolytic enzymes trypsin and pepsin. Both phytases PhyA and appA displayed differential sensitivity to trypsin and pepsin; PhyA from *A. niger* is more resistant to trypsin proteolysis while appA from *E. coli* showed better resistance to pepsin. Interestingly, appA showed a 30% increase in phytase activity and also released 30% more iP from soybean meal in the presence of pepsin. A likely explanation for this nutritionally beneficial property of appA is that potential pepsin-resistant polypeptides with higher phytase activity are released from the intact appA protein by partial pepsin digestion. Resistance to pepsin inactivation of the *E. coli* phytase was also confirmed in a subsequent study [38] where native appA proteins have been isolated from the endogenous host.

Engineering the intrinsic proteolytic stability of phytases has been a subject of research in the recent years [42, 130–133] with various degree of success. Resistance to proteolysis is not an absolute but relative property and is only a part of the required high phytase activity under the conditions of the upper digestive tract. Temperature stability, speed, and kinetics of stepwise phytate digestion under the same conditions are another important properties of the feed phytases and are discussed in detail.

3.3.9 Temperature Stability

The need for feed pelleting at 65–80 °C exerts high demands for stability on the enzyme producers and developer, respectively. The purpose of feed pelleting by using heat, moisture, and pressure is to take finely divided, sometimes dusty, feed particles to form into moderately cooked larger particles. While intrinsic heat stability of the enzyme is of utmost importance during feed pelleting distinction needs to be made between heat stability of the enzyme in the laboratory (in solution) and heat stability during feed pelleting where the enzyme is added to the feed and also moisture and pressure play a role. In general, phytate-degrading enzymes of microbial origin are more pH- and thermostable than their plant counterparts. A natural heat-stable phytase (PhyA) has been isolated from *Aspergillus fumigatus*, AfPhyA, able to withstand temperatures up to 100 °C over a period of 20 minutes, with loss of only 10% of the initial enzymatic activity [134]. Later, it was shown that the *A. fumigatus* enzyme is not thermostable, but has the property of being able to refold completely into native-like, fully active conformation after heat denaturation [135]. Irrespective of this remarkable property, thus far AfPhyA-based enzymes did not make it to commercial application. Other thermostable phytases were also identified in the β propeller family, or from *Pichia anomala*, or from Lactobacilli [76, 136–138]. However, from the concept of ideal phytase discussed above [98] thermal stability is only one of the three desired properties. Thermal stability needs to be combined with the ability to effectively hydrolyze phytate phosphorus in the upper digestive tract of the animal and have low production costs. Ideal phytases, therefore, can rarely be found in nature and the newly identified thermostable phytases often have trade-offs with efficient phytate hydrolysis in the upper digestive tract and/or cheap production costs. The contrary is also sometimes valid as highly efficient phytases in animal performance trials such as *E. coli* appA-based phytases or other HAPhy show reduced thermal stability. Therefore, many researchers have focused on improving thermal stability via protein engineering of existing phytases with proven nutritional efficacy and performance in the animals and established cheap production platforms. The publication of Shivange and Schwaneberg [139] provides an excellent review with the different approaches of protein engineering of the different phytases. A combination of rational design strategies with directed phytase evolution often aims to improve one property at a time. A multifactorial mutagenesis and screening method is desired in order to improve one property, such as the thermal stability, without affecting another, such as specific activity, K_m, k_{cat}, or resistance to proteases. Improving the thermal stability often affects the activity of the enzyme by changing the rigidity of the loops, and thereby as a trade-off of the increased thermal stability specific activity drops [140]. Furthermore, often increasing the thermal stability of the enzyme leads to also increased temperature optimum, which by itself makes it less efficient at the temperature optimum of the animals, 38–39 °C (Outchkourov, unpublished data). Examples from the literature show that improved thermal resistance can also lead to improved catalytic efficacy or improved gastric stability. The effect of different N-linked glycosylations on thermal stability partly described in the study of Rodriguez et al. [37] could be appreciated in Figure 3.3.4 (Outchkourov et al. unpublished data) where the same appA enzyme is expressed in three different expression systems, *E. coli*, *T. reezei*, and *P. pastoris*. Interestingly, *E. coli* appA is not glycosylated in its endogenous host but it carries four potential N-linked glycosylation sites (NXS/T) that signal for

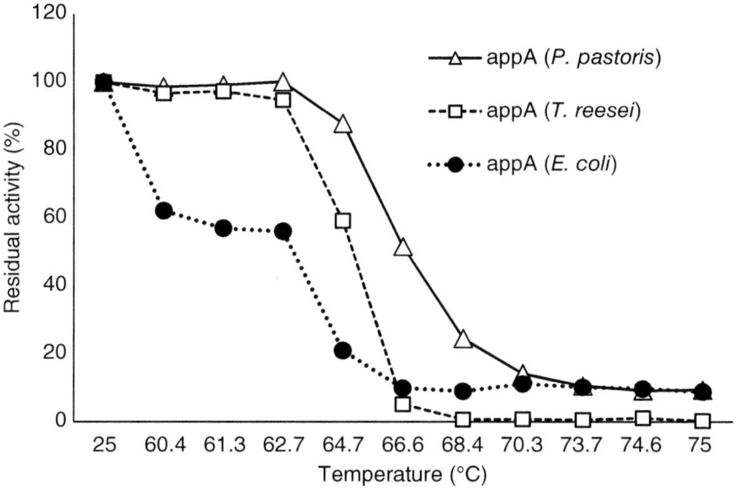

Figure 3.3.4 Impact of expression host and post-translational modification such as glycosylation on the thermal stability of *appA* encoded phytases. Example of thermal stability of appA phytase expressed in three different hosts. Phytase aliquots were incubated for 15 minutes at the indicated temperatures, subsequently phytases were chilled on ice and the remaining phytase activities were assayed at 37 °C.

N-linked glycosylation when secreted in eukaryotic hosts. Both glycosylation from *Pichia* and *Trichoderma* improve the thermal stability of appA phytase while addition of *Pichia* type N-linked glucans yields a more pronounced effect on the temperature stability. Another study of Rodriguez et al. [141] shows that rational design of additional N-glycosylation sites can lead to even further improved catalytic efficiency and thermostability of *E. coli* appA phytase mutants expressed in *P. pastoris*. Other fungal and plant phytate-degrading enzymes have been found to be glycosylated. The N-linked mannose and galactose of native enzyme from *A. niger* NRRL 3135 account for 27.3% of its molecular mass [142]. Glycosylation may have an effect on the catalytic properties, the stability, or the isoelectric point of an enzyme; e.g. complete deglycosylation of the phytate-degrading enzyme from *A. niger* resulted in 34% reduction in thermostability [143]. Study of Lehmann et al. [144] illustrates the possibility for exchanging the active site between phytases and rational design of thermostabilized variant of an enzyme with the most favorable catalytic properties. The study of Garrett et al. [42] illustrates successful saturation mutagenesis technology for the selection of new *E. coli*-based appA variant with eight individual mutations (Phy9X) with improved thermal and gastric stability. With the current developments of advanced high-throughput mutagenesis, screening, and sequencing technologies, new possibilities for the development of novel thermally and kinetically improved phytases exist [145–148]. Phytase development programs will depend entirely on the need and requirements of the phytase market and on the ability to define criteria for new phytase screening.

Application of coating on the surface of phytase granules has also been used to improve the thermostability of commercial phytases, e.g. the fat coating technology applied to OptiPhos CT or salt coating as applied to Ronozyme CT [149]. Coating technology has the advantage that it does not change the properties such as specific activity of the phytase itself but leads to relatively little thermostability improvement during feed pelleting. Another alternative strategy is to spray liquid phytase products post pelleting on cooled feed, which is successfully used today in many of the feed factories. Liquid spraying post pelleting requires stability of the phytase product in liquid form upon storage and transportation. Post pelleting applications of liquid enzymes also will require specific equipment and not all of the feed mills are equipped for this process.

3.3.10 In lieu of Conclusion: Lessons from Phytase Super Dosing Trials

While discussing all of the details of phytase superdosing is beyond the scope of this review, superdosing trials might provide an interesting clue for the mode of action of microbial phytases and point toward directions for future developments of next-generation phytases, more efficient in the upper digestive tract and with better animal performance.

The study of Shirley and Edwards [150] shows that unconventionally high doses of microbial phytases added to the feed provide additional substantial performance increase in terms of improved weight gain and lower feed conversion ratio in broiler chicks. Other studies [151–154] confirm the beneficial phytase superdosing effects in both poultry and pigs. As discussed earlier most of the monogastric diets contain phytate concentrations in the range of 5–25 g/kg, which consist of 2–4 g/kg phytate-bound phosphorus. Addition of a phytase dose of 500 phytase units/kg releases approximately 0.8 g of digestible phosphorus equivalent to replacement of 1.0 or 1.3 g/kg of phosphorus from mono- or dicalcium phosphate, which is sufficient to sustain growth of the animal. Why then are the effects of phytase superdosing observed? Apparently and as discussed by Cowieson et al. [12] and Lee et al. [153], this is likely due to phytate destruction rather than satisfying phosphorus requirements. Principle mechanisms of how high phytase doses may elicit beneficial effect have been outlined previously [12] as follows:

1. More liberated phosphate or restoration of P/Ca proportionate release.
2. Less residual phytate, i.e. destruction of the anti-nutritive effect and increased generation of more (persistently) soluble lower esters.
3. Generation of myoinositol with vitamin-like/lipotropic effects.

Cumulatively, most studies agree that inclusion levels of phytase at 500 phytase units/kg hydrolyzes about 45–60% of the phytate (20–48% above control). Phytase superdosing trial in pigs [154] show that diet supplemented with 15 000 FTU/kg (30 × the normal dose) can digest up to 85% of the phytate phosphorus, compared with 15% in the basal diet. In poultry, a similar situation is observed; addition of 12 000 FTU/kg (24×) leads to 95% of phytate phosphorous disappearance compared to 42% of the control with 93 FTU/kg [150]. Regular phytase doses leave a considerable proportion of IP6 remaining intact at the ileal level [8, 155, 156] and only high doses of phytase are required to partially digest them. Thus, complete phosphorus destruction is hard to achieve and excessive phytase doses can only partially dephosphorylate phytate in feed diets. Apparently, there is a fraction of the phytate that is highly resistant to phytase digestion. Furthermore, most of the current commercial phytases are 6-phytases, which, given the stomach (proventriculus, gizzard) residence time, digest IP6 *in vivo* through IP5 and yield accumulation of IP4 and IP3 molecules at the level of ileum [8, 156, 157]. While the IP3 and IP4 have much reduced chelating ability, their anti-nutritional properties might still be significant at the small intestines as there the pH rapidly returns from about 3 to about 6.5 and at the pH the solubility of IP6, IP5, IP4, IP3, and IP2 is respectively 2%, 7%, 8%, 31%, and 71% [53]. The low solubility of IP4 and IP3 indicates ability to precipitate out of solution a number of divalent cations, enzymes, and other nutrients, and thus antinutritional properties.

While it is unclear if super dosing effect are due to elimination of the anti-nutritional properties of the remaining IP6, IP4, IP3, or dephosphorylated (completely) *myo*inositol possess vitamin-like/lipotropic properties interesting directions for future phytase enzyme development will be to target the phytase-resistant IP6 part in the stomach and/or to generate phytases with high affinity for the IP4 and IP3 species generated by the stepwise action current 6-phytases with the structure of Ins(2,3,4,5)P_4 and Ins(2,3,4)P_3.

References

1 Greiner, R. and Konietzny, U. (2010). Phytases: biochemistry, enzymology and characteristics relevant to animal feed use. In: *Enzymes in Farm Animal Nutrition*, 2e, 96–128. Wallingford, CT: CABI.

2 Selle, P.H., Ravindran, V., Cowieson, A.J., and Bedford, M.R. (2010). Phytate and phytase. In: *Enzymes in Farm Animal Nutrition*, 2e, 160–205. Wallingford, CT: CABI.

3 Lei, X.G., Weaver, J.D., Mullaney, E. et al. (2013). Phytase, a new life for an "Old" enzyme. *Annu. Rev. Anim. Biosci.* 1 (1): 283–309.

4 Ashley, K., Cordell, D., and Mavinic, D. (2011). A brief history of phosphorus: from the philosopher's stone to nutrient recovery and reuse. *Chemosphere* 84 (6): 737–746.

5 Correll, D.L. (1998). The role of phosphorus in the eutrophication of receiving waters: a review. *J. Environ. Qual.* 27 (2): 261–266.

6 Reddy, N.R., Sathe, S.K., and Salunkhe, D.K. (1982). Phytates in legumes and cereals. *Adv. Food Res.* 28: 1–92.

7 Morgan, N.K., Walk, C.L., Bedford, M.R., and Burton, E.J. (2015). Contribution of intestinal- and cereal-derived phytase activity on phytate degradation in young broilers. *Poult. Sci.* 94 (7): 1577–1583.

8 Zeller, E., Schollenberger, M., Kühn, I., and Rodehutscord, M. (2015). Hydrolysis of phytate and formation of inositol phosphate isomers without or with supplemented phytases in different segments of the digestive tract of broilers. *J. Nutr. Sci.* 4: https://doi.org/10.1017/jns.2014.62.
9 Eeckhout, W. and De Paepe, M. (1994). Total phosphorus, phytate-phosphorus and phytase activity in plant feedstuffs. *Anim. Feed Sci. Technol.* 47 (1): 19–29.
10 Leske, K.L. and Coon, C.N. (1999). A bioassay to determine the effect of phytase on phytate phosphorus hydrolysis and total phosphorus retention of feed ingredients as determined with broilers and laying hens. *Poult. Sci.* 78 (8): 1151–1157.
11 Ravindran, V., Ravindran, G., and Sivalogan, S. (1994). Total and phytate phosphorus contents of various foods and feedstuffs of plant origin. *Food Chem.* 50 (2): 133–136.
12 Cowieson, A.J., Wilcock, P., and Bedford, M.R. (2011). Super-dosing effects of phytase in poultry and other monogastrics. *World's Poult. Sci. J.* 67 (2): 225–236.
13 Kraham, S.J. (2017). Environmental impacts of industrial livestock production. In: *International Farm Animal, Wildlife and Food Safety Law* (ed. G. Steier and K. Patel), 3–40. Cham: Springer.
14 Lei, X., Ku, P.K., Miller, E.R. et al. (1993). Supplemental microbial phytase improves bioavailability of dietary zinc to weanling pigs. *J. Nutr.* 123 (6): 1117–1123.
15 Lei, X.G., Ku, P.K., Miller, E.R. et al. (1994). Calcium level affects the efficacy of supplemental microbial phytase in corn-soybean meal diets of weanling pigs. *J. Anim. Sci.* 72 (1): 139–143.
16 Stahl, C.H., Han, Y.M., Roneker, K.R. et al. (1999). Phytase improves iron bioavailability for hemoglobin synthesis in young pigs. *J. Anim. Sci.* 77 (8): 2135–2142.
17 Lott, J.N.A., Ockenden, I., Raboy, V., and Batten, G.D. (2000). Phytic acid and phosphorus in crop seeds and fruits: a global estimate. *Seed Sci. Res.* 10 (1): 11–33.
18 Elser, J. and Bennett, E. (2011). Phosphorus cycle: a broken biogeochemical cycle. *Nature* 478 (7367): 29–31.
19 Adeola, O., Lawrence, B.V., Sutton, A.L., and Cline, T.R. (1995). Phytase-induced changes in mineral utilization in zinc-supplemented diets for pigs. *J. Anim. Sci.* 73 (11): 3384–3391.
20 Word Health Organization (2012). *Safety Evaluation of Certain food Additives*, WHO Food Additives Series 67. Geneva: WHO Press.
21 Nielsen, A.V.F., Tetens, I., and Meyer, A.S. (2013). Potential of phytase-mediated iron release from cereal-based foods: a quantitative view. *Nutrients* 5 (8): 3074–3098.
22 Schlemmer, U., Frølich, W., Prieto, R.M., and Grases, F. (2009). Phytate in foods and significance for humans: food sources, intake, processing, bioavailability, protective role and analysis. *Mol. Nutr. Food Res.* 53 (Suppl. 2): S330–S375.
23 Hartig, T. (1855). Uber das Klebermehl. *Bot. Z.* 13: 881–882.

24 Hartig, T. (1856). Weitere Mitteilungen, das Klebermehl (Aleuron) betreffend. *Bot. Z.* 14: 257–269.

25 Suzuki, U., Yoshimura, K., and Takaishi, M. Ueber ein Enzyme phytase das anhydrooxy-methylen disphosphorasure spaltet. *Collect. Agri. Bull. Tokyo Imp. Univ.* 7: 495–512.

26 Irving, G.C. and Cosgrove, D.J. (1972). Inositol phosphate phosphatases of microbiological origin: the inositol pentaphosphate products of *Aspergillus ficuum* phytases. *J. Bacteriol.* 112 (1): 434–438.

27 Mullaney, E.J., Gibson, D.M., and Ullah, A.H. (1991). Positive identification of a lambda gt11 clone containing a region of fungal phytase gene by immunoprobe and sequence verification. *Appl. Microbiol. Biotechnol.* 35 (5): 611–614.

28 van Hartingsveldt, W., van Zeijl, C.M., Harteveld, G.M. et al. (1993). Cloning, characterization and overexpression of the phytase-encoding gene (*phyA*) of *Aspergillus niger*. *Gene* 127 (1): 87–94.

29 Zhang, Z.B., Kornegay, E.T., Radcliffe, J.S. et al. (2000). Comparison of phytase from genetically engineered *Aspergillus* and canola in weanling pig diets. *J. Anim. Sci.* 78 (11): 2868–2878.

30 Kemme, P.A., Jongbloed, A.W., van der Klis, J.D., and Institute for Animal Science and Health (ID-DLO), and Department Nutrition of Pigs and Poultry, The Netherlands (1997). The impact of microbial phytase on the nutrition of monogastrics and the environment. In: *Feed Manufacturing in Southern Europe: New Challenges*, Cahiers Options Méditerranéennes; No. 26 (ed. P. Morand-Fehr), 135–147. Zaragoza: CIHEAM.

31 O'Quinn, P.R., Knabe, D.A., and Gregg, E.J. (1997). Efficacy of natuphos in sorghum-based diets of finishing swine. *J. Anim. Sci.* 75 (5): 1299–1307.

32 Cromwell, G.L., Coffey, R.D., Parker, G.R. et al. (1995). Efficacy of a recombinant-derived phytase in improving the bioavailability of phosphorus in corn-soybean meal diets for pigs. *J. Anim. Sci.* 73 (7): 2000–2008.

33 Tetu, C., Dassa, E., and Boquet, P.L. (1979). Unusual pattern of nucleoside polyphosphate hydrolysis by the acid phosphatase (optimum pH = 2.5) of *Escherichia coli*. *Biochem. Biophys. Res. Commun.* 87 (1): 314–322.

34 Dassa, E. and Boquet, P.L. (1985). Identification of the gene appA for the acid phosphatase (pH optimum 2.5) of *Escherichia coli*. *MGG Mol. Gen. Genet.* 200 (1): 68–73.

35 Dassa, J., Marck, C., and Boquet, P.L. (1990). The complete nucleotide sequence of the *Escherichia coli* gene appA reveals significant homology between pH 2.5 acid phosphatase and glucose-1-phosphatase. *J. Bacteriol.* 172 (9): 5497–5500.

36 Greiner, R., Konietzny, U., and Jany, K.D. (1993). Purification and characterization of two phytases from *Escherichia coli*. *Arch. Biochem. Biophys.* 303 (1): 107–113.

37 Rodriguez, E., Han, Y., and Lei, X.G. (1999). Cloning, sequencing, and expression of an *Escherichia coli* acid phosphatase/phytase gene (appA2) isolated from pig colon. *Biochem. Biophys. Res. Commun.* 257 (1): 117–123.

38 Golovan, S., Wang, G., Zhang, J., and Forsberg, C.W. (2000). Characterization and overproduction of the *Escherichia coli* appA encoded bifunctional

enzyme that exhibits both phytase and acid phosphatase activities. *Can. J. Microbiol.* 46 (1): 59–71.
39 Augspurger, N.I.L., Webel, D.M., Lei, X.G., and Baker, D.H. (2003). Efficacy of an *E. coli* phytase expressed in yeast for releasing phytate-bound phosphorus in young chicks and pigs. *J. Anim. Sci.* 81 (2): 474–483.
40 Kerr, B.J., Weber, T.E., Miller, P.S., and Southern, L.L. (2010). Effect of phytase on apparent total tract digestibility of phosphorus in corn-soybean meal diets fed to finishing pigs. *J. Anim. Sci.* 88 (1): 238–247.
41 Lei, X. (1999). Overexpression of phytase genes in yeast systems. EP1688500 B1, filed 23 June 1999 and issued June 1999.
42 Garrett, J.B., Kretz, K.A., O'Donoghue, E. et al. (2004). Enhancing the thermal tolerance and gastric performance of a microbial phytase for use as a phosphate-mobilizing monogastric-feed supplement. *Appl. Environ. Microbiol.* 70 (5): 3041–3046.
43 Kim, H.-W., Kim, Y.-O., Lee, J.-H. et al. (2003). Isolation and characterization of a phytase with improved properties from *Citrobacter braakii*. *Biotechnol. Lett.* 25 (15): 1231–1234.
44 Lassen, S.F., Breinholt, J., Østergaard, P.R. et al. (2001). Expression, gene cloning, and characterization of five novel phytases from four basidiomycete fungi: *Peniophora lycii*, *Agrocybe pediades*, a *Ceriporia* sp., and *Trametes pubescens*. *Appl. Environ. Microbiol.* 67 (10): 4701–4707.
45 Angel, R., Tamim, N.M., Applegate, T.J. et al. (2002). Phytic acid chemistry: influence on phytin-phosphorus availability and phytase efficacy. *J. Appl. Poult. Res.* 11 (4): 471–480.
46 Cheryan, M. (1980). Phytic acid interactions in food systems. *Crit. Rev. Food Sci. Nutr.* 13 (4): 297–335.
47 O'Dell, B.L. and De Boland, A. (1976). Complexation of phytate with proteins and cations in corn germ and oil seed meals. *J. Agric. Food Chem.* 24 (4): 804–808.
48 Graf, E. (1983). Applications of phytic acid. *J. Am. Oil Chem. Soc.* 60 (11): 1861–1867.
49 Agranoff, B.W. (2009). Turtles all the way: reflections on *myo*-inositol. *J. Biol. Chem.* 284 (32): 21121–21126.
50 Barrientos, L.G. and Murthy, P.P. (1996). Conformational studies of *myo*-inositol phosphates. *Carbohydr. Res.* 296: 39–54.
51 Costello, A.J., Glonek, T., and Myers, T.C. (1976). ^{31}P nuclear magnetic resonance-pH titrations of *myo*-inositol hexaphosphate. *Carbohydr. Res.* 46 (2): 159–171.
52 Bebot-Brigaud, A., Dange, C., Fauconnier, N., and Gérard, C. (1999). ^{31}P NMR, potentiometric and spectrophotometric studies of phytic acid ionization and complexation properties toward Co^{2+}, Ni^{2+}, Cu^{2+}, Zn^{2+} and Cd^{2+}. *J. Inorg. Biochem.* 75 (1): 71–78.
53 Schlemmer, U., Jany, K.D., Berk, A. et al. (2001). Degradation of phytate in the gut of pigs – pathway of gastro-intestinal inositol phosphate hydrolysis and enzymes involved. *Arch. Tierernahr.* 55 (4): 255–280.

54 Cowieson, A.J., Acamovic, T., and Bedford, M.R. (2004). The effects of phytase and phytic acid on the loss of endogenous amino acids and minerals from broiler chickens. *Br. Poult. Sci.* 45 (1): 101–108.
55 Deshpande, S.S. and Cheryan, M. (1984). Effects of phytic acid, divalent cations, and their interactions on α-amylase activity. *J. Food Sci.* 49 (2): 516–519.
56 Singh, M. and Krikorian, A.D. (1982). Inhibition of trypsin activity in vitro by phytate. *J. Agric. Food Chem.* 30 (4): 799–800.
57 Knuckles, B.e. (1988). Effect of phytate and other *myo*-inositol phosphate esters on lipase activity. *J. Food Sci.* 53 (1): 250–252.
58 Cowieson, A.J., Bedford, M.R., Selle, P.H., and Ravindran, V. (2009). Phytate and microbial phytase: implications for endogenous nitrogen losses and nutrient availability. *World's Poult. Sci. J.* 65 (3): 401–418.
59 Selle, P.H. and Ravindran, V. (2007). Microbial phytase in poultry nutrition. *Anim. Feed Sci. Technol.* 135 (1): 1–41.
60 Vincent, J.B., Crowder, M.W., and Averill, B.A. (1992). Hydrolysis of phosphate monoesters: a biological problem with multiple chemical solutions. *Trends Biochem. Sci.* 17 (3): 105–110.
61 Rigden, D.J. (2008). The histidine phosphatase superfamily: structure and function. *Biochem. J.* 409 (2): 333–348.
62 Oh, B.-C., Choi, W.-C., Park, S. et al. (2004). Biochemical properties and substrate specificities of alkaline and histidine acid phytases. *Appl. Microbiol. Biotechnol.* 63 (4): 362–372.
63 Wodzinski, R.J. and Ullah, A.H. (1996). Phytase. *Adv. Appl. Microbiol.* 42: 263–302.
64 Mullaney, E.J. and Ullah, A.H.J. (2003). The term phytase comprises several different classes of enzymes. *Biochem. Biophys. Res. Commun.* 312 (1): 179–184.
65 Kostrewa, D., Grüninger-Leitch, F., D'Arcy, A. et al. (1997). Crystal structure of phytase from *Aspergillus ficuum* at 2.5 Å resolution. *Nat. Struct. Biol.* 4 (3): 185–190.
66 Lim, D., Golovan, S., Forsberg, C.W., and Jia, Z. (2000). Crystal structures of *Escherichia coli* phytase and its complex with phytate. *Nat. Struct. Biol.* 7 (2): 108–113.
67 Ostanin, K. and Van Etten, R.L. (1993). Asp304 of *Escherichia coli* acid phosphatase is involved in leaving group protonation. *J. Biol. Chem.* 268 (28): 20778–20784.
68 Lindqvist, Y., Schneider, G., and Vihko, P. (1994). Crystal structures of rat acid phosphatase complexed with the transition-state analogs vanadate and molybdate. Implications for the reaction mechanism. *Eur. J. Biochem.* 221 (1): 139–142.
69 Porvari, K.S., Herrala, A.M., Kurkela, R.M. et al. (1994). Site-directed mutagenesis of prostatic acid phosphatase. Catalytically important aspartic acid 258, substrate specificity, and oligomerization. *J. Biol. Chem.* 269 (36): 22642–22646.

70 Wyss, M., Brugger, R., Kronenberger, A. et al. (1999). Biochemical characterization of fungal phytases (*myo*-inositol hexakisphosphate phosphohydrolases): catalytic properties. *Appl. Environ. Microbiol.* 65 (2): 367–373.
71 Powar, V.K. and Jagannathan, V. (1982). Purification and properties of phytate-specific phosphatase from *Bacillus subtilis*. *J. Bacteriol.* 151 (3): 1102–1108.
72 Oh, B.C., Chang, B.S., Park, K.H. et al. (2001). Calcium-dependent catalytic activity of a novel phytase from *Bacillus amyloliquefaciens* DS11. *Biochemistry (Moscow)* 40 (32): 9669–9676.
73 Fu, S., Sun, J., Qian, L., and Li, Z. (2008). Bacillus phytases: present scenario and future perspectives. *Appl. Biochem. Biotechnol.* 151 (1): 1–8.
74 Ha, N.C., Oh, B.C., Shin, S. et al. (2000). Crystal structures of a novel, thermostable phytase in partially and fully calcium-loaded states. *Nat. Struct. Biol.* 7 (2): 147–153.
75 Shin, S., Ha, N.C., Oh, B.C. et al. (2001). Enzyme mechanism and catalytic property of beta propeller phytase. *Structure (Lond. Engl. 1993)* 9 (9): 851–858.
76 Kerovuo, J., Rouvinen, J., and Hatzack, F. (2000). Analysis of *myo*-inositol hexakisphosphate hydrolysis by *Bacillus phytase*: indication of a novel reaction mechanism. *Biochem. J.* 352 (Pt. 3): 623–628.
77 Elkhalil, E.A.I., Männer, K., Borriss, R., and Simon, O. (2007). In vitro and in vivo characteristics of bacterial phytases and their efficacy in broiler chickens. *Br. Poult. Sci.* 48 (1): 64–70.
78 Lohse, D.L., Denu, J.M., Santoro, N., and Dixon, J.E. (1997). Roles of aspartic acid-181 and serine-222 in intermediate formation and hydrolysis of the mammalian protein-tyrosine-phosphatase PTP1. *Biochemistry (Moscow)* 36 (15): 4568–4575.
79 Yanke, L.J., Selinger, L.B., and Cheng, K.J. (1999). Phytase activity of *Selenomonas ruminantium*: a preliminary characterization. *Lett. Appl. Microbiol.* 29: 20–25.
80 Puhl, A.A., Greiner, R., and Selinger, L.B. (2008). A protein tyrosine phosphatase-like inositol polyphosphatase from *Selenomonas ruminantium* subsp. *lactilytica* has specificity for the 5-phosphate of *myo*-inositol hexakisphosphate. *Int. J. Biochem. Cell Biol.* 40 (10): 2053–2064.
81 Hegeman, C.E. and Grabau, E.A. (2001). A novel phytase with sequence similarity to purple acid phosphatases is expressed in cotyledons of germinating soybean seedlings. *Plant Physiol.* 126 (4): 1598–1608.
82 Dionisio, G., Madsen, C.K., Holm, P.B. et al. (2011). Cloning and characterization of purple acid phosphatase phytases from wheat, barley, maize, and rice. *Plant Physiol.* 156 (3): 1087–1100.
83 Barrientos, L., Scott, J.J., and Murthy, P.P. (1994). Specificity of hydrolysis of phytic acid by alkaline phytase from lily pollen. *Plant Physiol.* 106 (4): 1489–1495.
84 Mehta, B.D., Jog, S.P., Johnson, S.C., and Murthy, P.P.N. (2006). Lily pollen alkaline phytase is a histidine phosphatase similar to mammalian multiple inositol polyphosphate phosphatase (MINPP). *Phytochemistry* 67 (17): 1874–1886.

85 Skoglund, E., Näsi, M., and Sandberg, A.-S. (1998). Phytate hydrolysis in pigs fed a barley-rapeseed meal diet treated with *Aspergillus niger* phytase or steeped with whey. *Can. J. Anim. Sci.* 78 (2): 175–180.

86 Dvořáková, J., Kopecký, J., Havlíček, V., and Křen, V. (2000). Formation of *myo*-inositol phosphates by *Aspergillus niger* 3-phytase. *Folia Microbiol. (Praha)* 45 (2): 128–132.

87 Greiner, R., da Silva, L.G., and Couri, S. (2009). Purification and characterisation of an extracellular phytase from *Aspergillus niger* 11T53A9. *Braz. J. Microbiol.* 40 (4): 795–807.

88 Cosgrove, D.J. (1970). Inositol phosphate phosphatases of microbiological origin. Inositol phosphate intermediates in the dephosphorylation of the hexaphosphates of *myo*-inositol, *scyllo*-inositol, and D-*chiro*-inositol by a bacterial (*Pseudomonas* sp.) phytase. *Aust. J. Biol. Sci.* 23 (6): 1207–1220.

89 Konietzny, U. and Greiner, R. (2002). Molecular and catalytic properties of phytate-degrading enzymes (phytases). *Int. J. Food Sci. Technol.* 37 (7): 791–812.

90 Hayakawa, T., Suzuki, K., Miura, H. et al. (1990). Myo-inositol polyphosphate intermediates in the dephosphorylation of phytic acid by acid phosphatase with phytase activity from rice bran. *Agric. Biol. Chem.* 54 (2): 279–286.

91 Stahl, C.H., Roneker, K.R., Pond, W.G., and Lei, X.G. (2004). Effects of combining three fungal phytases with a bacterial phytase on plasma phosphorus status of weanling pigs fed a corn-soy diet. *J. Anim. Sci.* 82 (6): 1725–1731.

92 Gentile, J.M., Roneker, K.R., Crowe, S.E. et al. (2003). Effectiveness of an experimental consensus phytase in improving dietary phytate-phosphorus utilization by weanling pigs. *J. Anim. Sci.* 81 (11): 2751–2757.

93 Zimmermann, B., Lantzsch, H.-J., Mosenthin, R. et al. (2002). Comparative evaluation of the efficacy of cereal and microbial phytases in growing pigs fed diets with marginal phosphorus supply. *J. Sci. Food Agric.* 82 (11): 1298–1304.

94 Greiner, R. (2017). Activity of *Escherichia coli*, *Aspergillus niger*, and rye phytase toward partially phosphorylated *myo*-Inositol phosphates. *J. Agric. Food Chem.* 65 (44): 9603–9607.

95 Revy, P.S., Jondreville, C., Dourmad, J.Y., and Nys, Y. (2006). Assessment of dietary zinc requirement of weaned piglets fed diets with or without microbial phytase. *J. Anim. Physiol. Anim. Nutr.* 90 (1–2): 50–59.

96 Selle, P.H. and Ravindran, V. (2008). Phytate-degrading enzymes in pig nutrition. *Livest. Sci.* 113 (2): 99–122.

97 Rutherfurd, S.M., Chung, T.K., Thomas, D.V. et al. (2012). Effect of a novel phytase on growth performance, apparent metabolizable energy, and the availability of minerals and amino acids in a low-phosphorus corn-soybean meal diet for broilers. *Poult. Sci.* 91 (5): 1118–1127.

98 Lei, X.G. and Stahl, C.H. (2001). Biotechnological development of effective phytases for mineral nutrition and environmental protection. *Appl. Microbiol. Biotechnol.* 57 (4): 474–481.

99 Peers, F.G. (1953). The phytase of wheat. *Biochem. J.* 53 (1): 102–110.

100 Pointillart, A., Fourdin, A., and Fontaine, N. (1987). Importance of cereal phytase activity for phytate phosphorus utilization by growing pigs fed diets containing triticale or corn. *J. Nutr.* 117 (5): 907–913.

101 Madsen, C.K., Dionisio, G., Holme, I.B. et al. (2013). High mature grain phytase activity in the *Triticeae* has evolved by duplication followed by neofunctionalization of the purple acid phosphatase phytase (*PAPhy*) gene. *J. Exp. Bot.* 64 (11): 3111–3123.

102 ISO 30024:2009 (2009). Animal Feeding Stuffs -- Determination of Phytase Activity, ISO.

103 Patwardhan, V.N. (1937). The occurrence of a phytin-splitting enzyme in the intestines of albino rats. *Biochem. J.* 31 (4): 560–564.

104 Maenz, D.D. and Classen, H.L. (1998). Phytase activity in the small intestinal brush border membrane of the chicken. *Poult. Sci.* 77 (4): 557–563.

105 Hu, H.L., Wise, A., and Henderson, C. (1996). Hydrolysis of phytate and inositol tri-, tetra-, and penta-phosphates by the intestinal mucosa of the pig. *Nutr. Res.* 16 (5): 781–787.

106 Abudabos, A.M. (2012). Phytate phosphorus utilization and intestinal phytase activity in laying hens. *Ital. J. Anim. Sci.* 11 (1): 41–46.

107 Moore, R.J. and Veum, T.L. (1983). Adaptive increase in phytate digestibility by phosphorus-deprived rats and the relationship of intestinal phytase (EC 3.1.3.8) and alkaline phosphatase (EC 3.1.3.1) to phytate utilization. *Br. J. Nutr.* 49 (1): 145–152.

108 Raghavendra, P. and Halami, P.M. (2009). Screening, selection and characterization of phytic acid degrading lactic acid bacteria from chicken intestine. *Int. J. Food Microbiol.* 133 (1–2): 129–134.

109 Marounek, M., Skřivan, M., Rosero, O., and Rop, O. (2010). Intestinal and total tract phytate digestibility and phytase activity in the digestive tract of hens fed a wheat-maize-soyabean diet. *J. Anim. Feed Sci.* 19 (3): 430–439.

110 Walker, A.R., Fox, F.W., and Irving, J.T. (1948). Studies in human mineral metabolism: 1. The effect of bread rich in phytate phosphorus on the metabolism of certain mineral salts with special reference to calcium. *Biochem. J.* 42 (3): 452–462.

111 Sandberg, A.S., Larsen, T., and Sandström, B. (1993). High dietary calcium level decreases colonic phytate degradation in pigs fed a rapeseed diet. *J. Nutr.* 123 (3): 559–566.

112 Pileggi, V.J., De Luca, H.F., and Steenbock, H. (1955). The role of vitamin D and intestinal phytase in the prevention of rickets in rats on cereal diets. *Arch. Biochem. Biophys.* 58 (1): 194–204.

113 Wise, A. and Gilburt, D.J. (1982). Phytate hydrolysis by germfree and conventional rats. *Appl. Environ. Microbiol.* 43 (4): 753–756.

114 Jongbloed, A.W., Mroz, Z., and Kemme, P.A. (1992). The effect of supplementary *Aspergillus niger* phytase in diets for pigs on concentration and apparent digestibility of dry matter, total phosphorus, and phytic acid in different sections of the alimentary tract. *J. Anim. Sci.* 70 (4): 1159–1168.

115 Yi, Z., Kornegay, E.T., Ravindran, V., and Denbow, D.M. (1996). Improving phytate phosphorus availability in corn and soybean meal for broilers using microbial phytase and calculation of phosphorus equivalency values for phytase. *Poult. Sci.* 75 (2): 240–249.

116 Kemme, P.A., Jongbloed, A.W., Mroz, Z., and Beynen, A.C. (1998). Diurnal variation in degradation of phytic acid by plant phytase in the pig stomach. *Livest. Prod. Sci.* 54 (1): 33–44.

117 Yu, B., Jan, Y.C., Chung, T.K. et al. (2004). Exogenous phytase activity in the gastrointestinal tract of broiler chickens. *Anim. Feed Sci. Technol.* 117 (3): 295–303.

118 Maenz, D.D., Engele-Schaan, C.M., Newkirk, R.W., and Classen, H.L. (1999). The effect of minerals and mineral chelators on the formation of phytase-resistant and phytase-susceptible forms of phytic acid in solution and in a slurry of canola meal. *Anim. Feed Sci. Technol.* 81 (3): 177–192.

119 Persson, H., Türk, M., Nyman, M., and Sandberg, A.-S. (1998). Binding of Cu^{2+}, Zn^{2+}, and Cd^{2+} to inositol tri-, tetra-, penta-, and hexaphosphates. *J. Agric. Food Chem.* 46 (8): 3194–3200.

120 Dersjant-Li, Y., Awati, A., Schulze, H., and Partridge, G. (2015). Phytase in non-ruminant animal nutrition: a critical review on phytase activities in the gastrointestinal tract and influencing factors. *J. Sci. Food Agric.* 95 (5): 878–896.

121 Lönnerdal, B., Sandberg, A.S., Sandström, B., and Kunz, C. (1989). Inhibitory effects of phytic acid and other inositol phosphates on zinc and calcium absorption in suckling rats. *J. Nutr.* 119 (2): 211–214.

122 Sandström, B. and Sandberg, A.S. (1992). Inhibitory effects of isolated inositol phosphates on zinc absorption in humans. *J. Trace Elem. Electrolytes Health Dis.* 6 (2): 99–103.

123 Driver, J.P., Pesti, G.M., Bakalli, R.I., and Edwards, H.M. (2005). Effects of calcium and nonphytate phosphorus concentrations on phytase efficacy in broiler chicks. *Poult. Sci.* 84 (9): 1406–1417.

124 Pointillart, A., Fontaine, N., and Thomasset, M. (1984). Phytate phosphorus utilization and intestinal phosphatases in pigs fed low phosphorus: wheat or corn diets [phosphorus availability]. *Nutr. Rep. Int.* 22: 473–483.

125 Pointillart, A., Fontaine, N., Thomasset, M., and Jay, M.E. Phosphorus utilization, intestinal phosphatases and hormonal control of calcium metabolism in pigs fed phytic phosphorus: soyabean or rapeseed diets. *Nutr. Rep. Int.* 32: 155–156.

126 Berg, J.M., Tymoczko, J.L., and Stryer, L. (2002). The Michaelis–Menten model accounts for the kinetic properties of many enzymes. In: *Biochemistry*, 5e. Hoboken, NJ: W.H. Freeman https://www.ncbi.nlm.nih.gov/books/NBK22430/.

127 Huang, H., Luo, H., Wang, Y. et al. (2008). A novel phytase from Yersinia rohdei with high phytate hydrolysis activity under low pH and strong pepsin conditions. *Appl. Microbiol. Biotechnol.* 80 (3): 417–426.

128 Phillippy, B.Q. (1999). Susceptibility of wheat and *Aspergillus niger* phytases to inactivation by gastrointestinal enzymes. *J. Agric. Food Chem.* 47 (4): 1385–1388.

129 Rodriguez, E., Porres, J.M., Han, Y., and Lei, X.G. (1999). Different sensitivity of recombinant *Aspergillus niger* phytase (r-PhyA) and *Escherichia coli* pH 2.5 acid phosphatase (r-AppA) to Trypsin and Pepsin in Vitro. *Arch. Biochem. Biophys.* 365 (2): 262–267.

130 Niu, C., Yang, P., Luo, H. et al. (2017). Engineering of Yersinia phytases to improve pepsin and Trypsin resistance and thermostability and application potential in the food and feed industry. *J. Agric. Food Chem.* 65 (34): 7337–7344.

131 Niu, C., Luo, H., Shi, P. et al. (2015). N-glycosylation improves the pepsin resistance of histidine acid phosphatase phytases by enhancing their stability at acidic pHs and reducing pepsin's accessibility to its cleavage sites. *Appl. Environ. Microbiol.* 82 (4): 1004–1014.

132 Tan, H., Wu, X., Xie, L. et al. (2016). A novel phytase derived from an acidic peat-soil microbiome showing high stability under acidic plus pepsin conditions. *J. Mol. Microbiol. Biotechnol.* 26 (4): 291–301.

133 Zhao, Q., Liu, H., Zhang, Y., and Zhang, Y. (2010). Engineering of protease-resistant phytase from *Penicillium* sp.: high thermal stability, low optimal temperature and pH. *J. Biosci. Bioeng.* 110 (6): 638–645.

134 Pasamontes, L., Haiker, M., Wyss, M. et al. (1997). Gene cloning, purification, and characterization of a heat-stable phytase from the fungus *Aspergillus fumigatus*. *Appl. Environ. Microbiol.* 63 (5): 1696–1700.

135 Wyss, M., Pasamontes, L., Rémy, R. et al. (1998). Comparison of the thermostability properties of three acid phosphatases from molds: *Aspergillus fumigatus* phytase, *A. niger* phytase, and *A. niger* PH 2.5 acid phosphatase. *Appl. Environ. Microbiol.* 64 (11): 4446–4451.

136 Boukhris, I., Farhat-Khemakhem, A., Blibech, M. et al. (2015). Characterization of an extremely salt-tolerant and thermostable phytase from *Bacillus amyloliquefaciens* US573. *Int. J. Biol. Macromol.* 80: 581–587.

137 Vohra, A. and Satyanarayana, T. (2002). Purification and characterization of a thermostable and acid-stable phytase from *Pichia anomala*. *World J. Microbiol. Biotechnol.* 18 (7): 687–691.

138 De Angelis, M., Gallo, G., Corbo, M.R. et al. (2003). Phytase activity in sourdough lactic acid bacteria: purification and characterization of a phytase from *Lactobacillus sanfranciscensis* CB1. *Int. J. Food Microbiol.* 87 (3): 259–270.

139 Shivange, A.V. and Schwaneberg, U. (2017). Recent advances in directed phytase evolution and rational phytase engineering. In: *Directed Enzyme Evolution: Advances and Applications* (ed. M. Alcalde), 145–172. Cham: Springer.

140 Shivange, A.V., Serwe, A., Dennig, A. et al. (2012). Directed evolution of a highly active *Yersinia mollaretii* phytase. *Appl. Microbiol. Biotechnol.* 95 (2): 405–418.

141 Rodriguez, E., Wood, Z.A., Karplus, P.A., and Lei, X.G. (2000). Site-directed mutagenesis improves catalytic efficiency and thermostability of *Escherichia coli* pH 2.5 acid phosphatase/phytase expressed in *Pichia pastoris*. *Arch. Biochem. Biophys.* 382 (1): 105–112.

142 Ullah, A.H. (1988). Production, rapid purification and catalytic characterization of extracellular phytase from *Aspergillus ficuum*. *Prep. Biochem.* 18 (4): 443–458.

143 Han, Y. and Lei, X.G. (1999). Role of glycosylation in the functional expression of an *Aspergillus niger* phytase (phyA) in *Pichia pastoris*. *Arch. Biochem. Biophys.* 364 (1): 83–90.
144 Lehmann, M., Lopez-Ulibarri, R., Loch, C. et al. (2000). Exchanging the active site between phytases for altering the functional properties of the enzyme. *Protein Sci. Publ. Protein Soc.* 9 (10): 1866–1872.
145 Greiner-Stöffele, T. and Struhalla, M. (2003). Method from the selection of biomolecules from biomolecule variant libraries. Patent No. WO2005040376.
146 Greiner-Stoeffele, T., Feller, C., and Struhalla, M. (2008). Method for creating a variant library of DNA sequences. Patent No. EP2130918.
147 Champion, E., Vogel, A., Bartsch, S., and Dekany, G. (2014). Mutated fucosidase. Patent No. US2017313996.
148 Wijma, H.J., Fürst, M.J.L.J., and Janssen, D.B. (2018). A computational library design protocol for rapid improvement of protein stability: FRESCO. *Methods Mol. Biol. (Clifton, NJ)* 1685: 69–85.
149 Becker, N., Clarkson, K., Dale, D. et al. (2017). Stable, Durable Granules with Active Agents. MY161449 (A), issued 14 April 2017.
150 Shirley, R.B. and Edwards, H.M. (2003). Graded levels of phytase past industry standards improves broiler performance. *Poult. Sci.* 82 (4): 671–680.
151 Augspurger, N.R. and Baker, D.H. (2004). High dietary phytase levels maximize phytate-phosphorus utilization but do not affect protein utilization in chicks fed phosphorus- or amino acid-deficient diets. *J. Anim. Sci.* 82 (4): 1100–1107.
152 Pirgozliev, V., Oduguwa, O., Acamovic, T., and Bedford, M.R. (2007). Diets containing *Escherichia coli*-derived phytase on young chickens and turkeys: effects on performance, metabolizable energy, endogenous secretions, and intestinal morphology. *Poult. Sci.* 86 (4): 705–713.
153 Lee, S.A., Nagalakshmi, D., Raju, M.V.L.N. et al. (2017). Effect of phytase superdosing, *myo*-inositol and available phosphorus concentrations on performance and bone mineralisation in broilers. *Anim. Nutr.* 3 (3): 247–251.
154 Kies, A.K., Kemme, P.A., Sebek, L.B.J. et al. (2006). Effect of graded doses and a high dose of microbial phytase on the digestibility of various minerals in weaner pigs. *J. Anim. Sci.* 84 (5): 1169–1175.
155 Rapp, C., Lantzsch, H.J., and Drochner, W. (2001). Hydrolysis of phytic acid by intrinsic plant and supplemented microbial phytase (*Aspergillus niger*) in the stomach and small intestine of minipigs fitted with re-entrant cannulas. 3. Hydrolysis of phytic acid (IP6) and occurrence of hydrolysis products (IP5, IP4, IP3 and IP2). *J. Anim. Physiol. Anim. Nutr.* 85 (11–12): 420–430.
156 Beeson, L.A., Walk, C.L., Bedford, M.R., and Olukosi, O.A. (2017). Hydrolysis of phytate to its lower esters can influence the growth performance and nutrient utilization of broilers with regular or super doses of phytase. *Poult. Sci.* 96 (7): 2243–2253.
157 Pontoppidan, K., Pettersson, D., and Sandberg, A.-S. (2007). Interaction of phytate with protein and minerals in a soybean–maize meal blend depends on pH and calcium addition. *J. Sci. Food Agric.* 87 (10): 1886–1892.

Part IV

Enzymes for Biorefinery Applications

4.1

Enzymes for Pulp and Paper Applications

Debayan Ghosh, Bikas Saha, and Baljeet Singh

Epygen®Labs FZ LLC, D-021/022, Dubai Science Park, 485018 Dubai, UAE

Despite reduced usages of paper due to electronic media adaptation in society, the global demand for paper grows between 2% and 3% every year. The ever-increasing global demand for paper has been fueling the pulp and paper industry to grow. Although paper is considered biodegradable, the industry is dealing with constant challenges. Technologists can turn these challenges into opportunities to reduce the usage of virgin pulp, reduce energy consumption, reduce harsh chemical usage, reduce carbon footprint, and, finally, reduce the cost of paper making, for which enzymes play a crucial role.

The diminishing forest line has a negative association with the pulp and paper industry. It is also amongst the world's major generators of air and water pollutants and waste products, and contributes to the emission of greenhouse gases that cause climate change. While usage of recycled fiber to replace virgin fiber greatly reduces the impact on environment, it throws a new set of issues to the paper mills. On the other hand, enzymes, being biological and hence multifunctional in nature, can be engineered either at the genetic level or at the protein level based on the desired functionality. This is preferred over harsh chemicals and presents future directions to the pulp and paper industry. The potential use of enzymes in the pulp and paper industry was realized long ago by researchers. The main constituent of paper, pulp composed of natural polymers, i.e. cellulose, hemicellulose, and lignin, could be potentially modified by natural enzymes. In the initial stages, acceptability of enzymes in the pulp and paper industry appeared to be low. The key reasons could be attributed to the complexity and diverse nature of the substrate itself, i.e. wood and pulp. This dictated differentiation in solutions depending on the raw material used in different mills. It was also difficult for biotechnologists to design enzymes when the substrate–pulp composition was ever changing depending on its origin. A few other factors that hindered rapid adaptation of enzyme application in the pulp and paper industry could be the lack of methods to assess performance, poor enzyme stability under harsh process conditions, a narrow operating window of pH and temperature, and, finally, the relatively high cost of enzymatic treatment. Pulp and paper biotechnologists today have come out with innovative solutions to address these problems based on the tremendous progress in

Industrial Enzyme Applications, First Edition. Edited by Andreas Vogel and Oliver May.

enzyme technology and molecular biology in the last decades, as described in this chapter.

Our purpose in this chapter is to provide an overview of the various enzyme applications that are used in the pulp and paper industry today. There are more products in the horizon as biotechnology-based solutions fill future enzyme pipeline to address the needs of this industry and overcome the challenges that it still faces.

During the past two decades, the number of applications of enzymes in pulp and paper manufacture has grown steadily and received wide acceptability. Table 4.1.1 outlines some of the current usages.

4.1.1 Refining and Fiber Development Enzyme

Despite pioneering studies describing the interaction of cellulases and cellulose fibers by Mandels and Reese [1] half a century ago, there has been limited success in enzymatic modification of cellulosic fibers in the paper industry.

The more successful use of commercial cellulases has been in the textiles area [2] where enzymes have been adopted as a part of the garment wet processing industry to provide special effects in finished garments such as the stone washing of blue jeans or polishing and fuzz removal of knitted or woven fabric. This application has been enabled by the nature of activity of the cellulases produced by commercially exploited *Trichoderma* or *Aspergillum* based fungal systems.

On the other hand, despite the potential promise of the application of cellulases to pulp and paper, there have been only few reports about the use of these

Table 4.1.1 Different enzymatic applications in pulp and paper manufacturing.

Sr. no.	Enzyme application	Category of paper industry
1	Refining and fiber development enzyme	Writing and printing, tissue making
2	Pulp freeness or drainage improvement enzyme	Board making from recycled/virgin furnish
3	Stickies control enzyme	Liner, fluting making from recycled furnish, tissue making from de-inked pulp
4	Deinking enzyme	Produce de-inked pulp for writing and printing, tissue, boards, newsprint
5	Hard wood vessel breaking enzyme	Writing and printing
6	Native starch conversion enzyme	Starch processing for all type of paper
7	Bleach boosting enzyme	ECF bleaching
8	Effluent treatment enzyme	All types of paper mill effluent
9	Slushing enzyme	Various recycle furnish difficult to slush, e.g. sack Kraft, PE coated paper, hard-sized paper, etc.

enzymes to address manufacturing issues in the industry: Goma and coworkers [3] describe freeness/drainage improvements using the commercial cellulases available in the late 1980s. A more recent report [4] describes work at laboratory and pilot plant scale of pulp treatment using a *Trichoderma* sp. preparation. These authors observed pulp fiber shortening; they also report lower refining energy requirements to achieve the given drainage targets. Other published uses of cellulases in pulp and paper include deinking [5, 6] and pulp fiber modification to achieve production of a softer tissue [7, 8].

The recent availability of cellulase preparations produced by nonconventional cellulose-producing organisms such as *Chrysosporium* sp. [9] has expanded the library of cellulase systems available for application in the pulp and paper industry. It has been used for the pretreatment of fibers, rendering a downstream advantage of energy reduction and/or reduced steam requirement in drying of the treated pulp.

A refining and fiber developing carbohydrase enzyme complex is produced by genetic engineering and controlled fermentation of recombinant organisms. It is designed to enhance fibrillation and achieve drainage improvement of the paper web, by selective hydrolysis of xylan linkages on fiber surfaces and amorphous regions in fines. All these functions are orchestrated to a unique combination of exo- and endo-glucanase in a well-thought-about combination of several hemi-cellulase components.

As has been cited in earlier literature, cellobiohydrolase (CBH), which is an exo-glucanase, tends to degrade crystalline cellulose from the non-reducing end, developing microfibrils by cutting one side and raising them loose on the fiber surfacc [10a, b]. Other cellulases such as endo-glucanases (EG) are able to degrade amorphous cellulose (loosely packed, outer surface). Therefore, a predetermined design of a multienzyme complex of EGI, EGII, EGIII; CBH I, CBH II, and CBH III; and xylanases I, III, etc. is constructed to act in developing fibers, perforation, hydration, swelling of fiber, fibrillation (resulting in inter-fiber bonding), and brushing of fibers. Simultaneously, these enzyme complexes partially hydrolyze very small colloidal organic trash [11a, b], enhancing drainage.

The refining and fiber development enzymes described above were studied by three methods:

a. Microscopic evaluation of fiber treated with enzyme
b. Evaluation of handsheet prepared with enzyme-treated pulp in laboratory trials
c. Plant application trials

4.1.1.1 Microscopic Evaluation

The experiments were carried out with dried bleached kraft softwood and hardwood pulp. After incubation with cellulase enzymes, the fiber was studied under microscope as shown in Figure 4.1.1.

From the above microscopic analysis, using a digitized diameter-measuring method, it is evident that enzymatic refining resulted in substantial swelling of both hardwood and softwood fibers. The result is summarized in Table 4.1.2.

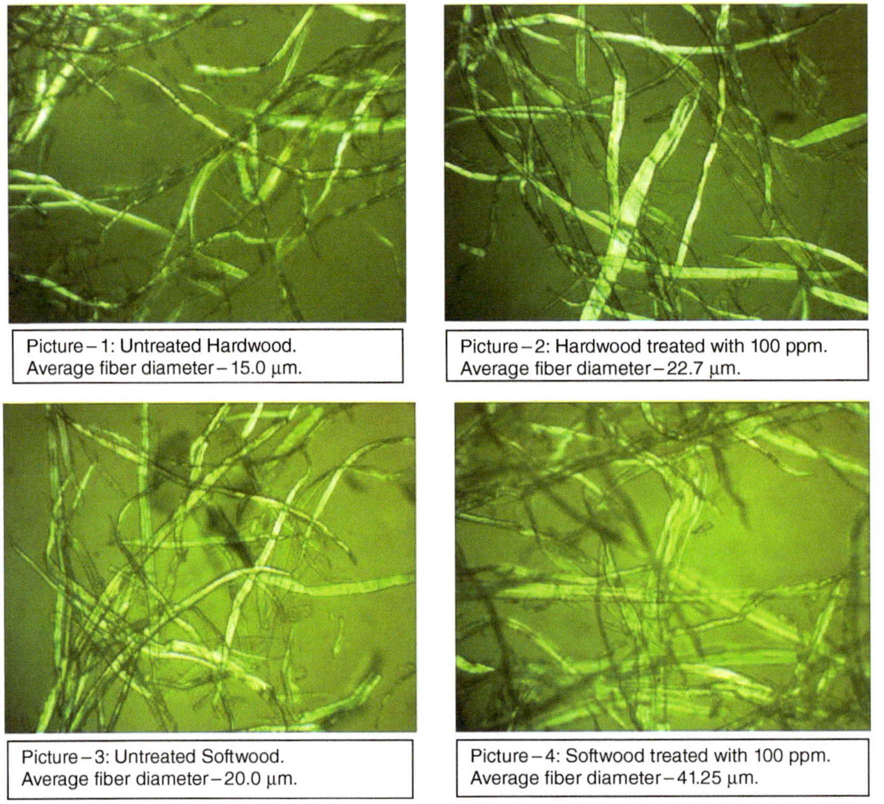

Figure 4.1.1 Comparison of microscopic pictures of hardwood and softwood using enzyme.

Table 4.1.2 Swelling effect of enzyme treatment on hardwood and softwood fiber.

Pulp type	Enzyme dose (ppm)	Average fiber diameter (μm)	Comparative fiber swelling (%)
Hardwood	0 (blank)	15.25	
	100	22.70	48.86
Softwood	0 (blank)	20.75	
	100	41.78	101.35

Since interfiber bonding is inversely proportional to interfiber distance, it is expected that swelling would result in higher strength properties such as tensile strength, burst, and wax pick number. Being a significant step of the fiber refining mechanism, enzymatic swelling would reduce the demand for mechanical refining for further fibrillation and development of fibers. In mills, where refiners are not used, this enzymatic swelling is expected to develop higher strength properties in paper.

4.1.1.2 Evaluation of Enzyme-Treated Handsheets

Laboratory evaluation of handsheets prepared with enzyme-treated fiber was carried out on many occasions (Epygen's, unpublished data). In one such laboratory evaluation, softwood pulp was treated with 150 ppm of enzyme and subjected to beating for various times. Hardwood was treated with enzyme only (100 ppm dosages). Beating was not done for hardwood. Handsheets were prepared with a blend of 80% hardwood and 20% softwood pulp. The results are shown in Table 4.1.3.

The laboratory test establishes significant improvement in tensile strength with the specific enzyme, compared to blank. At zero minute beating, the tensile strength increase was about 40% compared to blank. Pulp incubation with enzyme without any mechanical refining action helped the fiber to swell, resulting in a substantial increase in tensile strength. This was also evident from pulp freeness drop measured in CSF (Canadian Standard Freeness). Furthermore, the study indicates that after 20 minutes of beating, tensile strength showed 30% increase compared to blank; after 25 minutes, the increase was about 25%; after 30 minutes the increase was about 15%. Interestingly, the difference between the tensile strength of the blank sample after 30 minutes of beating and the tensile strength of the enzyme-treated sample without beating was close (7%). This indicates that a tensile strength of 19.55 can be achieved with very little refining of enzyme-treated pulp, indicating significant reduction in the required refining power and therefore substantial cost saving. In addition, mills can benefit from the increased tensile strength by reducing expensive softwood usage. Based on similar laboratory experiments, several mill trials were conducted with refining and fiber developing enzymes as outlined below.

The real-life test data provided here were supplied by the respective mills and were not necessarily performed under TAPPI (Technical Association of the Pulp and Paper Industry) standard test conditions.

4.1.1.2.1 Case Study 1

The product manufactured during this trial was facial tissue using a furnish mix of softwood and hardwood. Under pretrial conditions, only softwood was mechanically refined through two refiners, Refiner#1 and Refiner#2 (R1, R2), whereas hardwood was not refined. These two types of pulps were mixed at a certain ratio using a blending chest and subsequently passed through a mechanical refiner, Refiner#3 (R3). During the trial, Epygen Papyrase® group enzyme was added at the pulper to both pulps. As outlined in Figure 4.1.2, the dry tensile strength of the tissue increased significantly once an optimum dosage of the enzyme was identified. This development led to reduction in the overall applied load (power/energy in kilowatt) in mechanical refining and finally one refiner had to stop to get back the normal tensile strength. Potentially, refiner load reduction can lead to increase in caliper because of lesser cutting/collapsing of fibers, which has important commercial interest.

Apart from refining power reduction, softwood usage could be stepwise reduced from 35% to 25% depending on the target dry tensile strength of the tissue as shown in Figure 4.1.3. This reduction of softwood not only had huge

Table 4.1.3 Enzyme laboratory evaluation results with softwood and hardwood pulp.

Sl no.	Furnish composition (%)	Enzyme dose (softwood, hardwood)	Softwood beating time (min)	Average GSM	Freeness (CSF)	Tensile index (grf/25 mm)	Tear index (mN m^3/gm)	Thickness (μm)	Bulk (cm^3/g)
1	Blank (20% Softwood + 80% Hardwood)	Softwood – 0 ppm, Hardwood – 0 ppm	0	66.08	645	12.93	2.50	104.44	1.58
			20	59.90	603	17.16	4.36	88.94	1.48
			25	60.39	560	19.18	4.69	85.4	1.41
			30	60.49	539	19.55	5.25	84.64	1.40
2	Enzyme (20% Softwood + 80% Hardwood)	Softwood – 150 ppm, Hard-wood – 100 ppm	0	57.81	603	18.21	4.53	94.04	1.39
			20	67.89	593	22.43	5.34	84.00	1.35
			25	60.16	581	23.90	5.95	81.88	1.33
			30	65.16	550	22.50	5.59	80.94	1.36

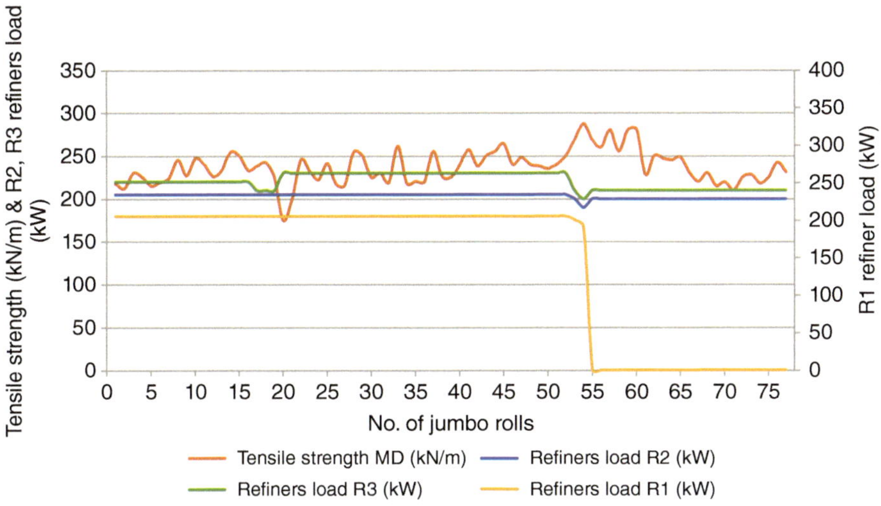

Figure 4.1.2 Effect of enzyme on refiner load and tensile strength.

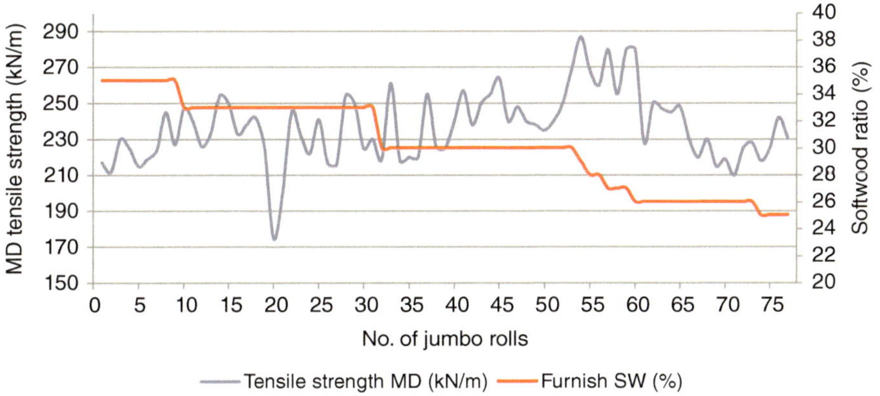

Figure 4.1.3 MD (machine direction) tensile strength and softwood ratio during enzyme usages.

commercial impact but also resulted in improved softness of tissue due to higher usage of hardwood. Few other notable changes were observed, for example, reduction of dust generation and improved runnability of tissue machine and winder.

4.1.1.2.2 Case Study 2

This mill trial was conducted with refining and fiber development enzyme with an objective to improve printability of online-coated paper measured by IGT printability tester. The rationale behind this was that Epygen Papyrase enzyme-treated pulp would improve the *z*-direction strength of the base paper, resulting in higher IGT values of coated paper. Hardwood and softwood pulps were treated separately with the enzyme at the pulper and followed by mechanical refining. Treated

Table 4.1.4 Impact of enzyme usage on IGT printability values.

Process condition	IGT test values	
	Machine direction	Cross direction
Blank	113	101
With enzyme	129	117

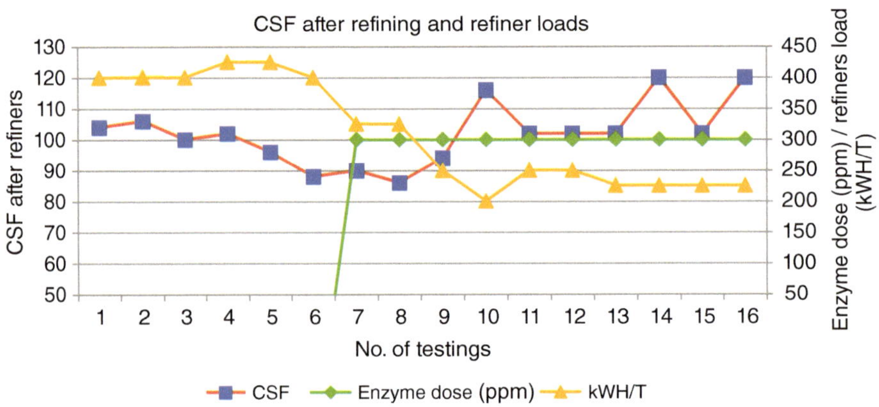

Figure 4.1.4 Comparison of after refining CSF and refiners loads (kilowatt hours per ton) during enzyme usages on NBKP pulp.

pulp was blended at a specific ratio at the subsequent stage. Table 4.1.4 shows improvement in IGT values of 65 GSM coated paper after Janus calendaring.

Few other highlights of this trial were substantial increase in ash loading in base paper as well as through coating and reduction of applied refining energy for NBKP (Northern bleached kraft pulp) as shown in Figures 4.1.4 and 4.1.5.

4.1.2 Drainage Improvement Enzyme

In the paper making process, the time needed to form a sheet increases with the square of the basis weight/grammage and is proportional to the drainage constant of the pulp [12a, b]. It is well known that drainage characteristics of the pulp are influenced by the percentage of fines present in the pulp slurry. Sometimes, the combination of higher basis weight with low drainage pulp is necessary, which can have a negative impact on the paper making process. In such cases, drainage characteristics of the pulp are improved by minimizing the ultra-fines components of the pulp mix [13]. Another method to improve the drainage characteristics is to add polymeric chemicals, which, however, often deliver inconsistent results and affect the formation of paper negatively. Even with good retention of

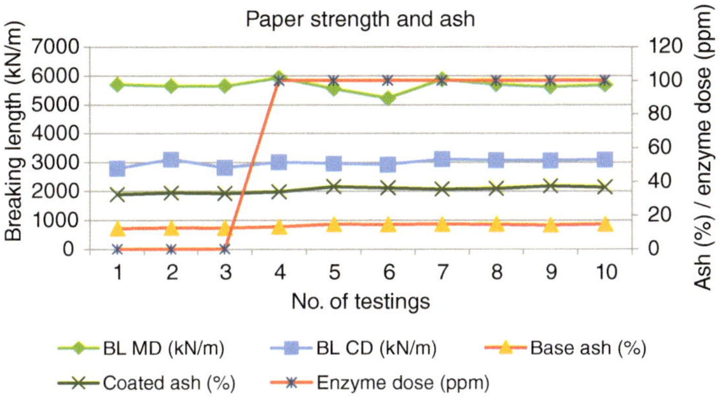

Figure 4.1.5 Comparison of breaking length (BL) and ash (base paper and coated paper) during enzyme usage.

fibers and fines, a significant fraction of smaller size fines fraction (ultra-fines) cannot be separated by paper web and therefore circulate in white water, rendering hindrance to drainage. Fines and ultra-fines are shown in Figure 4.1.6.

An enzymatic way to improve the drainage of pulp is by applying a recombinant carbohydrase enzyme, specially designed for selective hydrolysis of amorphous cellulose/hemicellulose contents of ultra-fines in virgin and recycled pulp furnish [10a, b]. During the entire process of pulping, refining, papermaking, and broke/recycle processing, long-chain carbohydrate depolymerization is influenced by mechanical disruption of internal bonds. Misalignment of naturally compact crystalline cellulose creates loosely packed amorphous cellulosic debris, which can be easily hydrolyzed by cellobiohydrolases. With abundant exo-ends being available for enzymatic hydrolysis, it seems that endo-enzyme activity limits the conversion of these ultra-fines cellulosic debris. The enzyme converts amorphous regions in ultra-fines to soluble reducing sugar, thus enhancing the drainage of pulp [11a]. This specially designed enzyme was tested at laboratory scale with various commercial pulps to observe how this synergetic design impacts fines of hardwood and softwood nature. Completely delignified

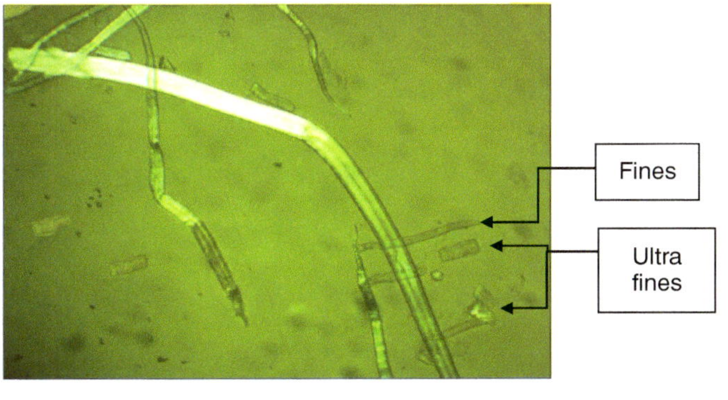

Figure 4.1.6 Polarized light microscopy image of fines and ultra-fines; magnification: 400 fold.

pulp fines bear characteristic differences compared to fines from wood or kraft pulp in terms of availability and preferential hydrolysis by this uniquely designed enzyme.

It has been hypothesized that hornified or repeatedly dried wood pulp debris contains less amorphous regions than delignified fines from various bleached pulps [11a]. Indeed, a difference in amorphous regions (availability of preferential enzyme sites) was confirmed by treating a wide range of furnish with a drainage enhancing enzyme (Tables 4.1.4–4.1.6). In case of leaf bleached kraft pulp (LBKP), the quantum of drainage improvement is about 30% with 400 ppm of enzyme dosages compared to blank, as indicated by increase in CSF values, whereas recycled fiber mix (MW/ONP/Dx) has shown improvement of 24% and CTMP has shown 20% improvement in CSF values at 400 ppm dosages compared to blank.

To prove the impact of a drainage enzyme, laboratory tests were conducted mimicking large-scale industrial conditions. Taking this hypothesis of differential fine hydrolysis impact, it has been shown in various plant-scale applications that fines contribution from a wide array of furnish can be predictably influenced by the introduction of this enzyme, providing a tool for the paper makers to de-bottleneck this costly limitation of commercially viable furnish mix.

The results are tabulated in Tables 4.1.5–4.1.7.

To establish drainage improvement, several plant applications were carried out. Results from a selected few are summarized below (Figure 4.1.7).

Table 4.1.5 Comparison of CSF and drainage test with different enzyme dosages on CTMP (chemi-thermomechanical pulp).

Enzyme dose (ppm)	GSM (g/m^2)	Temperature (°C)	CSF
0 (Blank)	56	35	400
100	55	35	440
200	56	35	450
300	54	35	460
400	54	35	480

Table 4.1.6 Comparison of CSF and drainage test with different enzyme dosages on LBKP (leaf bleached kraft pulp).

Enzyme dose (ppm)	GSM (g/m^2)	Temperature (°C)	CSF
0 (Blank)	57	35	310
100	54	35	360
200	54	35	380
300	53	35	380
400	53	35	400

Table 4.1.7 Comparison of CSF and drainage test with different enzyme dosages on MW/ONP/Dx (mixed waste/old newsprint/duplex).

Enzyme dose (ppm)	GSM (g/m^2)	Temperature (°C)	CSF
0 (Blank)	55	35	290
100	54	35	320
200	54	35	320
300	55	35	340
400	53	35	360

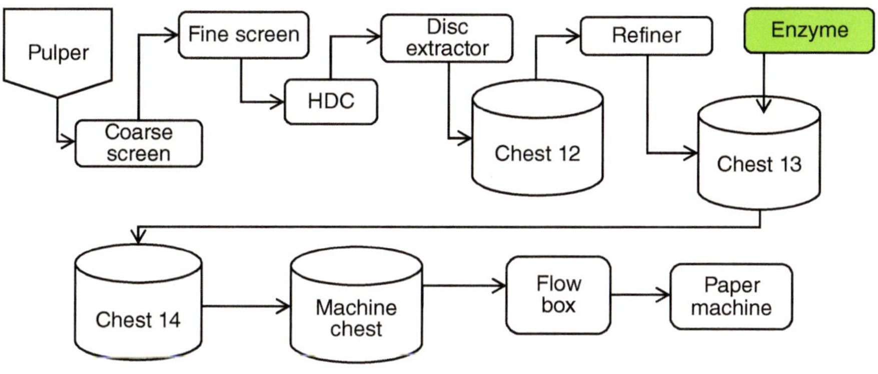

Figure 4.1.7 Flow chart of stock preparation of middle layer, where enzyme addition is after the refiner.

4.1.2.1 Case Study 3

Mill: A

Grade produced: Duplex Board

Capacity: 6000 MT/month

Enzyme used: Epygen Papyrase Drainage enzyme

Enzyme dosage: 300–500 ppm

Enzyme dosing location: Post refiners, chest 13.

Trial duration: 15 days

Middle layer furnishes: OCC* and mixed waste

Bottom layer furnish: ONP#

In this mill, paperboard consists of three layers: top, middle, and bottom. In the middle layer, old corrugated carton (OCC) and mixed waste (MW) are used in the ratio 2 : 1, limitation being the steep reduction in freeness (CSF drop) in case mixed waste is increased. It was shown during drainage enzyme trial (refer Figure 4.1.8) for 400 GSM that the CSF trend was significantly higher, even with inclusion of higher percentage of mixed waste in the pulp mix.

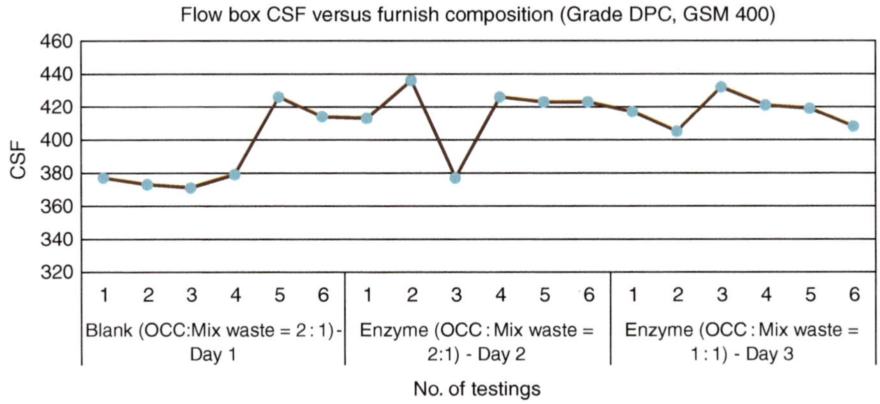

Figure 4.1.8 Improvement in CSF while using high fines (MW) furnish during enzyme usages.

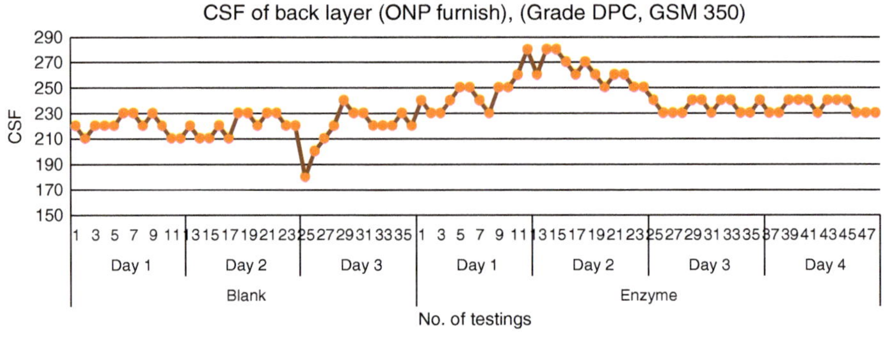

Figure 4.1.9 Improvement in CSF in ONP furnish during enzyme usages.

At a different instance, while producing 350 GSM for bottom layer furnish of old newsprint (ONP), it was observed that CSF values were higher during usage of enzyme compared to blank, which is shown in Figure 4.1.9.

4.1.2.2 Case Study 4

Mill: B

Grade produced: Medium Liner

Capacity: 8500 MT/month

Enzyme used: Epygen Papyrase Drainage enzyme

Enzyme dosage: 300–500 ppm

Enzyme dosing location: As shown in flowchart, Figure 4.1.10

Trial duration: Three days

Middle layer furnishes: OCC

In this case, drainage enzymes resulted in significant improvement in paper formation while producing medium liner owing to better dilution in head box, without any drainage constraint in the two-wire paper machine. The improvement in formation pattern during the trial period is showcased in Figure 4.1.11.

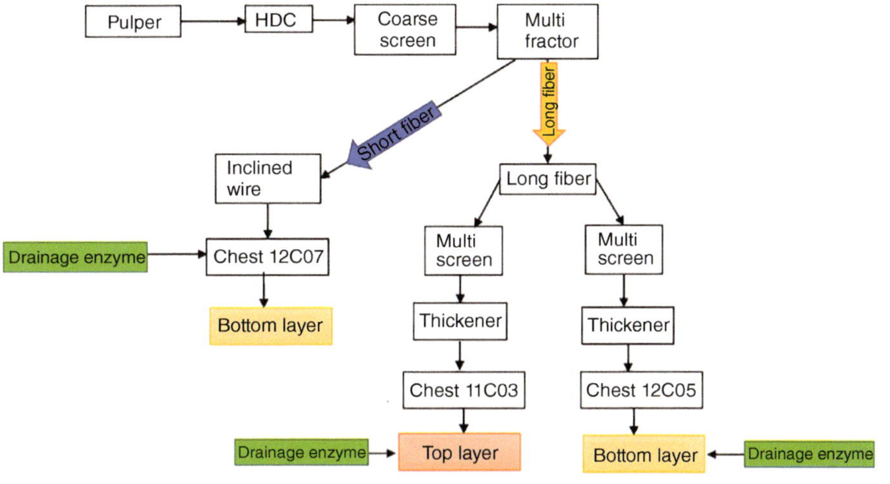

Figure 4.1.10 Flow chart of stock preparation.

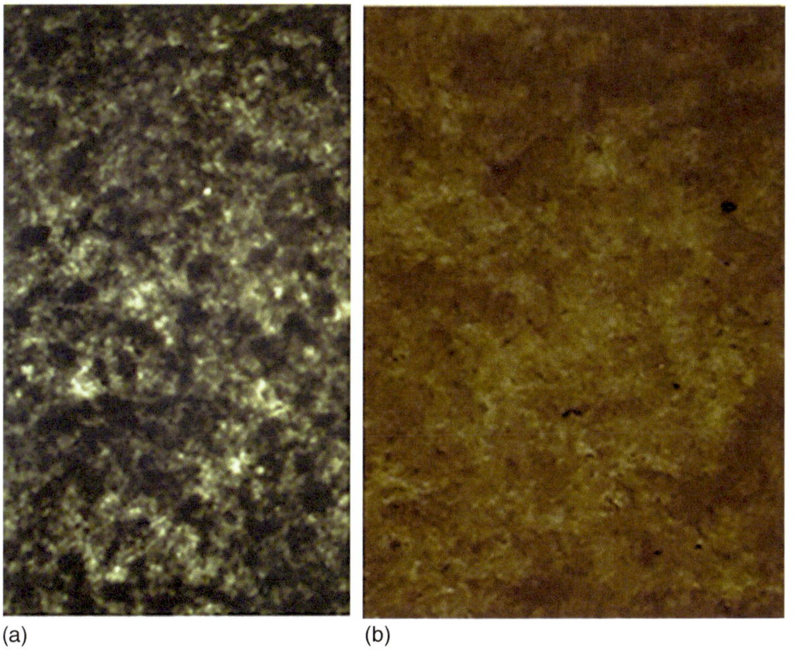

Figure 4.1.11 Paper formation before (a) and after (b) drainage enzyme usages.

4.1.3 Stickies Control Enzyme

Fluff, dust, lint, and stickies pose one of the biggest challenges in using recycled fiber in papermaking [14]. Fluff and stickies cause product quality issues, such as holes and specks in paper, often leading to paper rejection. Furthermore, the buildup of black and sticky or gummy deposits on the wire, felts, and dryer fabric leads to downtime of manufacturing equipment [15, 16]. Fluff or lint along with stickies creates also serious issues at the dryer section as the quality of raw materials deteriorates due to higher times of recycling. While doctoring the cylinders, it is essential to keep the cylinder surface clean; accumulation of fluff and stickies or a combination of these two at doctor blades becomes a serious problem. Owing to the washing of fabrics and cleaning of doctor blades, frequent interruption of the paper making operation becomes necessary to maintain good runnability at various stages of the paper making process, for example, at paper machines, winders, converting machines, and box making machines. All of these problems cause loss of production time, resulting in higher cost for the papermaker and converter. It is often difficult to estimate the exact loss caused by stickies deposit, as other factors (such as fluff or lint) also result in interruption of the operation [17, 18].

Typically, a stickies sample could consist of cellulosic materials, extractable organic substances, and inorganic components, e.g. talc, calcium carbonate, and silicate. It is therefore relevant to understand stickies and fluff or lint issues independently and then find an innovative solution for this compound and complex problem called "Stickies." Lint is usually defined as poorly bonded surface material from the surface of the paper sheet. Loosely bonded materials are, for example, fiber, fines, ray cells, fillers, and vessels.

Lint control strategies can be broadly categorized into three. One commonly used practice is reducing dryer temperatures in the first few dryers and ensuring clean smooth dryer surfaces to minimize surface disruptions and hence dusting. Increased press loadings help reduce dusting. The second option is a chemical route in which retention aids and other lint control additives are added to improve retention and thereby control lint or dusting. Internal sizing, starch addition to the wet end, defoamers, and, more recently, specialized roll release aid chemicals, which contain hydrocarbon materials, reduce surface disruption. However, one of the most effective strategies for lint control is to select furnish and develop the fiber in a desired manner with improved fines distribution.

Generally, the term "stickies" refers to sticky material in the recovered paper but excludes wood extractives. The most common sources of stickies are adhesives used in attaching labels, advertisements, CDs, and any other additional material in newspapers and magazines, for binding catalogs, for envelope sealing, stamps, and many more; the other sources of stickies are ink binders and coating binders, i.e. latex. Wax used in carton for packaging is also a significant source of stickies in recycling of old corrugated containers (OCC) [19]. Another factor contributing to the fluff and stickies issue has been the ever-increasing closed water loop in paper mills. This leads to fines buildup in the system, resulting in fluff or lint issue at the initial dryers of the drying system. On top of it, the stickies issue gets compounded as some mills put back the recovered fiber/sludge from the white water system or mill effluent water treatment plant into the pulper or pulp system to achieve high yield.

To deal with the stickies issues, a chemical option exists, which uses additives (such as Talc) for fixation and detackification that cause less number of issues. Another way to address the problem caused by stickies is the mechanical route in which centricleaners are applied for cleaning and screening of pulp. Despite the application of these methods, a substantial amount of stickies is carried forward to the paper making process. The reason is that stickies are flexible in nature, which reduces the effectiveness of screening systems by allowing these contaminants to be extruded. In addition to this characteristic, the fact that stickies' specific gravity is close to that of water and fiber, which causes them to be accepted by cleaners that are designed to permit water and fiber into the system. Next to the chemical and mechanical options, hot dispersion is applied to make any thermolabile sticky melt and wash away [17, 18].

Unfortunately, even with the best of all of these methods, too many of the stickies are still present in the final product to cause paper defects and increase machine downtime and interruptions at the rewinder.

To further reduce the stickies issue, an enzymatic solution was developed for recycled hornified OCC pulp, which enhances swelling and fibrillation of hornified fiber by hydrolyzing the lignocellulosic components of the highly crystalline outer layer of this fiber, even in the absence of mechanical refining. Simultaneously, bringing about a unique synergy between de-crystallization of the lignocellulose and hydrolysis of selected esters, these enzymes improve machine productivity by minimizing lint and fluff generation and agglomeration of stickies. Based on merits, many liner making mills apply enzymes to deal with the stickies issue, which is exemplified by the results shown in the following case study of a mill.

4.1.3.1 Case Study 5

A typical waste paper processing is shown in Figure 4.1.12.

Paper Machine Details:

Machine capacity: 12 000 ton/mo

Grade: Fluting and test liner

Furnish mix: Mixed OCC

Machine type: Fourdrinier; Speed: 600 m/min

Size press type: Metering size press

Mechanical refiner: No

Hot disperser: Yes.

Results show improvements for the paper making process for pulp treated with enzyme as breaks at dryer and rewinder are reduced significantly. Fluff and stickies accumulation is reduced as well.

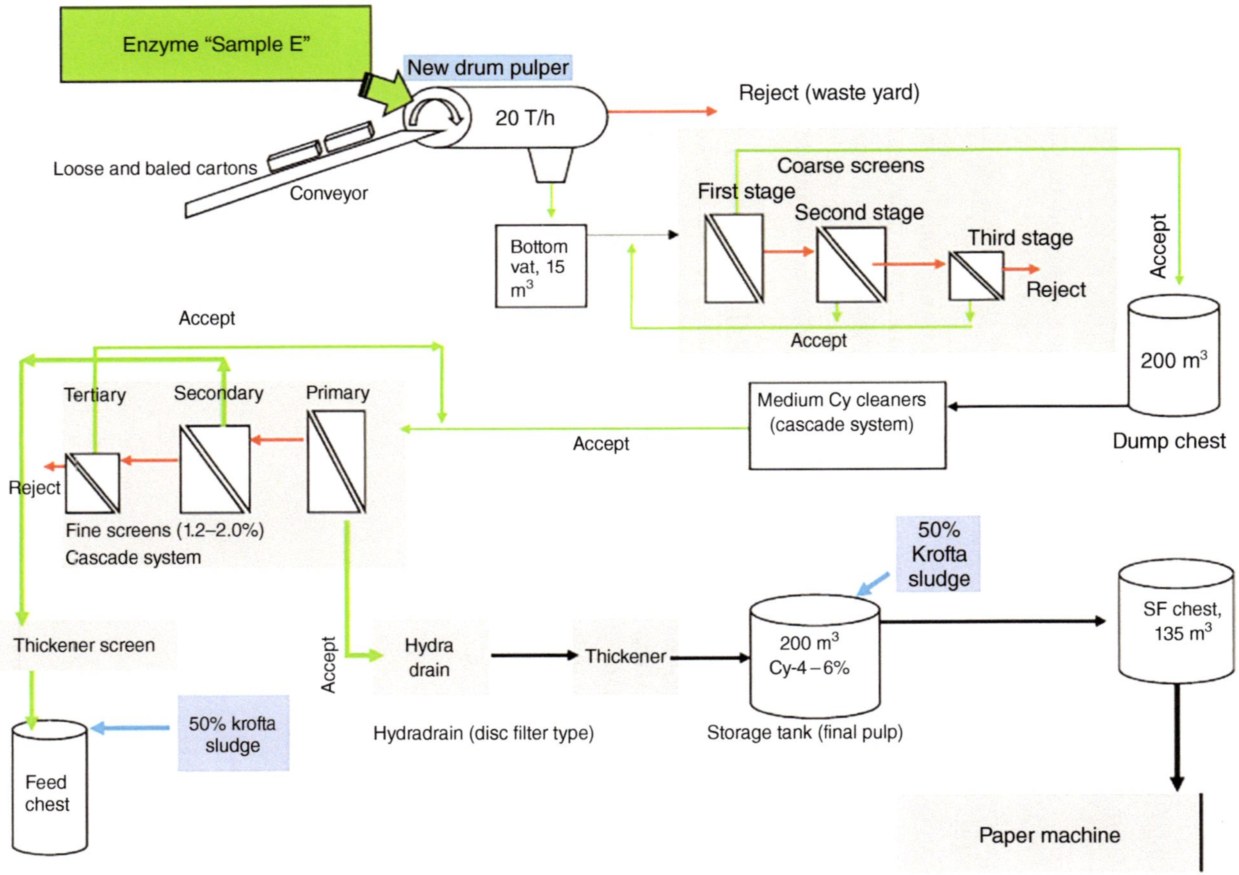

Figure 4.1.12 A typical waste paper processing plant.

Table 4.1.8 Reduction of paper breaks in dryers and rewinder in 112 GSM.

Sl no.	Time period	Paper quality	No. of paper breaks in dryers	No. of paper joints at rewinder	Percentage of paper breaks in dryers (%)	Percentage of paper joints at rewinder (%)
1	Blank	FL 112*	7	33	8.4	39.8
2	Enzyme	FL 112*	0	0	0.0	0.0

Table 4.1.9 Reduction of paper breaks in dryers and rewinder in 125 GSM.

Sl no.	Time period	Paper quality	No. of paper breaks in dryers	No. of paper joints at rewinder	Percentage of paper breaks in dryers (%)	Percentage of paper joints at rewinder (%)
1	Blank	TL 125*	23	8	19.0	6.6
2	Enzyme	TL 125*	8	0	9.0	0.0

Stickies breaks at paper machine dryers and rewinder:

(a) Results-01 (FL 112*: Fluting, 112 GSM) (Table 4.1.8)
(b) Results-02 (TL 125*: Testliner, 125 GSM) (Table 4.1.9)

Results of stickies and fluff weight collected from paper machine doctor blades (Table 4.1.10):

Table 4.1.10 Reduction of stickies and fluff accumulation at dryers.

Sl no.	Time period	Stickies and Fluff weight (kg)	
		First dryer group	Second dryer group
1	Day 1 (blank)	0.730	0.130
2	Day 2 (blank)	0.860	0.150
3	Day 3 (blank)	0.450	0.060
4	Day 4 (blank)	0.450	0.060
5	Day 5 (blank)	0.540	0.110
6	Day 6 (blank)	0.450	0.070
Average (blank)		*0.580*	*0.097*
7	Day 7 (with enzyme)	0.250	0.050
8	Day 8 (with enzyme)	0.210	0.045
Average (enzyme)		*0.230*	*0.0475*
Percentage reduction in stickies and fluff weight		60.34%	50.86%

4.1.4 Deinking Enzymes

While deinking is considered one of the most important steps in recycling paper, it is very difficult to remove contaminants, particularly ink. Although a conventional alkaline deinking process is still used by the majority of mills, they encounter difficulties with ink removal, depending primarily on the ink type, the printing process, and the fiber type. Nonimpact-printed papers are more difficult to deink and the quantity of such papers continues to grow as a proportion of the total volume of recovered paper. Similarly, the cross-linking of colored inks used in offset lithography or xerographic ink are also difficult to remove. On the other hand, newspapers printed with oil-based inks can be deinked relatively easily by conventional deinking processes. Many of the conventional deinking processes require large quantities of chemicals, resulting in high wastewater treatment costs to meet environmental regulations. Deinking processes are also substantial sources of solid and liquid waste [20, 21]. Enzymatic deinking, while increasing ink removal efficiency without using harsh chemicals, seems to be an effective solution to these problems. Enzymatic deinking produces white water (recycled process water) with a lower chemical oxygen demand (COD) than that from a conventional alkaline deinking process, thus reducing the load on wastewater treatment systems.

Enzymatic deinking involves dislodging ink particles from fiber surfaces and then separating the dispersed ink from the fiber suspension by washing or flotation. Enzymes attack either the ink or the fiber surfaces anchoring the ink, depending on the nature of the enzyme. Lipases and esterases can degrade vegetable-oil-based inks. Pectinases, hemicellulases, cellulases, and ligninolytic enzymes alter the fiber surface or bonds in the vicinity of the ink particles, thereby freeing ink for removal by washing or flotation [22, 23].

Different mechanisms have been proposed for enzyme deinking. Korean researchers pointed out that enzymes partially hydrolyze and depolymerize cellulose between fibers, freeing them from one another. Ink particles are dislodged as the fibers separate during pulping. Researchers also believe that enzyme treatment weakens the internal bonding of fibers, probably by increasing fibrillation or removing surface layers of individual fibers. It is suggested that extensive fiber hydrolysis may not be essential, since enzymes can remove ink under nonoptimal conditions by partially de-polymerizing micro fines attached to ink particles [24, 25].

A specially designed mix of cellulase and hemicellulases can sufficiently disrupt the fiber surface to release ink particles during pulping. It is also reported that cellulases peel fibrils from fiber surfaces, thereby freeing ink particles for dispersal in suspension. Often enzymes are deployed to smoothen the ink surface by hydrolyzing the fines attached to it, thus facilitating the release. Enzymatic treatment of nonimpact-printed paper has been consistently reported to remove material from ink particles, thereby increasing particle hydrophobicity and facilitating separation during flotation.

Over a period of time, the enzymatic deinking process has gained wide acceptance over conventional deinking, across mills and countries. Highlights of enzymatic deinking at one such mill is outlined below (Tables 4.1.11 and 4.1.12).

Table 4.1.11 Different chemicals dosages with and without enzyme.

Parameters	Blank	With enzyme
Furnish	45% Indian white cutting 45% POC (printed offset cutting) 10% Text books and magazines	45% Indian white cutting 45% POC (printed offset cutting) 10% Text books and magazines
Chemicals at pulper		
Caustic	0.50%	0
Sod Silicate	1.60%	1.20%
Surfactant	0.20%	0.20%
Pulper pH	9.0–9.5	7–7.5
Enzyme	0	0.01%

Table 4.1.12 Pulp brightness with and without enzyme at different stages.

Parameter	Blank	With enzyme
Pulper brightness (Br%)	61.7–65.2	62.7–65.2
Pulper pH	9.0–9.5	6.8–7.6
Pulper temperature	60 °C	55 °C
Deinking cell-I inlet Br%	59.2–61.7	59–62
Deinking cell-I outlet Br%	63–65	62–66
Gain across first cell	3.2–3.8	3.2–4.0
Deinking cell-II outlet Br%	67–70	66.5–70.2
Gain across second cell	4–5	4.5–5.2
Bleach tower brightness%	69–70	69–72
MC pump outlet Br%	70.6–72	69.6–72.8
HD tower brightness%	73.4–75	73–75
Final paper brightness%	75–76	75–76

4.1.4.1 Case Study 6

Paper Mill details: 200 TPD, writing printing paper

Furnish: Waste paper

Machine: Fourdrinier MF with size press

Make: Voith/Servall

Speed: 600 m/min

Deckle: 4.5 m

Grades: Creamwove, Maplitho

Deinking process: Voith (Double loop cell)

To summarize the benefits, the caustic was eliminated 100% during pulping. Sodium silicate was reduced by 33% during pulping. All other parameters were unchanged. The brightness gain across the flotation cell remained normal. Reduction in specks level was observed in both pulp and the final paper. Reduction in stickies was noticed, resulting in reduced paper breaks. The final paper brightness was maintained as normal. It is worth noting the substantial cost savings that were achieved from reduced chemical usage, higher yield, and improved machine performance by adopting enzymatic deinking.

4.1.5 Hardwood Vessel Breaking Enzyme

Hardwood pulp is used in the paper industry to produce a variety of products. Some of these are produced for the printing and book publishing segment. Vessels (non-pulpable pectin-rich material originating from connective tissue of plants) of hardwood create problem in printing, which is commonly known as "Vessel Pick." The printing press picks out large unbounded vessel elements on the surface of the sheet during the printing operation. This results in the ink not being applied to all parts of the paper where it is intended to be applied.

So far, vessel-picking problem has been addressed using sizing, coating, or refining technologies [26, 27]. The first two approaches have been unsuccessful in combating this problem. Refining tends to yield better results, though not sufficient, but requires significant amount of energy and capital, since dealing with vessel pick has become an issue that is not only costly, both to tolerate and to prevent, but often also goes unaddressed [28, 29]. Needless to mention here that this unaddressed issue hinders press room efficiency and causes substantial financial losses not only to the printer but also to the paper mill.

The objective of introducing a specially designed vessel breaking enzyme in bleached pulp is to selectively soften the hardwood vessel structure, rendering it susceptible to breaking under normal mill refining. Owing to selective hydrolysis of xylan linkages on vessel structure, vessel elements become susceptible to being disintegrated under normal mill refining. Once bigger and wider vessels are broken down to a smaller size, these are easily trapped by fiber network during paper formation. This phenomenon would result in substantial reduction in the vessel pick problem [30].

Laboratory experiments were carried out with an industrial ECF bleached kraft pulp. Pulp was produced from eucalyptus with little amount of casuarina. These experiments confirmed that enzyme-treated pulp, once refined by PFI, generates more broken vessels, establishing enzyme efficacy. Microscopic and fiber tester image study shown in Figure 4.1.13 further confirms these results.

4.1.5.1 Fiber Tester Image Analysis

Figure 4.1.14 shows fiber tester images of vessels for blank sample. Here, most of the vessel elements are unbroken. The number of broken vessels is 10 out of 36, which is about 27%.

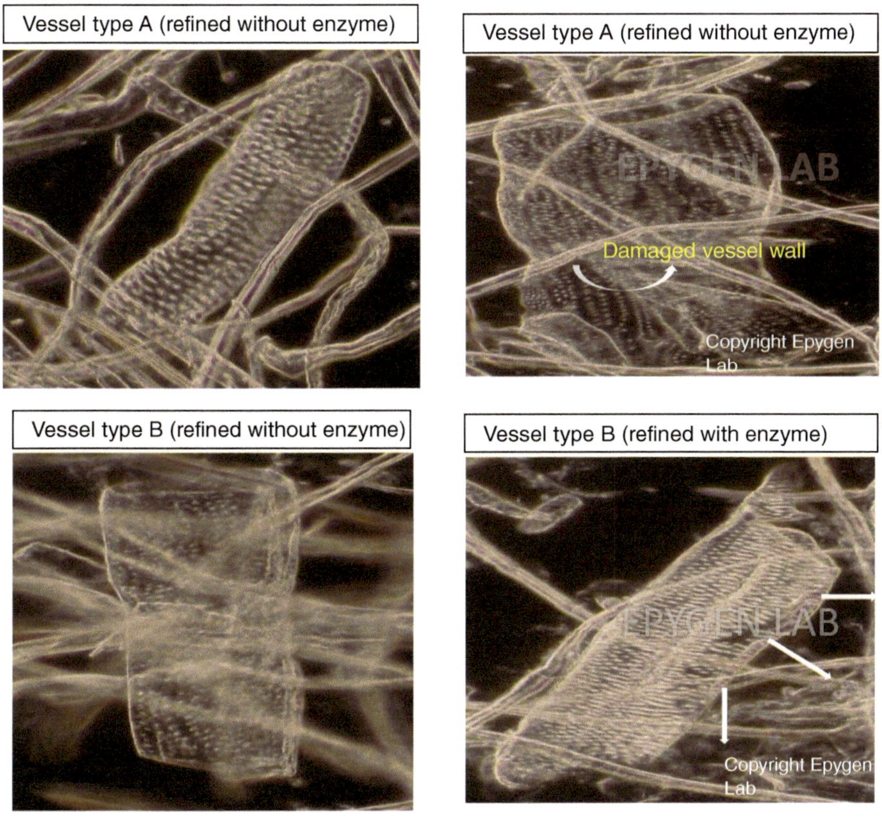

Figure 4.1.13 Visual analysis of microscopic pictures.

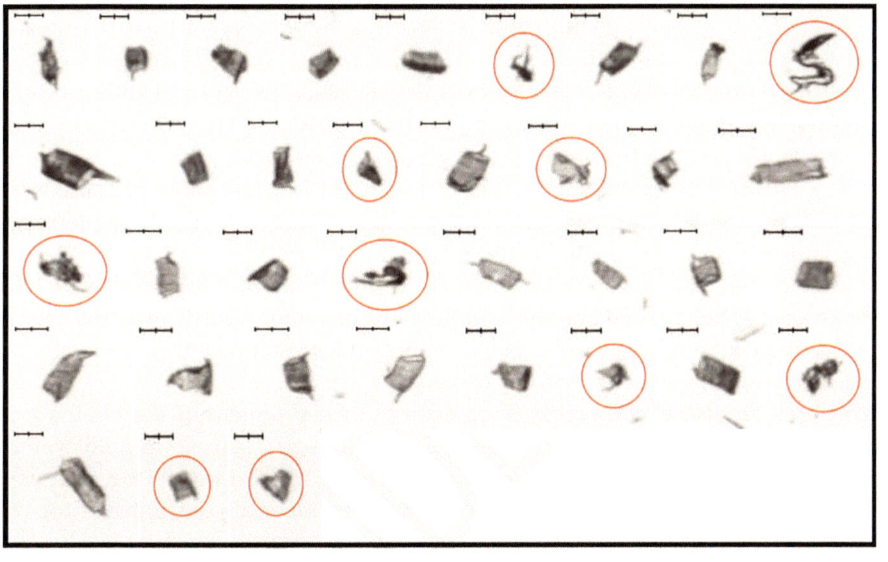

Figure 4.1.14 Fiber tester images of vessels for blank sample.

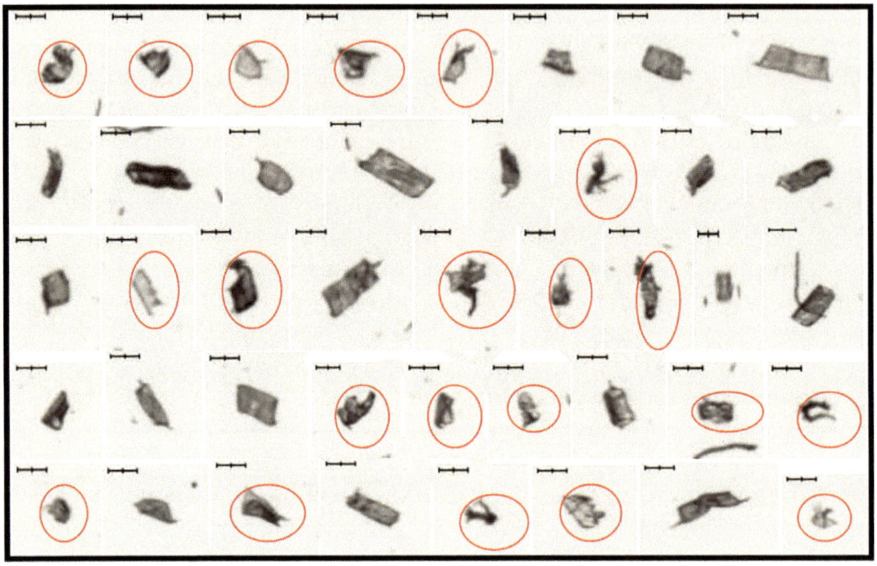

Figure 4.1.15 Fiber tester images of vessels of enzyme-treated sample.

Figure 4.1.15 shows that once the pulp sample is treated with enzyme, the number of broken vessels was 21 out of 42, which is about 50%.

In one mill application of vessel breaking enzyme, a significant reduction in vessel pick was observed, which was expressed by IGT Pick count. Vessel pick count reduced to 3 to 5 from 20 to 25 once the pulp was treated with enzyme.

4.1.6 Native Starch Conversion Enzyme

Starch is used in the pulp and paper industry in bulk quantity for various purposes. Usually, oxidized starch is being used for surface application in paper through size press. However, in recent times, in place of oxidized starch, mills use native starch resulting in substantial cost saving. Native starch is cooked and converted to a starch solution suitable for surface application.

Starch is a highly organized mixture of two carbohydrate polymers, amylose and amylopectin, which are synthesized by plant enzymes and simultaneously packed into dense water-insoluble granules.

Starch amylose is primarily a linear chain of glucose units. Amylose chains can coil into double helices and become insoluble in cold water. Amylose forms about 20–30% of the starch structure. D-Glucose molecules are linked together to form a large linear chain collectively to form amylose [31].

Amylopectin also is composed of chains of glucose units, but the chains are branched. This branched structure renders amylopectin soluble in cold water. The molecular architecture of the amylopectin and amylose within the granules is not entirely understood, but the granules are insoluble in cold water. It forms about 70–80% of the starch structure. D-Glucose molecules that are linked together to form a large branched chain collectively form amylopectin. These glucose molecules are linked together by α-1,4-glycosidic bonds in a linear manner and have α-1,6-glycosidic bonds in the branch chain. Each branch is attached to 20–30 units. The number of glucose units in amylopectin is 2000–200 000. These glucose molecules form long chains linearly as C1 attaches to other C4 and C1 and C6 as a branch (Figures 4.1.16 and 4.1.17).

Gelatinization of starch is the irreversible swelling of the granules when a suspension of the starch powder in water is heated. The temperature at which this occurs is called the gelling point. Uncooked starch has no water binding property. The binding power only develops during gelatinization.

Because of the larger size of the swollen granules compared to the size of amylose and amylopectin, the viscosity of the swollen granule mixture is much higher than the viscosity of the amylose/amylopectin mixture. In order to apply cooked starch on paper surfaces, swollen (gelled) molecules are to be broken down to reduce viscosity. Conventionally, strong oxidizing chemicals such as sodium peroxide or ammonium peroxide are applied to reduce viscosity at the gelling point. These chemicals are harsh and hazardous in nature and reportedly negatively affect starch molecules and quality. An alternative method to hydrolyze the starch αamylase enzymes can be used in place of harsh chemicals.

The native starch conversion enzyme is **EC 3.2.1.1** (commonly known as Alpha Amylase; BRENDA Comprehensive Enzyme Information System) or an endo amylase, which breaks the starch molecule down by acting on the 1,4 glucosidic bonds. This endo amylase hydrolyzes the 1,4-α-glucosidic bonds in amylose and amylopectin at random, resulting in rapid viscosity reduction of gelatinized starch.

Figure 4.1.16 Structure of amylose.

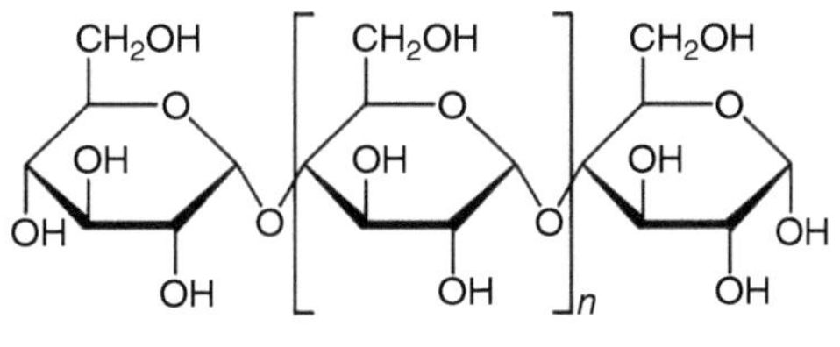

Figure 4.1.17 Structure of amylopectin.

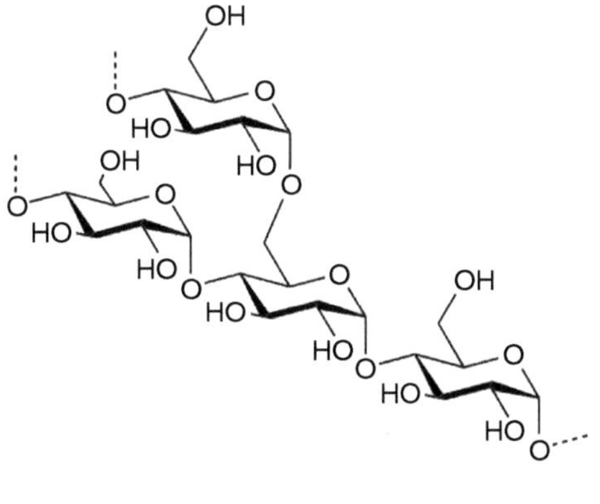

Owing to the distinctive advantages, enzymatic cooking is widely being practiced throughout the paper industry. Apart from cost benefits and being environment friendly, it is also reported to enhance paper strength properties in linerboard making. By adopting enzymes, mills also take advantage of starch cooking at higher solid concentrations. Recent developments of amylase enzyme could overcome earlier limitations of special requirement of enzyme storage and shelf life. Nowadays, these new generation enzymes can be stored at ambient temperature having a shelf life of more than one year.

4.1.7 Bleach Boosting Enzyme

In conventional kraft pulping, most part of lignin is removed from wood by cooking it in alkali liquor. However, after being washed with water, the cooked material still has some residual lignin. A multistage bleaching process is used to obtain a bright, stable final product by minimizing the remaining lignin [32]. Despite this process having many shortcomings, it is still the most cost effective, versatile, and efficient wood delignification method available.

4.1.7.1 Common Bleaching Agents

One of the main drawbacks of such bleaching process is its effluent containing high AOX levels (Table 4.1.13). Chlorine-based bleaching process contains several classes of toxic compounds, namely organochlorines. These compounds are formed when chlorine reacts with lignin in the first bleaching stage. The amount of organochlorines discharged from a pulp mill is closely related to the bleaching process used and, in particular, to the amount of chlorine used for bleaching [33].

Increased environmental awareness and demand for cleaner technology exert tremendous pressure to develop bleaching technologies with low AOX levels. During the last two decades, in response to these demands, various technologies are adopted to reduce or eliminate the use of chlorine and chlorine-based chemicals during bleaching. Application of enzymatic pre-bleaching has proved to be an attractive option for the paper industry, which has received considerable

Table 4.1.13 Common bleaching agents.

Sl no.	Bleaching agent	Bleaching chemicals	Symbols
1	Elemental chlorine	Cl_2	C
2	Alkali extraction	NaOH	E
3	Hypochlorite	NaOCl + NaOH	H
4	Chlorine dioxide	ClO_2	D
5	Peroxide	H_2O_2 + NaOH	P/E
6	Oxygen	O_2 + NaOH	O
7	Ozone	O_3	Z

attention. The enzymes used commercially in pulp bleaching are hemicellulases, which selectively affect the accessible hemicellulose fraction of the pulps. Among various commercially available hemicellulases, xylanases have been found to be the most effective as a pre-bleaching agent due to the following reasons:

- Enhances lignin extraction
- Allows for less cooking and preparation of higher Kappa pulps
- Reduction in use of chlorine and chlorinated bleach chemicals
- Reduction in discharge of AOX in bleach effluents
- Gain in final brightness and improved pulp properties

The ability of xylanases to facilitate the bleaching of kraft pulp was first reported in 1986 by the Finnish group led by Dr. Lisa Viikari. Crude xylanase preparations from *Aspergillus awamori* and *Streptomyces olivochromes* were shown to enhance the peroxide delignification of birch and (to a lesser extent) pine kraft pulp. Apparently, treating unbleached kraft pulp with xylanase does not, in itself, release significant amounts of lignin. However, the enzyme appears to render the lignin easier to extract in the subsequent bleaching steps.

The proposed mechanism accounts for the action of xylanases.

During the alkaline cooking stage, xylan gets dissolved. However, this dissolved xylan precipitates back on to the crystalline cellulose fibers as alkali concentration decreases. This re-precipitated, insoluble xylan hinders the process of solubilization of residual lignin. Selective hydrolysis of hemicellulose bonds near points of attachment between lignin and hemicellulose loosens up the pulp matrix, which facilitates penetration of bleaching chemicals, thereby improving the conversion of lignin to chloro-lignin, a soluble component [34, 35].

Since the effectiveness of xylanase enzyme mixes may vary with respect to activity, purity (particularly in terms of cellulase content), pretreatment conditions, and the type of pulps, evaluation studies for each type of enzyme are required to be carried out separately. Based on laboratory evaluation and plant applications, it has been reported to reduce chlorine demand to a level of 15–20% during bleaching with corresponding reduction in AOX level (20%) and improved pulp brightness to a level of 2–3% ISO.

Details of one such laboratory evaluation are outlined in Table 4.1.14. To evaluate the efficacy of bleaching enzymes, the parameters studied were Kappa reduction and brightness improvement of unbleached pulp, at different enzyme dosages. For this laboratory study, pulp sample was collected from the final unbleached washer outlet.

Results shown in Table 4.1.14 show a 38% and 17% reduction in Kappa number at 300 and 400 g/MT dosages of Epygen Papyrase bleach boosting enzyme respectively, when compared to control. Brightness improvement was also in line with expectation, 1% ISO and 2% ISO at 300 and 400 g/MT dosages respectively, when compared to control. Brightness was improved by 4.1% and 6.9% respectively with an enzyme concentration of 300 and 400 ppm.

4.1.7.1.1 Case Study 7

In one mill application, the reported benefits from bleaching enzyme application were significant as shown in Table 4.1.15.

Table 4.1.14 Lab evaluation of bleaching enzyme.

Parameter	Control	Set #1 (300 ppm)	Set #2 (400 ppm)
Pulp pH (PO1 O/L)	9.6	9.6	9.6
Pulp Cy% (PO1 O/L)	13.22	13.22	13.22
Amount of OD pulp taken	70 g	70 g	70 g
Cy% adjusted to	10%	10%	10%
Volume of 2N·H_2SO_4 added	6.5 ml	6.5 ml	6.5 ml
Pulp pH after acid addition	7.9	7.9	8.1
Dosage of Epygen Papyrase enzyme	**Nil**	**300 gr/MT**	**400 gr/MT**
Retention time at 65 °C	1 h	1 h	1 h
Retention time at room temperature	1 h	1 h	1 h
Pulp pH after retention	8.4	8.3	8.4
Washing of pulp	Done	Done	Done
Gentle disintegration of pulp	Done	Done	Done
Burette reading for K no.	34.8 ml	40.4 ml	37.3 ml
Kappa no.	**14.5**	**9.0**	**12.0**
Percentage of reduction in K no		**37.93**	**17.24**
Brightness%	**29.16**	**30.36**	**31.16**

Values in bold are so specific interest of the experiment.

Table 4.1.15 Benefits of bleaching enzyme application.

Description	Units	Before trials	During trials	Reduction in %
Raw chip	tons	4.01	3.7	7.73
WL charge	%	18.35	17.77	3.16
ClO_2	kg	28.33	22	22.35
NaOH	kg	24.22	17.42	28.08
SO_2	kg	0.005	0	—
O_2	kg	9.82	7.56	23.02
Xylanase	kg	0	0.185	

Mill produces eucalyptus pulp 1200 MT/d.
Analysis for 1 ADT final production (note calculations on wet chip):
To summarize, the benefits were as below:

- Pulp yield increases
- 22% reduction in chlorine dioxide consumption
- 28% reduction in sodium hydroxide consumption
- 23% reduction in oxygen consumption
- 11% reduction in COD for acid fiber saver
- 20% reduction in COD for alkali fiber saver
- Significant increase in viscosity readings

4.1.7.2 Overcoming Challenges Faced by Bleaching Enzymes in Pulp and Paper industry

Epygen Papyrase xylanase enzyme used in the above laboratory evaluation and mill application belongs to the category of genetically developed robust enzyme, designed to be stable at substantially higher temperatures and pH. These enzymes are active at 75 °C and above pH 9 for up to 1.5 h. This is a paradigm shift compared to earlier generation xylanases from mesophilic organisms. Earlier versions of xylanases used to rapidly lose activity at temperatures above 50 °C and at pH above 7. However, the ever-evolving bleaching sequence throws a highly challenging situation for bleaching enzyme to be effective at temperature and pH nearing 90 °C and 10 respectively. Therefore, to utilize bleaching enzymes in the most effective manner, the process might require cooling and pH adjustment. Understandably, it requires planning and would need suitable equipment or setup to embrace this enzymatic technology. The race is on for finding new enzymes, which will alleviate at least some of these problems. Research is in progress to produce xylanases by thermophilic organisms suitable at a pH of 10 and in boiling water.

4.1.8 Paper Mill Effluent Treatment Enzymes

Paper making process is one of the most water-intensive industrial production processes and hence it is needless to highlight the importance of the treatment of wastewater from paper mills. While regulatory authorities are tightening the norms of discharge water, people residing near or around paper mills are increasingly exerting pressure to treat mill effluent in the most desired manner [36]. At many places, rapid urbanization has stepped up further pressure on paper mills not only to maintain discharged water quality but also to maintain odor-free clean air. To meet these demands, mills are spending huge amounts in capital investments to upgrade their effluent treatment process or they are adopting various chemical routes. One of the most desired options is to utilize enzymes to improve the process efficiency [37].

Often, it is seen despite having all equipment of effluent treatment plant in place that the final outcome is not of the desired level due to inconsistent efficiency of bioreactors in aerobic or anaerobic processes. One of the main reasons is the nature of the wastewater of the pulp and paper industry, which often contains a complex mixture of solids and dissolved components, namely cellulose, hemicellulose, sulfur compounds, pulping chemicals, organic acids, chlorinated lignin, resin acids, and phenolics. Degradation of fibrous materials by aerobic/anaerobic bacteria is difficult. Hydrolysis of cellulosic debris is the rate-limiting step in digesters. During the regular operation of an effluent treatment plant, cellulosic materials enter into the anaerobic reactor or aeration chamber clogging up space for bacteria to colonize [38, 39], rendering bioreactors seriously compromised. Specially designed enzymes provide the necessary support to hydrolyze long carbohydrate polymers to readily degradable materials in aerobic environments. This speeds up hydrolysis of the waste material, improving the overall treatment efficiency and gas (methane) production by improving growth of activated sludge. This action brings significant COD reduction along with improved biodegradability of wastewater [40].

Wastewater treatment enzymes are robust fungal carbohydrases, comprising of a broad range of different enzymes for efficient hydrolysis of cellulose and hemicellulose. The enzymes operate between pH 5.5 and 8.5 and temperatures of 25–60 °C.

Laboratory studies are carried out by simulating conditions similar to industrial operations using wastewater, activated sludge, and compressed air. Experiments are run for 72 h. Samples collected at certain interval and COD are checked and compared with blank sample. During such laboratory studies, a 20–30% reduction in COD is reported (unpublished data).

4.1.8.1 Case Study 8

Mill application using Epygen Papyrase paper mill wastewater-treating enzyme results in outcome as shown in Table 4.1.16.

It may be noted that significant reduction of COD of river effluent was observed despite higher inlet COD at Bar screen.

Table 4.1.16 Mill trial of effluent enzyme.

	Blank			With enzyme			
Parameter	**Min**	**Max**	**Average**	**Min**	**Max**	**Average**	**Change in %**
TSS at bar screen	277.0	5370.0	2208.7	156.0	7050.0	3288.4	49
TSS at primary clarifier	15.0	95.0	42.1	20.0	97.0	47.6	13
MLSS (ppm) at V7	1706.0	2718.0	2187.8	2072.0	3094.0	2559.8	17
MLVSS (ppm) at DOD V7	1488.0	2286.0	1908.6	1824.0	2744.0	2247.6	18
TSS at effluent to river	9.0	34.0	13.4	6.0	34.0	10.6	−21
COD at bar screen	1015.2	8429.8	3111.8	1283.4	8090.4	3496.5	12
COD at primary clarifier	583.4	1263.1	915.3	573.6	1218.2	909.2	−1
BOD at primary clarifier	148.9	408.9	287.0	331.3	358.2	344.8	20
COD at effluent to river	32.7	160.7	97.4	70.2	112.7	80.9	−17
BOD at effluent to river	5.3	10.0	7.7	9.1	9.4	9.3	20

4.1.9 Slushing Enzyme

Despite paper products being biodegradable, sadly large amounts of paper end up at landfills. It becomes more painful when recycling centers show reluctance to collect a specific type of paper product, which paper mills do not accept because of their difficult pulp characteristic. For example, used paper cups, liquid packaging paper, cement bags, and paper core, though having long fiber or virgin fiber, are often faced with this problem. However, limited acceptability and attempts to recycle these types of paper products have been observed by a selective segment of paper mills despite the low cost, since actually the potential cost of pulping could be substantially high. In another situation, while producing high-strength paper from recycled furnish, mills often encounter hurdles at the pulping stage. Mills have difficulty recycling papers, e.g. strong kraft liner, paper having high amount of sizing chemicals and wet strength resins apart from sack bags, wax or PE-coated papers, mentioned earlier based on available pulping technology and quality demand of the final output. In dealing with these furnish, usually a drum pulper would throw higher rejects in the form of unslushed paper, leading to loss in yield and precious long fiber. On the other hand, hydro pulpers would take much more time and energy to produce suitable pulp [41]. Apart from higher time and energy, steam and sodium hydroxide are regularly being used as a pulping aid for these tough papers. Needless to mention, this means higher cost, higher effluent, and loss of fiber.

To aid the pulping process with the above paper products, a new enzyme system has been developed that helps to break the bonds of wet strength resins and waxy or PE layer, leading to faster hydration and subsequently ease of pulping. This enzyme is a ready to use liquid formulation, which can be used with or without dilution in the pulping process at pH between 6.0 and 8.0 and temperature between 25 and 70 °C.

Efficacy of this enzyme can be understood at the laboratory stage by soaking the sample paper at certain conditions and by measuring the increase in hydration compared to the blank sample. In addition, freeness can be measured and compared once the pulp is prepared and incubated under certain conditions. In one laboratory evaluation, it was observed that hydration of the target sample was increased by 20% after five minutes of soaking and freeness was increased by 15% compared to blank.

4.1.9.1 Case Study 9

The mill has a four-cylinder mold-former producing liner and fluting. It uses starch and dye as per quality requirement at the size press and maintains the top layer GSM at about 15–20% of the overall GSM and this layer furnish is from recycled cement bags. The top layer pulp is subjected to two refiners in series. The furnish of the other three layers is mixed OCC (Table 4.1.17).

Comparison of paper strength properties (Table 4.1.18).

Table 4.1.17 Mill trial of slushing enzyme.

Sl no.	Parameters	Blank	With enzyme
1	Pulper capacity	1.2 tons/batch	1.2 tons/batch
2	Raw material	Used cements bag	Used cements bag
3	pH	10–11	9–9.5
4	Temperature	40–45 °C	40–45 °C
5	Caustic	10 kg/ton	No caustic
6	Slushing time	120 min	85 min
7	Slushing enzyme	0	1.0 kg/ton of raw material

Table 4.1.18 Paper properties with slushing enzyme.

Sl no.	Parameters	Blank	With enzyme
1	GSM	138	134
2	CMT	107	106
3	RCT	141	142
4	Burst	184	182
5	SCT-MD	3.53	3.55
6	SCT-CD	1.73	1.76

4.1.9.2 Role of Enzymes in Pulp and Paper Industry – End Note!

Despite the fact that the pulp and paper industry plays a key role in the economy of many countries, the paper maker's bottom line seems eternally under pressure. For inherent reasons, the cost of production remains driven by the cost of raw material, i.e. cost of fiber accounting for 50% to 70%. In view of this, mills seem to be going out of their way in taking care while procuring raw material, be it pulp, secondary fiber, or wood. Surprisingly, when it comes to utilization of the fiber for making paper, it seems to adapt a less sensitive approach in developing the fiber suitable for this kind of paper making. Till recent times, mechanical refining was the only practiced tool for fiber development, and mills were compelled to rely on this operation despite several negative impacts and limitations that came along. In fact, a recycled mill could hardly use a mechanical refiner to redevelop fiber as the adverse effects often outweigh the benefits. In this challenged space, enzymes have been offered as innovative solutions for a wide range of paper mills and raw materials across the industry. As described earlier and demonstrated through various case studies, some of which have been cited, these new-age enzyme applications seem to be receiving wider acceptance. Based on heightened efforts on research and development of pulp and paper biotechnology enzymes, in the coming days, it could potentially bridge the gap between conservation of natural resources and utilization of the same in the best possible way. By harnessing the powerhouse of nature utilizing protein science and engineering, pulp and paper biotechnologists are bringing about a paradigm shift in approach of how fiber properties can be redeveloped, rather than being allowed to degenerate!

References

1 Mandels, M. and Reese, E. (1964). Fungal cellulases and the microbial decomposition of cellulosic fabric, developments. In: *Developments in Industrial Microbiol*, vol. 5. 5–20, Washington, DC: American Institute of Biological Sciences.

2 Kochavi, D., Videbaek, T., and Cedroni, D. (1990). *Text. Chem. Color.* 79 (9): 24.

3 Pommier, J., Fuentes, J.L., and Goma, G. (1989). Using enzymes to improve the process and the product quality in the recycled paper industry. *TAPPI J.* 72 (6): 187.

4 Mohlin, U.-B. and Petersson, B. (2002). Improved papermaking by cellulase treatment before refining. In: *Biotechnology in the Pulp and Paper Industry: Progress in Biotechnology*, vol. 21 (ed. L. Viikari and R. Lantto), 291. Elsevier.

5 Heise, O., Unwin, J.P., Klungness, J.H. et al. (1996). Industrial scaleup of enzyme-enhanced deinking of nonimpact printed toners. *TAPPI J.* 79 (3): 207–212.

6 Jobbins, J. and Franks, N. (1997). Enzymatic deinking of mixed office waste: process condition optimization. *TAPPI J.* 80 (9): 73.

7 US Patent 6,146,494 (Modified cellulosic fibers and fibrous webs containing these fibers).

8 US Patent 6,808,595 (Soft paper products with low lint and slough).

9 US Patent 5,811,381 (Cellulase compositions and methods of use).

10 (a) Wilson, D.B. (2008). Three microbial strategies for plant cell wall degradation. *Ann. N.Y. Acad. Sci.* 1125: 289–297. (b) Wilson, D.B. (2011). Microbial diversity of cellulose hydrolysis. *Curr. Opin. Microbiol.* 14: 1–5.

11 (a) Teeri, T.T. (1997). Crystalline cellulose degradation: new insight into the function of cellobiohydrolases. *Trends Biotechnol.* 15: 160–167. (b) Beguin, P. (1990). Molecular biology of cellulose degradation. *Annu. Rev. Microbiol.* 44: 2194–2248.

12 (a) Liimatainen, H., Haapala, A., Tomperi, J., and Niinimäki, J. (2009). Fibre floc morphology and dewaterability of a pulp suspension: role of flocculation kinetics and characteristics of flocculation agents. *Bioresources* 4 (2): 640–658. (b) Kalliokoski, J. (2011). Models of Filtration Curve as a Part of Pulp Drainage Analyzers. Acta Universitatis Ouluensis.

13 Liimatainen, H., Haavisto, S., Haapala, A., and Niinimäki, J. (2009). Influence of adsorbed and dissolved carboxymethyl cellulose on fibre suspension dispersing, dewaterability, and fines retention. *Bioresources* 4 (1): 321–340.

14 Doshi, M.R. (1991). Properties and control of stickies. *Prog. Paper Recycl.* 1 (1): 54.

15 Sharma, S., Kortmeyer, J.C., Spivey, A., and Lasmarias, V.B. (2002). A cost effective talc solution to stickies control in OCC pulps. *2002 Fall Technical Conference and Trade Fair*, San Diego, CA, USA.
16 US Patent 4,923,566 (Method of Pacifying Stickies in Paper).
17 Soung-Bae, P., Jung Myoung, L., and Tae-Jin, E. (2004). The control of sticky contaminants with enzymes in recycling of waste paper. *J. Ind. Eng. Chem.* 10 (1): 72–77.
18 Jones, D.R. and Fitzhenry, J.W. (2003). Esterase- type enzymes offer recycled mills an alternative approach to stickies control. *Pulp Pap.* 77 (2): 28–31.
19 Bossaer, P. (1999). The control of stickies by the means of dispersants and fixation aids. *1st Packaging Paper Recycling Symposium*, Grenoble, France.
20 US Patent 5,622,597 (Process for Deinking of Recycled Paper).
21 Shrinath, A., Szewczak, J.T., and Bowen, I.J. (1991). A review of ink-removal techniques in current deinking technology. *TAPPI J.* 74: 85–93.
22 Lee, C.K., Ibrahim, D., and Omar, I.C. (2013). Enzymatic deinking of various types of waste paper: efficiency and characteristics. *Process Biochemistry.* 48: 299–305.
23 Pathak, P., Nishi, K., and Singh, A. (2011). Optimization of chemical and enzymatic deinking of photocopier waste paper. *Bioresources* 6 (1): 447–463.
24 US Patent 6,767,728 B2 (Composition for Enzymatic Deinking of Waste Paper).
25 Kim, T.-J., Ow, S., and Eom, T.-J. (1991). Enzymatic deinking method of wastepaper. *TAPPI Pulping Conference Proceedings*, 1023–1027. Atlanta, GA: TAPPI Press.
26 Rakkolainen, M., Kontturi, E., Isogai, A. et al. (2009). Carboxymethyl cellulose treatment as a method to inhibit vessel picking tendency in printing of Eucalyptus pulp sheets. *Ind. Eng. Chem. Res.* 48: 1887–1892.
27 Jeffries, T.W. (1992). *Enzymatic Treatment of Pulps: Opportunities for the Enzyme Industry in Pulp and Paper Manufacture*, ACS Symposium Series, Chapter 18, vol. 476. https://doi.org/10.1021/bk-1992-0476.ch018.
28 Celso Foelkel (2007). Vessel Elements and Eucalyptus Pulps. Eucalyptuson line Book& Newsletter. ABTCP.
29 Mukoyoshi, S.-I., Komatsu, Y., and Ohsawa, J. (1986). Prevention of vessel picking trouble in tropical hardwood pulps, III Effect of pulp preparation on vessel picking. *Jpn. TAPPI J.* 40 (5): 51–57.
30 Ohsawa, J., Ohtake, T., Komatsu, Y., and Yoneda, Y. (1982). Prevention of vessel picking trouble in tropical hardwood pulps, I Effective vessel separation methods. *Jpn. TAPPI J.* 1982, 57 (10): 975–984.
31 Srichuwong, S., Sunarti, T.C., Mishima, T. et al. (2005). Starches from different botanical sources I: contribution of amylopectin fine structure to thermal properties and enzyme digestibility. *Carbohydr. Polym.* 60: 529–538.
32 Bajpai, P. and Bajpai, P.K. (1996). Realities and trends in enzymatic prebleaching of Kraft pulp. In: *Advances in Biochemical Engineering/Biotechnology*, vol. 56 (ed. T. Schepper), 1–31. Berlin: Springer-Verlag.
33 Bajpai, P., Anand, A., Sharma, N. et al. (2006). Enzymes in ECF bleaching of pulp. *BioResources* 1 (1): 34–44.

34 Call, H.P. (1999). New enzymatically mediated delignification and bleaching systems. *10th International Symposium on Wood and Pulping Chemistry – 10th Biennial ISWPC – Main Symposium*, Vol. 1, Yokohama, Japan, 540–545.

35 Kanosh, A.L. and Nagieb, Z.A. (2004). Xylanase and mannanase enzymes from *Streptomyces galbus NR* and their use in biobleaching of softwood Kraft pulp. *Antonie Van Leeuwenhoek* 85 (2): 103–114.

36 Eikelboom, D.H. and van Buijsen, H.J.J. (1981). *Microscopic Sludge Investigation Manual*. Delft: TNO Research Institute for Environmental Hygiene.

37 Jenkins, D., Richard, M.G., and Daigger, G.T. (1993). *Manual on the Causes and Control of Activated Sludge Bulking and Foaming*, 2e. Boca Raton, FL: Lewis Publishers, 2003, 3e.

38 Richard, M.G. (1989). *Activated Sludge Microbiology*. Alexandria, VA: Water Environment Federation.

39 Palm, J.C., Jenkins, D., and Parker, D.S. (1980). Relationship between organic loading, dissolved oxygen concentration and sludge settleability in the completely-mixed activated sludge process. *J. Water Pollut. Control Fed.* 52: 2484.

40 Walker, I. and Davis, M. (1977). The relationship between viability and respiration rate in the activated sludge process. *Water Res.* 11: 575.

41 Gangwar, A.K. (2016). Fiber from laminated cups. *BioResources* 11 (3): 5658–5659.

4.2

Enzymes in Vegetable Oil Degumming Processes

Arjen Sein [1], *Tim Hitchman* [2,*], *and Chris L.G. Dayton* [3]

[1] DSM Biotechnology Center, Alexander Fleminglaan 1, 2613 AX Delft, 2600 MA, Delft, the Netherlands
[2] DSM Food Specialties USA Inc, South Bend, IN 46628, USA
[3] Bunge Limited, 50 Main Street, Sixth Floor, White Plains, NY 10606, USA

4.2.1 Introduction

In 2016/17, the overall production of vegetable oils and fats was around 189 million metric tonnes (MMT) (Figure 4.2.1) [1]. The largest volume was from production of palm oil (a fruit oil), and more than 50% of the remaining volume was accounted for by seed oils, i.e. soy bean, canola/rapeseed, and sunflower oils. Oils and fats from vegetable or animal origin are used in many food and non-food applications. These include direct applications such as cooking and salad oil, but also indirect food applications such as margarines, dressings and emulsified sauces, bakery applications, chocolate and confectionary, dairy products, and ice cream. A major non-food application is biodiesel, accounting for 30% of the total domestic use of soy bean oil in the United States (2016/17) [2]. In addition, vegetable oils are used for cosmetics and pharmaceutical applications and can be a good renewable source for oleochemicals [3] and detergents [4].

In this chapter, the focus will be on enzymatic degumming (EDG) and refining of seed oils. Crude oil is obtained from seeds via solvent extraction and/or pressing. Crude oil contains minor components such as phospholipids and free fatty acids (FFAs) that some may find objectionable in food applications or hinder their use in non-food applications. The process of removing the minor components in a crude oil is called refining. During the refining process, some of the target neutral oil product is lost to the various refining steps, reducing the overall oil yield and costing the refiner a portion of profits. Nevertheless, removal of the "impurities" may be necessary to meet product or process requirements. Thus, oil processors may adopt processing methods to minimize losses.

The first major processing step is the removal of phospholipids or "degumming," which accounts for a significant proportion of total oil losses in refining. Use of enzymes as processing aids can greatly reduce the oil loss during degumming and refining. Application of these phospholipases in the degumming

* Tim Hitchman is no longer with DSM and can be contacted at thitchman@gmail.com.

Industrial Enzyme Applications, First Edition. Edited by Andreas Vogel and Oliver May.

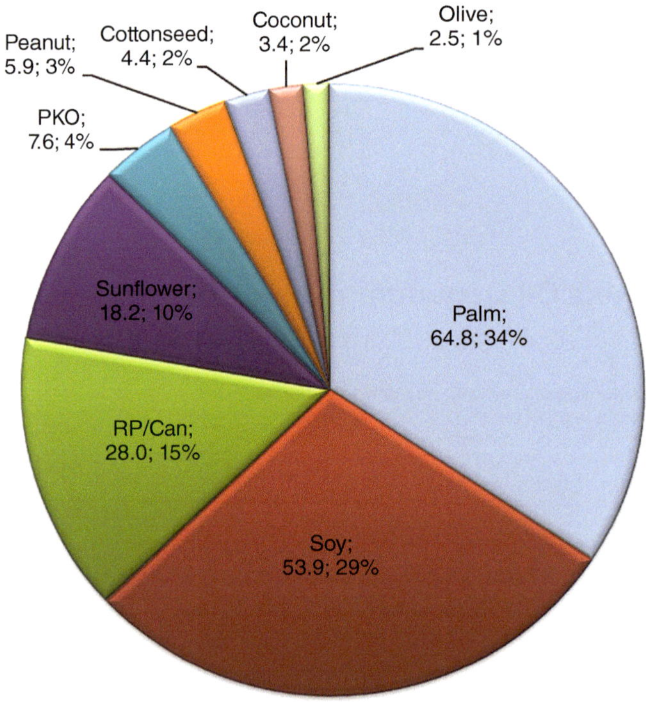

Figure 4.2.1 The 2016/2017 overall production of vegetable oils [1].

process leads to considerable yield improvements and cleaner processes with reduced use of chemicals [5].

In order to understand the role of enzymes in oil processes, it is first necessary to describe the general process for seed oil production, focusing on standard water degumming, and the chemistry involved. Then the role of phospholipases in the degumming process will be detailed, starting with description of the enzymology of the phospholipase application, followed by describing how improvements in yield and process efficiency can be obtained, and other implications for industrial processes. The chapter will close with a section on other enzymatic processes used in vegetable oil processing and some potential future applications.

4.2.2 General Seed Oil Processes

The general flow scheme of vegetable oil processing is given in Figure 4.2.2; more details can be found in the literature [6].

Soy beans contain about 19–20% of oil, based on total dry matter, the remainder being mostly protein and carbohydrates. In the vast majority of cases, the crude oil is separated by solvent extraction. The beans are mechanically pretreated to improve the extraction efficiency, by either flaking leading to thin flakes, or expanding (extrusion), which leads to more porous material (collets) with a higher surface area and hence a higher extraction efficiency.

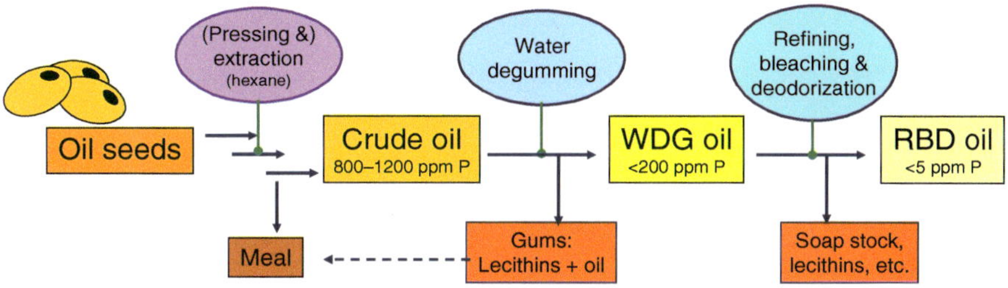

Figure 4.2.2 General flow scheme of seed oil processing

The crude "flake oils" and "expander oils" resulting from two processes differ in composition, specifically in their phospholipid content (Table 4.2.2).

The oil content in rapeseed/canola and sunflower seeds is around 40%. Typically, at first a pressing step releases roughly 2/3 of the oil, and the remainder is obtained by solvent extraction of the press cake. Canola is bred from rapeseed, leading to a plant lower in glucosinolates and oil that is low in erucic acid (C22:1ω9), which was present in large levels in the original seed. The varieties of rapeseed planted in the rest of the world for food consumption are also low in erucic acid. Generally, canola oil is the North American name for the same species. The pressed oil fraction is relatively low in phospholipids (1 wt% or less, and rich in phosphatidic acid; see Table 4.2.2), while the extracted oil fraction has a higher phospholipid content (2–3 wt%, and rich in phosphatidyl choline; see Table 4.2.2). These two fractions may be processed separately, but are usually combined before further processing, commonly referred to as the "crude" oil. Similarly, crude sunflower oil exists, but that is low in phospholipids, less than 1 wt%.

In addition to the seed of origin and oil extraction process, conditions such as geographical source of beans, weather conditions during growth and harvest, storage conditions, and upstream processing conditions can also affect the composition of the final crude oils.

4.2.2.1 Phospholipids

Phospholipids are polar lipids: molecules capable of interacting with polar (e.g. aqueous) phases due to a hydrophilic headgroup and nonpolar (e.g. neutral oil) phases due to a hydrophobic tail. Owing to their amphiphilic nature, phospholipids are excellent emulsifiers [7]. Phospholipids comprise the majority of polar lipids in crude vegetable oils (see Figure 4.2.3), but smaller amounts of glycolipids (e.g. sterol glucosides and mono- and di-galactolipids) are also present.

The phospholipids commonly found in seed oils vary in their polar, hydrophilic headgroup, and in the composition of the hydrophobic moiety, consisting of two fatty acid chains. The fatty acid composition simply reflects the composition of the source oil. The structures of the four most common polar headgroups of phospholipids are shown in Table 4.2.1, and the overall structure of these phospholipids is shown in Figure 4.2.3 [7–9].

The polar headgroup consists of a phosphate group with a further polar moiety attached, as indicated in Table 4.2.1.

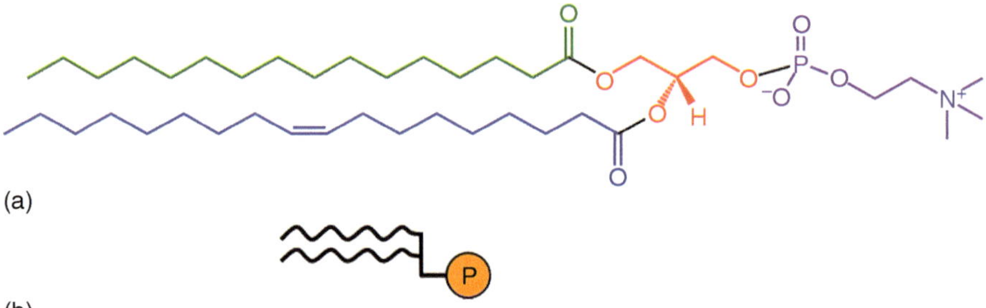

Figure 4.2.3 (a) Molecular structure of phosphatidyl choline (here L-α-1-palmitoyl-2-oleoyl phosphatidylcholine; POPC; PC (16 : 0/18 : 1 (9Z)) taken from Lipid Maps [8]) – purple, the phosphocholine headgroup; red, the glycerol backbone; green, the fatty acid on the sn1 position; blue, the fatty acid on the sn2 position. (b) Cartoon representation of a phospholipid.

Table 4.2.1 Overview of the four most common phospholipids.

Phospholipid name (abbreviation)	Headgroup structure
Phosphatidyl choline (PC)	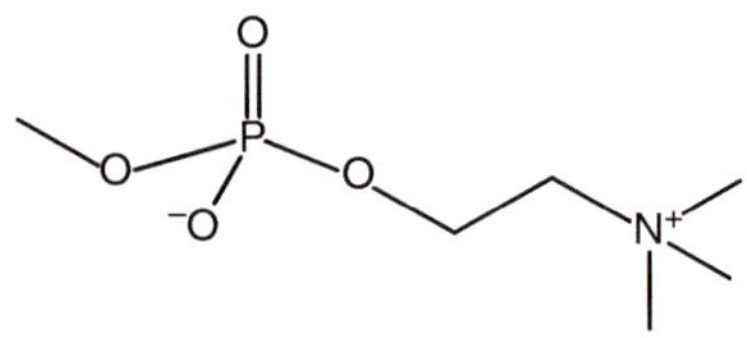
Phosphatidyl inositol (PI)	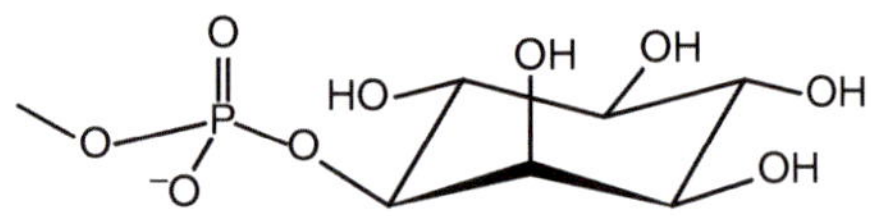
Phosphatidyl ethanolamine (PE)	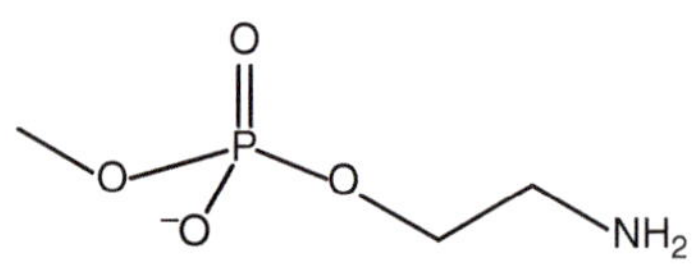 [or NH_3^+]
Phosphatidic acid (PA)	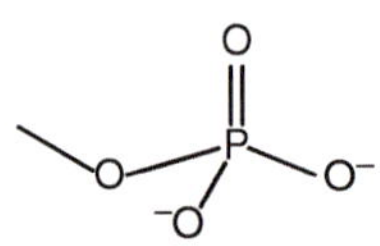

The total amount and composition of the polar lipids can be highly variable, and are impacted by a variety of influences as described for the extraction process. This is important to understand, because the polarity of the phospholipids is correlated with the ease of removal in the degumming process. Another key factor in ease of removal is the extent to which the polarity of the headgroup is modified by binding to metal cations, specifically calcium (Ca) and magnesium (Mg) ions. The approximate phospholipid and metal compositions of high volume seed oils are given in Table 4.2.2. The table is based on a large data set obtained over the years by Verenium and DSM. Atomic phosphorus (P) and metals (M) were measured

Table 4.2.2 Phospholipid composition of several common seed oils.

	P (ppm)	M[a] (ppm)	P/M					
	Range	Ca + Mg + Fe		PC	PE	PI	PA	Lyso PL
Expander soy	800–1200	160–250	5.6	37	24	20	13	6
Flake soy	600–1000	110–210	4.8	33	26	18	18	5
Rapeseed	400–800	160–300	2.3	28	16	17	33	9
Canola	400–900	180–440	1.9	32	14	20	28	6
Sunflower	200–400	60–200	2.3	43	16	2	16	—
Pressed rapeseed	200–400		1.0	6	13	4	71	6
Extracted rapeseed	600–1200		2.1	25	16	16	32	11

Individual phospholipids are given as percentage of total amount of phospholipids, measured by ^{31}P NMR.
Values given are obtained from 500 crude soy bean oils, 40 crude canola oils, 20 crude rapeseed oils, 7 sunflower oils, and 7 pressed and 7 extracted rapeseed oils.
a) M is metal ions. Large variations seen, especially for Ca.

by inductively coupled plasma (ICP). The amount of phospholipids is commonly expressed in terms of atomic phosphorus (P). This measure of phosphorus includes not only the phospholipids but also free phosphates, and may result in an overestimate of the actual amount of phospholipids present in oil. The table also mentions the P/M ratio, which is an indication for oil quality, as will be explained below. More accurate methods to quantify and characterize the phospholipid levels are ^{31}P NMR [10] and HPLC [11]. The values given in the table of phospholipid composition were measured by ^{31}P NMR in that same large set of samples. The numbers in the table should not be seen as absolute but can be compared relative to each other to show what the differences in profiles are. There can be substantial variation in the numbers, especially for the canola, rapeseed, and sunflower oils. As these numbers are measured in the unrefined oils, the values differ from what is commonly reported as phospholipid composition in lecithins, which are processed from gums obtained from a water-degumming process [7].

4.2.2.2 A Molecular View of the Degumming Process

During the water-degumming process, the addition of water to the oil results in hydration of the hydrophilic headgroups of the polar lipids. The hydrated polar lipids associate and form a dispersed phase that due to its higher density can be removed by centrifugation. This heavy phase is gummy or slimy by nature and is hence called the "gum phase." This process is, however, not optimal for separation of polar lipids from the neutral oil:

i. The amphiphilic nature of the polar lipids causes neutral oil to be "dragged" to the interface between the gum and the oil phases. Owing to the crude nature of the industrial process this results in separation of oil along with the phospholipids in approximately a 1 : 2 ratio, the so-called neutral oil loss (NOL).

Typically, this is assumed to be a result of the formation of emulsion in the gum phase, but molecular evidence suggests otherwise (see below).

ii. The varying polarity of the headgroups of phospholipids impacts the ease with which they are hydrated. As a result, not all polar lipids are effectively "brought" to the heavy gum phase and stay behind in the oil.

iii. Water-degumming is a relatively low-tech process, and where the coalescence of the hydrated polar lipids and water with the majority of the "heavy phase" is not achieved, these remain behind in the oil.

The various phospholipids are not all equally hydrated. The amount of water that is attracted by the hydrophilic headgroup depends on the structure of the headgroup and on environmental factors, such as the following:

- The pH of the aqueous surroundings. At low pH, all phosphate groups will be protonated; at high pH, all phosphate groups will be de-protonated. At neutral pH, PC and PE will be zwitterionic, and PI and PA will be anionic.
- The ionic strength of the aqueous surroundings, with additional acid and caustic pretreatments (see below); salting out conditions can be created competing with the phosphate headgroups for water.
- The total amount of water present, relative to the phospholipids.
- The presence of polyvalent cations such as calcium, magnesium, or iron ions. These ions will associate with the phosphate group, making the groups in total less hydratable. Phosphatidic acid has the highest preference for metal ions, followed by PI and PE when high levels of metal ions are present.

All these factors influence each other. The association of phospholipids in bilayers as occurs in aqueous environment (see below) influences the state of hydration too. The level of hydration is also reflected in the efficiency of removal of the phospholipid subclasses: the better the hydratability, the better their removal. Practically, the order of hydratability is PC > PI > PE > PA > metal salts of PA, as was once observed by the relative rates of hydration of specific phospholipids in oil at 80 °C [12]. Studies on hydratability of specific phospholipids in biomembranes or in binary phospholipid–water phases confirm this trend: [13, 14] PC is the most hydratable and can bind twice as many water molecules as PE [15]. One reason might be that the PE can form a cyclic structure by intramolecular hydrogen bonding between the P=O and N—H [16–18]. According to molecular dynamics simulations, PI is even more hydratable than PC [19]. The headgroup of PA is the smallest of all; therefore, hydration is expected to be the lowest of all four major phospholipids.

A closer look at the hydrated phospholipid molecule shows that it has a roughly cylindrical shape, the diameter of the hydrated headgroup being about the same as the cross section of the apolar moiety. This molecular shape of the polar lipid determines its aggregation behavior, as described by the "packing parameter concept." [16, 20]. On detailed view, subtle variations exist between the molecular shapes of hydrated PC, PE, PI, and PA but a mixture of these four will on average have cylindrical molecular shape (packing parameter of around 1). The consequence of this cylindrical shape is that the molecules will form a lamellar liquid-crystalline phase: stacking of planar, bilayer-like lamellae with headgroups facing the water and fatty acids tails pointing inward. This is called *liquid*-crystalline, because intramolecular mobility represents a liquid state such as cis–trans transitions in the fatty acid tails, molecular rotations, high molecular mobility within the bilayer, lateral mobility in the monolayer half, on the interface, and flip-flop from one side of the bilayer to the other [13].

For PA with its small headgroup – even in hydrated state – this packing parameter will be even larger than one, leading to inverse hexagonal liquid-crystalline structures as can be seen in the water–PA phase diagram [21], showing occurrence of inverse structures at water levels higher than 50% (w/w).

Lysophospholipids have a roughly truncated cone form [16], as the hydrophobic moiety contains now only one fatty acid tail. On their own, such molecules would rather form cylindrically-shaped aggregates, also called worm-like micelles, in excess water and no oil. However, mixed with excess double-tailed phospholipids, bilayers will be formed, but now these bilayers tend to curve to accommodate for the molecular structure.

Thus, during oil degumming, when water is added to crude oil containing phospholipids in roughly equal amounts to the amount of phospholipids, the hydration of phospholipids promotes formation of bilayers, and the resultant lamellar liquid-crystalline phase. In this structure, water will be present as layers between headgroups, and oil will be entrained between the tails of the phospholipids [13], as graphically represented in Figure 4.2.4. These structures can be visualized by light microscopy, as shown in the same figure. Especially in the crossed polar settings on light microscopes anisotropic phases appear by their birefringence. This pattern (Figure 4.2.4b, top) is typical for a planer lamellar liquid-crystalline phase [22]. Molecular dynamics simulations further illustrate the same microstructural arrangement [19].

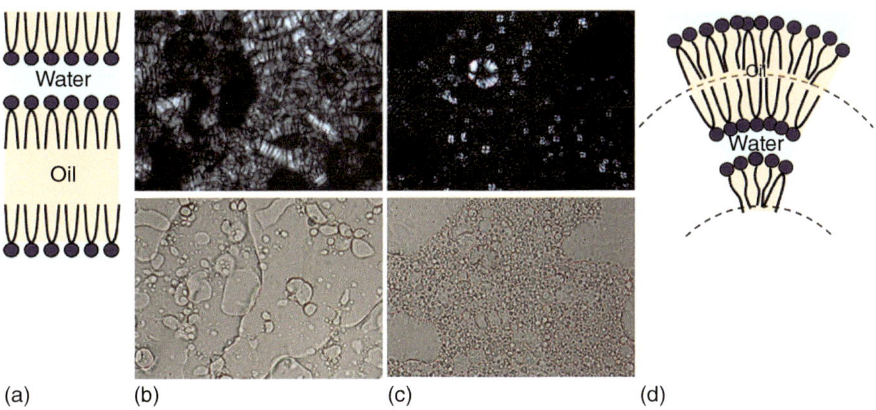

Figure 4.2.4 (a) Cartoon of lamellar liquid-crystalline state of phospholipids during the oil degumming state; (b) light microscopy image of dispersed phase in oil, bottom bright field, top crossed polars, showing planar lamellar liquid-crystalline arrangement; (c) microscopic images of the dispersed phase after treatment with Purifine 3G, showing highly curved lamellar liquid-crystalline arrangement; (d) cartoon representing the state of the polar lipids left after treatment with Purifine 3G (see below).

The thickness of the water layer depends first on how much water is present relative to amphiphiles and further on how well the polar lipids are hydrated and on the repulsive forces between the layers. The overall charges of the phospholipids and the ionic strength of the water phase determine these repulsive forces: the more ions dissolved in the water layer, the lower the repulsion between the bilayers and the thinner the water layers.

Metal ions will most preferably associate with PA, and such structures tend to form inverse structures, as is also observed in the crystalline state [23]. These might average out in the planar liquid-crystalline state, but as in practice phosphatidic acid is not removed with the degumming but remains present in the oil; it is expected that the PA with associated metal ions does not participate in the lamellar liquid-crystalline state, and instead forms separate non-hydrated inverse structures in the oil phase.

In essence then, water degumming of crude oils is a process whereby phospholipids are hydrated, forming a lamellar liquid-crystalline phase that is dispersed in oil. That dispersed phase can, due to its higher density, be removed by centrifugation (or sedimentation). However, as neutral oil is entrained between the phospholipids, much oil is lost during this process step. This amount of neutral oil can be about two-thirds of the total weight of polar lipids removed, running up to over 2% of the original amount of crude oil. Furthermore, not all the phospholipids are removed in this processing step.

This conclusion diverges from the conventional model of phospholipids acting as emulsifying agents, forming a water-in-oil emulsion with phospholipids only on the oil–water interface. If the amount of water would be far higher, such an emulsion would indeed be formed. In the degumming process, however, with about equal amounts of polar lipids and water in the crude oil, a dispersion of lamellar liquid-crystalline phase in oil is formed, confirmed by microscopy (see Figure 4.2.4b).

4.2.3 Enzymatic Degumming

Enzymatic degumming (EDG) is a relatively recent adaptation of the standard water-degumming process that offers the benefit of increasing the efficiency, i.e. reducing losses, during the degumming step. The enzymatic process variation uses phospholipases to separate polar and nonpolar moieties of phospholipids, thereby improving the removal of the polar components and reducing the amount of oil that is entrained in the gum phase [5]. In an EDG process, the enzyme is added to the oil with the hydration water, followed by incubation for an extended period (one to eight hours) to allow for conversion of the phospholipids. After the reaction, the aqueous gum phase, with the enzyme and water-soluble and polar reaction products, can be removed by centrifugation.

In general, there are two types of phospholipase classes used for EDG – see also Figure 4.2.5 – and are explained in detail below:

- Phospholipase C (PLC) enzymes convert the phospholipids into a diglyceride (DAG) and a phosphoryl moiety.
- Phospholipase A (PLA1 and PLA2) enzymes yield lysophospholipids and free fatty acids (FFA).

Two other classes of phospholipases are known, which are not used in industrial processes today. These are phospholipase B (PLB) and phospholipase D (PLD). Phospholipase B will remove the remaining fatty acid present on the lysophospholipid creating a glycerophosphate, also seen as a lysophospholipase (see Figure 4.2.5c). PLD is an enzyme that cleaves the choline or ethanolamine group of the phosphate, leading to formation of phosphatidic acid, generally an unwanted reaction in oilseed processing.

4.2.3.1 Phospholipase C

PLC is, strictly speaking, a phosphatase: it cleaves an organic group (the DAG) from a phosphate. Most commonly used phospholipase C enzymes are from bacterial origin and are expressed in a variety of hosts [24]. Typically, these enzymes are highly active on PC, somewhat less active on PE, hardly active on PA, and not at all on PI. The active site of bacterial PLC enzymes contains metal ions such as zinc to direct the phosphate group such that the cleaving of the DAG is well facilitated (see Figure 4.2.6a) [25]. These metal ions are tightly bound in the active site of the enzyme at neutral to basic conditions. Once brought into a more acidic environment, one or more of the active-site zinc ions might dissociate, inactivating the enzyme.

For PI, phosphatidyl inositol-specific phospholipase C enzymes (PI-PLC) have been developed [26], which are used in combination with normal PLC enzymes in EDG in commercial products such as Purifine. The inositol headgroup structure of PI is different from those of PC and PE. The larger inositol moiety does not fit in the normal PLC active site, and the polarity is distributed over a bigger volume, so that a different enzyme active site structure is needed, as indicated in Figure 4.2.6b [27]. It is also remarkable that the reaction takes place through a cyclic intermediate, which is slowly converted to inositol phosphate [27]. This cyclic inositol phosphate is also clearly visible by ^{31}P NMR [10a].

Recently, also several enzymes with a broader substrate specificity have been discovered, but are not yet commercially available [28].

The primary application of phospholipase C enzymes is for the replacement of water degumming with an enzymatic process, leading to improvement of the oil yield during the degumming process. Substantial yield gain is obtained via (i) breaking the emulsifying character of the phospholipid leading to reduced loss of neutral oil from entrainment in gums, and (ii) the additional neutral oil yield derived from the diglyceride oil obtained from hydrolysis of phospholipids. After PLC treatment, the total volume of the gum phase is significantly reduced, and gum fluidity is increased, facilitating separation of the reacted gums from the neutral oil.

While PLC-based processes substantially improve yield of the water-degumming processes, they may be less effective at enabling removal of all phospholipids from the oil. Where degumming is integrated with downstream processes, in a so-called deep-degumming or full-refining process, residual phosphorous levels must be reduced to below 10 ppm atomic P, preferably even lower. To reach this goal by PLC-based concepts is difficult, as it requires a successful combination of enzymatic reaction and removal of P-containing

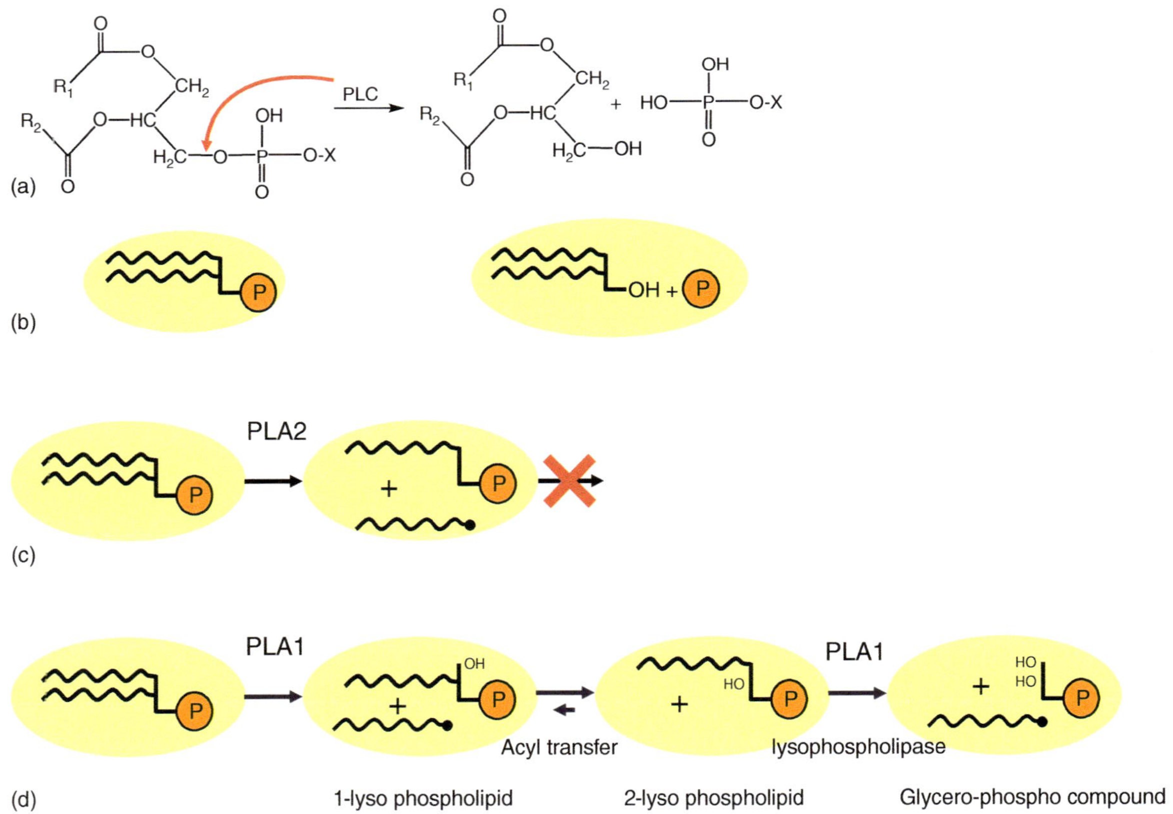

Figure 4.2.5 Schematic representation of the relevant phospholipase reactions. (a) Phospholipase C; (b) phospholipase A2 leading to a 2-lysophospholipid; and (c) phospholipase A1 leading to a 1-lysophospholipid, and after subsequent acyl transfer, certain PLA1 enzymes can hydrolyze the second fatty acid too, called lysophospholipase activity or PLB.

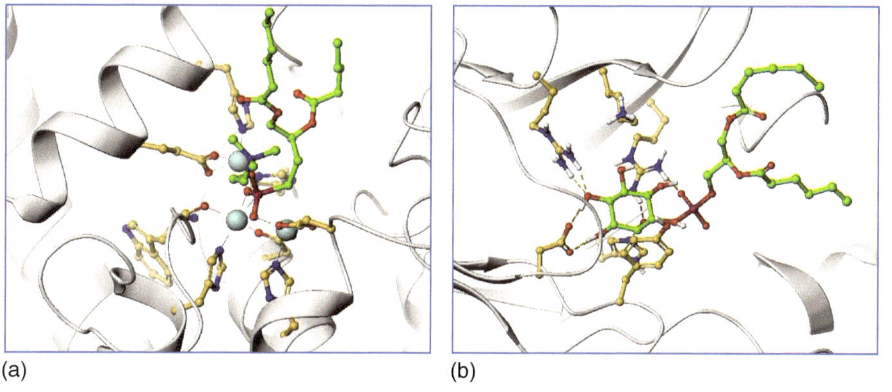

Figure 4.2.6 Active sites of (a) PLC showing the three zinc atoms coordinating with the phosphate, and (b) PI-PLC showing the hydrogen bonds between amino acids and inositol and the tyrosine that has pie-stacking interaction with the inositol.

compounds. Under these conditions, the PLC-catalyzed reactions are difficult to drive to completion. PA is hardly converted during normal operations and PE may be also difficult to hydrolyze completely, depending on the type and quality of the oil, as will be explained in the next section.

One solution to achieve better conversion and more efficient degumming is to combine a PLC-based process with use of a PLA, such as occurs in Purifine 3G, a combination of PLC, PI-PLC, and a minor amount of PLA2. The PLA2 (see also below) converts specifically the PA into lysoPA, which allows for better removal of P-containing molecules during degumming.

The microstructure of the dispersed gum phase after treatment with PLC differs from untreated gum. This is seen by the topology of the liquid-crystalline structures in the gum phase, as shown in Figure 4.2.4d – in this case by Purifine 3G. The structure is still lamellar, but now in the form of spherulites that consist of curved bilayers, often termed giant multilamellar vesicles or "onions" – as is graphically shown in Figure 4.2.4d. Now that most of the planar-bilayer-promoting phospholipids (PC, PE, and PI) are reacted away, the bilayers are formed by the polar lipids that are not converted, mostly residual PA, glycolipids, and lysophospholipids already present in the oil before degumming, and – in the case of Purifine 3G – the phospholipids that are converted to lysophospholipids by the PLA2. Lysophospholipids with their truncated-cone molecular shape promote positive curvature of the bilayers, located in the outer half of the bilayer. On the other hand, the unreacted PA, promoting inversely curved structures, may reside in the inner half of the bilayer.

4.2.3.2 Ways to Cope with Poor Conversion/Poor Quality Oils in PLC-Based Processes

The efficiency of the conversion by PLC depends on the quality of the crude oil (see also Table 4.2.2). Oil quality depends (at least) on the following factors:

- Total amount of phospholipids.

- Composition of the phospholipids: good quality oil is high in PC and low in PA; poor quality oil is high in PA.
- Free fatty acids: making the oil relatively acidic, poor quality oil is high in FFA.
- Total amount of metal ions. This usually goes hand in hand with high PA levels where they associate to. A high metal is related to poorer quality. An indicator of quality is the P/M ratio – the value of atomic P over the sum of atomic metals, all by ICP – as shown in Table 4.2.2. For good quality oils the ratio is far larger than 2; for poor quality this is around or even below 2. The latter indicates a high metal content relative to the phosphorus content.

Good quality oils are, for instance, North American or Latin American expander soy bean oils, high in phosphorous-containing components, and within this group high in PC (see Table 4.2.2). Flake oils typically have lower PC content as well as lower total phospholipid content than expander oils. While flake oils may be considered good quality, the phospholipid profiles may challenge the PLC enzyme.

Poor quality soy bean oil arises from beans that experienced poor growing conditions and/or suboptimal geography, or beans that were improperly conditioned (heat treated beans *activating* endogenous PLD instead of denaturing it), or have been stored long under poor conditions or on long and poor conditioned transport. Crude rapeseed/canola oils (combination of expeller and extracted oils) are usually lower in quality – see Table 4.2.2. The extracted oil stream in rapeseed/canola oil processes have profiles closer to crude soy bean oils. Pressed oils are low in P and relatively high in PA (see Table 4.2.2). High FFA levels may arise from chemical or endogenous enzymatic hydrolysis of triglyceride oils.

Conversion of PC by PLC occurs efficiently under most conditions and oil qualities. PLC has, however, some difficulty converting PE in poor quality oil. This is illustrated in Figure 4.2.7, showing laboratory-scale results of incubation of oils of different quality with Purifine PLC in a relatively low dose. The figure shows the residual intact PC and PE levels after 90 minutes of reaction at 55 °C with various caustic pretreatments and 3% of water. The low enzyme dose is used to better illustrate the effect of oil quality and conditioning measures on the individual phospholipids. The various caustic pretreatments lead to varying alkalinity of the aqueous surroundings in which the enzyme operates. With such low water content, one cannot speak of the pH value of the water phase: almost all the water is bound to a substrate, barely enough to fully hydrate the phospholipids. There is no "free water" to make a thermodynamically ideal solution that can be expressed by a pH value. The figure shows that the aqueous environment with increased alkalinity leads to increased conversion of PC. PE conversion is more influenced by the nature of the oil: a more alkaline environment improves conversion, but in "bad" quality oil, the conversion still lags. In practice, with high enzyme dosages and longer reaction times most of the PC and PE will be fully converted. In addition to a better environment for the PLC to work in, caustic preconditioning also neutralizes a part of the FFAs, and thereby increases the charges on the bilayer surfaces, making the water layers swell.

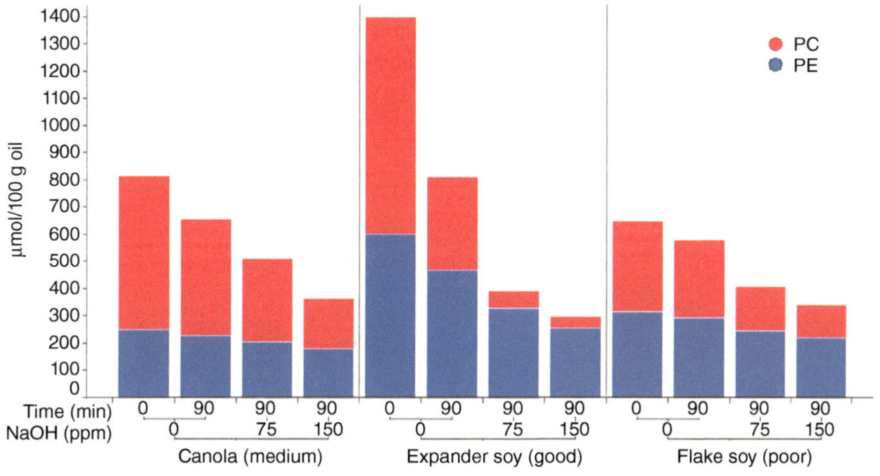

Figure 4.2.7 Intact phosphatidyl choline and phosphatidyl ethanolamine levels of oils of varying quality (poor–good–medium), treated with a *low dose* PLC to show the different response of the oil quality and the nature of the aqueous environment on conversion (after 90 minutes at 55 °C and with 3% water added). With a normal dose of PLC, the conversion would have progressed further. Values obtained by ^{31}P NMR; detection limit is at 40 µmol/100 g oil for both PC and PE.

PI is not shown in the data in Figure 4.2.7 as in all cases the PI conversion by PI-PLC is near complete. The quality of the oil does not hamper PI-PLC, and operates well under almost all conditions.

Why is PE conversion so difficult? Several potential reasons are proposed:

- The PE headgroup is more poorly hydrated compared to that of PC, which might translate into poorer participation in the O/W interface, and hence less contact opportunities with the enzyme.
- It has been shown that intramolecular ring formation can occur [16, 18], which may make the molecule less open to interaction with the active site. Although there is enough room in the active site (as it accommodates the bigger choline too), the intramolecular ring formation may reduce binding in the active site: the ring should first open up before it can bind, which is another kinetic hurdle in the enzyme reaction as shown in this schematic reaction:

$$\text{(cyclic PE)} \rightleftarrows \text{(open PE)} + \text{enzyme} \rightleftarrows [\text{ES}] \rightleftarrows [\text{EP}] \rightleftarrows \text{E} + \text{ethanolamine phosphate} + \text{diglyceride}$$

  in which [ES] is the enzyme–substrate (phosphatidyl ethanolamine) complex and [EP] the enzyme–product complex.

A common method to improve hydration of phospholipids is to pretreat the oil with acid (e.g. phosphoric or citric), followed by addition of caustic to increase the basicity of the aqueous environment. The response of PLC and PI-PLC to this is illustrated in Figure 4.2.8, showing that acid–caustic pretreatments reduce conversion by PLC of PC and PE. With so much sodium citrate – an excellent salting out agent – being present, the PLC enzyme suffers from dehydration. Also, oil

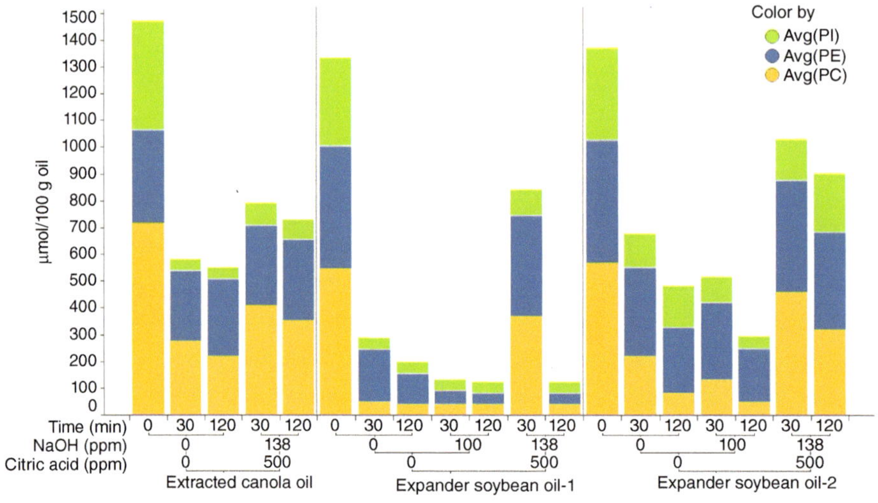

Figure 4.2.8 Reactions with PLC+PI-PLC on different oils with only water or preconditioned with only caustic (100 ppm) or 500 ppm citric acid and adding 138 ppm NaOH just before addition of the enzymes.

quality plays a role: good quality oil (see Figure 4.2.8, middle) still gets mostly converted with citric/caustic preconditioning, but not as well as with only caustic or only water. For oils of lesser quality, the difference gets increasingly pronounced. Noteworthy is that the activity of PI-PLC is not influenced by this variation in environmental conditions.

4.2.3.3 Phospholipase A

Two types of phospholipase A exist: Phospholipase A (PLA1 and PLA2) enzymes yield lysophospholipids and FFAs (see Figure 4.2.5). The lysophospholipids entrain substantially less oil. The FFAs may also be partially removed in the gum phase.

PLA2 hydrolyzes the fatty acid at the sn2 position making a 1-acyl-2-lysophospholipid. A commercial PLA2 (derived from pancreatic PLA2, produced by the fungus *Aspergillus niger*) [29] can hydrolyze all relevant phospholipids, with preference in the order PA > PC > PE > PI. The enzyme has a calcium ion in its active site as cofactor to bind the phosphate of the phospholipid. It prefers to operate under close to neutral to basic conditions. The reaction product, 2-lysophospholipid, is not a substrate for further reaction by this enzyme. Recently, an acidic PLA2 was brought to market [30].

PLA1 enzymes hydrolyze the fatty acid at the sn1 position, leading to a 1-lysophopholipid. This is however not a thermodynamically stable species: the acyl in the 2-position can migrate to the more stable sn1 position, leading to 2-lysophospholipid. In practice, the 1-lyso/2-lyso equilibrium is around 1/10, and the acyl migration is acid or base catalyzed. In aqueous surroundings, the minimum migration rate occurs at pH 4–5 [31]; in practice this occurs already under relatively acidic processing conditions. After migration to the sn1 position, the fatty acid can be cleaved off too, leading to a glycerophosphate compound. In this way, the fatty acids can be fully removed; however, full conversion to the glycerophosphate does not occur under practical conditions (see below).

Most PLA1 enzymes – and the acidic PLA2 enzyme – can operate in acidic conditions that improve hydratability of certain phospholipids, especially by creating an environment where the metal ion can more easily dissociate from the phospholipids, which is particularly the case for PA. Hence the reaction can be driven almost to completion under acidic reaction conditions. This can be advantageous in several processes:

- In degumming of oils, particularly poor quality and low P oils, such as crude sunflower oil or pressed rapeseed (see Table 4.2.2).
- In deep degumming processes, going in one process from crude oils to oil ready for bleaching and deodorization. In this type of process, the FFAs are removed by physical refining in the deodorizer. Deep degumming can, in principle, be applied in any type of seed oil and without the use of enzymes, but only with the use of a lot of chemicals. However, with high levels of phospholipids the neutral oil loss can be unacceptably high. Since enzymes greatly reduce these losses, they are the key enabling technology for deep degumming and physical refining of seed oils. Another consideration in the presence of high phospholipids is the lysophospholipase activity of the PLA1 enzymes resulting in glycerophosphates, and a high total amount of FFAs generated. In this case, it is essential that a well-dimensioned and large deodorizer is present.
  The effect is illustrated in Figure 4.2.9 for a range of soy bean oils with varying P content and varying relative PA levels, indicative of oil quality variations. It shows how 25 ppm PLA1 (Lecitase Ultra) converts all intact phospholipids in water-degummed oil within four hours at 55 °C after preconditioning with acid and caustic. After 120 minutes, all PA and PE are converted, but some PC and most PI still need to be converted. In some cases, even glycerophosphates were formed after four hours.
- Further processing of water-degummed oils that still contain those phospholipids that were not removed during water degumming, i.e. the poorly hydratable phospholipids such as PA and metal-associated PA. This process may be seen in plants where commodity degummed oil is brought in for full refining. The initial "water degumming" step is aimed at producing oil low in P, ready for the downstream process. The commodity oil with <200 ppm P contains < 0.5% of phospholipids, and so there is not much yield gain to be obtained from use of a PLC. This is illustrated in Figure 4.2.10, showing how 25 ppm PLA1 can convert all intact phospholipids in water-degummed oil within four hours at 55 °C, preconditioned with citric acid and caustic. Here too, some glycerophosphocholine could be formed after two hours.

4.2.4 Enzymatic Degumming in Industrial Practice

Laboratory-scale EDG is done in batch style and under well-controlled conditions, with proper high-shear mixers, continuous stirring, and good temperature

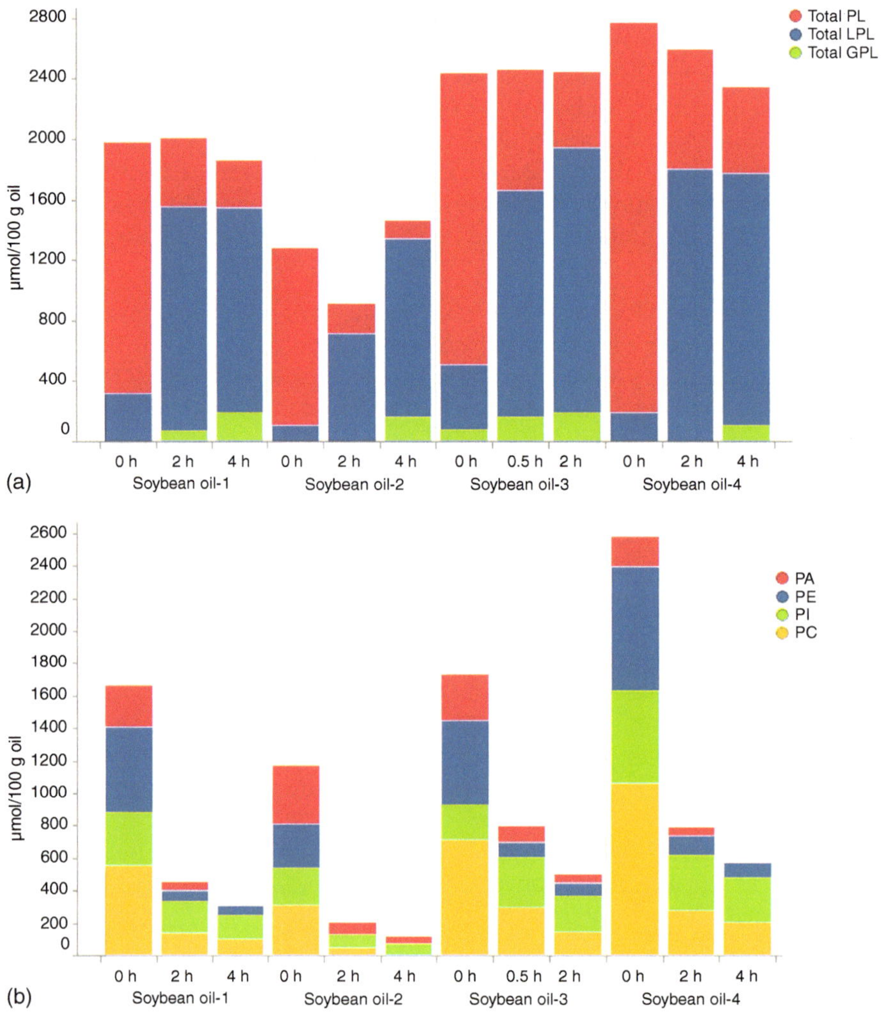

Figure 4.2.9 Levels of phosphorus-containing compounds versus time in four different crude soy bean oils during treatment with PLA1 (25 ppm, after preconditioning the oil with 500 ppm citric acid and 138 ppm NaOH). Laboratory-scale experiments, measured by ^{31}P NMR; detection limit is 40 μmol/100 g oil per individual compound. (a) Sum of intact phospholipids, sum of lysophospholipids, and sum of glycerophosphates. (b) Levels of intact phospholipids. (c) Levels of lysophospholipids. (d) Levels of glycerophosphate. Note that the *Y*-axis varies for each figure.

control. In industrial practice with continuous processing, ideal conditions are usually harder to achieve. Figure 4.2.11 shows a simplified industrial process, indicating what enzymatic processes can best be applied.

In an industrial process with good quality oil in its simplest form, the crude oil coming from the desolventizer is cooled to 55 °C (enzyme-dependent), and then water and PLC enzymes are added to the oil with intense mixing to produce the lamellar liquid-crystalline phase. Since the enzyme needs time to

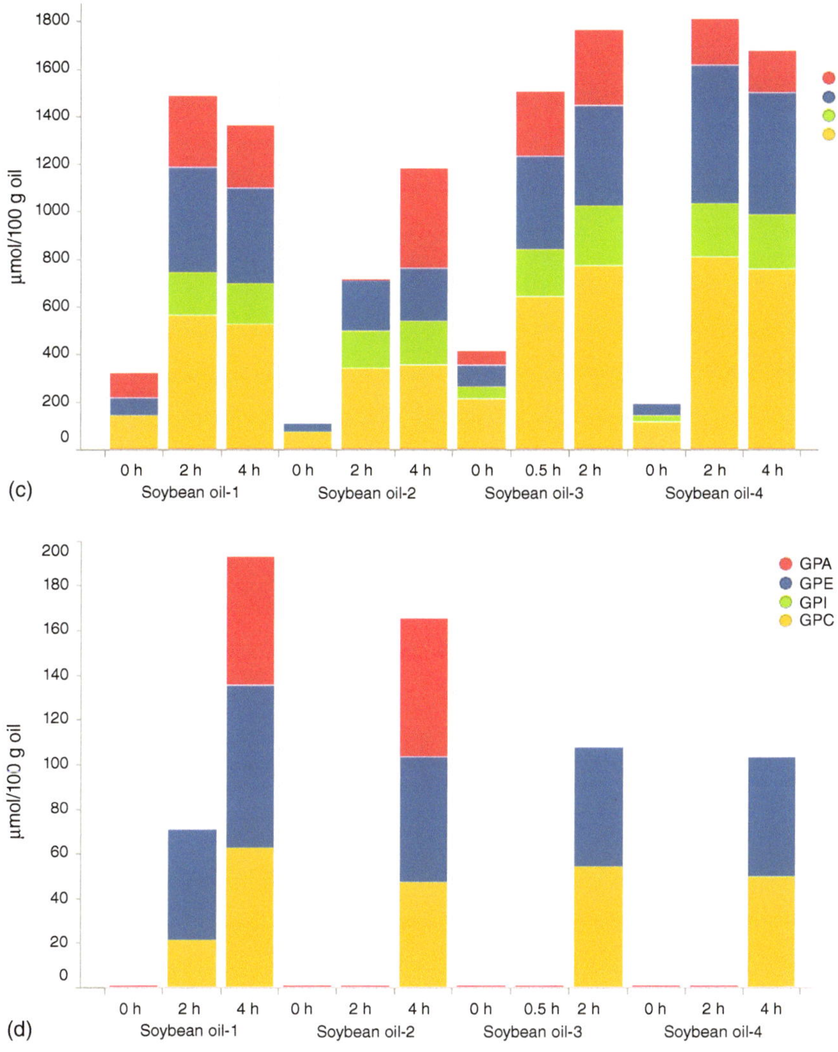

Figure 4.2.9 *(Continued)*

convert the phospholipids, the reaction mixture is passed into a continuous reactor, preferably maintained at a temperature of around 55 °C, and a flow rate ensuring a residence time of two to four hours. During this stage, good agitation in the reaction tank is required, to improve enzyme–substrate contact for an efficient conversion and to prevent the hydrated phospholipid phase from sedimenting from the oil after reaction. The oil is heated up to 70–80 °C, and the aqueous gum phase is separated by centrifugation. Because of the conversion of the phospholipids, the total volume of gum phase is substantially reduced, and is less "gummy" since the fluidity has increased relative to gums from a water-degumming process. This process works well for good quality oils in combination with phospholipase C enzymes. However, when the quality of the

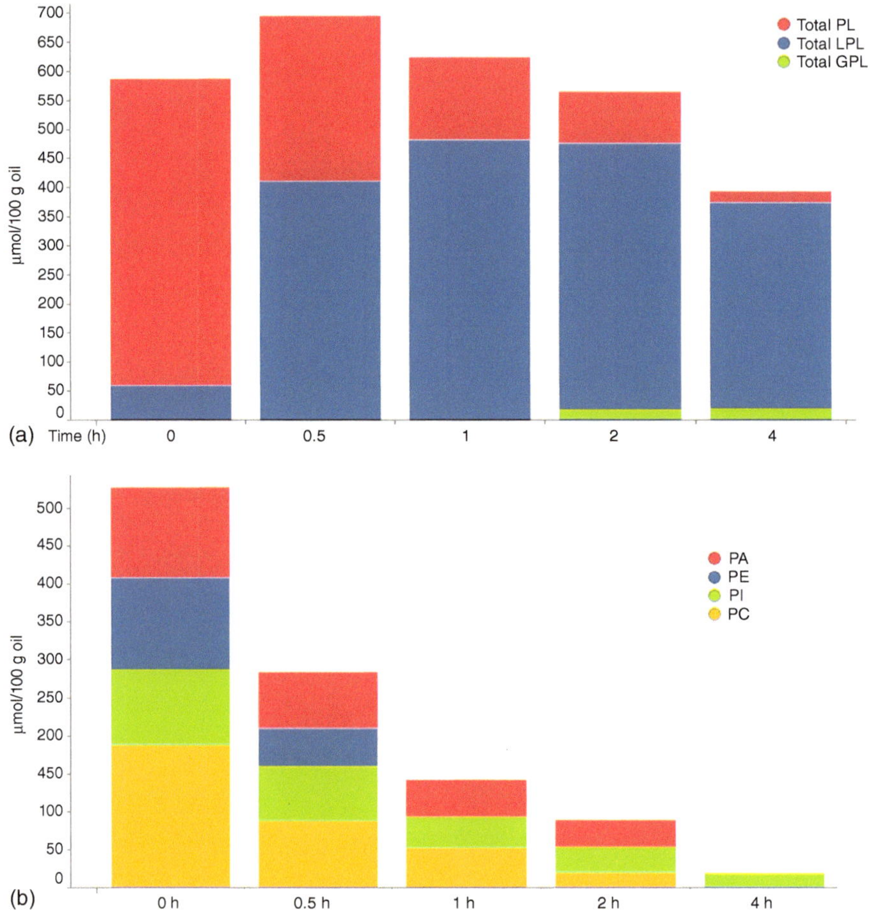

Figure 4.2.10 Levels of phosphorus-containing compounds versus in water-degummed soy bean oil versus time during treatment with PLA1 (25 ppm, after preconditioning the oil with 500 ppm citric acid and 138 ppm NaOH) laboratory-scale experiments, measured by ^{31}P NMR; detection limit is 40 μmol/100 g oil per individual compound. (a) Sum of intact phospholipids, sum of lysophospholipids, and sum of glycerophosphates. (b) Levels of intact phospholipids. (c) Levels of lysophospholipids. (d) Levels of glycerophosphate. Note that the *Y*-axis varies for each figure.

oil is lower, mild preconditioning with caustic helps achieve good results (see Section 4.2.3.2).

In the case of deep degumming using an acidic PLA enzyme, the oil must be preconditioned with acid alone for certain PLA, which prefers an acidic environment, or acid followed by addition of caustic to establish the right acidity conditions for the enzyme. The acid is added in highly concentrated form, for instance, a 50% citric acid solution, at 70 °C and mixed well, and kept for around 30 minutes. After cooling to the required temperature (45–60 °C, depending on type of enzyme), caustic (optionally) is added, together with or followed by addition of water with enzyme. During subsequent incubation of two to four hours,

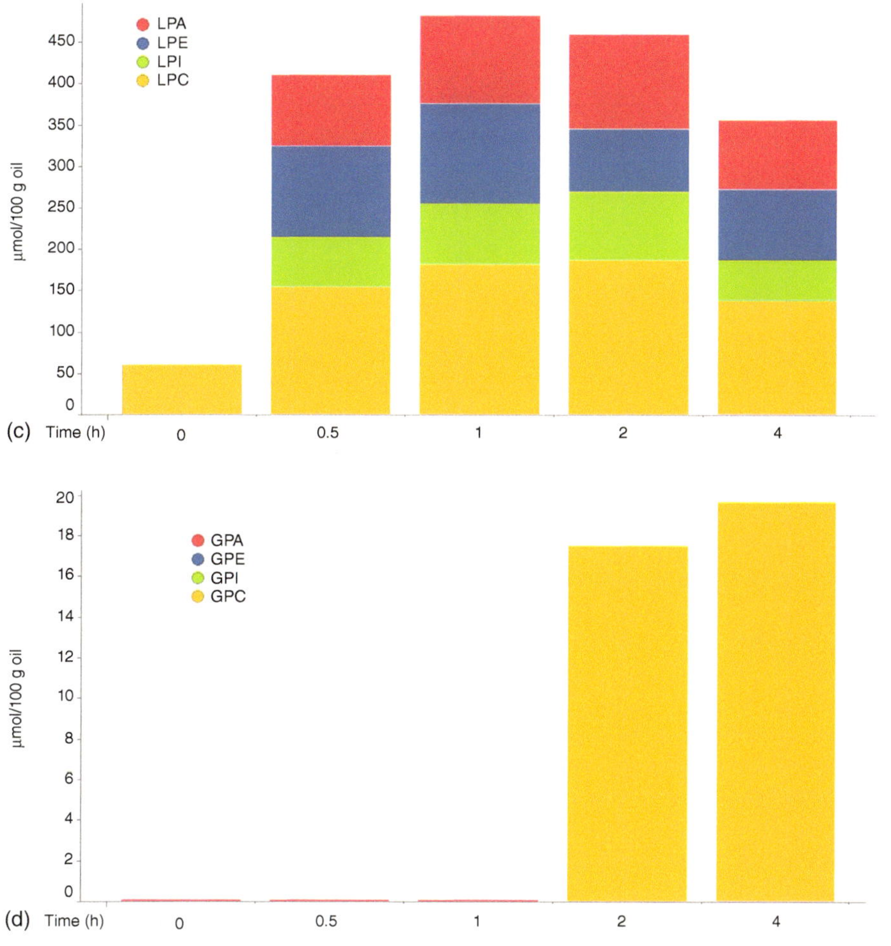

Figure 4.2.10 (*Continued*)

agitation is needed. After reaction, the residual "gum" phase is separated by centrifuge and the oil can be washed once more with water or acid. The latter can be done also to prevent scaling of the stacks in the centrifuge [32]. This should lead to oils of less than 5 ppm P. A similar process is applied for *refining* already water-degummed oil with an acidic PLA.

4.2.4.1 Introduction Hurdles

Given the advantages of EDG and refining, what prevents the industry from adoption of this technology *en masse*? Several reasons can be given:

- (Competing priorities for) Capital expenditures (CAPEX) of the EDG process: to introduce such a process, the processing line needs to be adapted by extra shear equipment and reaction tanks.

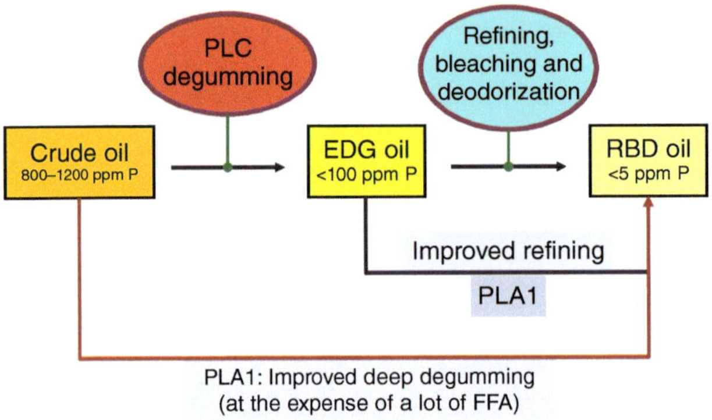

Figure 4.2.11 Enzymatic processes, degumming, refining, or deep degumming. EDG, enzymatically degummed; RBD, refined, bleached, and deodorized.

- When not properly designed, the process is not easy to control: a smooth operation needs careful control of the dispersion of the water with enzyme and to keep it dispersed throughout the incubation. The most optimal reaction conditions (enzyme-dependent preconditioning, time) need to be found for the oil processed in the particular plant, and adjusted for varying quality.
- The battle between high yield and low P. Especially for lower quality oils, the PLC concept can be less efficient: low level of phospholipids convertible by PLC enzymes, leading to low DAG increase and insufficient P reduction. Solutions are found in combinations of PLC enzymes and PLA1 enzymes, either added together or sequentially.
- The gum phase has a value:
  - In most processes the gum is added onto the meal: true cost yield in the process is therefore the difference between oil price and meal price. However, due to global reduction in protein values in the oilseed crops, it may not be possible to add the gums back to the meal.
  - Some processors (about 10–15% of the worldwide volume) use the gums as a source for lecithins, E322, emulsifier for food, for which a higher price can be obtained than for the oil. Gums from a PLA catalyzed process can have added value compared to non-enzymatic processes, as they can be rich in lysophospholipids; or in the case of a PLC-only process (without PI-PLC), the gums will be rich in PI and some PA, making a relatively low hydrophilic-lipophilic balance (HLB) emulsifier that may do well in W/O emulsions or chocolate applications [33]. Such lecithin fraction has higher commercial value than common untreated lecithin, and the producers gain extra oil from the process.
  - Gums sold separately as feed additive can also give higher value, based on the thought that it has a high nutritional value. However, a gum from a PLC+PI-PLC process is rich in choline phosphate, ethanolamine phosphate, and inositol phosphate, and these have a higher nutritional value as feed additive than the components in a normal gum phase, apart from the energy value. The volume might be lower, but the financial value of such a side stream will be higher.

4.2.5 Other Applications of Enzymes in Oil – Outlook

This chapter has dealt with the use of phospholipases in degumming and refining. Enzymes can play a wider role in oils and fats processes [3, 34]. Several of those are listed below; some of these are already well established, while others are still in development.

4.2.5.1 Enzymatic Interesterification of Triglyceride Oils

The physical, chemical, and nutritional properties of triglycerides (or triacylglycerides, TAGs) depend largely on the type of fatty acids and their distribution on the glycerol backbone. For various applications this profile needs to be adapted by a process called interesterification. This process can be improved, or controlled better by the specific use of TAG lipases [35]. This technology has been around already for many years and is often reviewed. Generally, two cases are identified: randomization of triglycerides and structured lipids.

Randomization: For fats with specified melting profiles special fat blends are designed, for use in, for instance, table or baking margarines and spreads. Today, these fats are made by combining low melting oils and high melting fats of certain specified fatty acid compositions (saturated and unsaturated), and under the influence of a catalyst fatty acids can be exchanged, randomizing the fatty acid composition of the blend. This leads to fats with a melting profile better suited for the applications such as margarine and bakery products. This is an alternative to partial hydrogenation, which is an unacceptable process leading to high levels of trans fats.

Commonly, a base catalyst is used (sodium (m)ethanoate), but this "chemical interesterification" has several disadvantages: The process takes place at elevated temperatures (energy use); catalysts and feedstocks are sensitive to moisture; sodium (m)ethanoate is a hazardous chemical; The reaction is prone to side products such as soaps, Me-esters, MAG (monoglyceride), and DAG (diglyceride); it is a batch process that needs postprocessing to kill and wash out the catalyst and to remove color; total randomization occurs, which is not always wanted; and finally there is the risk of flavor reversion and reduced oxidative stability (antioxidant losses).

An alternative process is found in enzymatic interesterification, where a TAG lipase on a silica carrier is added to the anhydrous mixture of oil and fat, facilitating the exchange of fatty acids under far milder conditions, and leading to similar interesterified blends, but now more controlled with less side reactions [36]. This process runs already on large scale at some major oil processors.

Structured lipids: In applications such as human milk fat analogues or cocoa butter replacement, distribution of fatty acids on the glycerol backbone should be controlled. This involves specific design of triglycerides called structured lipids. Human milk fat analogues are characterized by a saturated fatty acid on the middle, sn2 position of the triglyceride, and unsaturated ones on the sn1 and sn3 positions. For cocoa butter equivalents, the fatty acid on the middle position should be an unsaturated fatty acid, and the ones on the sn1 and sn3 positions are saturated fatty acids. To make these structured lipids from vegetable oils, specific hydrolysis and re- or transesterification processes are developed, making use of position-specific TAG lipases immobilized on, for instance, resin carriers [3, 37, 38].

A third trend in this field – still not commercial yet – is the development of fatty-acid-specific lipases, which are targeted to hydrolyze these specific fatty acids for increasing the wanted fatty acid [3, 39], or decreasing unwanted fatty acids [3, 40, 41] in the fat or oil.

4.2.5.2 Biodiesel

Another interesterification process using lipases relates to the production of biodiesel from triglycerides and methanol (or ethanol), resulting in fatty acid methyl esters (FAMEs) [42]. Although quite attractive on paper, it is not easy to convert this to a commercially viable process. Solutions are sought in reusing soluble lipases rather than immobilized lipases, where the lipase can be recovered from the non-FAME phase for reuse [43].

A further issue in biodiesel production that can be solved by enzymes is the removal of sterol glucosides using sterol glucosidases. Sterol glucosides have a low solubility in FAME and precipitate from cold biodiesel, causing clogging and engine failure. By hydrolyzing the sterol glucoside in the biodiesel [44] or even in the crude oil where the biodiesel is made from [45], the precipitation does not occur.

4.2.5.3 Enzyme-Assisted Decoloring

A crucial step in the refining of vegetable oils is the bleaching step. In this step usually activated bleaching earth is added to remove chlorophyll, pheophytin and pyropheophytin. An additional advantage is that residual phospholipids and metals are caught in this step. During the bleaching step further oil loss can occur, and handling of bleaching earth is cumbersome. When chlorophyll-derived pigments can be converted to the more water-soluble heme group and the phytol fatty alcohol tail, the colored heme compound can be removed during washing or by alternative absorbents that absorb less oil. Such a process can be catalyzed by chlorophyllases or pheophytinases [46], resulting in hydrolysis of the phytol tail from the heme group. This would circumvent the use of bleaching earth.

4.2.5.4 Enzyme-Assisted Oil Extraction

Finally, the field of improving oil yield from the extraction step using enzymes is briefly touched upon. Fruit oils such as palm oil are processed in a different way than seed oils. For palm oil, the oil should be released from the fruit pulp. Carbohydrases improve the recovery process [47], although this is not yet a generally applied process on large scale.

Also in the extraction of olive oil, a mixture of carbohydrases helps increase the yield and improve the quality of the oil[1] [48].

1 Maxoliva product data sheet DSM Food Specialties, the Netherlands.

Improving the yield of seed oils from pressing and extraction by enzymes is still only studied on the laboratory scale. The rapeseed oil press step can be improved by pretreating the seeds with carbohydrases [49].

Replacing hexane extraction for seed oils is identified as a high need, for food and process safety, and environmental reasons. Enzyme-assisted aqueous extraction of seed oils receives quite some attention [50]; however, this process uses far too much water to be competitive. Biorefinery-type solutions employing less water should lead to oil extraction, at the same time yielding nutritionally and functionally improved proteins [51].

4.2.6 Conclusion

This chapter has shown how the efficiency of the oil degumming and refining processes can be improved using phospholipases A and C. In a world where environmental concerns and pressure on prices become increasingly important, the use of enzymes in such processes is highly needed to obtain increase in yields and reduction in chemicals use. Beyond degumming and refining, many more processes in oils and fats profit from the use of enzymes.

Acknowledgments

The authors wish to acknowledge Remco Muntendam, Ying Zha, and René de Jong for their input in this manuscript.

References

1 USDA (2017). Oilseeds: World Markets and Trades – December 2017. https://apps.fas.usda.gov/psdonline/circulars/oilseeds.pdf (accessed 14 May 2018).
2 USDA (2017). Mark Ash, Oil Crops Outlook – 14 August 2017. https://www.ers.usda.gov/webdocs/publications/84678/17h.pdf?v=42961 (accessed 14 May 2018).
3 (a) Zorn, K., Oroz-Guinea, I., Brundiek, H., and Bornscheuer, U.T. (2016). Engineering and application of enzymes for lipid modification, an update. *Prog. Lipid Res.* 63: 153–164. (b) Biermann, U., Bornscheuer, U.T., Meier, M.A.R. et al. (2011). Oils and fats as renewable raw materials in chemistry. *Angew. Chem. Int. Ed.* 50: 3854–3871. (c) Bornscheuer, U.T. (2018). Enzymes in lipid modification. *Annu. Rev. Food Sci. Technol.* 9: 85–103.
4 Foley, P., Kermanshahi Pour, A., Beach, E.S., and Zimmerman, J.B. (2012). Derivation and synthesis of renewable surfactants. *Chem. Soc. Rev.* 41: 1499–1518.
5 (a) Dayton, C.L.G. and Galhardo, F. (2012). Enzymatic degumming. In: *Green Vegetable Oil Processing*, Chapter 6 (ed. E. Farr and A. Proctor), 105–146. Urbana IL: AOCS Press. (b) Dayton, C.L.G. and Galhardo, F. (2012). Enzymatic degumming AOCS lipid library. http://lipidlibrary.aocs.org/OilsFats/

content.cfm?ItemNumber=40324 (accessed 14 May 2018). (c) Galhardo, F. and Dayton, C.L.G. (2010). Desgomado enzimático de aceites de vegetales. *Acietas Grasas* 79: 204–212. (d) De Maria, L., Vind, J., Oxenbøll, K.M. et al. (2007). Phospholipases and their industrial applications. *Appl. Microbiol. Biotechnol.* 74: 290–300.

6 (a) Dijkstra, A.J. and Segers, J.C. (2007). Production and refining of oils and fats. In: *The Lipid Handbook*, 3e (ed. F.D. Gunstone, J.L. Harwood and A.J. Dijkstra), 143–262. Boca Raton, FL: CRC Press. (b) Dijkstra, A.J. (2014). AOCS lipid library. http://lipidlibrary.aocs.org/OilsFats/content.cfm?ItemNumber=40318&navItemNumber=19214 (14 May 2018). (c) Gupta, M. (ed.) (2017). *Practical Guide to Vegetable Oil Processing*, 2e. London: Academic Press and AOCS Press.

7 (a) van Nieuwenhuyzen, W. and Tomás, M.C. (2008). Update on vegetable lecithin and phospholipid technologies – review. *Eur. J. Lipid Sci. Technol.* 110: 472–486. (b) van Nieuwenhuyzen, W. (2014). The changing world of lecithins. *AOCS Inform* 25: 254–259.

8 The LIPID MAPS® Lipidomics Gateway (2017). http://www.lipidmaps.org/ (accessed 14 May 2018).

9 (a) Cevc, G. (1993). *Phospholipids Handbook*. New York: CRC Press. (b) Szuhaj, B.F. (2005). Lecithins. In: *Bailey's Industrial Oil and Fat Products*, Chapter 13 (ed. F. Shahidi), 361–456. Hoboken, NJ: Wiley.

10 (a) van Rijn, J., Groen, P., Lankhorst, P. et al. (2019) submitted). Robust and reliable quantification of phospholipids in edible oils using ^{31}P NMR spectroscopy. *J. Am. Oil Chem. Soc.* (b) Yang, Y., Hiserodt, R., and Li, J. (2017). Rapid identification and relative quantification of the phospholipid composition in commercial lecithins by ^{31}P NMR. *J. Am. Oil Chem. Soc.* 94: 885–892. (c) Diehl, B.W.K. (2015). NMR of polar lipids. In: *Polar Lipids: Biology, Chemistry, and Technology* (ed. M.U. Ahmad and X. Xu), 391–438. Urbana, Il: AOCS Press.

11 (a) Brouwers, J.F. (2011). Review – liquid chromatographic-mass spectrometric analysis of phospholipids. Chromatography, ionization and quantification. *Biochim. Biophys. Acta* 1811: 763–775. (b) Yamada, T., Nagasawa, Y., Taguchi, K. et al. Polar lipid profiling by supercritical fluid chromatography/mass spectrometry method. In: *Polar Lipids: Biology, Chemistry, and Technology* (ed. M.U. Ahmad and X. Xu), 439–462. Urbana, Il: AOCS Press.

12 Sen Gupta, A.K. (1986). Neue Entwicklungen auf dem Gebiet der Raffination der Speiseöle. *Fette Seifen Anstrichm.* 88: 79–86.

13 Seddon, J.M. and Cevc, G. (1993). Lipid polymorphism: structure and stability of lyotropic mesophases of phospholipids. In: *Phospholipids Handbook* (ed. G. Cevc), 403–454. New York: CRC Press (See also [9a]).

14 Disalvo, E.A., Lairion, F., Martini, F. et al. (2008). Structural and functional properties of hydration and confined water in membrane interfaces. *Biochim. Biophys. Acta* 1778: 2655–2670.

15 Marsh, D. (1989). Water adsorption isotherms and hydration forces for lysolipids and diacyl phospholipids. *Biophys. J.* 55: 1093–1100.

16 Israelachvili, J.N., Marčelja, S., and Horn, R.G. (1980). Physical principles of membrane organization. *Q. Rev. Biophys.* 13: 121–200.

17 Bouchet, A.M., Frías, M.A., Lairion, F. et al. (2009). Structural and dynamical surface properties of phosphatidylethanolamine containing membranes. *Biochim. Biophys. Acta* 1788: 918–925.
18 Suits, F., Pitman, M.C., and Feller, S.E. (2005). Molecular dynamics investigation of the structural properties of phosphatidylethanolamine lipid bilayers. *J. Chem. Phys.* 122: 244714.
19 Henneré, N., Prognon, P., Brion, F. et al. (2009). Molecular dynamics simulation of a mixed lipid emulsion model: influence of the triglycerides on interfacial phospholipid organization. *J. Mol. Struct. THEOCHEM* 901: 174–185.
20 Israelachvili, J.N., Mitchell, D.J., and Ninham, B.W.J. (1976). Theory of self-assembly of hydrocarbon amphiphiles into micelles and bilayers (review). *J. Chem. Soc., Faraday Trans. 2* 72: 1525–1568.
21 Lindblom, G., Rilfors, L., Hauksson, J.B. et al. (1991). Effect of head-group structure and counterion condensation on phase equilibria in anionic phospholipid–water systems studied by ^{2}H, ^{23}Na, and ^{31}P NMR and X-ray diffraction. *Biochemistry* 30: 10938–10948.
22 (a) Sein, A., Engberts, J.B.F.N., van der Linden, E., and van de Pas, J. (1996). Lyotropic phases of dodecylbenzenesulfonates with different counterions in water. *Langmuir* 12: 2913–2923. (b) Shearman, G.C., Ces, O., Templer, R.H., and Seddon, J.M. (2006). Inverse lyotropic phases of lipids and membrane curvature. *J. Phys. Condens. Matter* 18: S1105–S1124. (c) Tiddy, G.J.T. (1980). Surfactant-water liquid crystal phases. *Phys. Rep.* 57: 1–46.
23 Harlos, K., Eibl, H., Pascher, I., and Sundell, S. (1984). Conformation and packing properties of phosphatidic acid. *Chem. Phys. Lipids* 1883: 115–126.
24 (a) Gramatikova, S., Hazlewood, G., Lam, D., and Barton, N.R. (2005). Phospholipases, nucleic acids encoding them and methods for making and using them. US7226771 – US2005108789 (patent of Diversa). (b) Xu, J., Xuan, Y., and Li, J. (2015). Phospholipase C mutant and use thereof. WO2015067161 (patent of Wilmar). (c) Castelli, M.E., Menzella, H., Peiru, S., and Vetcher, L. (2015). Modified *Bacillus cereus* phospholipase C protein and method of processing vegetable oil. WO2015017045 (patent of Keclon). (d) Segura, D.R., Olinski, R.P., Borch, K. et al. (2015). Polypeptides having phospholipase C activity and polynucleotides encoding same. WO2015140275 (patent of Novozymes).
25 Hansen, S., Hough, E., Svensson, L.A. et al. (1993). Crystal structure of phospholipase C from *Bacillus cereus* complexed with a substrate analog. *J. Mol. Biol.* 234: 179–187.
26 Barton, N.R., Hitchman, T.S., Lyon, J.D. et al. (2011). Phospholipases, nucleic acids encoding them and methods for making and using them. WO2011046812 (patent of Verenium).
27 (a) Griffith, O.H. and Ryan, M. (1999). Bacterial phosphatidylinositol-specific phospholipase C: structure, function and interaction with lipids. *Biochem. Biophys. Acta* 1441: 237–254. (b) Heinz, D.W., Essen, L.-O., and Williams, R.L.C. (1998). Structural and mechanistic comparison of prokaryotic and eukaryotic phosphoinositide-specific phospholipase. *J. Mol. Biol.* 275:

635–650. (c) Moroz, O.V., Blagova, E., Lebedev, A.A. et al. (2017). The structure of a calcium-dependent phosphoinositide-specific phospholipase C from *Pseudomonas* sp. 62186, the first from a Gram-negative bacterium. *Acta Crystallogr., Sect. D: Struct. Biol.* 73: 32–44.

28 (a) Sein, A., De Jong, R.M., Van Rij, E.T. et al. (2016). Phospholipase C. WO2016162456 (patent of DSM). (b) Tang, L., Liu, Y., Borch, K., and Brask, J. (2012). Polypeptides having phospholipase C activity and polynucleotides encoding same. WO2012062817 (patent of Novozymes).

29 DSM (2018). Application Datasheet Maxapal A2 PLA2.

30 AB Enzymes (2016). Application Datasheet Rohalase® PL-XTRA.

31 Okudaira, M., Inoue, A., Shuto, A. et al. (2014). Separation and quantification of 2-acyl-1-lysophospholipids and 1-acyl-2-lysophospholipids in biological samples by LC-MS/MS. *J. Lipid Res.* 55: 2178–2192.

32 Dayton, C.L.G., Staller, K., and Berkshire, T.L. (2005). Process for improving enzymatic degumming of vegetable oils and reducing fouling of downstream equipment. WO2005063950 (patent of Bunge).

33 Kindlein, M., Greiner, M., Elts, E., and Briessen, H. (2015). Interactions between phospholipid head groups and sucrose crystal surface at the cocoa butter interface. *J. Phys. D: Appl. Phys.* 48: 384002.

34 Bornscheuer, U.T. (ed.) (2018). *Lipid Modification by Enzymes and Engineered Microbes*. London: Academic Press and AOCS Press.

35 (a) Schmidt, R.D. and Verger, R. (1998). Lipases: interfacial enzymes with attractive applications. *Angew. Chem. Int. Ed.* 37: 1609–1633. (b) Casas-Godoy, L., Duquesne, S., Bordes, F. et al. (2012). Lipases: an overview. In: *Lipases and Phospholipases: Methods and Protocols, Methods in Molecular Biology*, vol. 861 (ed. G. Sandoval), 3–30. New York: Springer-Verlag. (c) Naika, S., Basua, A., Saikiaa, R. et al. (2010). Lipases for use in industrial biocatalysis: specificity of selected structural groups of lipases. *J. Mol. Catal. B: Enzym.* 65: 18–23.

36 (a) Dayton, C.L.G. (2012). Enzymatic interesterification. In: *Green Vegetable Oil Processing* (ed. E. Farr and A. Proctor), 205–224. Urbana IL: AOCS Press. (b) Holm, H.C. and Cowan, W.D. (2008). The evolution of enzymatic interesterification in the oils and fats industry. *Eur. J. Lipid Sci. Technol.* 110: 679–691. (c) Cowan, W.D. (2018). *Enzymatic Interesterification*. AOCS Lipid Library. http://lipidlibrary.aocs.org/OilsFats/content.cfm?ItemNumber=40329. (d) Dijkstra, A.J. (2007). Interesterification. In: *The Lipid Handbook*, 3e (ed. F.D. Gunstone, J.L. Harwood and A.J. Dijkstra), 285–300. Boca Raton, FL: CRC Press.

37 (a) Xu, X. (2000). Production of specific-structured triacylglycerols by lipase-catalyzed reactions: a review. *Eur. J. Lipid Sci. Technol.* 102: 287–303. (b) Şahin-Yeşilçubuk, N. and Akoh, C.C. (2017). Biotechnological and novel approaches for designing structured lipids intended for infant nutrition. *J. Am. Oil Chem. Soc.* 94: 1005–1034.

38 (a) Kadivar, S., De Clercq, N., Dewettinck, K., and Danthine, S. (2016). Crystallization and polymorphic behavior of enzymatically produced sunflower oil based cocoa butter equivalents. *Eur. J. Lipid Sci. Technol.* 118: 1521–1538.

(b) Kim, S., Kim, I.-H., Akoh, C.C., and Kim, B.H. (2014). Enzymatic production of cocoa butter equivalents high in 1-palmitoyl-2-oleoyl-3-stearin in continuous packed bed reactors. *J. Am. Oil Chem. Soc.* 91: 747–757.

39 Suarez, E.R., Mugford, P.F., Rolle, A.J. et al. (2010). ^{13}C-NMR regioisomeric analysis of EPA and DHA in fish oil derived triacylglycerol concentrates. *J. Am. Oil Chem. Soc.* 87: 1425–1433.

40 Brundiek, H.B., Evitt, A.S., Kourist, R., and Bornscheuer, U.T. (2011). Creation of a lipase highly selective for trans fatty acids by protein engineering. *Angew. Chem. Int. Ed.* 51: 412–414.

41 (a) Dayton, C.L.G. (2013). Process for production of low saturate oils. WO2013188615 (patent of Bunge Global Innovations). (b) Halich, R., Kline, K., Shanahan, D., and Ciofalo, V. (2012). Safety evaluation of a lipase enzyme (BD29241 Palmitase) preparation, expressed in *Pseudomonas fluorescens*, intended for removing palmitic acid from triacylglycerol. *Regul. Toxicol. Pharmacol.* 64: 87–94.

42 (a) Adamczak, M., Bornscheuer, U.T., and Bednarski, W. (2008). The application of biotechnological methods for the synthesis of biodiesel. *Eur. J. Lipid Sci. Technol.* 111: 806–813. (b) Norjannah, B., Ong, H.C., Masjuki, H.H. et al. (2016). Enzymatic transesterification for biodiesel production: a comprehensive review. *RSC Adv.* 6: 60034–60055.

43 (a) Nielsen, P.M., Brask, J., and Fjerbaek, L. (2008). Enzymatic biodiesel production: technical and economical considerations. *Eur. J. Lipid Sci. Technol.* 110: 692–700. (b) Cesarini, S., Falkenberg Haller, R., Diaz, P., and Nielsen, P.M. (2014). Combining phospholipases and a liquid lipase for one-step biodiesel production using crude oils. *Biotechnol. Biofuels* 7: 29–41.

44 Peiru, S., Aguirre, A., Eberhardt, F. et al. (2015). An industrial scale process for the enzymatic removal of steryl glucosides from biodiesel. *Biotechnol. Biofuels* 8: 223–235.

45 (a) Münger, L.H. and Nyström, L. (2014). Enzymatic hydrolysis of steryl glycosides for their analysis in foods. *Food Chem.* 163: 202–211. (b) Münger, L.H. and Nyström, L. (2015). Enzymatic hydrolysis of acylated steryl glycosides and methods for treating biofuel. WO2015022367 (patent of ETH Zürich).

46 (a) Bitar, M., Karboune, S., Bisakowski, B., and Kermasha, S. (2004). Chlorophyllase biocatalysis in an aqueous/miscible organic solvent medium containing canola oil. *J. Am. Oil Chem. Soc.* 81: 927–932. (b) Lam, D., Weiner, D., Hitchman, T. et al. (2006). Compositions and methods for enzymatic decolorization of chlorophyll. WO2006009676 (patent of Diversa). (c) Soe, J.S., Poulsen, C.H., Zargahl, M.S. et al. (2010). Method. WO2010143149 (patent of Dupont). (d) Sein, A., van Rij, T.J., van der Laan, J.M., and Dayton, C.L.G. (2018). Enzymatic removal of chlorophyll substrates from triacylglycerol-based oils. WO2019063613 (patent of DSM).

47 (a) Teixeira, C.B., Macedo, G.A., Macedo, J.A. et al. (2013). Simultaneous extraction of oil and antioxidant compounds from oil palm fruit (*Elaeis guineensis*) by an aqueous enzymatic process. *Bioresour. Technol.* 129: 575–581. (b) Soe, J.S., Mikkelsen, R., Lee, L.S., and Kjaer, K.H. (2016).

Method of improving palm oil yields from palm fruit or palm fruit liquids. WO2016097264 (patent of Dupont).
48 Clodoveo, M.L. (2012). Malaxation: influence on virgin olive oil quality. Past, present and future – an overview. *Trends Food Sci. Technol.* 25: 13–23.
49 Szydłowska-Czerniak, A., Karlovits, G., Hellner, G. et al. (2010). Effect of enzymatic and hydrothermal treatments of rapeseeds on quality of the pressed rapeseed oils. *Process Biochem.* 45: 7–17.
50 (a) Yusof, M.M., Gordon, M.H., and Niranjan, K. (2015). Aqueous enzyme assisted oil extraction from oilseeds and emulsion de-emulsifying methods: a review. *Trends Food Sci. Technol.* 41: 60–82. (b) Campbell, K.A., Glatz, C.E., Johnson, L.A. et al. (2011). Advances in aqueous extraction processing of soybeans. *J. Am. Oil Chem. Soc.* 88: 449–465.
51 (a) Loman, A.A. and Ju, L.-K. (2016). Optimization of enzymatic process condition for protein enrichment, sugar recovery and digestibility improvement of soy flour. *J. Am. Oil Chem. Soc.* 93: 1063–1073. (b) Loman, A.A., Mahfuzul Islam, S.M., Li, Q., and Ju, L.-K. (2017). Enzyme recycle and fed-batch addition for high-productivity soybean flour processing to produce enriched soy protein and concentrated hydrolysate of fermentable sugars. *Bioresour. Technol.* 241: 252–261.

Part V

Enzymes used in Fine Chemical Production

5.1

KREDs: Toward Green, Cost-Effective, and Efficient Chiral Alcohol Generation

Chris Micklitsch, Da Duan, and Margie Borra-Garske

Codexis Inc., Departments of Biochemistry and Process Chemistry, 200 Penobscot Drive, Redwood City, CA 94063, USA

5.1.1 Introduction

Reduction of ketones to generate chiral secondary alcohols is one of the most predominant asymmetric transformations in pharmaceutical manufacturing. Enantiopure secondary alcohols have been exploited as building blocks to generate a wide range of pharmaceutically relevant products [1]. Traditional synthetic routes are commonly used, although the use of biocatalysts in such transformations is becoming more prevalent.

Two traditional synthetic methodologies toward asymmetric induction are considered in modern organic synthesis, and these are dependent on the chirality of the substrate molecule (Figure 5.1.1). In the case of substrate-guided synthetic routes, the reducing agent is guided by the chirality already present in the substrate to yield the desired diastereoisomer. In this manner, achiral sodium borohydride provides a straightforward and preferred way for large-scale asymmetric reductions. Sodium borohydride is commercially inexpensive and readily available in bulk and in various forms. Reactions are typically carried out using sodium borohydride in combination with additives such as cerium chloride, triethylborane, or diethylmethoxyborane. An alternative to sodium borohydride is lithium aluminum hydride (LAH), which has also been used for large-scale ketone reduction. However, LAH has lower chemical selectivity, and solid LAH is highly flammable and may ignite in moist or heated air, making it a less desirable reagent. Aluminum salts often complicate reaction workup and product isolation. Another alternative is the aluminum isopropoxide/isopropanol mixture for the so-called Meerwein–Ponndorf–Verley reduction, which is a greener and more cost-effective process compared to sodium borohydride and LAH methods. However, Meerwein–Ponndorf–Verley reactions have seen limited adoption to this point [1].

In the case of reagent-guided synthetic routes, the substrates lack chirality but have prochiral centers that allow for the formation of new chiral center(s) upon

Industrial Enzyme Applications, First Edition. Edited by Andreas Vogel and Oliver May.

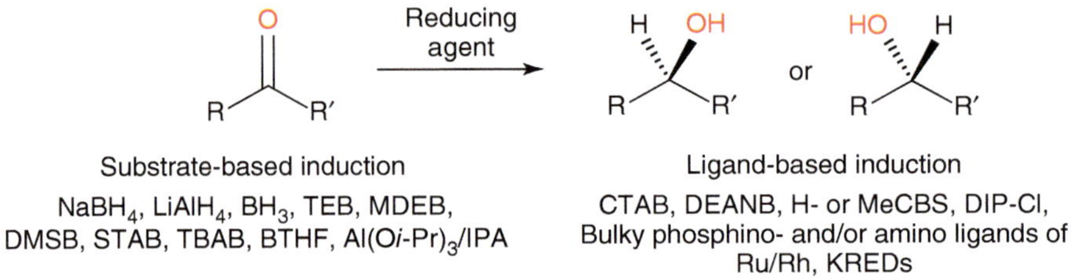

Figure 5.1.1 Approaches to asymmetric induction of ketones by reducing agents. Examples of reducing agents are divided between those that rely on previously established molecular chirality and those that use the ligand framework to direct the reduction and establish a particular stereoisomer. TEB, triethylborane; MDEB, methoxydiethylborane; DMSB, dimethylsulfide borane; STAB, Sodium triacetoxyborohydride; TBAB, tetrabutylammonium borohydride; BTHF, borane-tetrahydrofuran complex; CTAB, catecholborane; DEANB, diethylaniline borane; MeCBS, Methyl-Corey–Bakshi–Shibata ligand.

reduction. The prochiral faces of the substrate can be differentiated by chiral reducing agents that deliver the hydride selectively and provide the desired stereoisomer. There are two reagent classes that have shouldered the lion's share of the work in this type of reduction and have been so important that they have been recognized with Nobel Prizes (see Figure 5.1.4). Oxazaborolidine-mediated reagents make use of catalytic amounts of Corey–Bakshi–Shibata (CBS) oxazaborolidines, along with stoichiometric application of the borane for the synthesis of alcohols in high enantiomeric ratio. The Noyori class of ruthenium catalysts are complexed with bulky chiral phosphino and/or amino ligands to guide ketone hydrogenation in the presence of H_2. Another historically successful approach employed stoichiometric addition of an asymmetric borane reagent such as β-chlorodiisopinocampheylborane, or DIP-Cl. This method is now disfavored, replaced by modern chemical and biochemical asymmetric catalytic methods [1].

The need to have efficient, scalable, cost-effective, and green processes to generate pharmaceutical intermediates has underscored the importance of biocatalysis in the pharmaceutical industry. Enzymes with improved and novel properties have been great alternatives for equivalent or less efficient chemical processes. Biocatalysis has been increasingly utilized in manufacturing intermediates to support pharmaceutical development, from safety evaluations to development of commercial processes to clinical trials. Throughout these stages of pharmaceutical development, speed, production cost, robustness of process, and safety are major drivers for manufacturing and commercialization. In this review, we will focus on the biocatalytic routes for ketone reduction to generate chiral secondary alcohols. We will delve into the substrate scope, as well as stereoselectivity of the biocatalytic process. We will discuss the resulting process improvements and beneficial environmental consequence of using ketoreductase (KRED) enzymes. In addition, we will highlight protein engineering efforts that are driving the transformation of KRED enzymes to fit the demands of modern pharmaceutical manufacturing processes.

5.1.2 Ketoreductases

The application of isolated ketoreductases for commercial synthesis of chiral alcohols has been on the rise in the last few decades [2]. Many of these industrially relevant enzymes have been derived from bacteria and yeast, such as *Rhodococcus*, *Lactobacillus*, *Pseudomonas*, *Thermoanaerobium*, *Candida*, and *Saccharomyces* species. *Saccharomyces cerevisiae*, Baker's yeast, was first used to demonstrate biocatalytic ketone reduction a century ago [3]. Merck, with parallel fermentation optimization efforts, has successfully employed whole cell yeast for chiral alcohol production [2]. In conjunction with technology improvements, ketone reduction has been accomplished using isolated KRED enzymes coupled with cofactor recycling systems. Engineered KREDs have been used in isolated form to commercially generate chiral intermediates for the production of atorvastatin (Lipitor®), montelukast (Singulair®), and atazanavir (Reyataz®), to name a few. KREDs have been shown to reduce a wide range of substrates and with high enantioselectivity (>99% enantiomeric excess [ee]) and high yield (>90%) [2].

KREDs are part of the short-chain dehydrogenases/reductases enzyme family. Many homologs exist, and although sequence identity among this class of enzymes is low, three-dimensional structures determined so far show similar α/β folding pattern [4–8]. Crystal structures show homotetrameric enzymes, with bound Mg^{2+} ions (Figure 5.1.2a). One of the conserved domains is on the interior surface of an α-helix, containing the active-site catalytic triad of serine, tyrosine, and lysine residues (Figure 5.1.2b). An active-site asparagine residue was shown to be important for stabilizing the active-site lysine in many homologs. Typical KRED reactions proceed by the following steps: (1) Cofactor, NAD(P)H, binding to the active site of the enzyme, (2) hydride transfer from the cofactor to the carbonyl carbon of the substrate and concurrent proton transfer from the active-site tyrosine to the substrate carbonyl oxygen atom, (3) release of the chiral alcohol product from the enzyme-oxidized cofactor complex, and (4) release of the oxidized cofactor from the enzyme. A proton relay is formed during the reaction, involving the 2′-hydroxyl of the ribose group of NAD(P)H, the active-site lysine, and a water molecule [9].

Ketoreductases are active on a wide range of substrates and act with exquisite regio- and stereoselectivity. For example, KREDs are able to distinguish subtle steric and electronic differences in substituted diaryl ketone substrates to produce a myriad of diaryl methanol products [10]. A flexible substrate-binding loop, the most variable portion of the enzyme, plays a key role in determining the reactivity and stereoselectivity of the KREDs. Mutations in the loop affect the reactivity and stereoselectivity of the enzyme [8, 11]. The orientation of the substrate-binding loop opens or closes the small binding pocket (SBP), facilitating effective binding and catalysis of different substrates.

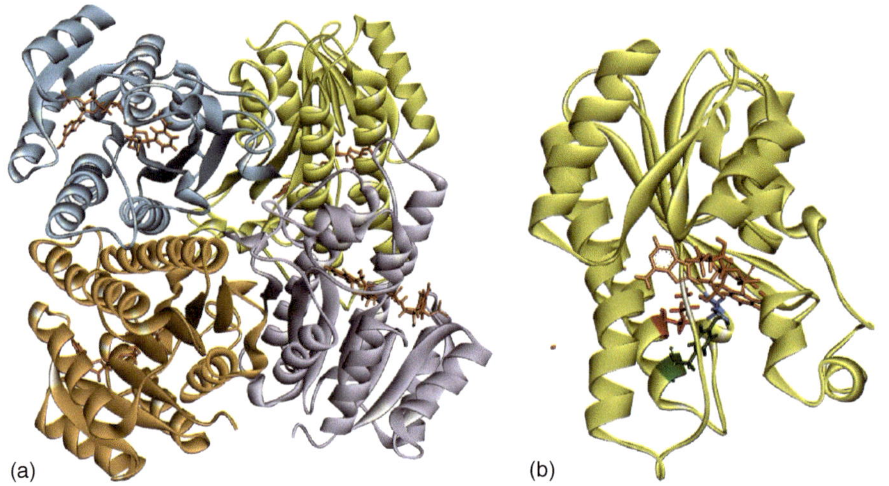

Figure 5.1.2 *Lactobacillus kefir* KRED crystal structure with bound NADP$^+$ (4RF2) [8]. (a) The structures show a homotetrameric enzyme (each monomer is shown in a different color), with α/β folding pattern similar to other short-chain dehydrogenases/reductases. (b) Each monomer is composed of two domains, forming a large binding pocket (LBP) and small binding pocket. The cofactor (shown in brown) and the substrate bind in these binding pockets where catalysis is carried out by the catalytic triad (serine in blue, tyrosine in green, and lysine in red).

5.1.3 Cofactor Recycling

KRED enzymes require NADH or NADPH as hydride sources during the reaction. NADH and NADPH are costly; therefore, recycling of these cofactors is desirable at all scales. A few recycling enzymes are commonly used (Table 5.1.1): glucose dehydrogenase (GDH), formate dehydrogenase (FDH), phosphite dehydrogenase (PDH) (for additional information, see the reviews [2, 9, 12]). GDH oxidizes glucose to gluconolactone, which spontaneously hydrolyzes to gluconic acid, while reducing NAD(P)$^+$ to NAD(P)H. GDH is highly active; however, formation of gluconic acid reduces the pH and requires pH control to carry out in large scale. pH control is typically performed by incorporating a pH titrator with the reaction vessel. When the pH drops, due to acid formation, sodium hydroxide (or any desired base) is added into the reaction to maintain the pH within the desired range. FDH irreversibly converts formate to carbon dioxide. While most wild-type FDHs have low activity and stability, FDH from *Candida boidinii* was utilized for large-scale synthesis of L-*tert*-leucine [13]. PDH irreversibly converts sodium phosphite to sodium phosphate. The resulting high phosphate content, however, can make the workup more cumbersome.

In addition to the dehydrogenases mentioned above, KREDs can also catalyze the reverse reaction, converting an alcohol substrate to the corresponding ketone and reducing the cofactor in the process. In a substrate-coupled approach, KRED oxidizes, for example, isopropanol to acetone, and reduces NAD(P)$^+$ back to NAD(P)H [2, 9, 12, 14].

Table 5.1.1 Commonly used recycling systems to generate NAD(P)H. Most of the recycling systems require an additional enzyme in the process.

Enzyme	Reaction	Comments
Glucose dehydrogenase (GDH)	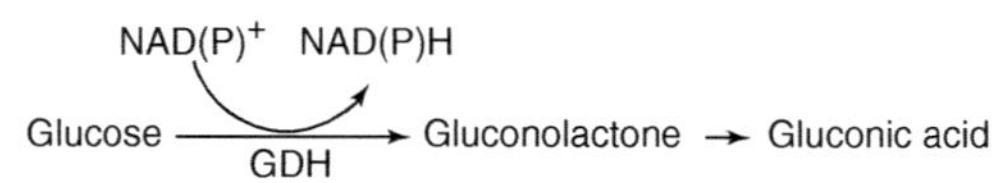	Requires a pH control system to compensate for the gluconic acid production
Formate dehydrogenase (FDH)	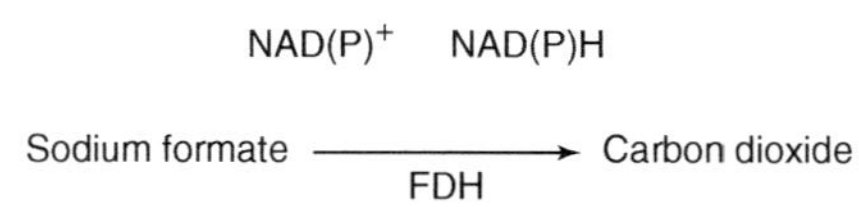	Product is easily removed. However, enzyme has low activity and stability, undesirable for large-scale manufacturing
Phosphite dehydrogenase (PDH)	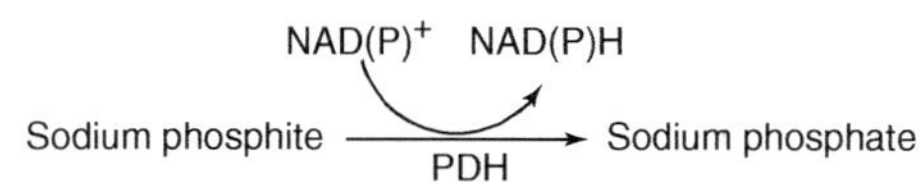	Reaction is irreversible. Reaction components are exposed to high ionic strength and high phosphate concentration
KRED	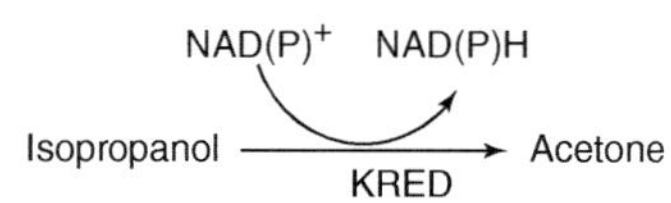	The same enzyme can be used for both the asymmetric ketone reduction and cofactor recycling. The co-product is volatile and can be removed with engineering controls

5.1.4 CodeEvolver® Protein Engineering Technology

With increasing demand for biocatalysts in pharmaceutical space, engineering of commercially and industrially relevant enzymes has become critical. Companies, such as Codexis, c-LEcta, Johnson Matthey, and Almac, have a wide array of KREDs that are commercially available. The Codex® KRED panel plates and kits, for example, consist of engineered enzyme variants that were selected to perform under a wide range of functional space. Some of the variants were chosen due to their high likelihood of activity toward a broad range of possible substrates. Others were chosen because of their ability to function under typical process conditions, including tolerance to high temperature and high solvent levels. Additional ones were chosen for their enantioselectivity, cofactor utilization, expression level, compatibility with high substrate load, enhanced overall robustness, in process stability, and long shelf life.

Depending on the substrates and process conditions, many of the enzymes in the Codexis collection can be used in a "plug-and-play" manner; meaning the enzymes may have the desired characteristics to allow for immediate utilization. In cases where initial activity is insufficient and enzyme improvement is desired, Codexis' CodeEvolver® protein engineering technology allows for a rapid directed evolution approach to improve biocatalysts toward the desired reaction and manufacturing process. The CodeEvolver protein engineering technology involves generation of variants of the desired enzyme, identifying variants with the necessary improvements for process application, and using Codexis' proprietary statistical analysis software to deconvolute the roles of each mutation (Figure 5.1.3, see also Chapter 1.2). In a typical round of evolution, a few thousand variants, containing single or a combination of mutations, are generated. High-throughput screens, which target enzyme activity (increased activity, pH stability, thermostability, etc.) in actual or process-like conditions, are developed and utilized in order to screen variants for the desired improvements. Using Codexis' ProSAR™ or MOSAIC® statistical analysis software, the contributions of individual amino acid substitutions are deconvoluted, and mutations are categorized as beneficial, neutral, or deleterious. Beneficial mutations are recombined onto the most improved variant. Evolution is repeated until an enzyme that is fit for purpose is identified.

5.1.5 Reduction of a Wide Range of Ketones/Aldehydes

As evident in recent literature, KREDs have become a major class of enzymes for the manufacture of chiral alcohols at industrial scale. A number of previous reviews highlight the utilization of these enzymes at very low catalyst loadings and harsh process conditions to successfully produce enantiopure chiral alcohol intermediates in drug manufacturing processes [12, 15–18]. Table 5.1.2 is an up-to-date list of the continuously growing number of disclosed substrate and product pairs transformed by KRED enzymes with high productivity, as well as high regio- and stereoselectivity.

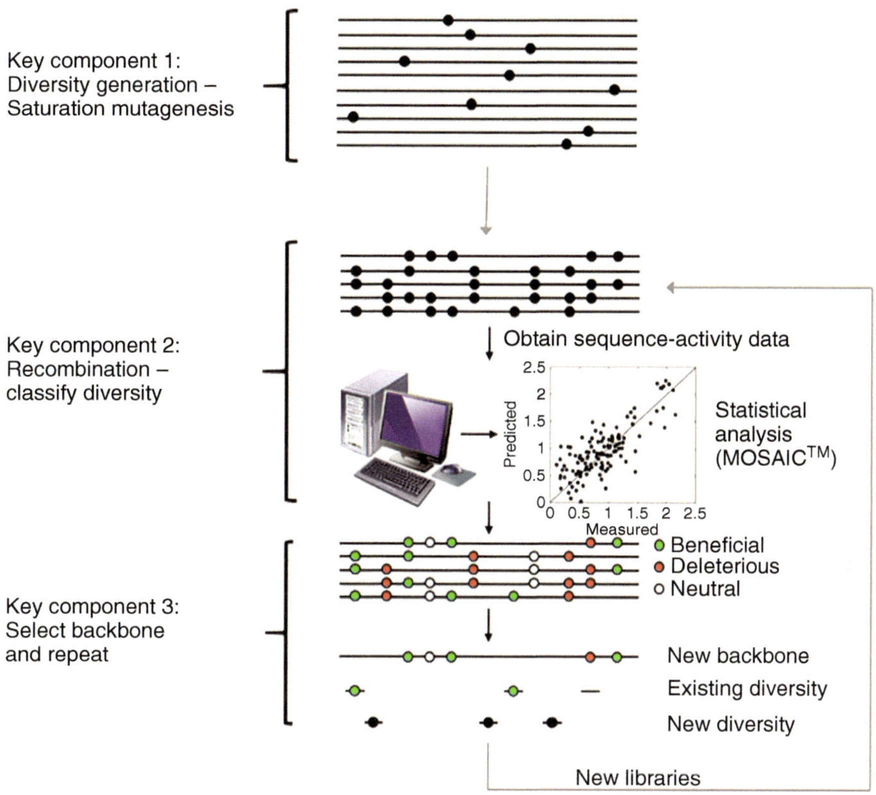

Figure 5.1.3 Codexis' CodeEvolver® protein engineering technology. The technology involves generation of variants with single or multiple mutations, screening the variants for the desired characteristics, deconvoluting the effects of individual amino acid substitutions, and combining beneficial mutations onto a new and improved variant.

As shown in Table 5.1.2 and in other reviews, it is obvious that KREDs can be engineered to work on substrates with diverse chemical properties. KREDs are able to transform substrates of different sizes; in particular, one of the smallest substrates **2B** is almost five times smaller than that of the Montelukast intermediate **2O**. Using the CodeEvolver protein engineering technology, Codexis engineered highly productive KREDs for these distinct substrates: **2B, 2D, 2L, 2M, and 2O**. In addition, KREDs can reduce substrates with a large range of hydrophilicity (*A*log *P* of −0.30 to 7.76).

The ability of KREDs to reduce difficult substrates is exemplified by (*S*)-1-(2,6-dichloro-3-fluorophenyl)ethanol (DCFPE) (**2G**). Despite many efforts to perform asymmetric reduction of DCFPE by traditional chemistry means, there has been only one report of successful chemical enantioselective reduction of **2G** to the *R* enantiomer of the alcohol product with 95% yield and 96% ee [34, 35]. However, the *R*-enantiomer was not the desired enantiomeric product. The challenge in selectivity may be attributed to the 2,6-dichloro substitution pattern on the phenyl ring. Steric repulsion between

Table 5.1.2 KRED Substrates described in recent literature.

Entry	Substrate	Product	Application	MW	Alog *P*	References
A	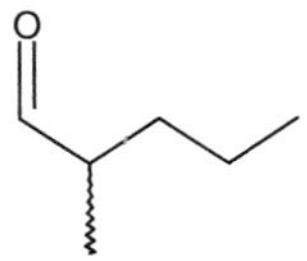	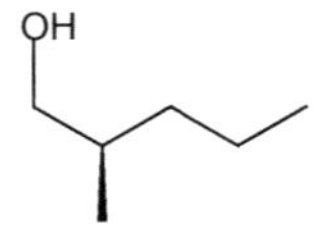	Intermediate for pharmaceuticals and liquid crystals	100	1.86	[19]
B	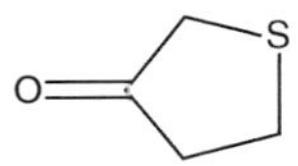	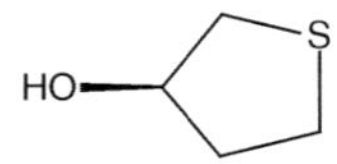	Sulopenem intermediate	102	0.31	[20]
C	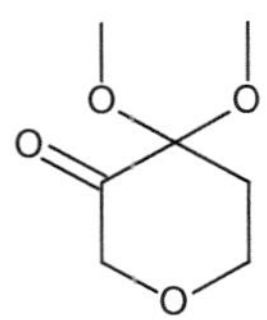	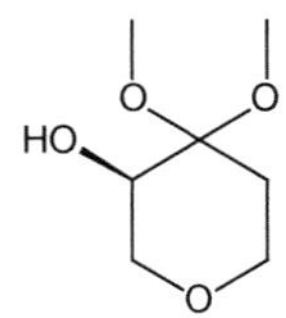	Chemokine receptor inhibitor intermediate	160	−0.30	[21]
D	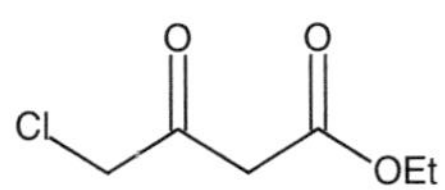	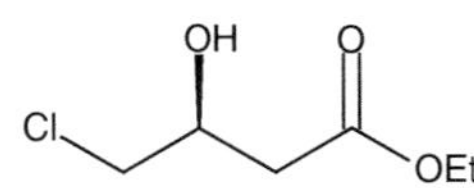	Atorvastatin intermediate	164	0.81	[22]
E	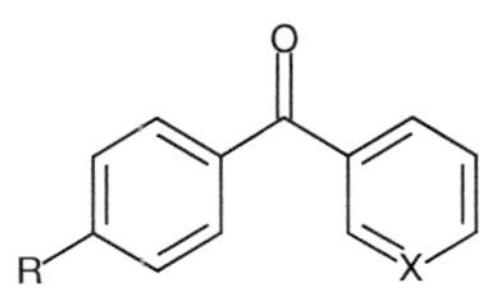	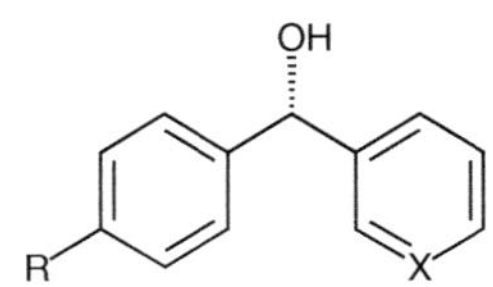	Drug API intermediate	>182	—	[10]
F	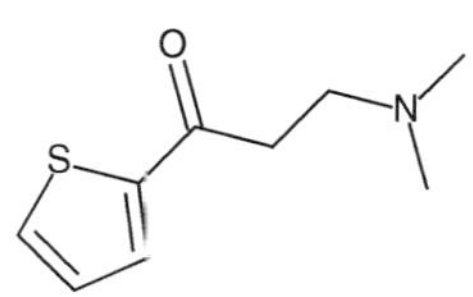	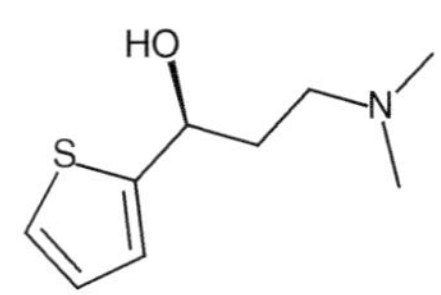	Duloxetine intermediate	183	1.78	[23]

G	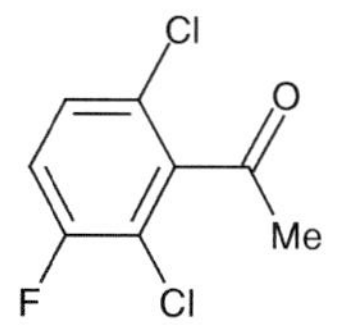	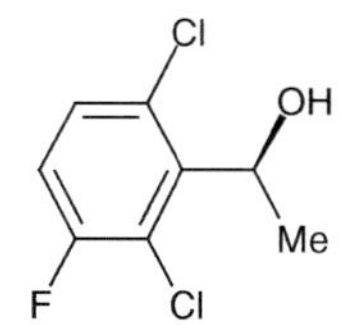	HGFR-antagonist (PF-2341066) intermediate	206	3.10	[24]
H	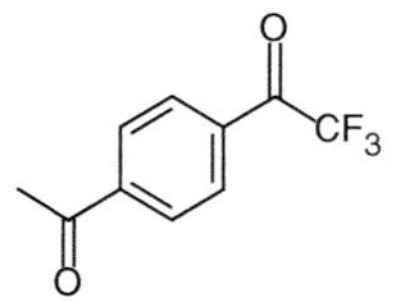	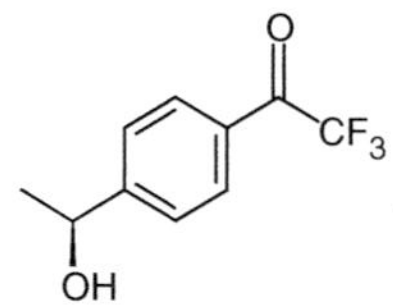 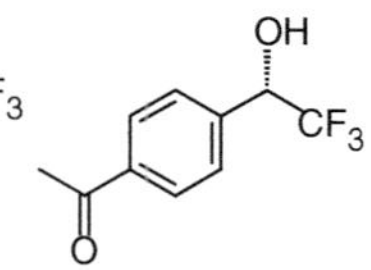	Drug API intermediates	216	2.40	[25]
I	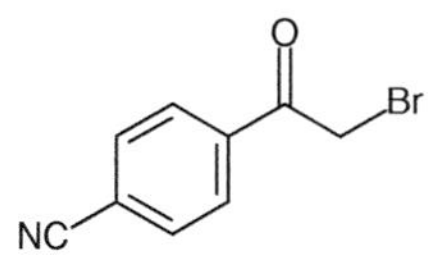	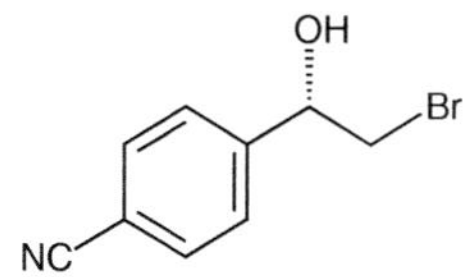	SIP receptor agonist (BMS-960) intermediate	223	2.13	[26]
J	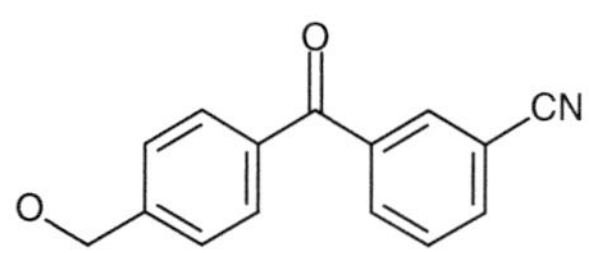	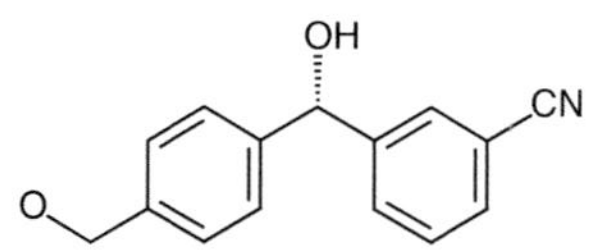	mGlu2 potentiator intermediate	237	2.51	[27]
K	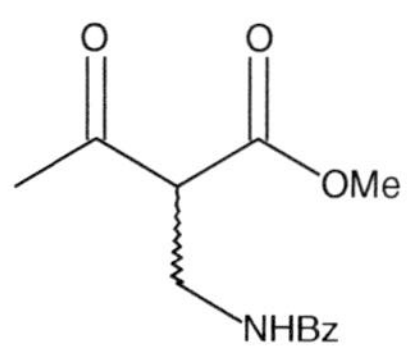	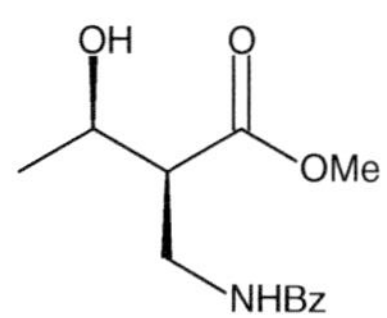	Azetidinone intermediate	251	—	[28]

(Continued)

Table 5.1.2 (Continued)

Entry	Substrate	Product	Application	MW	Alog *P*	References
L	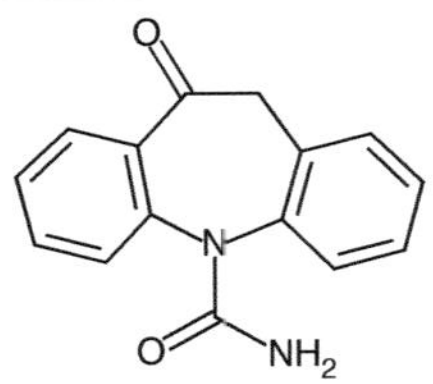	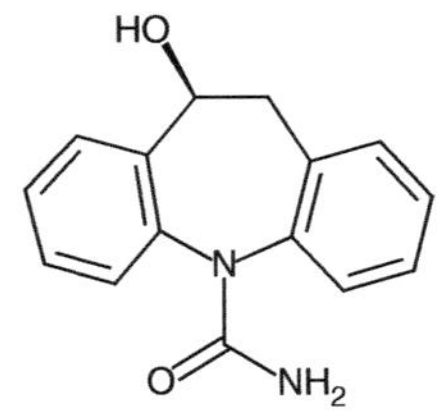	(*S*)-Licarbazepine	252	2.13	[29]
M	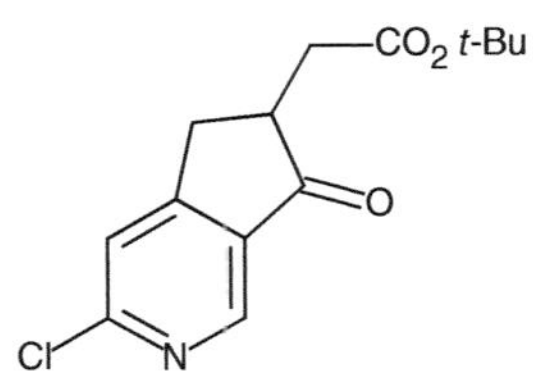	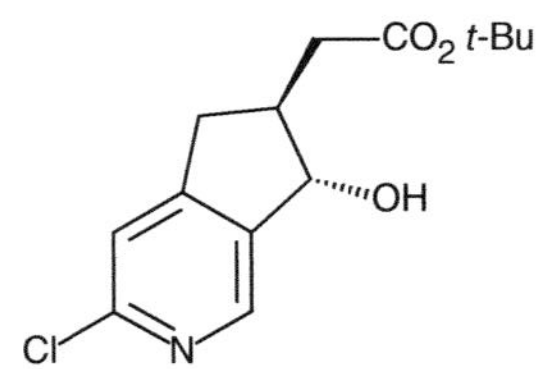	MK-8666 intermediate	296	—	[30]
N	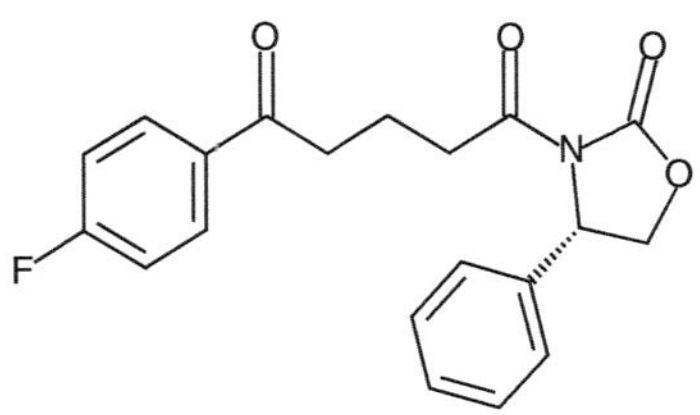	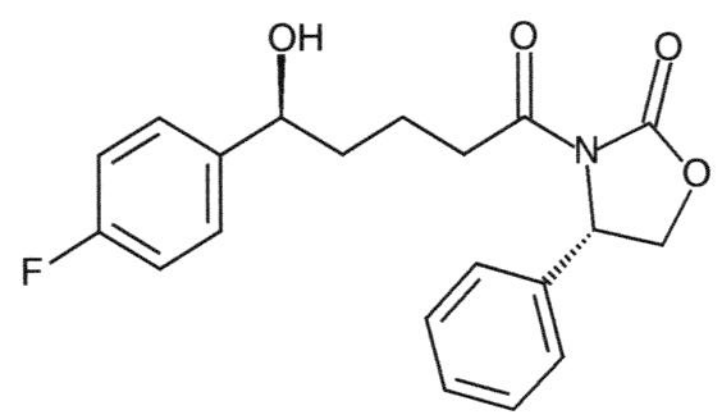	Ezetimibe intermediate	355	3.72	[31, 32]
O	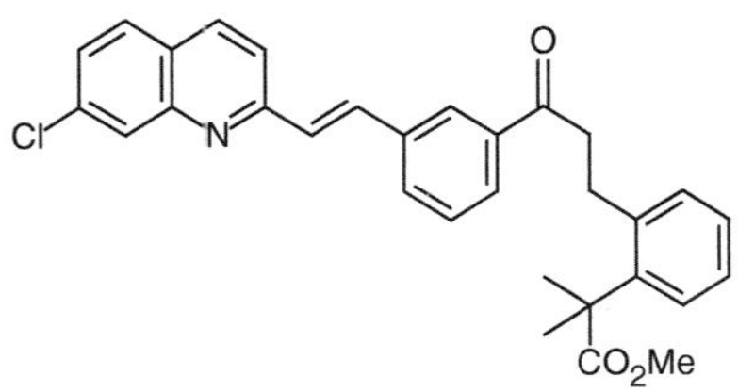	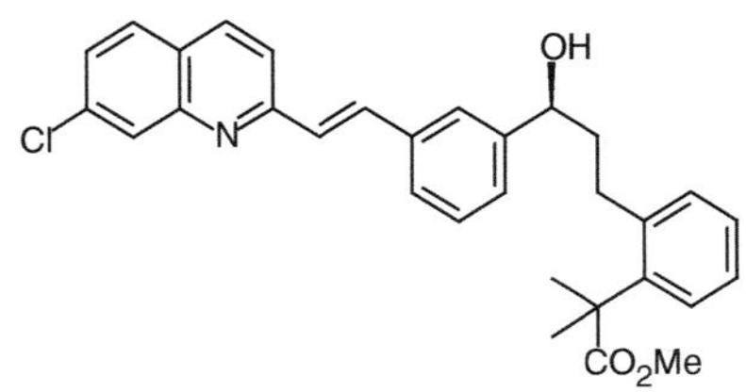	Montelukast intermediate	498	7.76	[33]

the two *ortho*-chloro substituents and the methyl ketone functionality leads to the orthogonal positioning between the plane of the aromatic ring and the plane of the methyl ketone. Unlike typical acetophenone derivatives, the predicted ground state conformation of (2,6-dichloro-3-fluorophenyl) ethanone is not planar but rather perpendicular. In this conformation, the chlorine atoms effectively "shield" the carbonyl carbon from the incoming hydride, making 1-(2,6-dichloro-3-fluorophenyl) ethanone significantly more bulky ("3-dimensional") than at a first glance ("2-dimensional"). Utilizing the CodeEvolver protein engineering technology, Codexis evolved a KRED to have the opposite, but desired, selectivity. After three rounds of evolution, an evolved KRED showed >99% conversion of 100 g/l of the substrate at 1% enzyme in 24 hours. The isolated yield was 93%, and the chemical and optical purities were both >99.5%.

In traditional chemical asymmetric reductions, it is desirable to use substrates with electronically or sterically differentiated groups flanking the target carbonyl in order to impart a high degree of enantioselection. However, in biocatalytic transformations, KREDs do not require high degrees of differentiation between flanking groups in order to exert very high enantioselectivity, as demonstrated by Truppo et al., using a range of substituted benzophenones and benzoylpyridine substrates (**2E**) [10]. Additionally, substrates such as the atorvastatin intermediate (**2D**), fluorinated hydroxyketones (**2H**), and the ezetimibe intermediate (**2N**) have multiple reducible ketones, making regioselective reduction challenging via conventional chemical catalysts. KREDs, on the other hand, displayed exquisite stereoselectivity and regioselectivity for each of these substrates.

As mentioned in a prior section, the Codexis KRED panel is a collection of KREDs that had shown unique activity in Codexis KRED directed evolution programs over the years. KREDs evolved for activity toward a different substrate could be appropriated as a starting enzyme for an entirely different substrate under different process manufacturing conditions. This fact was elegantly demonstrated on KREDs from the *Lactobacillus kefir* lineage. In the initial stage of DCFPE manufacturing, none of the commercially available KREDs were active toward the DCFPE substrate **2G**. Codexis screened KRED variants, evolved originally for activity toward the sulopenem substrate **2B**, and identified a variant weakly active toward the DCFPE substrate. Using the CodeEvolver protein engineering technology, three rounds of directed evolution produced a progeny that was suitable for large-scale manufacturing of the DCFPE intermediate.

Not all KRED enzymes identified by panel screening require further improvement to fit the manufacturing process. A number of KREDs from the Codexis collection have served as "plug-and-play" enzymes. Two Codexis KRED enzymes displayed sufficient activity and selectivity toward **2I** and **2J** substrates in which no additional improvement was required.

5.1.6 Critical Selectivity Tools for Enantiopure Asymmetric Carbonyl Reduction

The source of the enantiodiscrimination of any enzyme is derived from the three-dimensional configuration of its binding pocket. The binding pocket orients the substrate with respect to the active site residues and the relative position of the cofactor, directing the chemical conversion spatially. Conversion proceeds in a manner in which one configuration of the product is preferred, such as in the case of KREDs, where the addition of the hydride from the NAD(P)H cofactor is to a specific face of the ketone. When it comes to generating chiral alcohols from ketones, the groups on either side of the carbonyl are key contributors in determining how the molecule orients in the binding pocket. In some cases, one side will contain a larger or bulkier group compared to the other side of the ketone. Depending on the size, shape, and polarity of the KRED active site, accommodation of the groups adjacent to the carbonyl determines the prochiral face of the ketone that is presented for reduction. In this scenario with one larger and one smaller binding site, the enzyme has clear landmarks by which to distinguish and orient the substrate.

Until recently, producing enantiopure compounds in high yield and efficiency had been one of the most challenging and desirable goals in pharmaceutical development [36]. Creating novel chirally active pharmaceutical ingredients (APIs) requires performing chemistry with precise spatial control and high efficiency [37, 38]. There are many stereoselective tools available to synthetic chemists for setting the stereochemistry of a desired pharmaceutical target. In this section, we will delve more into reagent-guided asymmetric reduction of ketones. Figure 5.1.4 displays the key reagents used in pharmaceutical production. Generally, the CBS catalyst is used along with boranes, complexed with ligands covering a spectrum from dimethyl sulfide, ethers (such as $BH_3 \cdot THF$), and amines, as well as catecholboranes [39]. $BH_3 \cdot SMe_2$ is the most commonly used stoichiometric reducing agent, despite the olfactory drawback associated with the process. Use of these reagents, which results in large volumes of waste

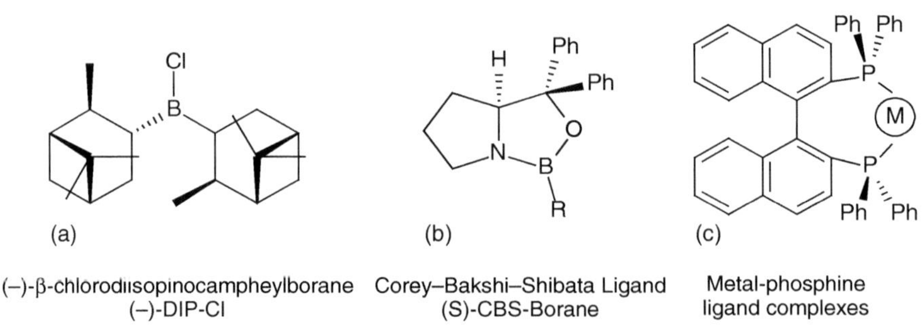

Figure 5.1.4 Typical chemical reagents and catalysts for asymmetric reduction of ketones and aldehydes. (a) Several types of hindered boranes are available for asymmetric hydroboration of carbonyl groups. (b) The R group of the CBS catalyst can be acquired with alkyl or aryl functionality. (c) The BINAP ligand shown is just one of several chiral bulky bis-phosphine ligands coordinated with a transition metal (M), primarily Ru, for catalytic hydrogenation.

due to the stoichiometric application of the borane ligands, requires significant care and handling. Similarly, Brown's DIP-Cl has seen reduced use due to similar concerns, as well as by-product formation associated with pinene, which requires extensive purification efforts [1, 39]. In addition, chiral phosphine architectures used along with the reductive power of transition metals, such as ruthenium, have also had good success [1, 40]. They are highly selective and efficient in controlling stereospecific installation of key alcohol functionality. All of these methodologies have been extremely effective for enantioselective reduction; however, Noyori himself stated that there are no universal catalysts because of the complexity of organic compounds [41], and so the search continues for more selective and efficient catalysts with greater flexibility of application.

With the turn to the more environmentally responsible, as well as economically competitive routes to produce new drug candidates, the search for precise biocatalytic agents has accelerated dramatically since the start of the new millennium [42–45]. The limitations of biocatalysis in pharmaceutical development have been discussed in the literature [44, 46]. However, with respect to stereochemical quality of the products obtained, biocatalysis often times offers the far superior route [47]. In the case of redox reactions, oxidoreductases present a target-rich environment, comprised of nearly a quarter of all known enzymes, for finding key tools to use in developing redox-based stereochemical routes. Many of these enzymes provide supreme chemo- and stereochemical control over the production and maintenance of metabolites essential to their hosts. Unfortunately, whole cell compositions or enzymes directly from Nature are, in large part, not sufficient without some level of engineering for application in pharmaceutical processes. This is why till recently they have been ignored by the industry. Many of these limitations have now been addressed by introducing recombinant DNA technology and enzyme engineering.

Enhanced by the power of directed evolution, KREDs are now looked at as the first choice in developing routes to novel therapeutic targets. The many advantages associated with incorporating KREDs in production of APIs are discussed throughout this chapter. It must be considered though that the primary benefit of KREDs, with or without protein engineering improvements, is the exquisite stereochemical control that they can provide for setting stereocenters in new chiral pharmaceutical alcohols [42, 45, 48].

There are two methodologies with which KREDs are applied to derive chiral alcohols, one by kinetic resolution of racemic mixtures and another by creation of one or more chiral centers from a prochiral keto substrate (Figure 5.1.5a). Recovery of an enantiomer via kinetic resolution is an inefficient route due to the inherent inability to reach greater than 50% recovery of the desired product. However, dynamic kinetic resolution (DKR) allows for 100% theoretical yield by providing a mechanism for rapid interchange between chiral configurations before conversion to the desired product is achieved. As can be seen in Figure 5.1.5b (iii), racemization of the target stereocenter is driven by either reaction conditions or use of a racemizing agent. What is critical for this methodology is the rapid interconversion of the racemate prior to conversion with the KRED that will provide

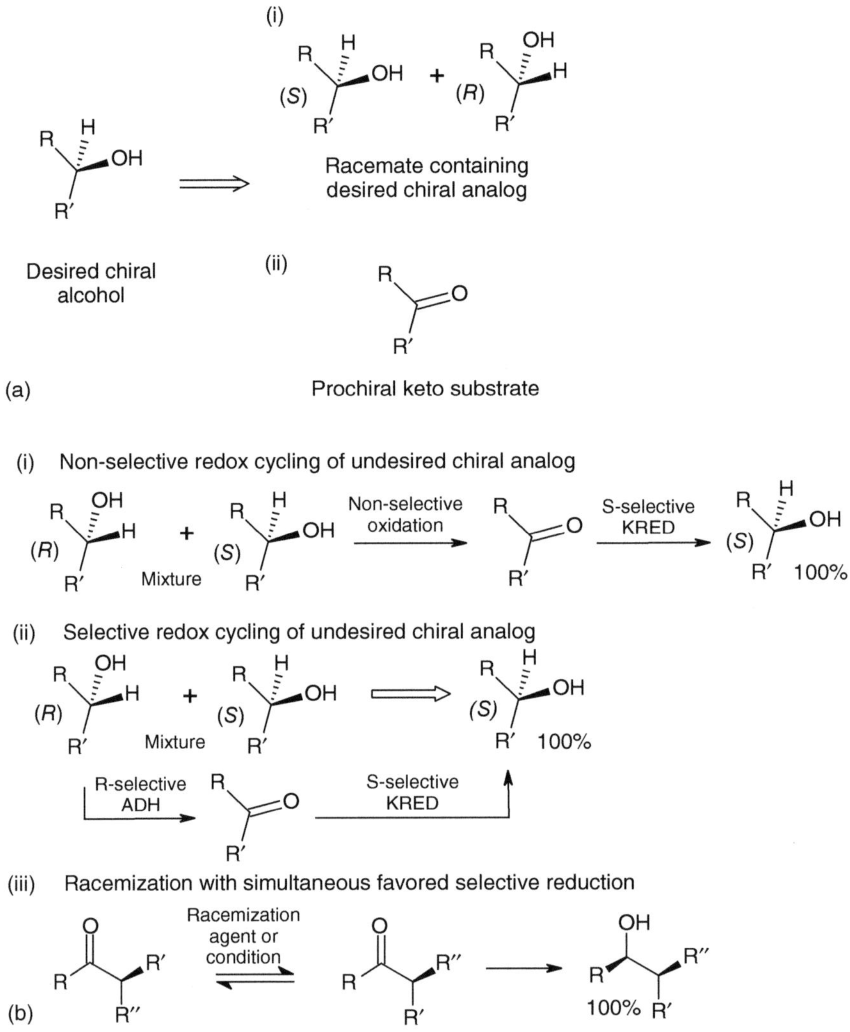

Figure 5.1.5 (a) Precursors to the preparation of chiral alcohols. (b) Approaches to performing dynamic kinetic resolution (DKR) to convert stereomeric mixtures to one specific stereoisomer.

only the desired enantiomer. A purely biocatalytic DKR requires a KRED to perform in its dehydrogenase (ADH) capacity in the oxidative direction, with a separate highly enantioselective KRED for performing the reduction to the desired final configuration.

Enzymatic DKR methodology has found application in pharmaceutical generation and will continue to grow as a vital tool for organic chemists [36]. For example, a KRED-driven DKR route was proposed to replace a traditional Noyori catalyst in the first step in the synthesis of the key intermediate of a β-lactam antibiotic [49]. The chiral starting material was prepared from the starting racemate via a DKR process driven by pH racemization of the cheap keto-ester (Figure 5.1.6), in combination with a selective KRED conversion to the desired chiral alcohol [28]. Remarkably, a KRED enzyme with the desired characteristics was rapidly engineered in just three rounds of directed evolution. The final KRED displayed a large increase in selectivity and productivity over the initial enzyme identified from the Codexis collection, providing an improved selectivity route over the Noyori catalytic reduction (Figure 5.1.7) [50].

Several chiral pharmaceutical targets have now been developed and produced with KREDs acting on prochiral substrates. For the antiasthmatic montelukast substrate (Table 5.1.2, entry O), the large central aromatic ring sits opposite the less sterically demanding alkyl tether of the methoxybenzoate group. There is a clear size distinction between the groups driving the orientation of the substrate

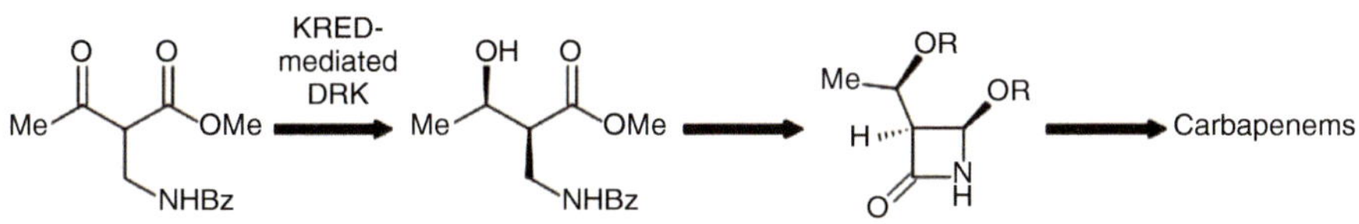

Stage	Loading (g/l) Substrate	Enzyme	NADP	Conversion	(2*S*,3*R*) Purity
Wild Type	20	5	0.1	<5%	~80% ee
Final KRED	250	1	0.01	100%	>99.9% ee

Figure 5.1.6 Carbapenem synthesis Incorporating KRED-mediated DKR. Reaction scheme that provides a means to apply KRED biocatalysis to set the desired diastereomer to produce carbapenem antibiotic. The associated table details the dramatic improvement in enzyme characteristics desired for process economics.

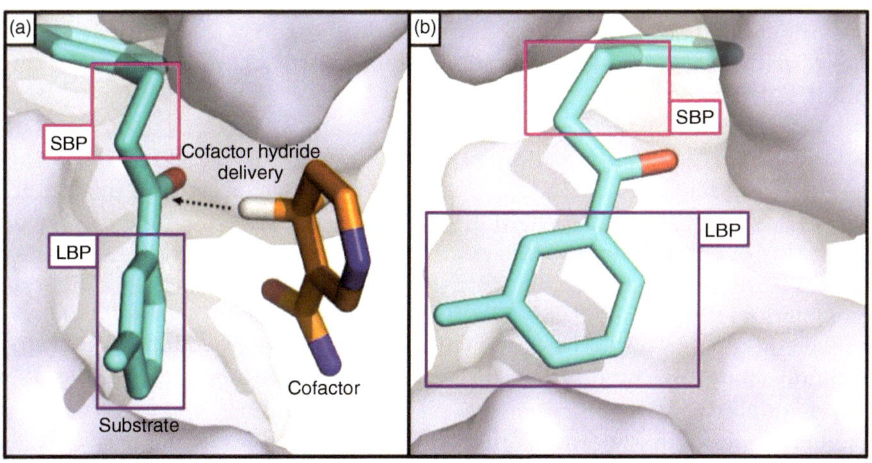

Figure 5.1.7 KRED binding pockets determine substrate orientation. (a) (top-down view) and (b) (side-on view) of a generic substrate fit in the large (LBP) and small (SBP) binding pockets around KRED's active site to demonstrate the manner in which the binding pockets direct the substrate's position and determine to which side of the ketone the cofactor's hydride is delivered.

in the binding pocket. In the initial enzyme selection at Codexis, a KRED variant from *L. kefir* showed great starting selectivity, with the substrate orientation that favored alcohol formation with the desired *S*-configuration. Using the CodeEvolver protein engineering technology, the selectivity was enhanced and essentially "locked in," to the point that additional rounds of evolution only required achiral analysis [33]. This arises as a result of elucidating key enzymatic residues essential to the development of a biocatalyst with the desired high degree of enantioselectivity. Once these positions are identified, they are rarely targeted for further DE probing. The final variant had excellent selectivity and the process was successfully scaled to a production process.

The extensive Codexis KRED collection, acquired through selection of a diverse pool of genes and directed evolution efforts, provides the opportunity to rapidly identify robust, highly active, and selective enzymes for identification of biocatalytic starting points for other drug candidates. In the case of ezetimibe, the API of Zetia®, a treatment for elevated cholesterol levels, an effort was made to apply the KRED methodology to its production. Application of the favored asymmetric hydroboration routes was capable, but provided room for improvement due to poor selectivity or over-reduction concerns [51]. In Codexis' effort to develop a KRED for this process, several rounds of evolution were carried out to generate a KRED with superior stereoselective control in the generation of the desired chiral alcohol.

More challenging substrates for pharmaceutical chemists are prochiral intermediates that have minimal structural differentiation or bulk to drive stereoselectivity (Table 5.1.2, entry L). Take, for example, eslicarbazepine, an anti-seizure treatment, which is relatively flat and symmetric with no distinguishing features by which to differentiate along its C_2 axis. A number of chemical methods were able to obtain decent to good selectivity; however, they suffered from productivity or other process concerns [51–58]. Codexis was able to develop a KRED along with a production process that yielded efficient production of the drug with excellent selectivity (>99% ee S) [29], despite the unbiased prochiral limitations of the substrate [29, 59].

Perhaps the most challenging substrates, at least with respect to a lack of prochiral bias, are symmetrical small molecule intermediates. Consider the cases of two precursors for the development of a protease inhibitor and an antibiotic (Figure 5.1.8) [20, 59]. Both pharmaceuticals require small, chiral, heterocyclic cyclopentanones, each with ring structures having a pseudo symmetry; a heteroatom versus a methylene is the only discernable structural element and is isosteric in the case of sulfur. In previous examples, the size of the groups on either side of the carbonyl was the primary feature for directing the orientation within the binding pocket of the KRED. However, here the KRED has to distinguish minor differences in the atomic volumes and electrostatics of the heteroatoms in the ring. What is understood now is that there are key positions around the binding site whose residues can be mutated to provide this sensitive chiral selectivity [8].

Looking at the crystal structure alone is not enough to understand the substrate/cofactor/ KRED relationship because enzymes are dynamic structures that are constantly "breathing." After molecular dynamics (MD) simulations [8] of

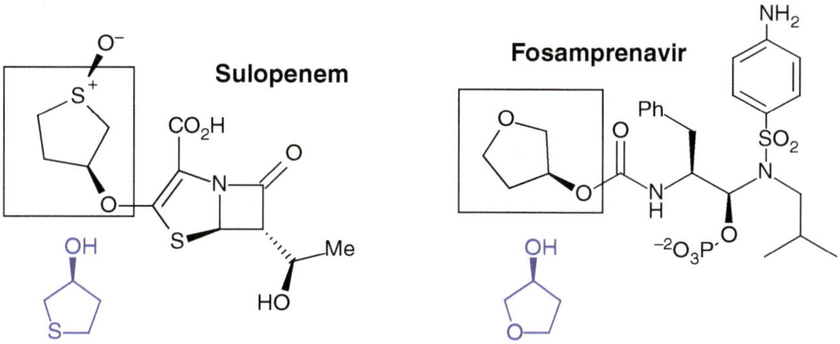

Figure 5.1.8 Application on small molecule intermediates. Small heterocyclic cyclopentanol components of antibiotic and protease inhibitor drug APIs produced from KREDs. Despite the small size and limited structural architecture to act on, KREDs can still provide high enantioselectivity.

the KRED with NADPH and the thio- or oxycyclopentanone substrates, it was clear that the most perturbed region was the substrate binding loop (residues ~190–210), where key contacts are made with the substrate and cofactor [20]. By tracking the distance of the residues within the binding loop, open and closed conformation of the binding pocket could be followed. By examining specific residue modifications to the KRED, it was observed that evolved KRED mutants in each case exhibited a surprisingly greater flexibility than the wild type that gave rise to the pro-R or S chirality with respect to the substrate. Depending on the substrate and which heteroatom was part of the ring, the enzyme flipped the orientation of the cyclopentanone ring along the C_2 axis. Interestingly, the modifications in the final KRED variant made key changes to either “side” of the binding pocket in order to be sensitive to slight spatial differences in each ring, leading to those orientation changes with each respective cyclopentanone substrate.

What is remarkable to note about all of the KREDs utilized for the critically selective reductions presented herein is that they are all of a lineage from the *L. kefir* organism. It is not the only KRED with the ability to offer effective asymmetric reduction, yet it highlights the substrate range and potential that can be developed from a single enzyme starting point. From substrates in which orientation is easily discerned due to clear difference in structure and size, to substrate with clear electrochemical differences, KREDs have been molded by directed evolution to provide exceptional stereoselectivity at the process level. This will continue to offer synthetic chemists a flexible tool with the means to effectively prepare highly enantiopure chiral alcohols.

5.1.7 Examples of Improved KREDs for Improved Manufacturing

Generation of pharmaceutical compounds using traditional chemical methods often requires multiple synthetic steps, and frequently results in the formation of extraneous compounds or impurities, which then require cumbersome purification. Chemical steps often involve protection and deprotection of functional groups or addition of groups to direct the required transformation to the desired location. These steps not only add cost and additional waste, but typically produce impurities, which can be unreacted starting materials, undesired stereoisomers or by-products of the reaction, and degradation products, all of which could have deleterious effects on the downstream process and effectiveness of the pharmaceutical ingredient, and affect the stability of the desired compound. Onerous purification workup is often required to isolate the desired product and to meet regulatory agency guidelines for purity. Extensive purification typically results in reduced product yield. In this section, we will highlight a few examples in which the use of KREDs has improved the manufacturing process.

Generation of duloxetine hydrochloride presents a good example of using a KRED enzyme to circumvent the cumbersome traditional synthesis and the subsequent purification steps to increase the yield. Duloxetine, a thiophene derivative, belongs to a class of heterocyclic antidepressants known as serotonin-norepinephrine re-uptake inhibitors. It was originally produced by Eli Lilly and sold under the name Cymbalta®. One of the chemical steps during the manufacturing of duloxetine includes reduction of *N*,*N*-dimethyl-3-keto-3-(2-thienyl)-1-propanamine to the (*S*)-*N*,*N*-dimethyl-3-hydroxy-3-(2-thienyl)-1-propanamine. Using a traditional synthetic route, reduction of the ketone is performed using sodium borohydride in sodium hydroxide solution [60]. The resulting alcohol is extracted with methylene chloride, after which the organic and aqueous layers are separated. Purification also includes distillation of the solvent, followed by addition of hexane and heating of the reaction to reflux. After cooling the reaction and precipitating the solids, the purified material is dried. All these cumbersome purification processes to isolate the desired alcohol product reduce the yield to only 75%.

To develop an economically competitive route for the formation of the duloxetine alcohol intermediate, an alternative biocatalytic route was developed. Codexis evolved a KRED enzyme to perform the reduction (Figure 5.1.9) [23]. Although the wild-type KRED from *L. kefir* did not have activity toward the desired substrate, initial screening of different variants identified a few that

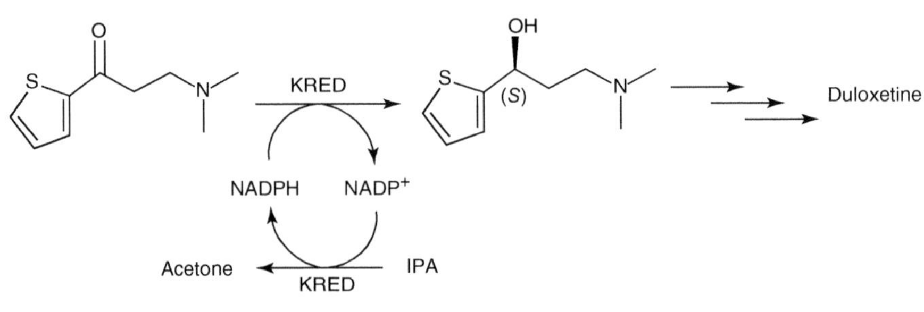

Figure 5.1.9 Biocatalytic reduction of a duloxetine intermediate. A KRED biocatalytic system that takes advantage of the irreversible formation of acetone, which can then be driven off by nitrogen or by distillation as a means to improve manufacturing processability.

produced the desired product. A variant containing 12 mutations relative to the wild type showed the best activity, with >90% of the product having the correct configuration, albeit having low conversion.

Initial process development showed that the enzyme was not active at the desired pH 11 when NaCl was present in the reaction; therefore, the reaction was carried out at pH 9. It was determined that the initial enzyme was not stable at temperatures greater than 25 °C, necessitating thermostability improvement, in addition to activity improvement. In order to improve cofactor recycling for the desired process conditions, higher isopropanol tolerance was required; the acetone produced is then easily removed by distillation or nitrogen sweeping. Using the CodeEvolver protein engineering technology and screening conditions employing high isopropanol concentrations, Codexis carried out three rounds of evolution to generate a final variant that was 150-fold improved in activity and thermostable. The final variant showed greater than 90% conversion, with greater than 90% of the product having the desired (*S*)-configuration [23].

In addition to duloxetine, generation of a key intermediate of atorvastatin also highlights the improvements gained in using biocatalysis. Atorvastatin calcium is an active ingredient in Lipitor, the first drug in the world with annual sales over $10 billion [22]. Because of high demand and high purity requirements, extensive efforts were carried out to make manufacturing cost-effective. The key chiral building molecule for atorvastatin is ethyl (*R*)-4-cyano-3-hydroxybutyrate (Figure 5.1.10). Several routes were developed to generate this key building block, including kinetic resolutions using microbes, use of (*S*)-hydroxy butyrolactone from chiral pool, and asymmetric reduction of ethyl 4-chloroacetoacetate derived from diketene. Other routes included the use of lipases and nitrilases, all of which required high enzyme loadings that were not commercially desirable, with the exception of an aldolase-based route.

Commercial chemical processes used to generate ethyl (*R*)-4-cyano-3-hydroxybutyrate include reaction of ethyl 3-hyroxy-4-halobutyrate with cyanide ion in alkaline solution at high temperatures [22]. However, this process resulted in by-product formation due to the base sensitivity of the substrate and the product. By-product formation and subsequent purification processes significantly decreased the yield. For example, Mitsubishi Chemical Corporation reported a yield of only 85% with chlorohydrin reaction at 80 °C and pH 10 [61]. Because the product and by-products have boiling points close to 80 °C, product recovery necessitated the use of high-vacuum fractional distillation. Further

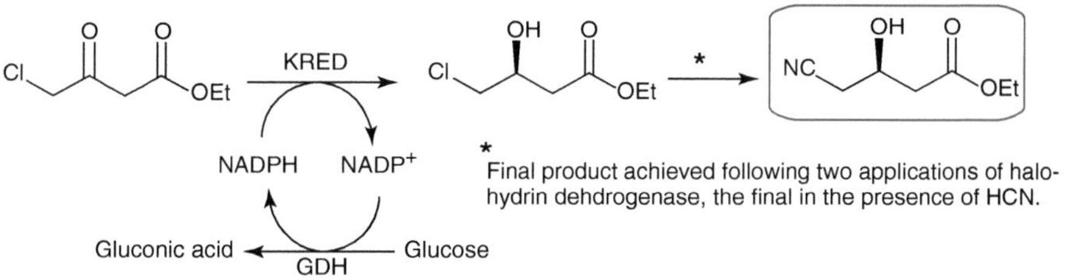

Figure 5.1.10 KRED route to ethyl (*R*)-4-cyano-3-hydroxybutyrate.

yield reduction from the distillation procedure made the process economically undesirable.

To circumvent this issue, a two-step, three-enzyme system was developed (see Figure 5.1.10) [22]. In the first step, ethyl 4-chloroacetoacetate was reduced to ethyl (*S*)-4-chloro-3-hydroxybutyrate using a KRED enzyme, with GDH used for recycling of the cofactor. The second step is the cyanation of the ethyl (*S*)-4-chloro-3-hydroxybutyrate to form the desired ethyl (*R*)-4-cyano-3-hydroxybutyrate building block.

The activity and stability of the wild-type KRED and GDH enzymes were insufficient to enable a scalable biocatalytic pathway. In initial reactions using 100 g/l of substrate, 6% (wt enzyme/wt substrate) of KRED and 3% (wt enzyme /wt substrate) of GDH were required for full conversion in 15 hours. The high enzyme loading also resulted in emulsion formation, which rendered product recovery less efficient. Directed evolution of the two enzymes was carried out using conditions as close to the process conditions as possible. After four rounds of evolution, GDH was improved 13-fold. Eight rounds of evolution were carried out to improve the KRED enzyme by sevenfold, sufficient for the desired process. Using the evolved enzymes, the substrate load was increased 60% (w/v), and the enzyme load was decreased by 17-fold for KRED and 12.5-fold for GDH to result in full conversion in about half the time. The lower enzyme load also diminished emulsions, resulting in higher product recovery.

The application of KREDs provides the possibility of aiding in product recovery as highlighted in a fairly recent report [62]. Typical biocatalytic reactions are carried out with enzymes dissolved in buffered aqueous conditions, at near neutral pH and relatively mild temperatures. However, reactions where enzymes are in solutions have a few disadvantages. Often, the desired product has to be purified from the reaction, and the presence of enzymes could present a challenge during the reaction workup due to emulsification and other issues. Reactions are also hampered by the instability of the enzyme in organic solvents and the loss of enzyme activity over time. To circumvent these drawbacks, immobilization of enzymes has provided a platform to increase the efficiency and speed of product purification, as well as to increase enzyme stability and recyclability. Benefits of enzyme immobilization technology in drug manufacturing have been demonstrated with a few enzyme classes, such as CalB, PenG-acylases, and transaminases, with the immobilized enzymes having the desired solvent tolerance and/or recyclability [63–66].

In 2015, Merck demonstrated that immobilization of organic solvent-tolerant KRED enzyme was industrially viable [62]. Merck identified a KRED from Codexis' enzyme collection that was capable of reducing 3,5-bis(trifluoromethyl) acetophenone to the desired product, (*R*)-1-(3,5-bis(trifluoromethyl) phenyl) ethanol. The process used isopropanol as a co-substrate to recycle the oxidized $NADP^+$ back to NADPH, again taking advantage of the ease of acetone removal to drive the reaction to completion. Merck assessed various resins and evaluated the activity of immobilized enzymes. Optimization of immobilization conditions included increased incubation time, washing the resin with cofactor-containing buffer, and optimization of water content of the reaction. The immobilized KRED displayed high activity (up to 98% conversion), adequate stability at 90% organic solvent (hexane and isopropanol), and good stability at 40 °C. The immobilized KRED, under optimized conditions, demonstrated improved recyclability, with the enzyme maintaining activity for 10 cycles in seven days.

This Merck study has demonstrated that, in addition to activity, stability, and stereoselectivity improvements, KRED enzymes are also amenable to immobilization process, allowing for ease of product recovery. Immobilization also significantly affects process feasibility, as it provides a major reduction in cost by being able to get multiple cycles from the same material. As we will discuss in the next section, the ability to enhance process efficiency is a facet of process development where KREDs and biocatalysts cause significant changes to how pharmaceuticals will be manufactured in the near future.

5.1.8 KREDs: Going Green and Saving Green

As always, the direction of pharmaceutical development is toward more efficient processes with improved cost-effectiveness. Manufacturing processes incorporating approaches to minimize waste and decrease environmental impact are becoming more prevalent [67, 68]. More than ever before, traditional chemical routes are being replaced with more eco-friendly and sustainable biocatalytic processes [69, 70]. Biocatalysis has already proven, and continues to demonstrate, that it can equal or surpass its traditional chemical counterparts.

Now that this technology has been proved to be effective, processes are being developed with "green-by-design" methodology from the initial phase of planning [44, 68, 71]. By employing a biocatalytic approach, renewable, eco-friendly, and biodegradable reaction components are sought in place of harmful and toxic reagents and/or solvents. The end result is processes that are cheaper to manage, due to reduced costs of handling and treatment of hazardous waste. A further benefit is the mild conditions under which these catalysts are effective, reducing energy costs associated with using traditional chemical steps and providing safer alternatives. In some instances, these processes allow for several steps from the established chemical routes to be removed; protection/deprotection steps could be eliminated.

As discussed in the previous section, new and improved means of immobilizing biocatalysts provide new approaches for recycling and or recovering the biocatalyst [62, 72–74]. Enzyme immobilization and recycling could further reduce the costs of biocatalysis. In some cases, adoption of this methodology is leading to a shift from large-scale reactors to smaller, modular multiuse flow-through reactors for near and continuous flow processes [75–80]. This shift allows for significantly smaller physical footprints and reduces the need for associated hardware to run large batch reactors. Smaller and modular reactors also use less energy for heating and temperature maintenance, resulting in fewer runaway reactions. All of these elements mean safer and cheaper processes as a result of requiring less time, materials, labor, waste, hazardous conditions, and energy.

Ketoreductases have challenged the paradigm of asymmetric reduction in pharmaceutical development. A way to evaluate the efficiency and "greenness" of a process is to look at the amount and type of components used to prepare a given amount of product formed. ACS's Green Chemistry Institute has formulated the Process Mass Index [81], or PMI, which compares the mass of ingoing components to the mass of product produced. Although there are critics to this method of scoring process efficiency, the PMI calculation allows a viable comparison of different process routes to the same product molecule. With this in mind, we compare below a few cases where traditional chemical and biocatalytic ketoreductase approaches have been applied to solve challenging transformations to provide the same intermediate or API.

Several commercial-scale productions requiring asymmetric reductions have demonstrated that engineered KREDs are suitable agents in pharmaceutical process development. For a specific example, KRED has replaced the use of DIP-Cl in montelukast (Singulair by Merck, chronic obstructive pulmonary disease) production. DIP-Cl was difficult to handle, atomically inefficient, and required stoichiometric reagent addition. Its use also necessitated extensive purification to recover the product, which resulted in suboptimal yield. A KRED was engineered to have activity toward the highly hydrophobic montelukast substrate, which greatly improved the production process [20]. This enzymatic route removed the need to use DIP-Cl, required 50 times less mass of reducing agent, and eliminated the need for post-process workup, as the process conditions led to product crystallization as the reaction proceeds.

Another tool for installing chiral alcohols is CBS oxazaborolidine chiral catalysts complexed with a borane reducing agent [39]. It is applied to prochiral alkenes followed by hydrolysis of the resultant boronic ester. This too suffers from similar drawbacks as DIP-Cl. In addition, it has been shown to be capable of over-reducing the intended target [82]. CBS catalyst has been applied to the production of eslicarbazepine acetate, (Aptiom®, (*S*)-licarbazepine, anti-seizure) and ezetimibe. In the case of eslicarbazepine, KRED substitution not only removed the requirement of the hydroboration reagent, but also removed a process step by telescoping the reduction and the esterification to produce the final API in one step [29]. Schering Plough's approach to ezetimibe found that application of the expensive reagent did not provide the desired stereocontrol and over-reduced the substrate [51]. KRED substitution for these chemical processes eliminated the associated concerns.

As can be seen in Table 5.1.3, these processes are evaluated according to the PMI guidelines in order to compare between corresponding KRED biocatalytic and current chemical routes. These have been further scrutinized by breaking out each general operation in the overall reaction procedure to look at the components involved in the conversion itself, the post-reaction processing to recover the crude product, also known as the workup, and the recrystallization, when necessary. In the case of montelukast, KRED conversion is an improvement on the traditional chemistry route by nearly 40%. The ketoreductase route for eslicarbazepine, however, is moderately less efficient than the Noyori route. Here, the primary difference between the methods is the mass requirements for the recrystallization in the biocatalytic route. It should be considered though that the PMI was calculated for the non-optimized KRED purification process and it contributes heavily to the PMI value used in Table 5.1.3 [29]. This demonstrates that

Table 5.1.3 PMI comparison table of developed production processes of some commercial APIs. This table draws on reports from scholarly and patent literature sources to provide an updated comparison of these processes. It provides a comprehensive breakdown of the operations of a transformation and the contributions of each to the overall PMI for that reaction.

API	Reducing agent type	Agent	Steps	Cumulative PMI	Step PMI		
					Reaction	Workup	Recrystallization
Montelukast	Hydroboration	DIP-Cl	1	60.5 [83]	12.0	46.6	
	Asymmetric Reduction	KRED-026	1	37.7 [33]	8.9	17.2	13.2
Eslicarbazepine	Hydrogenation	Ru – (*S*,*S*)-TsDAEN	1	27.8 [56]	11.3	7.4	8.9
	Asymmetric Reduction	KRED-021	1	40.5 [29]	12.2	8.1	19.7
Sulopenem	N/A	Multiple	5	406.4 [84]			
	Hydroboration	DIP-Cl	1	112.3 [85]	25.6	81.5	
	Asymmetric Reduction	KRED-033	1	30.2 [20]	10.1	20.2	

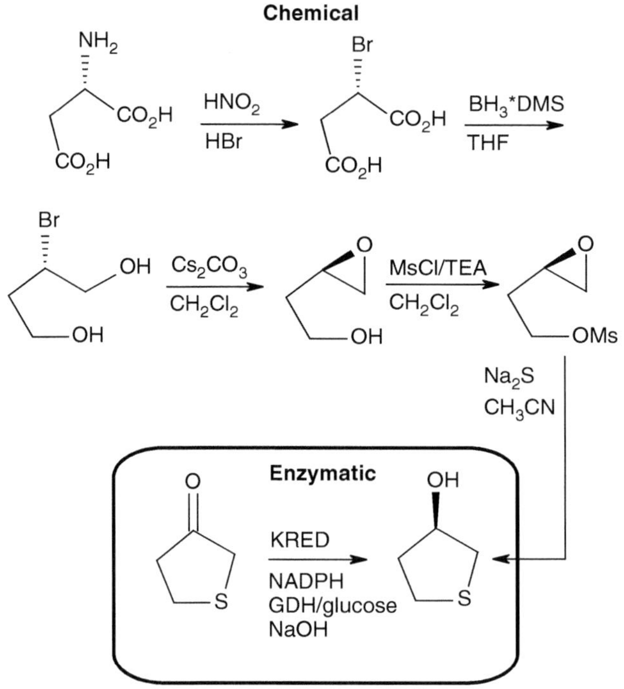

Figure 5.1.11 Chemical and enzymatic approach comparison to (*R*)-3-hydroxytetra-hydrothiophene.

the ideal situation occurs where the reaction conditions are such that the product phase separates from the reaction mass in order to simplify product recovery when possible.

One of the best examples of the environmental impact of ketoreductase application can be seen in an asymmetric reduction used to prepare the sulopenem (Pfizer/Iterum, antibiotic) intermediate (*R*)-3-hydroxytetrahydrothiophene [20]. Using Pfizer's original approach (Figure 5.1.11), the five-step process yielded 11% product overall with 96–98% ee [84]. Unsurprisingly, a comparison (Table 5.1.3) between the PMIs for this process shows more than an order of magnitude improvement by going to the single biocatalytic step. A more "apples-to-apples" comparison where the conversion is reduced to one step can be evaluated using the previously discussed CBS catalyst [86], or more closely by Brown's DIP-Cl [87], which provides excellent enantiomeric purity. The PMI for the DIP-Cl approach is about 2.5 times more than the PMI of the KRED approach. This stems from the use of a non-catalytic route requiring stoichiometric amounts of reagent and a large amount of mass for removing α-pinene from the crude product for recovery. In contrast, the workup for the biocatalytic route requires a quarter of the post-processing materials to reach a similar quality product. This PMI comparison puts into context the capabilities of KREDs in improving process efficiency. It is not always the superior route; however, it is competitive with traditional approaches and should be considered when formulating processes for new pharmaceutical candidates (Table 5.1.4).

Table 5.1.4 Comparison of stereoselective alcohol incorporation for (*R*)-tetrahydrothiophene-3-ol.

Route	Starting material	Reducing agent type	Reagent application	Steps	Yield	ee
Original synthesis	L-Aspartic acid	Hydroboration	Various/ stoichiometric	5	11%	96–98%
DIP-Cl	2,3-Dihydrothiophene	Hydroboration	Stoichiometric	1	92%	100%
CBS	Tetrahydrothiophen-3-one	Hydroboration	Stoichiometric	1	76%	23%
Ir-BINAP	Tetrahydrothiophen-3-one	Hydrogenation	Catalytic	1	87%	≤75%
KRED	Tetrahydrothiophen-3-one	Hydrogenation	Catalytic	1	85–88%	≥99%

Application of an evolved KRED to the one-step tetrahydrothiophenone conversion was run in aqueous buffer and provided the desired tetrahydrothiophenol in ≥99% ee [20]. The reduction used the GDH recycling system, which required sodium hydroxide addition to counter gluconic acid production and maintain the pH. Despite these requirements, the biocatalytic route is preferable compared to the highly inefficient six-step chemical route that required the use of hazardous bromic and nitrous acids, hygroscopic and hazardous borohydride dimethylsulfide, explosive epoxide intermediates, noxious sodium sulfide, and toxic cesium salts and chlorinated organic solvents. Some of the gains in removal of hazardous components and reduction of steps are offset by the energy needed to perform the reaction at 15 °C, but overall, the enzymatic route is significantly improved over the chemical route.

Changing the paradigm of traditional chemistry by the incorporation of biocatalytic processes is well on its way. What is important to note is that these changes are not just individual cases specific to KREDs. Adoption of biocatalysis provides many incentives over current traditional chemical routes, and investment into these processes provides financial and productivity windfalls to those who invest in them.

References

1 Magano, J. and Dunetz, J.R. (2012). Large-scale carbonyl reductions in the pharmaceutical industry. *Org. Process Res. Dev.* 16 (6): 1156–1184.

2 Moore, J.C., Pollard, D.J., Kosjek, B., and Devine, P.N. (2007). Advances in the enzymatic reduction of ketones. *Acc. Chem. Res.* 40 (12): 1412–1419.

3 Neuberg, C. and Lewite, A. (1918). Hydrogenation of ketone by yeast. *Biochem. Z.* 91: 257.

4 Schlieben, N.H., Niefind, K., Muller, J. et al. (2005). Atomic resolution structures of *R*-specific alcohol dehydrogenase from *Lactobacillus brevis* provide the structural bases of its substrate and cosubstrate specificity. *J. Mol. Biol.* 349 (4): 801–813.

5 Niefind, K., Muller, J., Riebel, B. et al. (2003). The crystal structure of *R*-specific alcohol dehydrogenase from *Lactobacillus brevis* suggests the structural basis of its metal dependency. *J. Mol. Biol.* 327 (2): 317–328.
6 Benach, J., Atrian, S., Gonzalez-Duarte, R., and Ladenstein, R. (1999). The catalytic reaction and inhibition mechanism of *Drosophila* alcohol dehydrogenase: observation of an enzyme-bound NAD-ketone adduct at 1.4 Å resolution by X-ray crystallography. *J. Mol. Biol.* 289 (2): 335–355.
7 Filling, C., Berndt, K.D., Benach, J. et al. (2002). Critical residues for structure and catalysis in short-chain dehydrogenases/reductases. *J. Biol. Chem.* 277 (28): 25677–25684.
8 Noey, E.L., Tibrewal, N., Jimenez-Oses, G. et al. (2015). Origins of stereoselectivity in evolved ketoreductases. *Proc. Natl. Acad. Sci. U.S.A.* 112 (51): E7065–E7072.
9 De Wildeman, S.M., Sonke, T., Schoemaker, H.E., and May, O. (2007). Biocatalytic reductions: from lab curiosity to "first choice". *Acc. Chem. Res.* 40 (12): 1260–1266.
10 Truppo, M.D., Pollard, D., and Devine, P. (2007). Enzyme-catalyzed enantioselective diaryl ketone reductions. *Org. Lett.* 9 (2): 335–338.
11 Zhang, R., Xu, Y., Sun, Y. et al. (2009). Ser67Asp and His68Asp substitutions in *Candida parapsilosis* carbonyl reductase alter the coenzyme specificity and enantioselectivity of ketone reduction. *Appl. Environ. Microbiol.* 75 (7): 2176–2183.
12 Huisman, G.W., Liang, J., and Krebber, A. (2010). Practical chiral alcohol manufacture using ketoreductases. *Curr. Opin. Chem. Biol.* 14 (2): 122–129.
13 Gröger, H., Werner, H., Altenbuchner, J. et al. (2005). Process for preparing optically active amino acids using a whole-cell catalyst. WO2005093081 A1.
14 Bradshaw, C.W., Hummel, W., and Wong, C.H. (1992). *Lactobacillus kefir* alcohol dehydrogenase: a useful catalyst for synthesis. *J. Org. Chem.* 57 (5): 1532–1536.
15 Bornscheuer, U.T., Huisman, G.W., Kazlauskas, R.J. et al. (2012). Engineering the third wave of biocatalysis. *Nature* 485 (7397): 185–194.
16 Solano, D.M., Hoyos, P., Hernaiz, M.J. et al. (2012). Industrial biotransformations in the synthesis of building blocks leading to enantiopure drugs. *Bioresour. Technol.* 115: 196–207.
17 Ni, Y. and Xu, J.-H. (2012). Biocatalytic ketone reduction: a green and efficient access to enantiopure alcohols. *Biotechnol. Adv.* 30 (6): 1279–1288.
18 Bezborodov, A.M. and Zagustina, N.A. (2016). Enzymatic biocatalysis in chemical synthesis of pharmaceuticals (review). *Appl. Biochem. Microbiol.* 52 (3): 237–249.
19 Gooding, O.W., Voladri, R., Bautista, A. et al. (2010). Development of a practical biocatalytic process for (*R*)-2-methylpentanol. *Org. Process Res. Dev.* 14 (1): 119–126.
20 Liang, J., Mundorff, E., Voladri, R. et al. (2010). Highly enantioselective reduction of a small heterocyclic ketone: biocatalytic reduction of tetrahydrothiophene-3-one to the corresponding (*R*)-alcohol. *Org. Process Res. Dev.* 14 (1): 188–192.

21 Kosjek, B., Nti-Gyabaah, J., Telari, K. et al. (2008). Preparative asymmetric synthesis of 4,4-dimethoxytetrahydro-2H-pyran-3-ol with a ketone reductase and in situ cofactor recycling using glucose dehydrogenase. *Org. Process Res. Dev.* 12 (4): 584–588.

22 Ma, S.K., Gruber, J., Davis, C. et al. (2010). A green-by-design biocatalytic process for atorvastatin intermediate. *Green Chem.* 12 (1): 81–86.

23 Savile, C., Gruber, J.M., Mundorff, E. et al. (2010). Ketoreductase polypeptides for the production of a 3-aryl-3-hydroxypropanamine from a 3-aryl-3-ketopropanamine. WO 2010025238 A2, March 4 2010.

24 Anke, K., Ching, C., Emily, M. et al. (2009). [EN] Ketoreductase polypeptides for the reduction of acetophenones. PN:CN101855342A PD:20101006;PN:CN101855342B PD:20130710;PN:DE602008028883D1 PD:20140116;PN:EP2198018A2 PD:20100623;PN:EP2198018B1 PD:20131120;PN:IN02039CN2010A PD:20100917;PN:JP2010538657T2 PD:20101216;PN:JP2015091269A2 PD:20150514;PN:JP2016105739A2 PD:20160616;PN:JP5973131B2 PD:20160823;PN:JP6137758B2 PD:20170531;PN:KR101586503B1 PD:20160118;PN:KR20100061525A PD:20100607;PN:US2009191605AA PD:20090730;PN:US2013078692AA PD:20130328;PN:US2013344552AA PD:20131226;PN:US2015017695AA PD:20150115;PN:US2017067032AA PD:20170309;PN:US8512973BB PD:20130820;PN:US8748143BB PD:20140610;PN:US8852909BB PD:20141007;PN:US9528131BB PD:20161227;PN:WO09036404A2 PD:20090319;PN:WO09036404A3 PD:20090716;PN:WO09036404A8 PD:20100506.

25 Grau, B.T., Devine, P.N., DiMichele, L.N., and Kosjek, B. (2007). Chemo- and enantioselective routes to chiral fluorinated hydroxyketones using ketoreductases. *Org. Lett.* 9 (24): 4951–4954.

26 Hou, X.P., Zhang, H.P., Chen, B.C. et al. (2017). Regioselective epoxide ring opening for the stereospecific scale-up synthesis of BMS-960, a potent and selective isoxazole-containing $S1P_1$ receptor agonist. *Org. Process Res. Dev.* 21 (2): 200–207.

27 Magnus, N.A., Coffey, D.S., DeBaillie, A.C. et al. (2011). Diarylketone ketoreductase screen and synthesis demonstration to access mGlu2 receptor potentiators. *Org. Process Res. Dev.* 15 (6): 1377–1381.

28 Campopiano, O., Mundorff, E., Borup, B., and Voladri, R. (2011). Ketoreductase polypeptides for the production of azetidinone. Google Patents WO 2009/046153 A4.

29 Modukuru, N.K., Sukumaran, J., Collier, S.J. et al. (2014). Development of a practical, biocatalytic reduction for the manufacture of (*S*)-licarbazepine using an evolved ketoreductase. *Org. Process Res. Dev.* 18 (6): 810–815.

30 Hyde, A.M., Liu, Z.J., Kosjek, B. et al. (2016). Synthesis of the GPR40 partial agonist MK-8666 through a Kinetically controlled dynamic enzymatic ketone reduction. *Org. Lett.* 18 (22): 5888–5891.

31 Edward, P., Homann, J., Homann Michael, J. et al. (1997). [EN] Stereoselective microbial reduction of 5-fluorophenyl-5-oxo-pentanoic acid and a phenyloxazolidinone condensation product thereof. PN:AT232239E PD:20030215;PN:AU199673617A1 PD:19970417;PN:CA2231808AA

PD:19980311;PN:CA2231808C PD:20070522;PN:DE69626128D1 PD:20030313;PN:DE69626128T2 PD:20031127;PN:DK0862645T3 PD:20030407;PN:EP0862645A1 PD:19980909;PN:EP0862645B1 PD:20030205;PN:ES2191771T3 PD:20030916;PN:JP10512454T2 PD:19981202;PN:JP3023179B2 PD:20000321;PN:MX9802377A1 PD:19980830;PN:US5618707A PD:19970408;PN:WO9712053A1 PD:19970403.

32 Emily, M., Erik, D.V., Mundorff, E. et al. (2010). [EN] Ketoreductase polypeptides for the stereoselective production of (4S)-3-[(5S)-5-(4-fluorophenyl)-5-hydroxypentanoyl]-4-phenyl-1,3-oxazolidin-2-one. PN:CN102186972A PD:20110914;PN:CN102186972B PD:20140820;PN:DE602009027373D1 PD:20141204;PN:EP2329014A2 PD:20110608;PN:EP2329014A4 PD:20120321;PN:EP2329014B1 PD:20141022;PN:IN02000CN2011A PD:20111202;PN:SG10201404330VA1 PD:20141030;PN:SI2329014T1 PD:20150130;PN:US2010062499AA PD:20100311;PN:US2012322136AA PD:20121220;PN:US2013210098AA PD:20130815;PN:US2015132806AA PD:20150514;PN:US2015337272AA PD:20151126;PN:US2016376565AA PD:20161229;PN:US8273554BB PD:20120925;PN:US8415126BB PD:20130409;PN:US8956840BB PD:20150217;PN:US9139820BB PD:20150922;PN:US9422530BB PD:20160823;PN:US9796964BB PD:20171024;PN:WO10025085A2 PD:20100304;PN:WO10025085A3 PD:20100415.

33 Liang, J., Lalonde, J., Borup, B. et al. (2010). Development of a biocatalytic process as an alternative to the (−)-DIP-Cl-mediated asymmetric reduction of a key intermediate of montelukast. *Org. Process Res. Dev.* 14 (1): 193–198.

34 Jiang, B., Feng, Y., and Zheng, J. (2000). Highly enantioselective reduction of achiral ketones with $NaBH_4/Me_3SiCl$ catalyzed by (*S*)-alpha,alpha-diphenylpyrrolidinemethanol. *Tetrahedron Lett.* 41 (52): 10281–10283.

35 Martinez, C.A., Keller, E., Meijer, R. et al. (2010). Biotransformation-mediated synthesis of (1S)-1-(2,6-dichloro-3-fluorophenyl)ethanol in enantiomerically pure form. *Tetrahedron: Asymmetry* 21 (19): 2408–2412.

36 Carey, J.S., Laffan, D., Thomson, C., and Williams, M.T. (2006). Analysis of the reactions used for the preparation of drug candidate molecules. *Org. Biomol. Chem.* 4 (12): 2337–2347.

37 Lu, H. (2007). Stereoselectivity in drug metabolism. *Expert Opin. Drug Metab. Toxicol.* 3 (2): 149–158.

38 Jamali, F., Mehvar, R., and Pasutto, F.M. (1989). Enantioselective aspects of drug action and disposition: therapeutic pitfalls. *J. Pharm. Sci.* 78 (9): 695–715.

39 Burkhardt, E.R. and Matos, K. (2006). Boron reagents in process chemistry: excellent tools for selective reductions. *Chem. Rev.* 106 (7): 2617–2650.

40 Sumi, K. and Kumobayashi, H. (2004). Rhodium/ruthenium applications. In: *Organometallics in Process Chemistry* (ed. R.E. Larsen), 63–95. Berlin, Heidelberg: Springer-Verlag.

41 Noyori, R., Kitamura, M., and Ohkuma, T. (2004). Toward efficient asymmetric hydrogenation: architectural and functional engineering of chiral molecular catalysts. *Proc. Natl. Acad. Sci. U.S.A.* 101 (15): 5356–5362.

42 Matsuda, T., Yamanaka, R., and Nakamura, K. (2009). Recent progress in biocatalysis for asymmetric oxidation and reduction. *Tetrahedron: Asymmetry* 20 (5): 513–557.
43 Applegate, G.A. and Berkowitz, D.B. (2015). Exploiting enzymatic dynamic reductive kinetic resolution (DYRKR) in stereocontrolled synthesis. *Adv. Synth. Catal.* 357 (8): 1619–1632.
44 Choi, J.-M., Han, S.-S., and Kim, H.-S. (2015). Industrial applications of enzyme biocatalysis: current status and future aspects. *Biotechnol. Adv.* 33 (7): 1443–1454.
45 Gavin, D.P. and Maguire, A.R. (2017). Chirality related to biocatalysis and enzymes in organic synthesis. In: *Chirality in Supramolecular Assemblies* (ed. F.R. Keene), 343–406. Wiley.
46 Koenig, S. (2013). *Scalable Green Chemistry: Case Studies from the Pharmaceuticals Industry*. Boca Raton, FL: CRC Press.
47 Torrelo, G., Hanefeld, U., and Hollmann, F. (2015). Biocatalysis. *Catal. Lett.* 145 (1): 309–345.
48 Hall, M.l. and Bommarius, A.S. (2011). Enantioenriched compounds via enzyme-catalyzed redox reactions. *Chem. Rev.* 111 (7): 4088–4110.
49 Berks, A.H. (1996). Preparations of two pivotal intermediates for the synthesis of 1-β-methyl carbapenem antibiotics. *Tetrahedron* 52 (2): 331–375.
50 Noyori, R., Ikeda, T., Ohkuma, T. et al. (1989). Stereoselective hydrogenation via dynamic kinetic resolution. *J. Am. Chem. Soc.* 111 (25): 9134–9135.
51 Fu, X., McAllister, T.L., Thiruvengadam, T.K. et al. (2003). Process for preparing ezetimibe intermediate by an acid enhanced chemo- and enantioselective CBS catalyzed ketone reduction. *Tetrahedron Lett.* 44 (4): 801–804.
52 Benes, J., Parada, A.n., Figueiredo, A.A. et al. (1999). Anticonvulsant and sodium channel-blocking properties of novel 10,11-dihydro-5H-dibenz[b,f] azepine-5-carboxamide derivatives. *J. Med. Chem.* 42 (14): 2582–2587.
53 Learmonth, D.A. (2002). Method for preparation of (s)-(+)- and (r)-(-)-10,11-dihydro-10-hydroxy-5h-dibenz/b,f/azepine-5-carboxamide. WO02092572.
54 Koilpillai, J.P., Kulkarni, P.B. et al. (2010). Process for purifying eslicarbazepine acetate. WO2010113179.
55 Mathes, C., Sedelmeier, G., Blatter, F. et al. (2004). Enantioselective process for the preparation of both enantiomers of 10,11-dihydro-10-hydroxy-5H-dibenz [b,f]azepine-5-carboxamide and new crystal forms thereof. WO2004031155.
56 Learmonth, D.A., Grasa, G.A., and Zanotti-Gerosa, A. (2007). Asymmetric catalytic reduction of oxcarbazepine. WO200712793.
57 Katkam, S., Sagyam, R.R., Buchikonda, R. et al. (2011). Eslicarbazepine acetate and its polymorphs. WO2011091131.
58 Husain, M. and Datta, D. (2011). Process for preparing (S)-(-)-10-acetoxy-10,11-dihydro-5H-dibenz[B,F]azepine-5-carboxyamide and its esters thereof. WO2011045648.
59 Liang, J., Jenne, S.J., Mundorff, E. et al. (2009). Improved ketoreductase polypeptides for the stereoselective production of (*R*)-3-hydroxythiolane. WO2009029554.

60 Reddy, M.S., Kumar, M.K., Rajan, S.T., and Prasad, D.S. (2007). Improved process for pure duloxetine hydrochloride. WO 2007077580 A2.
61 Matsuda, H., Shibata, T., Hashimoto, H., and Kitai, M. (1999). Method for producing (*R*)-4-cyano-3-hydroxybutyric acid lower alkyl ester. Google Patents USOO.5908953A.
62 Li, H., Moncecchi, J., and Truppo, M.D. (2015). Development of an immobilized ketoreductase for enzymatic (*R*)-1-(3,5-bis(trifluoromethyl)phenyl) ethanol production. *Org. Process Res. Dev.* 19 (7): 695–700.
63 Chen, B., Hu, J., Miller, E.M. et al. (2008). *Candida antarctica* lipase B chemically immobilized on epoxy-activated micro- and nanobeads: catalysts for polyester synthesis. *Biomacromolecules* 9 (2): 463–471.
64 Kallenberg, A.I., van Rantwijk, F., and Sheldon, R.A. (2005). Immobilization of penicillin G acylase: the key to optimum performance. *Adv. Synth. Catal.* 347 (7–8): 905–926.
65 Truppo, M.D., Strotman, H., and Hughes, G.J. (2012). Development of an immobilized transaminase capable of operating in organic solvent. *ChemCatChem* 4 (8): 1071–1074.
66 Laszlo, J.A., Jackson, M., and Blanco, R.M. (2011). Active-site titration analysis of surface influences on immobilized *Candida antarctica* lipase B activity. *J. Mol. Catal. B: Enzym.* 69: 60–65.
67 Woodley, J.M., Breuer, M., and Mink, D. (2013). A future perspective on the role of industrial biotechnology for chemicals production. *Chem. Eng. Res. Des.* 91 (10): 2029–2036.
68 Sheldon, R.A., Arends, I.W.C.E., and Hanefeld, U. (2007). *Green Chemistry and Catalysis*, 433. Weinheim: Wiley-VCH.
69 Philp, J.C., Ritchie, R.J., and Allan, J.E.M. (2013). Biobased chemicals: the convergence of green chemistry with industrial biotechnology. *Trends Biotechnol.* 31 (4): 219–222.
70 Wenda, S., Illner, S., Mell, A., and Kragl, U. (2011). Industrial biotechnology-the future of green chemistry? *Green Chem.* 13 (11): 3007–3047.
71 Liang, Y., Zhang, M.M., Ang, E.L., and Zhao, H. (2016). Biocatalysis for drug discovery and development. In: *Green Biocatalysis* (ed. R.N. Patel), 421–455. Hoboken, NJ: Wiley.
72 Sheldon, R.A. (2007). Enzyme immobilization: the quest for optimum performance. *Adv. Synth. Catal.* 349 (8–9): 1289–1307.
73 Azevedo, A.M., Prazeres, D.M.F., Cabral, J.M.S., and Fonseca, L.P. (2001). Stability of free and immobilised peroxidase in aqueous–organic solvents mixtures. *J. Mol. Catal. B: Enzym.* 15 (4): 147–153.
74 Khalaf, N., Govardhan, C.P., Lalonde, J.J. et al. (1996). Cross-linked enzyme crystals as highly active catalysts in organic solvents. *J. Am. Chem. Soc.* 118 (23): 5494–5495.
75 Porta, R., Benaglia, M., and Puglisi, A. (2016). Flow chemistry: recent developments in the synthesis of pharmaceutical products. *Org. Process Res. Dev.* 20 (1): 2–25.
76 Tamborini, L., Fernandes, P., Paradisi, F., and Molinari, F. (2018). Flow bioreactors as complementary tools for biocatalytic process intensification. *Trends Biotechnol.* 36 (1): 73–88.

77 Rossetti, I. and Compagnoni, M. (2016). Chemical reaction engineering, process design and scale-up issues at the frontier of synthesis: flow chemistry. *Chem. Eng. J.* 296 (Suppl. C): 56–70.

78 Bolivar, J.M., Eisl, I., and Nidetzky, B. (2016). Advanced characterization of immobilized enzymes as heterogeneous biocatalysts. *Catal. Today* 259 (Part 1): 66–80.

79 Gutmann, B., Cantillo, D., and Kappe, C.O. (2015). Continuous-flow technology – a tool for the safe manufacturing of active pharmaceutical ingredients. *Angew. Chem. Int. Ed.* 54 (23): 6688–6728.

80 Kockmann, N., Thenee, P., Fleischer-Trebes, C. et al. (2017). Safety assessment in development and operation of modular continuous-flow processes. *React. Chem. Eng.* 2 (3): 258–280.

81 Hughes, D. (2011). ACS Green Chemistry Institute: Pharmaceutical Roundtable's Process Mass Intensity Calculator. https://www.acs.org/content/dam/acsorg/greenchemistry/industriainnovation/roundtable/process-mass-intensity-calculation-tool.xls.

82 Shinkai, I., King, O.A., and Larsen, R.D. (1994). A practical asymmetric synthesis of LTD4 antagonist. *Pure Appl. Chem.* 66: 1551–1556.

83 King, A.O., Corley, E.G., Anderson, R.K. et al. (1993). An efficient synthesis of LTD4 antagonist L-699,392. *J. Org. Chem.* 58 (14): 3731–3735.

84 Volkmann, R.A., Kelbaugh, P.R., Nason, D.M., and Jasys, V.J. (1992). 2-thioalkyl penems: an efficient synthesis of sulopenem, a (5*R*,6*S*)-6-(1(*R*)-hydroxyethyl)-2-[(*cis*-1-oxo-3-thiolanyl)thio]-2-penem antibacterial. *J. Org. Chem.* 57 (16): 4352–4361.

85 Olah, G.A., Herges, R., Laali, K., and Segal, G.A. (1986). Onium ions. 34. The methoxydiazonium ion: preparation, proton, carbon-13, and nitrogen-15 NMR and IR structural studies, theoretical calculations, and reaction with aromatics. Attempted preparation and the intermediacy of the hydroxydiazonium ion. *J. Am. Chem. Soc.* 108 (8): 2054–2057.

86 Quallich, G.J. and Woodall, T.M. (1993). Enantioselective oxazaborolidine reduction of ketones containing heteroatoms. *Tetrahedron Lett.* 34 (5): 785–788.

87 Brown, H.C. and Prasad, J.V.N.V. (1986). Chiral synthesis via organoboranes. 6. Hydroboration. 74. Asymmetric hydroboration of representative heterocyclic olefins with diisopinocampheylborane. Synthesis of heterocyclic boronates and heterocyclic alcohols of very high enantiomeric purity. *J. Am. Chem. Soc.* 108 (8): 2049–2054.

5.2

An Aldolase for the Synthesis of the Statin Side Chain

Martin Schürmann

InnoSyn B.V., Urmonderbaan 22, 6167 RD, Geleen, the Netherlands

5.2.1 Introduction – Biocatalysis

5.2.1.1 Enzymes as Biocatalysts in Chemical Process

One of the most prominent industrial applications of enzymes is their use as biocatalysts in chemical processes as described in Chapter 1.1. One recent example of such chemo-enzymatic processes is the amination of a prochiral ketone substrate to the chiral amine product sitagliptin; for this active pharmaceutical ingredient (API) product the enzyme as well as the process has been developed by researchers at Merck and Codexis [1, 2]. But also at the other end of the value chain, in processes for large-volume/low-value bulk chemicals, enzymes are applied: for instance, nitrile hydratases for the production acrylamide from acrylonitrile, the former being used as monomer in various polymer applications [3].

The motivation to use enzymes as biocatalysts originates from a number of properties that make them superior to chemo-catalysts or conventional chemical processes: next to their ability to work efficiently under benign reaction conditions such as neutral pH, ambient temperature, and pressure as well as in water as solvent, it is especially their selectivity that is unrivaled in organic synthesis. While for the synthesis of chiral pharma molecules the usually excellent enantioselectivity of enzymes is the most important and impressive feature, their evenly superior regio- and chemoselectivity should not be underestimated (Figure 5.2.1), especially for non-chiral non-pharma molecules, which represent – volume wise – a much larger product class than pharma small molecules. The abovementioned nitrile hydratase used in acryl amide production, for instance, very cleanly stops at the target amide stage while the copper chemo-catalyst generates low – but still too high – acrylic acid concentrations for subsequent polymerization steps [4].

The chemo-, regio- and enantioselectivity of enzymes therefore not only enables the replacement of a chemical process step by a biocatalytic one, but even more importantly facilitates shortcuts in route design and route scouting

Industrial Enzyme Applications, First Edition. Edited by Andreas Vogel and Oliver May.

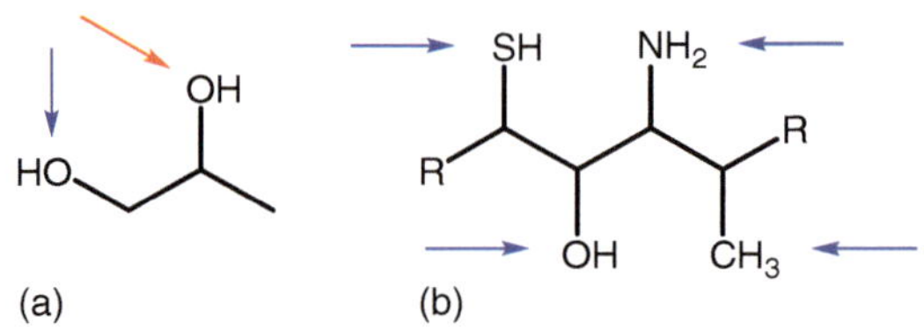

Figure 5.2.1 Regio- and chemoselectivity of enzymes: (a) Regioselective enzymes can typically discriminate very well between identical functional groups in one molecule, for instance, oxidation of a primary versus a secondary alcohol in 1,2-propanediol by an alcohol oxidase or dehydrogenase. (b) Chemoselective discrimination of different functional groups with similar chemical reactivities in the same (hypothetic) molecule by an enzyme. For methylations, for example, distinct methyl transferase classes exist in nature, each of them being selective for either *S*-, *N*-, *O*-, or *C*-methylation and generally not having side activities, even on chemically similar functional groups.

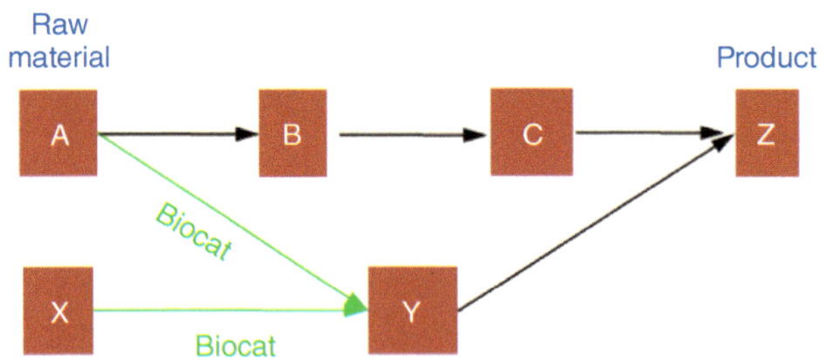

Figure 5.2.2 Schematic representation of a multistep chemical process from starting material **A** via intermediates (**B** and **C**) to a product molecule **Z**. Applying selective biocatalysts can reduce the number of process steps, e.g. by going via a different intermediate **Y**, which would not be accessible via conventional chemistry. Also, alternative raw materials **X** can be utilized using enzymes.

for multistep chemical processes, for instance, by omitting otherwise required protection/de-protection steps (Figure 5.2.2). Additionally, alternative (lower cost) raw materials, which otherwise could not be considered, can be unlocked by enzymes in biocatalytic process steps (Figure 5.2.2).

In combination, multistep chemical processes can be significantly shortened and therefore made more efficient with respect to costs and environmental impact. Just as an exemplary case, the production of the nonnatural cyclic amino acid (*S*)-2,3-dihydro-1*H*-indole-2-carboxylic acid was re-designed from a seven-step classical chemical resolution process with recycling of the undesired enantiomer to a two- to three-step chemo-enzymatic 100% ee/100% theoretical yield process [5]. The newly developed process included as the key step the enantioselective ammonia addition to a cinnamic acid derivative by phenylalanine ammonia lyase (PAL) furnishing an *ortho*-halo-L-phenylalanine, which could be efficiently ring-closed by the low concentrations of copper present in the tap-water used for the process solvent [5]. The positive effects of shortening and intensifying this process were analyzed in an environmental analysis (Eco-indicator 99) and – next to the cost advantage – mainly resulted in the reduction of inorganic wastes and lower fossil fuel consumption (Figure 5.2.3) [6].

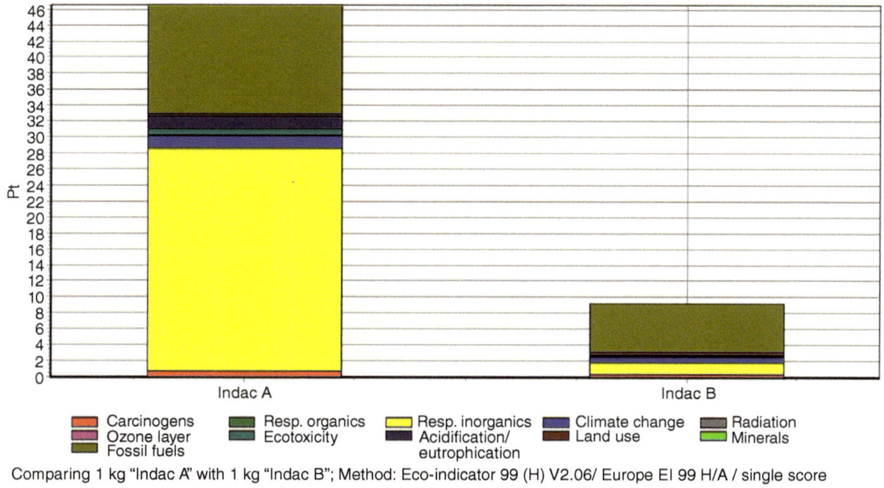

Figure 5.2.3 Improvement of ecological footprint of a pharma chemical by changing from a seven-step conventional chemical process (left) including a classical resolution step to a three-step biocatalysis-enabled process (right) as determined with Eco-indicator 99 (H) method (Version 2.06). The main improvement comes from significant reduction of inorganics use and wastes (yellow) as well as reduced consumption of fossil fuels, e.g. for transportation of goods (olive).

5.2.1.2 Biocatalytic Routes to the Statin Side Chain

One of the API product classes that has attracted much attention of academic and industrial scientists in the last decades is the cholesterol-lowering drugs of the statin type. Statin drugs such as atorvastatin and rosuvastatin are potent inhibitors of hydroxymethyl-glutaryl coenzyme A (HMG-CoA) reductase, a key step of cholesterol biosynthesis in humans, because of the structural reminiscence of their side chains to the product of HMG-CoA reductase mevalonic acid (Figure 5.2.4).

Apart from numerous synthesis routes based on conventional organic chemistry and chemo-catalysis, biocatalysis has been applied extensively by academic groups and chemical companies to develop highly selective and elegant routes to the statin side chains. The biocatalysis-enabled routes contain enzymes from a large variety of EC classes, e.g. oxidoreductases, transferases, hydrolases, and lyases. A selection of biocatalysis routes to statin side chains is depicted in Figure 5.2.5.

5.2.2 The Aldolase DERA in Application

5.2.2.1 DERA-Catalyzed Aldol Reactions

When we – almost 20 years ago (and at that time part of DSM Pharma Chemicals) – received a customer request to develop a cost-efficient route to a statin side chain intermediate, we quickly arrived at the pioneering publications by

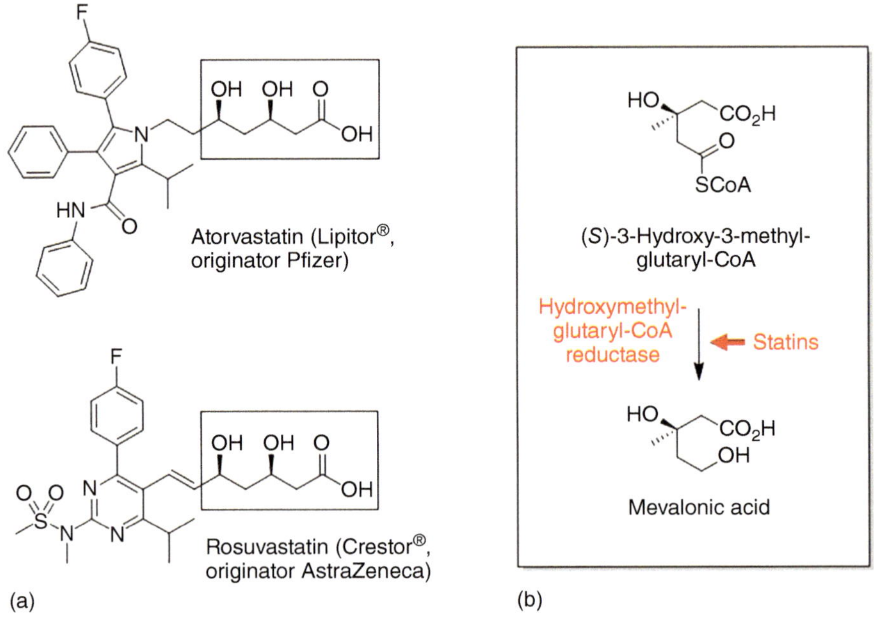

Figure 5.2.4 Structural comparison of (a) the two top-selling statin drugs atorvastatin and rosuvastatin and (b) the substrate and product of the enzyme the statin drugs inhibit, hydroxymethyl-glutaryl (HMG)-CoA reductase.

Gijsen and Wong [7], in which 2-deoxy-D-ribose 5-phosphate aldolase (DERA) had been used to synthesize a variety of hydroxy-aldehydes or deoxysugars from simple and cheap starting aldehydes. In its physiological role, DERA splits 2-deoxy-D-ribose 5-phosphate in a retro-aldol reaction into acetaldehyde (AA) and D-glyceraldehyde 3-phosphate as part of the recycling of DNA breakdown (Figure 5.2.6) [8]. In biocatalytic applications, on the other hand, DERA as well as other aldolases can also be used for *in vitro* aldol reactions with suitable carbonyl substrates. DERA was at that time the only aldolase reported to accept aldehydes as donor (nucleophile) as well as acceptor (electrophile) substrates (Figure 5.2.6) [9].

Still the aldolase discriminates very well between acetaldehyde **2** as donor substrate and acceptor substrate; while acetaldehyde is the most efficient donor substrate for DERA-catalyzed aldol reactions with a K_m value of 1.7 mM, it is not used as acceptor as long as a better aldehyde acceptor substrate **1** is present. The reported activities with chloroacetaldehyde (ClAA) as acceptor were rather low – only 0.3% compared to the activity on the physiological substrate D-glyceraldehyde 3-phosphate, but still among the highest reported for alternative acceptor substrates [10]. This, together with the long reaction times published originally, initially made us hesitant to follow up on a DERA-based synthesis route to the statin side chain intermediate. In their pioneering work, Gijsen and Wong reported for the target double aldol reaction of one molecule of chloroacetaldehyde **1** with two equivalents of acetaldehyde **2** reaction times of up to six days, low substrate concentrations (for an industrial process) of

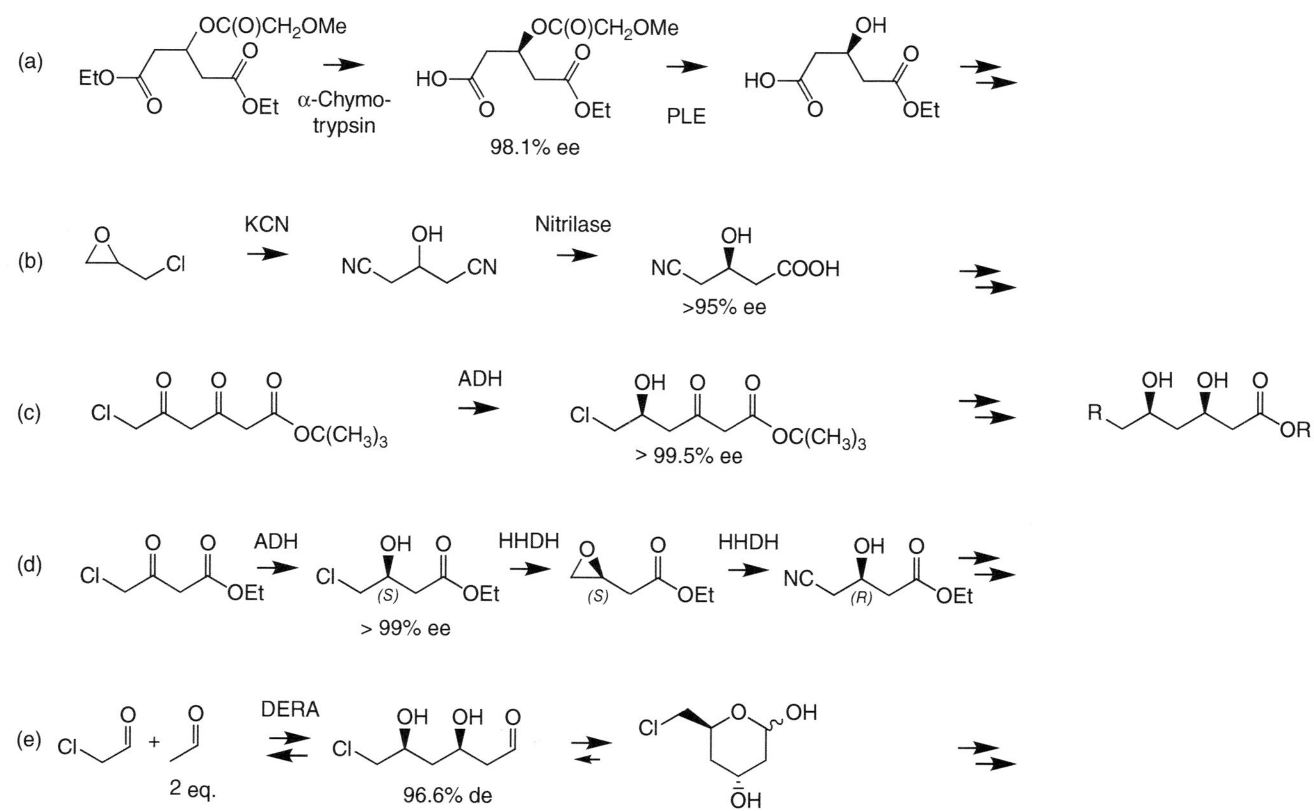

Figure 5.2.5 Overview of biocatalytic routes to statin side chains. PLE, pig-liver esterase; ADH, alcohol dehydrogenase; HHDH, halohydrin dehalogenase; DERA, 2-deoxy-D-ribose 5-phosphate aldolase.

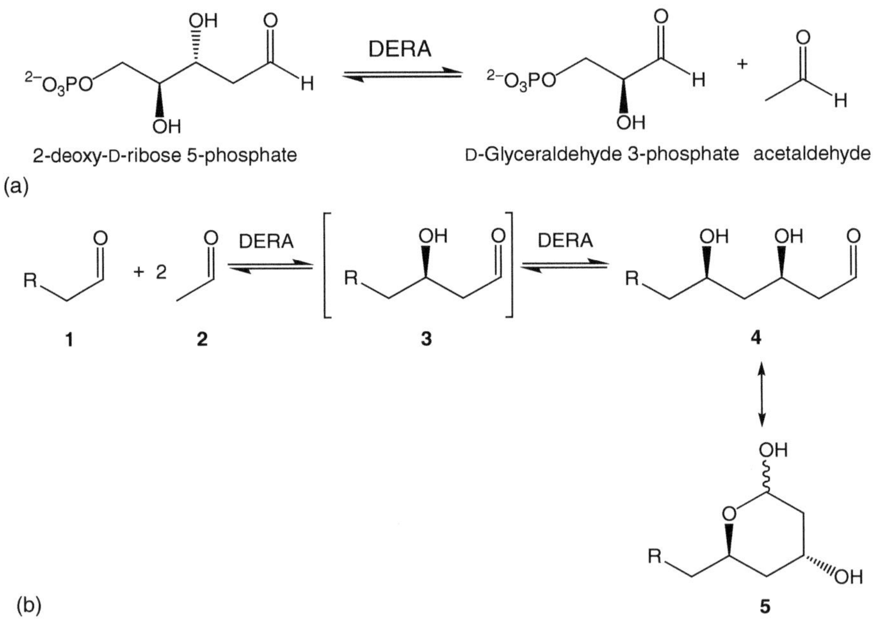

Figure 5.2.6 (a) Physiological reaction of DERA reversibly splitting 2-deoxy-D-ribose 5-phosphate into D-glyceraldehyde 3-phosphate and acetaldehyde and (b) biocatalytic application of DERA in the synthesis of lactols/deoxysugars from an acceptor aldehyde **1** such as chloroacetaldehyde (R = Cl) and two equivalents of acetaldehyde **2** via a monoaldol intermediate **3** and the open aldehyde form of the double aldol product **4**, which spontaneously ring-closes to the target product (3*R*,5*S*)-6-chloro-2,4,6-trideoxyhexapyranoside (CTHP) **5** (R = Cl).

100 mM ClAA and 300 mM AA, respectively, moderate product **5** yields of 70%, as well as little about side products and mass balances. The customer, however, demanded a project to develop a route; so we embarked on the DERA route and started the practical work in the laboratory.

5.2.2.2 Feasibility Phase of DERA-Enabled Statin Side Chain Process

The first step was to amplify the *deoC* gene encoding DERA [11] from chromosomal DNA by PCR and in parallel clone it into our laboratory and large-scale production *Escherichia coli* expression systems. Expression levels in both systems proved to be very good and led to the first successful fermentative production of recombinant DERA in our laboratories after three weeks, so that the experimental work on the enzymatic reaction itself could be started and the production strain was transferred to development phase.

It was straightforward to reproduce the results published by Gijsen and Wong [7] in the target tandem aldol reaction of ClAA and two equivalents of AA to (3*R*,5*S*)-6-chloro-2,4,6-trideoxyhexapyranoside (CTHP) **5** in our laboratories. But when trying to scale up from these conditions (100 mM ClAA, 300 mM AA and 50 U/ml DERA) to industrially required concentrations of ≥5% (w/v) or 300 mM of product, we quickly encountered the reaction stalling at very low

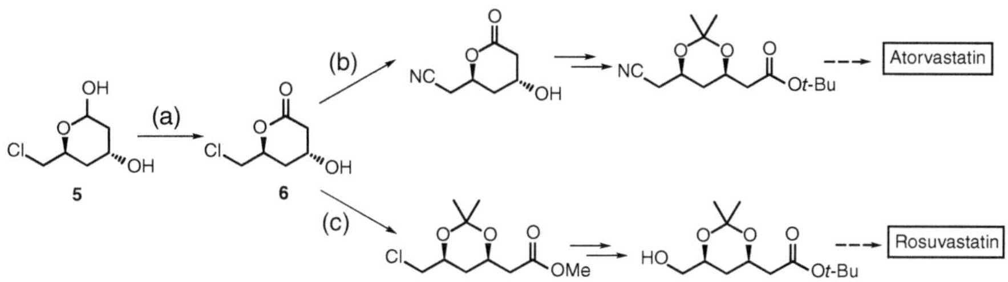

Figure 5.2.7 Functional group transformations from CTHP **5** via hydroxy-lactone **6** *en route* to atorvastatin and rosuvastatin. (a) Br_2, H_2O; (b) 1. CN^-, H_2O; 2. H_3O^+; (c) 2,2-dimethoxypropane, cat. H^+.

conversions of around 10–20%. This was accompanied by protein precipitation, which occurred rapidly in the reactions, strongly indicating the rapid denaturation of DERA (and other *E. coli* cell-free extract proteins) by at least one of the substrates, intermediates, or products of the reaction. As high product concentrations are a prerequisite for efficient product isolation and thus for a cost-efficient biocatalytic process, it was, nevertheless, required to achieve higher CTHP yields and concentrations in the target DERA reaction. Additionally, the reaction time of six days had to be reduced as this leads to low productivity (space–time yield), which is not compatible with a cost-efficient process.

The laboratory proof-of-principle for the target double aldol reaction of ClAA and AA to CTHP was obtained within five weeks after the start of the experimental work, but it took us another two months to develop the DERA reaction to an efficient biocatalytic reaction and the five subsequent process steps. As there is no alternative access to the DERA reaction product CTHP, the aldolase reaction had to deliver the starting material for the following chemical reactions. This starts with the oxidation of the CTHP lactol **5** to the corresponding hydroxy-lactone **6**, after which the routes to the different statin products diverge (Figure 5.2.7). Therefore, all experimental work had been dependent on quick and reliable access to larger quantities of CTHP, which we established by going to even higher substrate concentrations in combination with higher DERA amounts. This, together with a detailed analysis of the reaction course and a systematic optimization of the other reaction conditions such as pH and temperature, was key to developing the double aldol reaction of ClAA and AA to CTHP to a commercially viable process step for the production of statin side chains [12].

When increasing the substrate concentrations further from the original 0.1 M ClAA and 0.3 M AA to 0.6 M ClAA and 1.5 M AA, the reaction proceeded quickly via the monoaldol intermediate to the double aldol product CTHP and was finished within several hours, but only in case significantly increased concentrations of DERA were also applied. We tried to determine an apparent K_m value for ClAA based on the rate of formation of the monoaldol and double aldol product as these reactions could not be separated easily [12]. The value of approximately 150 mM indicates that even at 600 mM ClAA the DERA enzyme is still not completely saturated with the acceptor substrate, while there is no such limitation for the donor substrate AA (K_m = 1.7 mM) at elevated substrate concentrations. Thus, even from the beginning, the reaction does not proceed at its theoretical maximum rate. Additionally, both substrates and the monoaldol intermediate, which has to act as acceptor substrate as well (Figure 5.2.6), have a strong inactivating effect on DERA, the most pronounced being the inactivation/denaturation by ClAA and the monoaldol intermediate [12]. As this intermediate cannot be easily produced or isolated as pure compound, we could not determine kinetic data either for monoaldol as substrate for the second aldol reaction toward the target product CTHP or for the retro-aldol reaction back to ClAA and AA, which occurs simultaneously. The DERA reaction optimization was therefore more of an empiric approach than based on detailed kinetic data. Nevertheless, it was possible to fine-tune substrate concentrations and applied DERA amounts to finally arrive at product concentrations above 0.5 M with acceptable DERA consumptions. Notably, the (analytical) CTHP yield in the DERA reaction never really exceeded 85%, most probably due to the low affinity of DERA for ClAA, the resulting low rates at reduced substrate concentrations, and the fast inactivation of the aldolase.

Laboratory scale-up of these conditions delivered sufficient CTHP material to investigate and optimize the five subsequent chemical steps and finally to produce a 2 g sample of the required statin side chain intermediate for the customer. In total, it took only three months to prove the feasibility of six reaction steps, which was only possible because of the close interaction of the biologists and chemists in the department. Shortly after the positive customer feedback, we transferred the process for the production of the statin intermediate to the development phase at the different sites for enzyme and chemical production and supported the implementation of the whole process based on the wild-type DERA. Despite the suboptimal performance of the aldolase in the target reaction, a cost-competitive industrial process could be developed and implemented. The beauty and the advantage of the DERA-enabled process, finally, originate from the unique capability of aldolases (and similar C—C bond forming enzymes) to build up the carbon skeleton of a target molecule and introduce chiral hydroxy functionalities in one reaction and process step, thereby making use of simple and cheap starting materials and cutting down process steps.

5.2.3 Directed Evolution and Protein Engineering to Improve DERA

5.2.3.1 Rational Design

The poor stability in the presence of and low affinity for the aldehyde substrates and intermediate nevertheless called for an experimental program to improve ideally both these shortcomings of DERA in the application of statin side chain synthesis. This was initiated in 2002 in an EU-supported Marie Curie Industrial Host Fellowship Post-doc project drawn up by Sven Panke and Daniel Mink, both still working at DSM Pharma Chemicals (Geleen, the Netherlands) at that time. Combining the scientific competences in organic chemistry, microbiology, and molecular biology, we started out testing alternative prokaryotic DERAs, rational design based on the 3D structure of *E. coli* DERA (published by Heine et al. in 2001) [13], as well as immobilization. Immobilization of DERA proved to be straightforward in delivering highly active supported DERA, but after one reaction cycle the immobilized enzyme was as deactivated as the free enzyme, so we quickly gave up on this approach. In the rational design approach, we followed a similar strategy as DeSantis et al. [14] to exchange active site amino acids, mainly in the phosphate group binding pocket of the enzyme, to achieve better binding and conversion of the more hydrophobic acceptor substrate chloroacetaldehyde. The DERA variants we obtained were still active in the physiological and the target reaction with ClAA but not improved sufficiently compared to the wild-type enzyme. We achieved similar results testing alternative wild-type DERAs (see below).

However, we still wanted to solve the problem of low stability of DERA, and even more importantly, the overall low productivity in the target application. We expected the stability problem to be the easier one: we had learned from chloroacetaldehyde exposed *E. coli* DERA, which had been analyzed by our colleagues at DSM Biotechnology Center in Delft by peptic digestion and LC-MS analysis of obtained fragments, that several of the 19 lysine residues of *E. coli* DERA were covalently modified [15]. Molecular modeling studies based on the published DERA X-ray structure [13] showed that these modified lysine residues were at solvent-exposed surface positions, with lysine K172 even being located at the entrance of the substrate binding site and probably blocking it when being modified. To examine the role of the lysine residues in the inactivation of DERA by chloroacetaldehyde, we decided to target all DERA lysine residues, except lysine residue 167 (active site lysine involved in the catalytic mechanism), individually by site-directed mutagenesis. From the obtained active clones, the variant DERA enzymes were then tested for improved chloroacetaldehyde tolerance. For all lysine residues except lysines 167 and 201, amino acid substitutions were identified, resulting in enzymatically active DERA enzymes. However, no amino acid substitution was identified that led to a detectable improvement in chloroacetaldehyde tolerance of the tested variant DERA enzymes. Even the K172I amino acid substitution (located at the entrance of the substrate binding site, Figure 5.2.8) showed no improvement of chloroacetaldehyde resistance of the variant DERA enzyme compared to wild-type DERA [15]. Slightly discouraged from these results, we did not recombine the individual exchanges anymore.

In parallel, we had screened the sequence databases for DERAs with a low number of lysine residues. Remarkably, the *E. coli* and *Bacillus subtilis* DERA have evolved to contain 19 and 20 lysine residues, respectively, whereas the DERAs from *Burkholderia mallei*, *Magnetospirillum magnetotacticum*, and *Halobacterium* sp. NRC-1 can fulfill the same physiological function with only five and six lysine residues, respectively. We cloned and expressed the *deoC* genes of *B. mallei* and *Halobacterium* sp. NRC-1 recombinantly in *E. coli*, but again, despite being enzymatically active, these low-lysine DERAs were not more stable or productive than the *E. coli* wild-type DERA (M. Schürmann et al., unpublished results). The explanation for DERA inactivation in aldol synthesis reactions remained unsolved for more than a decade until the groups of Pietruszka and Willbold at Research Center Jülich and University of Düsseldorf joined forces and used NMR spectroscopy on both the protein and the small-molecule scale to show that the monoaldol intermediate binds to the inner part of the enzyme and

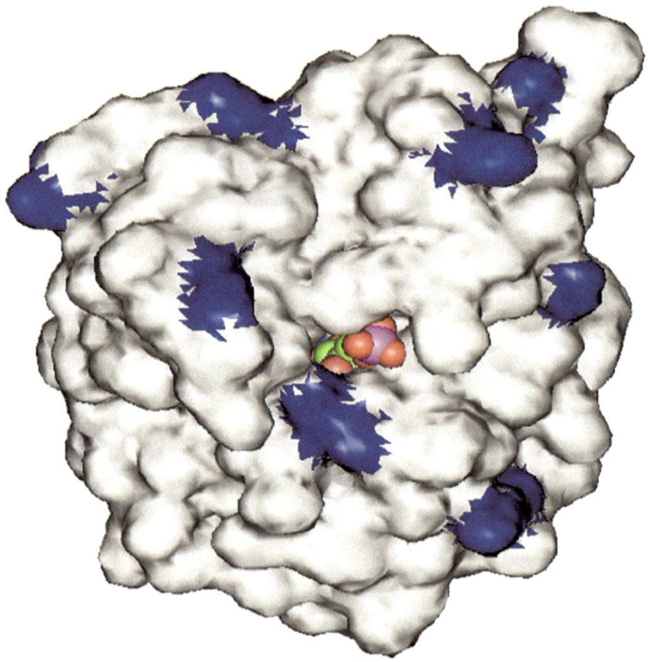

Figure 5.2.8 DERA X-ray structure-based model, highlighting solvent exposed lysine residues in blue and bound 2-deoxy-D-ribose 5-phosphate substrate in the active site. Lysine residue 172 is located at the substrate entry channel close to the active site.

not the outer [16]. The authors identified a covalently bound reaction product bridging the catalytically active lysine K167 to a nearby cysteine (C47) and not a lysine in the deactivated enzyme [16], in retrospect explaining very well why all our approaches focusing on lysine residues had failed.

5.2.3.2 Directed Evolution of DERA

While our knowledge-based and structure-guided approaches had not been successful in 2002/2003, the high-throughput (HTP)/brute-force directed evolution strategies had become very popular and successful in those days, so we decided to put all our efforts into these (see also Chapter 1.2). Again aiming at improved stability and productivity, we devised two microtiter plate-based HTP assays to probe an error-prone PCR library of the *E. coli deoC* gene. The chloroacetaldehyde tolerance screening was developed based on the spectrophotometric activity assay for DERA: (i) retro-aldol splitting of 2-deoxy-D-ribose 5-phosphate into acetaldehyde and D-glyceraldehyde 3-phosphate, (ii) isomerization of the latter by auxiliary enzyme triose phosphate isomerase (TPI) to dihydroxyacetone phosphate, and (iii) reduction of the previous under equimolar consumption of NADH by the second auxiliary enzyme glycerol 3-phosphate dehydrogenase. We adapted this established assay to MTP scale and included preincubation with increasing ClAA concentration per directed evolution round. The first screening round was performed on 10 000 DERA variants from the error-prone

PCR library by first measuring the individual activity of each variant, followed by two minutes incubation with 100 mM ClAA and remeasuring the remaining activity [15]. This way, we identified 63 DERA variants with at least twofold increased ClAA tolerance. These were subjected to a second and third round including further mutagenesis and recombination of the hits and challenged by increasing the ClAA concentration to 200 and 300 mM, respectively. The retesting of the third round screening hits identified four superior clones, which reproducibly showed the best tolerance to ClAA under the applied conditions and fell into two groups: two variants containing a M185V exchange and two containing mutations at the C-terminus of the *E. coli* DERA, namely, deletion of the terminal tyrosine Y259 and a mutation of the *deoC* stop codon. The latter mutation resulted in the extension of DERA by nine amino acids encoded by the vector DNA sequence until by chance a stop codon was reached [15]. These two C-terminally modified DERA variants only displayed significant activity in the spectrophotometric assay with 2-deoxy-D-ribose 5-phosphate as substrate when ClAA was added to the assay, while without addition of ClAA almost no activity could be detected in these assays. In small-scale aldol reaction application tests with AA and ClAA, however, both C-terminal variants exhibited two- to threefold increased CTHP product formation compared to wild-type DERA when comparable protein amounts were applied, while variant M185V even showed a fivefold improved product formation. The reasons and mechanisms of low activity in the physiological reaction combined with ClAA tolerance and even activation of ΔY259 and the C-terminally extended variant are not clear. While the 3D structure of *E. coli* DERA was solved at atomic resolution, the C-terminal helix was not resolved in the structure, most likely because of its flexibility [13]. It is, however, considered that it folds back into the active site and participates in (acceptor) substrate binding, which would explain our observations of reduced activity of the C-terminal variants with DERA's physiological substrate.

For the productivity screening we calculated that we would not be able to form considerable amounts of CTHP product from ClAA and AA with the DERA amounts, which could be produced on MTP scale. Therefore, we decided to focus on the formation and detection of the monoaldol intermediate by GC-MS. We anticipated that DERA variants that are improved in monoaldol formation could also be better in the second aldol reaction from monoaldol and AA to CTHP. This was a certain risk, taking the golden rule of directed evolution into account that "you get what you screen for." Nevertheless, we established a GC-MS-based medium- to HTP method for the detection of monoaldol intermediate and CTHP product having a cycle time of less than three minutes. As this represents a considerably lower throughput compared to the stability assay described above, we only screened 3000 randomly chosen variants of the same error-prone PCR library [15]. Indeed, we did not detect CTHP formation in this evolution round, but identified nine DERA variants with more than threefold increased monoaldol formation than the wild type. Retesting these primary hits as described above showed that the best variant was improved more than 15-fold in CTHP formation than the wild-type DERA. The corresponding F200I exchange had not occurred in the active site itself, but in a helix in the second shell "behind" the active site lysine K167 [15]. Again, the exact reason for this improvement could not be derived from the 3D structure, but the mutation of the bulky hydrophobic phenylalanine residue to a smaller hydrophobic isoleucine probably causes more flexibility in the active site to accommodate a bound monoaldol intermediate and the formation of the linear CTHP analogue product. Additionally, the adduct formation of the monoaldol intermediate to C47, similar to the effect described by Pietruszka and Willbold [16], might be disfavored.

Finally, we tried to combine the best mutations obtained from the stability and productivity screening and were pleased to see that the F200I exchange mutation and the mutations at the C-terminus indeed had an additive effect, which is not self-evident in such independent directed evolution approaches [13, 16]. Additionally, we observed in later projects with other aliphatic acceptor aldehydes that the F200I DERA variant was generally improved with respect to aldol synthesis rate compared to the wild type. In that sense, we were lucky that we not only "got what we screened for," but we also got a bit more.

5.2.3.3 Other Approaches to Suitable or Improved DERAs

Greenberg et al. from Diversa identified a DERA from metagenomes that proved to be suitable for catalyzing the target double aldol reaction of ClAA and AA already in 2004 [17]. The enzyme is probably obtained from a *Geobacillus* species and had about 30% sequence identity to the *E. coli* enzyme. It is striking that even such low sequence identities still result in enzymes that perform both their physiological reaction and the statin side chain synthesis reaction well. Under the applied reaction conditions this alternative wild-type DERA outcompeted the *E. coli* wild-type DERA in the required biocatalyst loading and the synthesis reaction rate, but also exhibited a somewhat higher enantio- and diastereoselectivity [17].

Probably stimulated by these and our positive results on applying DERA in the synthesis of statin side chains, researchers at Pfizer and Lek Pharmaceuticals also embarked on the aldolase technology. Hu et al. focused on an alternative nitrogen-containing acceptor substrate targeted at the atorvastatin side chain [18]. Researchers from Lek and University of Ljubljana also worked with chloroacetaldehyde, which delivered the generic building block for statin side chains CTHP and applied DERA in whole cell biocatalyst formulations [19], but also used acetoxy-acetaldehyde as substrate, which is tailored to rosuvastatin synthesis [20].

Additionally, several academic groups as the ones of Pietruszka, Xu, and Ohshima discovered a number of wild-type DERAs suitable to catalyze the target reaction and engineered them further [21–23]. Especially DERAs from thermophilic and hyperthermophilic species proved to be more stable than enzymes of mesophilic origin, also in the presence of organic (co-)solvents and aldehyde substrates. We had tested a selection of DERAs from thermophilic prokaryotes as well, such as the ones from *Aeropyrum pernix* and *Thermotoga maritima*. In the desired application for statin side chain synthesis the advantage of improved stability was, however, compromised by their lower intrinsic activity at the ambient reaction temperature as well as the lower recombinant expression level compared to the *E. coli* aldolase, for which we could achieve up to 50% of the total soluble protein content (M. Schuermann et al., unpublished data). More recently, Pietruszka and coworkers published a systematic study of DERAs from psychrophilic, mesophilic, and (hyper)thermophilic prokaryotes, which underlines that also cold-adapted and mesophilic DERA enzymes can show a better performance than thermophilic ones [24].

5.2.3.4 Other Applications of Process Intermediates and the DERA Technology

With the DERA reaction product CTHP **5** and the corresponding hydroxy-lactone **6** as its subsequent oxidized process intermediate at hand we reasoned that the latter would serve as an excellent starting material for the preparation of the α,β-unsaturated δ-lactone 7 (Figure 5.2.9). In general, β-hydroxy δ-lactones are known to be susceptible to dehydration, a fact we used to our advantage to eliminate water from hydroxy-lactone **6** to δ-lactone 7 by acid-catalyzed dehydration in toluene, aided by azeotropic removal of water [25]. The α,β-unsaturated δ-lactone is a characteristic structural motif in a large family of physiologically active secondary metabolites, widely distributed in nature [26]. Furthermore, these kinetically controlled conjugate additions to such chiral unsaturated δ-lactones such as 7 usually proceed with very high diastereoselectivity in favor of the *trans* configuration. We used this characteristic to prepare a range of base-catalyzed conjugate additions of various *C*-, *N*-,

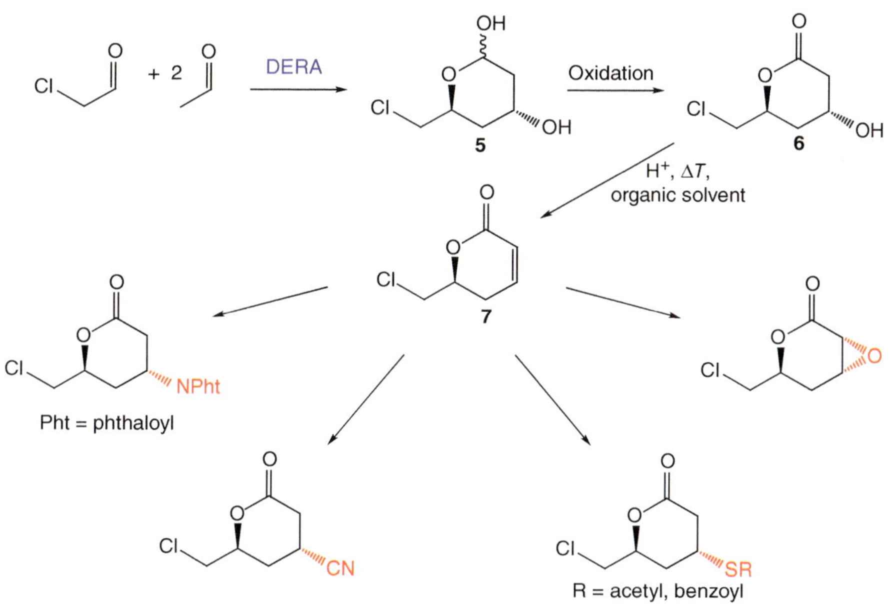

Figure 5.2.9 Reaction scheme from DERA-synthesized CTHP **5** via hydroxy-lactone **6** to unsaturated δ-lactone **7**, a precursor to several synthons obtained by conjugate additions with various *C*-, *N*-, *O*-, and *S*-nucleophiles.

O-, and *S*-nucleophiles to lactone **7** (Figure 5.2.9). The products were obtained usually as crystalline solids in high isolated yields of 67–92% and excellent diastereoselectivity of *trans:cis* of more than 95 : 5 (by NMR) [25]. Owing to their favorable set of functional groups, lactone **7** and the derivatives synthesized from it are versatile building blocks for the synthesis of pharmaceutically relevant compounds such as β-amino acids as well as natural products.

When we looked deeper into molecules having the saturated and unsaturated δ-lactone structural motif, we quickly arrived at a number of flavor and fragrance lactone products (http://www.leffingwell.com/chirality/lactones.htm), which can, in principle, be obtained by reacting an aliphatic unsubstituted aldehyde as DERA acceptor substrate instead of chloroacetaldehyde with two equivalents of acetaldehyde [27]. Oxidizing the resulting lactol to the corresponding hydroxy-lactone and eliminating water as described above result in several unsaturated δ-lactones (Figure 5.2.10) as they occur, for instance, in tropical plants such as massoialactone, which has been isolated from the bark of the massoia tree. The enantiopure saturated and unsaturated δ-lactones with medium side chain length (as they can be derived from DERA reactions with butanal to octanal as acceptor aldehydes) are typically applied as flavor enhancers in foods and beverages. Nowadays, these lactone products are, however, either derived from chemical synthesis (usually as racemates) or isolated from the tropical plants as enantiomerically pure products. Synthesizing them using the DERA technology as key chiral and C—C bond-forming step thus offers the advantage of saving scarce natural resources or producing them with high optical purity in an efficient process from cheap and readily available starting materials [28].

5.2.4 Conclusions

The DERA technology we developed based on the pioneering work of Wong and coworkers represents in our view the most elegant and efficient way to produce the statin side chain molecules and this technology is currently being applied by the DSM Sinochem Pharmaceuticals (DSP) joint venture to produce atorva-, rosuva- and pitavastatin at industrial scale. This advantage and elegance is based on DERA's capability to couple simple and cheap aldehydes enantioselectively and building up the complete carbon skeleton of the side chain in one step. This, overall, saves process steps and reagents and thus fixed costs (installations) and variable costs (consumables) compared to other statin side chain processes. Because of the lower number of process steps the employed DERA does not need to compete in initial activity and stability with other enzymes used in alternative processes such as lipases, esterases, alcohol dehydrogenases (ADHs), or nitrilases.

Although the process stability and activity of wild-type DERA in the target reaction was far from optimal, we could develop a sufficiently cost-efficient process and establish it as a first-generation process to a statin side chain and not "just" a second-generation process to replace a preexisting, suboptimal conventional chemical process, which is often the case for biocatalytic processes.

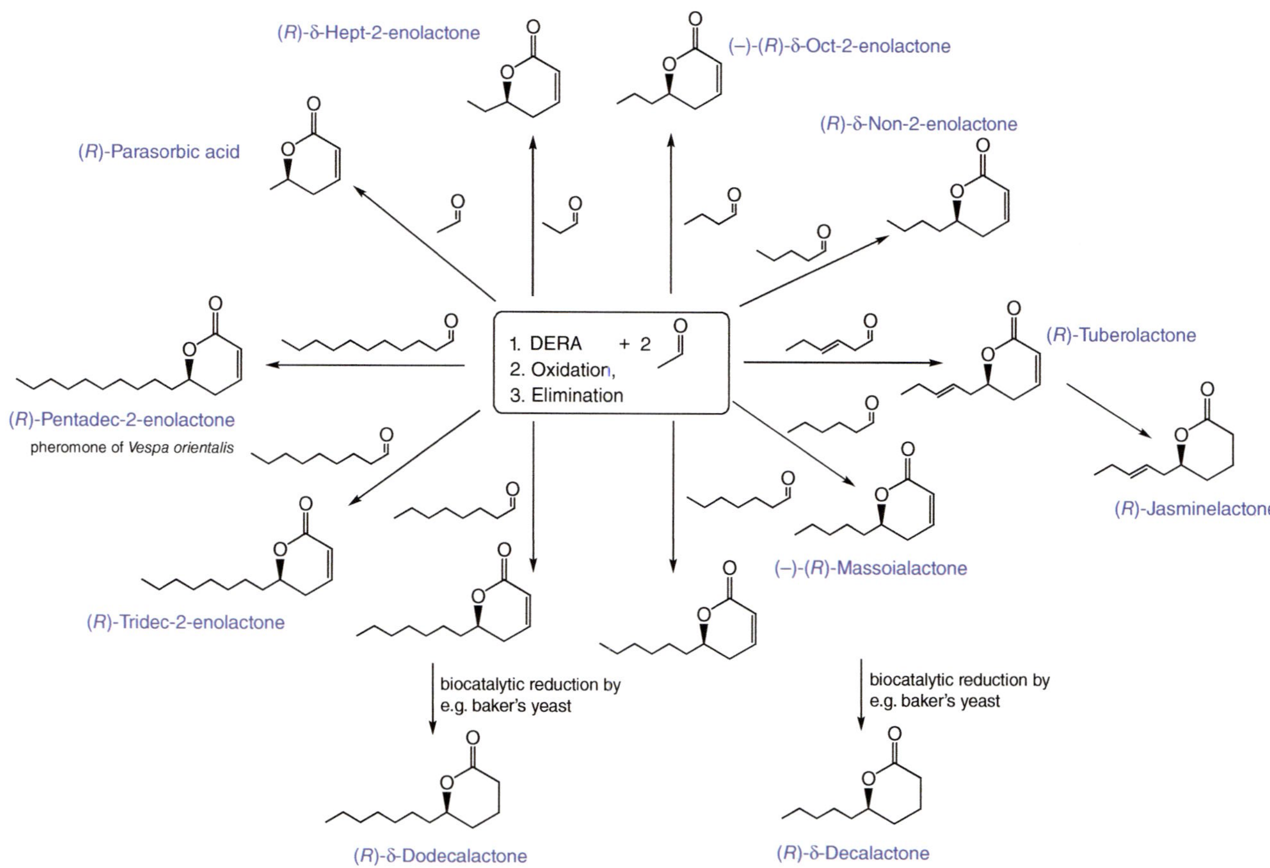

Figure 5.2.10 Reaction scheme of DERA-enabled product tree to unsaturated and saturated δ-lactones analogous to Figure 5.2.9 by 1. DERA double aldol reaction of an acceptor aldehyde and two molecules of acetaldehyde to a hydroxy-lactol, 2. Oxidation of the lactol product of the DERA reaction to a hydroxy-lactone and 3. Water elimination furnishing the unsaturated δ-lactone. By chemical or biocatalytic reduction also the saturated δ-lactones can be obtained.

One of the reasons for the underrepresentation of biocatalysis-enabled processes as first-generation production processes is that synthesis routes have often been registered based on the medicinal chemistry synthesis routes and therefore need to be produced via the same registered starting materials or process intermediates. As medicinal chemists in the past decades were frequently not aware of biocatalysis or did not have suitable enzymes available, their routes seldom contained biocatalysis. Additionally, in the past biocatalysis used to suffer from longer development times compared to chemo-catalysis and conventional chemistry. The short process development times described above and the presence of biology and chemistry competences to produce DERA and apply it in a chemical reaction in the same company and even department were key to this success. The detailed process knowledge we built up on the advantages, and even more about the shortcomings, of DERA enabled us to improve its stability and performance in a – nowadays almost old-fashioned, but still very efficient – directed evolution program. This, together with further improvements of the complete route to the statin side chain, helped remain competitive in the very dynamic market.

For us this original DERA process opened up a technology platform to produce more molecules and products other than statins, either derived from hydroxy-lactone **6** or from DERA reactions with acceptor aldehydes different than chloroacetaldehyde such as, for instance, flavor and fragrance lactones. Building on our good experience with DERA we later developed a new route to (+)-biotin, again based on an aldolase, namely D-fructose 6-phosphate aldolase (FSA), setting the C-skeleton and both stereocenters of the target molecules in one reaction step [28]. The number of enzymes with different substrate scope in sequence databases is continuously increasing and modern enzyme engineering techniques have resulted in aldolase variants with extended donor (nucleophile) acceptance. Just to name one example, Clapes and coworkers engineered an FSA, which efficiently catalyzes aldol reactions of glycolaldehyde (hydroxy-acetaldehyde) with a wide range of acceptor aldehyde, so that DERA is not the only aldolase anymore that converts aldehydes as acceptor and donor substrate [29]. In the future, we will see more and successful applications of aldolases and other carbon–carbon bond-forming enzymes such as hydroxy-nitrile lyases, the big group of thiamine diphosphate decarboxylases and carboligases, as well as other less explored C—C bond-forming enzymes in the production of pharma, Flavour & Fragrance (F&F), and other fine chemicals.

Acknowledgments

I would like to thank my former Marie Curie Post-Doc colleagues Michael Wolberg, Stefan Jennewein, and Iris Hilker as well as the colleagues currently or formerly at DSM, who made the Marie Curie Industrial Host Fellowship project possible, especially Daniel Mink, Marcel Wubbolts, Sven Panke, Theo Sonke (all formerly DSM Pharma Chemicals, Geleen, NL), and Ruud Luiten (DSM Biotechnology Center DBC, Delft, NL). Furthermore, the collaboration with the former colleagues of DSM Pharma Chemicals (DPC Venlo, NL), DSM Sinochem Pharmaceuticals (DSP, Deflt, NL), and colleagues at InnoSyn (Geleen, NL) involved over the years in the various projects on DERA technology for the synthesis of statin side chains is greatly acknowledged; especially Ben de Lange, Dennis Heemskerk, Peter Hermsen, Hans Kierkels (all InnoSyn), John Jansen, Thijs Kuilman, Robert Jan Vijn (all DPC), Jan-Metske van der Laan, Richard Kerkman (both DBC), Herman Slijkhuis (DSP), John Mommers (DSM Resolve), Dick Schipper (DBC), and Lucien Duchateu (DSM Resolve & DBC). We thank the European Commission for the industrial host fellowship grant HPMI-CT-2001-00142 to Michael Wolberg, Stefan Jennewein, Iris Hilker, and Martin Schürmann.

References

1 Savile, C.K., Janey, J.M., Mundorff, E.C. et al. (2010). Biocatalytic asymmetric synthesis of chiral amines from ketones applied to sitagliptin manufacture. *Science* 329: 305–309.

2 Truppo, M.D., Strotman, H., and Hughes, G. (2012). Development of an immobilized transaminase capable of operating in organic solvent. *ChemCatChem* 4: 1071–1074.

3 Kobayashi, M. and Shimizu, S. (2000). Nitrile hydrolases. *Curr. Opin. Chem. Biol.* 4: 95–102.

4 Yamada, H. and Kobayashi, M. (1996). Nitrile hydratase and its application to industrial production of acrylamide. *Biosci. Biotechnol., Biochem.* 60: 1391–1400.

5 de Lange, B., Hyett, D.J., Maas, P.J.D. et al. (2011). Asymmetric synthesis of (*S*)-2-indolinecarboxylic acid by combining biocatalysis and homogeneous catalysis. *ChemCatChem* 3: 289–292.

6 Poechlauer, P., Braune, S., de Vries, A.H.M., and May, O. (2010). Sustainable route design for pharmaceuticals – why, how and when. *Chem. Today* 28: 14–17.

7 (a) Gijsen, H.J.M. and Wong, C.-H. (1994). Unprecedented asymmetric aldol reactions with three aldehyde substrates catalyzed by 2-deoxyribose-5-phosphate aldolase. *J. Am. Chem. Soc.* 116: 8422–8423. (b) Gijsen, H.J.M. and Wong, C.-H. (1995). Sequential three- and four-substrate aldol reactions catalyzed by aldolases. *J. Am. Chem. Soc.* 117: 7585–7591. (c) Wong, C.-H., Garcia-Junceda, E., Chen, L. et al. (1995). Recombinant 2-deoxyribose-5-phosphate aldolase in organic synthesis: use of sequential two-substrate and three-substrate aldol reactions. *J. Am. Chem. Soc.* 117: 3333–3339.

8 (a) Ahmad, S.I. and Pritchard, R.H. (1969). A map of four genes specifying enzymes involved in catabolism of nucleosides and deoxynucleosides in *Escherichia coli*. *Mol. Gen. Genet.* 104: 351–359. (b) Valentin-Hansen, P., Hammer-Jespersen, K., and Buxton, R.S. (1979). Evidence for the existence of three promoters for the *deo* operon of *Escherichia coli* K12 *in vitro*. *J. Mol. Biol.* 133: 1–17.

9 Dean, S.M., Greenberg, W.A., and Wong, C.-H. (2007). Recent advances in aldolase-catalyzed asymmetric synthesis. *Adv. Synth. Catal.* 349: 1308–1320.

10 Chen, L., Dumas, D.P., and Wong, C.-H. (1992). Deoxyribose-5-phosphate aldolase as a catalyst in asymmetric aldol condensation. *J. Am. Chem. Soc.* 114: 741–748.
11 Valentin-Hansen, P., Boetius, F., Hammer-Jespersen, K., and Svendsen, I. (1982). The primary structure of *Escherichia coli* K12 2-deoxyribose 5-phosphate aldolase. *Eur. J. Biochem.* 125: 561–566.
12 Schürmann, M., Wolberg, M., Panke, S., and Kierkels, H. (2010). The development of short, efficient, economic, and sustainable chemoenzymatic processes for statin side chains. In: *Green Chemistry in the Pharmaceutical Industry* (ed. P.J. Dunn, A.S. Wells and M.T. Williams), 127–144. Weinheim: Wiley-VCH.
13 Heine, A., DeSantis, G., Lutz, J.G. et al. (2001). Observation of covalent intermediates in an enzyme mechanism at atomic resolution. *Science* 294: 369.
14 DeSantis, G., Liu, J., Clark, D.P. et al. (2003). Structure-based mutagenesis approaches toward expanding the substrate specificity of D-2-deoxyribose-5-phosphate aldolase. *Bioorg. Med. Chem.* 11: 43–52.
15 Jennewein, S., Schürmann, M., Wolberg, M. et al. (2006). Directed evolution of an industrial biocatalyst: 2-deoxy-D-ribose 5-phosphate aldolase. *Biotechnol. J.* 1: 537–548.
16 Dick, M., Hartmann, R., Weiergräber, O.H. et al. (2016). Mechanism-based inhibition of an aldolase at high concentrations of its natural substrate acetaldehyde: structural insights and protective strategies. *Chem. Sci.* 7: 4492–4502.
17 Greenberg, W.A., Varvak, A., Hanson, S.R. et al. (2004). Development of an efficient, scalable, aldolase-catalyzed process for enantioselective synthesis of statin intermediates. *Proc. Natl. Acad. Sci. U.S.A.* 101: 5788–5893.
18 Hu, S., Tao, J., and Xie, Z. (2006). Process for producing atorvastatin, pharmaceutically acceptable salts thereof and intermediates thereof. WO2006/134482A1.
19 Oslaj, M., Cluzeau, J., Orkic, D. et al. (2013). A highly productive, whole-cell DERA chemoenzymatic process for production of key lactonized side-chain intermediates in statin synthesis. *PLoS ONE* 8 (5): e62250. https://doi.org/10.1371/journal.pone.0062250.
20 Ručigaj, A. and Krajnc, M. (2013). Optimization of a crude deoxyribose-5-phosphate aldolase lyzate-catalyzed process in synthesis of statin intermediates. *Org. Process Res. Dev.* 17: 854–862.
21 Kullartz, I. and Pietruszka, J. (2012). Cloning and characterisation of a new 2-deoxy-D-ribose-5-phosphate aldolase from *Rhodococcus erythropolis*. *J. Biotechnol.* 161: 174–180.
22 Jiao, X.-C., Pan, J., Xu, G.-C. et al. (2015). Efficient synthesis of a statin precursor in high space-time yield by a new aldehyde-tolerant aldolase identified from *Lactobacillus brevis*. *Catal. Sci. Technol.* 5: 4048–4054.
23 Sakuraba, H., Yoneda, K., Yoshihara, K. et al. (2007). Sequential aldol condensation catalyzed by hyperthermophilic 2-deoxy-D-ribose-5-phosphate aldolase. *Appl. Environ. Microbiol.* 73: 7427–7434.
24 Dick, M., Weiergräber, O.H., Classen, T. et al. (2016). Trading off stability against activity in extremophilic aldolases. *Sci. Rep.* 6: 17908. https://doi.org/10.1038/srep17908.

25 Wolberg, M., Dassen, B.H.N., Schürmann, M. et al. (2008). Large-scale synthesis of new pyranoid building blocks based on aldolase-catalysed carbon–carbon bond formation. *Adv. Synth. Catal.* 350: 1751–1759.

26 (a) Collett, L.A., Davies-Coleman, M.T., and Rivett, D.E.A. (1998). Naturally Occurring 6-Substituted 5,6-Dihydro-α-Pyrones. In: *Fortschritte der Chemie organischer Naturstoffe/Progress in the Chemistry of Organic Natural Products*, vol. 75 (ed. W. Herz, H. Falk, G.W. Kirby, et al.). Vienna: Springer.
b) Davies-Coleman, M.T. and Rivett, D.E.A. (1989). Naturally Occurring 6-substituted 5,6-dihydro-α-pyrones. In: *Fortschritte der Chemie organischer Naturstoffe/Progress in the Chemistry of Organic Natural Products*, vol. 55 (ed. W. Herz, H. Grisebach, G.W. Kirby and C. Tamm). Vienna: Springer.

27 Mink, D., Wolberg, M., Schürmann, M., and Hilker, I. (2007). Process for the preparation of delta lactones. International Patent Application. WO2007068498.

28 Meyer, H.-P., Eichhorn, E., Hanlon, S. et al. (2013). The use of enzymes in organic synthesis and the life sciences: perspectives from the Swiss Industrial Biocatalysis Consortium (SIBC). *Catal. Sci. Technol.* 3: 29–40.

29 Garrabou, X., Castillo, J.A., Guérard-Hélaine, C. et al. (2009). Asymmetric self- and cross-aldol reactions of glycolaldehyde catalyzed by D-fructose-6-phosphate aldolase. *Angew. Chem. Int. Ed.* 48: 5521–5525.

Index

a

Abraham, Edward Penley 11
acetic acid bacteria 12
acetobacter enzymes 210
acid lactases 158, 160, 207–208
acid phosphatase 63, 257, 261, 263
acrylamide 13, 40, 110, 116, 129, 130, 135–136, 385
Acrylaway® 135–137
Acyltransferase 112, 162
Addgene 54, 55
aflatoxin B_1 241
aflatoxin contamination 221, 241
air bubbles 100, 105, 106, 109
alcohol dehydrogenases 37, 210, 389, 398
alcoholic beverage 4
alcoholic fermentation 10
alcohol oxidase 1 (AOX1) 53, 58
aldolases 12, 385–400
α-amylases 8, 13, 19, 20, 99, 100, 106, 113, 212, 260, 311
α-Galactosidase (ADG) 205, 207–209, 255
α2-6 sialyltransferase 187
α1-3/4 transfucosidase (α1-3/4 TF) 189, 195
α2-6 transsialidase 187, 192, 193, 195
amdS marker system 56, 61
American brewing industry 15
amino acids 13, 28, 30, 31, 33–36, 55, 73, 87, 110–112, 125, 127–130, 133–135, 138, 147, 148, 153, 185, 191, 206, 211, 230, 237, 238, 256, 257, 259, 260, 265, 333, 358, 359, 386, 393, 395, 398
amylase, bread making process
  dough rheological properties 99, 106, 108
  fermentation 106, 107
  flour quality and standardization 98–100
amylases 15, 16, 19, 20, 99, 100, 104–107, 110, 112–114, 117, 205, 206
amyloglucosidase 106, 107, 110–112
animalcules 5
AN-PEP enzyme 135
antibiotics
  prokaryotic and eukaryotic organisms 53, 54
antinutrient 259
anti-staling enzymes 113, 114, 116
AQUAZYME® 19
artificial chromosomes 51
Asahi Chemical Industry 13
l-ascorbic acid 10, 12
asexual evolution 31
asparaginase 116, 129, 135–137
Aspergillus flavus 61, 220
Aspergillus niger 18, 49, 50, 61, 63, 107, 129, 135, 151, 208, 209, 226, 241, 256, 258, 262, 270, 336
Aspergillus strains 4, 15
Association of Manufacturers and Formulators of Enzyme Products (AMFEP) 48, 61
atom economy 76, 92

Industrial Enzyme Applications, First Edition. Edited by Andreas Vogel and Oliver May.

atoxigenic strains 221
autonomously replicating sequence (ARS) 52
auxotrophic systems 55
Azospirillum brasilense 51

b

Bacillus circulans 160
Bacterium xylinum 10
baking enzymes 97, 100, 107, 111, 112, 114–118
batch reactor 87–89, 373
β-chlorodiisopinocampheylborane 354, 364
β-propeller phytase (BPP) 261–262
beverages 4, 7–9, 15, 131, 136, 143, 168, 169, 398
Bifidobacterium bifidum 160
Bifidobacterium longum subsp. *infantis* 187
Bimuno GOS 160
bioactive peptides 130, 135–137
biocatalysis 13, 71, 73, 76, 77, 80, 82, 84, 242, 354, 365, 371, 373, 377, 387, 400
biocatalyst
  assessment criteria for 90–92
  characteristics 71
  chemo-catalysts/conventional chemical processes 385, 386
  environmental analysis 386, 387
  enzyme classes, commercial availability and utility 73
  enzyme immobilization 74–76
  reaction media
    monophasic system 77–80
    multiphasic systems 80–87
  reactor types in 87
  regio- and chemoselectivity 385, 386
  resting and growing cells 73
  statin side chain 387
  whole-cells 72
biocatalyst consumption 91, 92
biodiesel 20, 323, 344
Biolacta 160
biomass degrading enzymes 62
BIOMIN 222, 225, 227, 228, 234, 235
biotransformation processes 12, 71
bleach boosting enzyme 290, 312–315
Bommarius group 38
bovine spongiform encephalopathy (BSE) 132
bread making process
  baking and oven spring 109–110
  baking enzymes in 114, 115
  fermentation and dough stability 105–109
  flour quality 98–100
  flour standardization 98–100
  mixing and dough handling 100–105
  phospholipases in 107
bread quality
  color and flavor 111–112
  shelf life 111–116
BRENDA database 26, 27, 311
Brewers Clarex® 61
Brewing 7, 8, 13, 15, 17
Browning 109–111, 135, 157, 161, 168, 169, 174
bubble column reactor 83, 89
bubbling water 5
BURNUS® 18, 19

c

cake baking process 116
calf rennet 131, 147–149, 151
Canadian Standard Freeness (CSF) 293
Candida antarctica 29, 63
carbohydrate-active enZYme (CAZy) 167
Carica papaya 132
CASTing method 34, 35
catalophore approach 26
cell factories 49, 186
cellobiohydrolases (CBH) 62, 63, 291, 297
cellobiose
  cellobiose phosphorylase enzymes 171
  chemical synthesis of 170
  enzymatic synthesis of 16–68

phosphorolysis units 173
process development
purification of 174–176
synthesis of 174
properties and applications 168–170
sucrose phosphorylase (SP) 171
synthesis units (SU) 173
cellobiose phosphorylases (CP) 167, 168, 170–173
cheese making process 131, 145, 146, 148, 151, 152
chemo-enzymatic processes 385
chlorine-based bleaching process 312
chloroacetaldehyde (CIAA) 388, 390, 393, 394, 396, 398, 400
(3*R*,5*S*)-6-chloro-2,4, 6-trideoxyhexapyranoside (CTHP) 390
Chorleywood bread process 101, 103
Chymax® 151, 152
Chymosin 9, 40, 63, 131, 147, 148, 150–152
chymus 222, 224, 228
CIAA aldol reaction 390–392, 396
classical enzymes 144, 150
Claviceps purpurea 220
c-LEcta 28, 35–38, 136, 139, 171, 172, 184, 189, 191–193, 358
coagulants
α-and β-caseins 146
aspartic endo-proteases 145
curd–whey separation 145
fermentation-produced chymosin 147
International Milk Clotting Unit method 145, 146
microbial rennets 146, 148–151
traditional rennet 147–148
coagulation *versus* proteolysis (C/P) activity 146, 147
CodeEvolver protein engineering technology 358, 359, 363, 368, 371
consensus approach 34
pyruvate decarboxylase promoter (P_{PDC}) 58
contaminated wastewater 76
continuously operated stirred tank reactor (CSTR) 87, 88
cookies 98, 116, 135
core promoter 52, 53, 57
Corey–Bakshi–Shibata (CBS) oxazaborolidines 354, 364
counter selection markers 55
CRISPR/Cas9 based technology 60
crude oil 323–325, 329, 330, 333, 337, 338, 344
crumb color 111
crumb setting 109
Cryphonectria parasitica 149
cysteine phytase 261, 263

d

dairy applications 15, 144, 163
dairy industry
acid lactases 158
acyltransferase 162
coagulants
α-and β-caseins 146
aspartic endo-proteases 145
curd-whey separation 145
fermentation produced chymosin 151
International Milk Clotting Unit method 145
microbial rennets 146, 148–151
traditional rennet 147–148
glucose oxidase 161
GOS production 158–160
hexose oxidase 161
lactose oxidase 161
limitation of 144
lysozyme 163
neutral lactases 156–158
new developments 163
oxidases/peroxidases 161
peroxidase 161
phopholipases 162
phospholipase A1 162
preservation 163
protein cross-linking 162
ripening enzymes
enzyme modified cheese 153

dairy industry (*contd.*)
lipases/esterases 154
proteases and peptidases 153
transglutaminase 162
damaged starch 99–101, 106, 111
delta-lactone 104, 161, 397–399
de novo design, enzymes 39
deoxycholic acid 12
2-deoxy-D-ribose 5-phosphate aldolase (DERA)
catalyzed aldol reactions 387–390
CTHP 397
delta-lactone 397
directed evolution of 394–396
hydroxy-lactone 397
rational design 392–394
statin side chain process 390–392
thermophilic prokaryotes 396
unsaturated and saturated delta-lactones 399
deoxynivalenol 220–223, 227, 241, 242
derepression effect 58
detergent applications 19
dextran 208
diastase 5, 6, 15, 16, 205, 206
diastatic enzyme 15
(S)-1-(2,6-dichloro-3-fluorophenyl) ethanol (DCFPE) 359
digestive enzymes
therapeutic 207
in United States 206–207
5,6-dihydroxyindole (DHI) 213, 214
lactose (Gal(β1-4)Glc) 180
dog dung 4
dough mixing methods 101
downstream processing (DSP) 26, 48, 76, 225
DSM Food-Specialties 130, 135, 144, 146, 150–154, 156, 158, 161, 344
dynamic kinetic resolution (DKR) 365, 366

e

effervescent tablets 170
egg white lysozyme 11
endo-acting hydrolases 153
endo-acting xylanases 109
endogenous enzymes 50, 99, 100, 117, 129, 224
endopeptidases 128, 129, 132, 135–137
endoplasmic reticulum (ER) 49
enzymatic degumming
in industrial practice 337–342
phospholipase A (PLA) enzymes 336
phospholipase C (PLC) enzymes 330, 331
enzyme-assisted decoloring 344
enzyme-assisted oil extraction 34–45
enzyme development technology
enzyme engineering
library design and generation 30–37
modifications 30
screening analytics 37–38
impact of 38–41
wild-type enzymes 26–30
enzyme dynamics 39
enzyme engineering
library design and generation 30–37
modifications 30
screening analytics 37–38
enzyme for ferments 10
enzyme immobilization method 74
enzyme membrane reactor (EMR) 13, 90
enzyme-modified-cheese (EMC) 132, 138, 144, 152
enzyme recovery 75
enzyme recycle 75
enzyme supplements 15
episomal vectors 51, 52
epoxide ring 241, 242
Epygen Papyrase group 293, 295, 313–316
ergopeptines 242
error prone PCR (epPCR) 31, 394, 395
Escherichia coli 29, 50, 51, 151, 171, 183, 231, 257, 258, 262, 270, 390
eslicarbazepine 368, 374, 375
ethanol production 20
eukaryotic expression systems
advantages and disadvantages 48

DNA 51
expression hosts 47
production 47
promoters 56
vector design 51–56
eukaryotic microorganisms 48
European Food Safety Authority (EFSA) 158, 239
exoglucanase 49, 58
exopeptidase 128, 129, 153
expander oils 325, 334
extensibility and surface-active properties 105
extraction rate 98, 99
extraphosphoric effects 260

f
facial tissue 293
fed batch 87, 89
feed industry 19, 221
fermentation produced chymosin (FPC) 147, 151–152
fermentative processes 4, 73
fermenting yeast 10
ferments 10
flake oils 325, 334
flour standardization, bread 99, 105
flower of vinegar 10
Fluorouracil 60
food for special medical purpose (FSMP) 184
FoodPro Cleanline 162
food proteins
acrylamide reduction 135–136
animal derived enzymes in 132
bioactive peptides 136
enzyme types 127
future needs enzymes 137
gluten modification 135
ingredients 125
plant-derived enzyme 132
plant protein hydrolysates 134–135
processing of 127
products of 125, 126
protein value chain 130–131
research and development 125
whey protein hydrolysate 134
formate dehydrogenase (FDH) 13, 356, 357
fruit processing 18
fucose 180, 190
l-fucose (Fuc) 180
fucosylation 180, 184, 187
2'-fucosyllactose (2'-FL) 180, 183, 184
FumD 230–235, 237–239
fumonisins 220, 221, 224–228, 230–238, 240, 242, 243
FUMzyme
enzyme activity assays 232–233
enzyme characterisation and evaluation 233–234
enzyme discovery 227–230
enzyme engineering 237–238
enzyme feeding trials and biomarker analysis 234–237
enzyme production 238–239
enzyme selection 230–232
fumonisins 225–227
registration 239–240
fungal enzyme 4, 128
fungicides 221
Fusarium resistance 221
Fusarium verticillioides 221, 225–227, 231, 232, 234, 235

g
galacto-oligosaccharides (GOS) 131, 145, 158, 181
d-galactose (Gal)180
gas/liquid mixtures 83–84
GenBank® 26
gene of interest (GOI) 52
genetically modified enzymes (GMO enzymes) 205
genetically modified microorganism (GMM) 61
Gliocladium roseum 241
d-glucose (Glc) 154, 184, 310
glucose dehydrogenase (GDH) 356, 357
recycling 377

glucose oxidase 7, 61, 63, 104, 108–110, 115, 117, 161, 163, 208–210
glutaminase enzyme 138
gluten-free bread 117
glyceraldehydes-3-phosphate dehydrogenase promoter (P_{GAP}) 58
glycerol kinase 1 (GUT1) gene 55
Glycom A/S
  c-LEcta GmbH 184
  2'-FL and LNnT 181–183
  in vitro diversification technology 187
  in vivo fermentation technology 184
  whole-cell microbial fermentation 18–86
glycosyltransferases (GTs) 184, 193
Golgi compartments 49
Good Manufacturing Practice (GMP) guidelines 205
green chemistry 76, 374
gum phase 328, 330–331, 333, 336, 339, 341, 342

h

Hansen, Christian 15, 144, 146, 149–151, 154, 161, 162
hardwood pulp 291, 292, 308
hardwood vessel breaking enzyme 308–310
heat treatment technique 9
hemicellulases 103, 115, 206, 306, 313
heterologous protein production 60, 62–63
hexose oxidase 7, 104, 161
histidine acid phosphatases (HAP) superfamily 258, 261
histidine acid phytase (HAPhy) 261
homologous protein production 61–62
Hot Spot Wizard 26
human digestion 6
human milk oligosaccharides (HMOs) 159, 179
  discovery and function of 179–180
  diversification concept 187–189
  Glycom A/S 184–189
  historical overview 181–184
  optimized enzymes 195
  structure of 180–181
human nutrition and healthcare
  acetobacter enzymes 210
  acid lactase 207–208
  α-Galactosidase (ADG) 208
  dextran 208
  digestive enzymes 207–208
  glucose oxidase 208–210
  laccase 211
  polyphenol oxidase 210
  problems of 205
  transglucosidase (TG) 211
hydratability 328, 337
hydrolase-catalyzed reactions 76
hydrolyzed FB_1 (HFB_1) 228, 232
hydrolyzed fumonisins 230–232, 235
hydroxymethyl-glutaryl coenzyme A (HMG-CoA) reductase 387, 388
(R)-3-hydroxytetrahydrothiophene 376

i

induced fit model 11
industrial biotechnology 47, 49, 62, 239
industrial biotransformation process 12
industrial enzyme applications
  aldolases 12, 385–400
  biocatalyst 72
  biotransformation process 12
  business 13, 14
  cell factories 49
  entrepreneurs 17, 20
  enzyme classes of 6–7
  feed industry 19
  fungal enzyme 4
  human nutrition and healthcare
    acetobacter enzymes 210
    acid lactase 207–208
    α-Galactosidase (ADG) 208
    dextran 208
    digestive enzymes 207–208

glucose oxidase 208–210
laccase 211–215
polyphenol oxidase 210
problems of 205
transglucosidase (TG) 211
prehistoric applications 3–5
prokaryotic and eukaryotic host organisms 48
starch processing 19
traditional leather processing 18
industrial enzyme production
CRISPR/Cas9 based technologies 60
multicopy integration 59
multiple integration, expression cassettes 59
posttransformational vector amplification (PTVA) method 59
QM9414 and RutC30 59
scale of
heterologous protein production 62–63
homologous protein production 61–62
suicide substrates 60
industrial protein expression 48
infant nutrition, HMOs 183
inositol hexakisphosphate 255
in situ product removal (ISPR) 80, 81
integrative systems expression 53
International Milk Clotting Unit (IMCU) method) 145
International Minerals & Chemicals (IMC) 256
iterative saturation mutagenesis (ISM) method 34

j

Jokichi Takamine's patent 4, 15, 16

k

ketoreductase (KRED) enzymes
CodeEvolver protein engineering technology 358
crystal structures 355, 356
DIP-Cl 365, 374
enantiopure asymmetric carbonyl reduction 364–369
environmental impact of 376
green-by-design methodology 373
ketones/aldehydes 358–363
manufacturing process 370
NADH/NADPH 356
PMI guidelines 374
regio-and stereoselectivity 355
short-chain dehydrogenases/reductases enzyme family 355
tetrahydrothiophenone 377
Kluyveromyces lactis 49, 52, 151, 154, 156, 160
Koji process 15
Kozak Consensus Sequence (KS) 57
kraft pulping 312
kraft softwood 291, 292
Kühne, Wilhelm Friedrich 10

l

laccase 7, 63, 104, 139, 163, 205, 210–215, 241
lactases
acid 158, 160, 161, 207–208
GOS production 158–160
neutral 156–158
Lactobacillus kefir 356, 363
lacto-*N*-biose (Gal(β1-3)GlcNAc) (Type I) 180
lacto-*N*-fucopentaose III (LNFP-III) 180, 184, 195–197
lacto-*N*-neotetraose (LNnT) 183, 192, 195–197
lacto-*N*-neo-tetraose (LNnT) 180, 181, 192
lactose-derived prebiotics 158
lactose-free milk 156, 157
lactose oxidase 161
laundry applications 18, 62
laundry enzymes 49
leaf bleached kraft pulp (LBKP) 298
leather processing 4, 18
Leloir GTs 184
L-(−)ephedrine 12
linoleic acid 168, 169

lipases 19, 20, 78, 99, 107, 108, 110, 112, 114, 147, 153, 154, 206, 306, 343, 344, 371, 398
lipoxygenases 7, 99, 111, 112
liquid-crystalline 328–330, 333, 338
liquid/liquid mixtures 80–83
liquid PTVA 59
lithium aluminum hydride (LAH) 353
living creatures 5
living microbes 5
lysophospholipids 162, 329–333, 336, 338, 340, 342
lysozyme 11, 63, 78, 163

m
magic forces 5
Maillard reaction 107, 110, 112, 116, 118, 129, 135, 156, 157, 161, 169, 174
maltogenic amylase 113, 114, 116
maltotetraose-producing amylase 114
Maxiren XDS 152
Meerwein–Ponndorf–Verley reduction 353
Melanin's biosynthesis pathway 213
MetaKids^{TM} Nutrition Powder 184
methanol oxidase promoter (P_{MOX}) 58
Michaelis–Menten kinetic properties 91, 145, 270
microbial 11-α-hydroxylation 12
microbial contamination 76, 160, 196
microbial eukaryotes 51
microbial fermentation 183, 185–186
microbial hydroxylation 12
microbial lipase 154
microbial rennets 146, 148–151, 154
microbial spoilage 112, 114
microemulsions 82, 83
milk fat globular membrane (MFGM) 162
milling process 98, 243
Ministry of Health, Labor and Welfare (MHLW) 205
mixed waste (MW) 299
monodirectional promoters (MDPs) 58
monogastric animals 224, 255, 266, 267
monophasic system 77–80
MTA144 228–232
multibillion Euro enzyme industry 12
multicopy integration 59
multi-dimensional mutagenesis (MDM) platform 35, 191
multienzyme reactions 75, 167
multiphasic systems
  gas/liquid mixtures 83–84
  liquid/liquid mixtures 80–83
  solid/liquid mixtures 84–87
Myceliophthora thermophila 50, 62, 63, 239
mycotoxins
  in animal 220
  contaminated feed/food 219
  enzyme applications
    chymus 224
    digestion 224
    FUMzyme 225
  filamentous fungi 219
  fungal molecules 219
  future aspects 240
  human nutrition and health 220
  mitigation strategy 220, 223
  research 220
Myrothecium verrucaria 215

n
N-acetylglucosamine (GlcNAc) 180
N-acetyllactosamine (Gal(β1-4)GlcNAc) (Type II) 180
National Food Research Institute 170
native starch conversion enzyme 290, 310–312
Natuphos® 257, 258, 266, 267
natural polymers 61, 289
NCBI databases 26
Neolithic Revolution 3
neutral lactases 15–58, 207
neutral oil loss (NOL) 327, 337
Nippon Petrochem 170
Nitto Chemical Industries 13
nixtamalization 231, 232
non-food application 131, 323

nonimpact-printed papers 306
Novozymes 17, 18, 47, 61, 63, 129, 134–136, 139, 144, 158, 258

o

ochratoxin A 220, 223, 227, 240, 241
Old corrugated carton (OCC) 299
old corrugated containers (OCC) 302
old newsprint (ONP) 300
oligonucleotides 33–35
OptiPhos phytase 267
organic breads 118
organic media 76–80
origin of replication (ORI) 51
OROPON^{TM} 14, 18
outflow conditions 88
over-the-counter (OTC) drugs 205
oxidative biotransformation 10

p

packed bed reactor (pbr) 75, 87, 88
palm oil 323, 344
pancreatic enzymes 18, 132
pancreatic proteases 18
pancreatin 128, 132, 133, 206
papain 128, 129, 131–133, 205, 206
paper mill effluent treatment enzymes 315–316
parbaked/frozen storage 118
partially hydrolyzed FB_1 ($pHFB_1$) 232
pasteurization 9, 149, 152, 156
Pasteur, Luis 9, 10
Payen, Anselme 5, 15
Peroxidise 7, 99, 104, 139, 161
Persoz, Jean-Francois 5, 15
Pfizer's original approach 376
phenylalanine ammonia lyase (PAL) 386
phopholipases 162
phosphite dehydrogenase (PDH) 356, 357
phospholipase A1 108, 162, 330, 332
phospholipase A (PLA) enzymes 108, 162, 324, 330, 332, 336–337
phospholipase B (PLB) 108, 331
phospholipase C (PLC) enzymes 108, 163, 330–333, 335, 336, 339
phospholipase D (PLD) 108, 331
phospholipids 107, 115, 162, 323, 325–340, 342, 344
phosphorolysis units (PU) 173
phosphorous 255, 275, 331, 334
phytase 20, 34, 40, 63, 99, 117, 224, 255–275
phytases, animal feed application
  characteristics 265
  kinetic description of 269–271
  low pH and proteases resistance 271
  market development 256–258
  nutritional values 265
  phosphorous
    β-propeller phytase 261
    cysteine phytases 263
    dephosphorylation 263, 264
    histidine acid phosphatases superfamily 258, 261, 262
    properties of 259–260
    purple acid phytase 263
  significance of 255
  super dosing 274–275
  temperature stability 271–274
  upper digestive tract 266
phytate 255, 256, 259–275
phytate-P content 255
phytic acid 255, 259, 268
phytin 259, 260
Piccantase KLC 154
piccante 154
Pichia pastoris 48–53, 58, 60, 63, 231–234, 238, 239, 257, 258, 270–273
pills 158, 168, 170
plant biotechnology 221
plant protein hydrolysates 134–135
plasmids 50, 51, 54–56, 185, 186
plug flow reactor (pfr) 30, 88
polypeptide 26, 30, 31, 39, 54, 271
polyunsaturated fatty acids (PUFA) 99, 111, 168
porcine trypsin 128, 134, 137
postharvest mycotoxin formation 219, 221

posttransformational vector amplification (PTVA) method 59
prebiotics 158, 159, 170, 206
pregastric esterases 154
preharvest mycotoxin contamination 220
prokaryotic expression hosts 47
  advantages and disadvantages 48
  in industry 47
  promoters 56
  vector design 51
prokaryotic host organisms 51–56
Promod™ 950L 133
promoter of the alcohol oxidase 1 gene (PAOX1) 58
ProSAR protocol 33
PROSS approach 26
proteases 104, 153–154, 271
protein cross-linking 139, 162
protein engineering methods 34
protein production
  heterologous 62–63
  homologous 61–62
protein value chain 130–131, 134, 137, 138
proteolytic activity 100, 147, 152, 271
pulp and paper enzymes market 20
pulp and paper industry
  bleach boosting enzyme 312–315
  deinking enzymes 306–308
  drainage improvement enzyme 296–301
  hardwood vessel breaking enzyme 308–310
  paper mill effluent treatment enzymes 315–316
  refining and fiber development enzyme
    handsheets 293–296
    microscopic evaluation 291–292
  slushing enzyme 317–319
  starch 310–312
  stickies control enzyme 301–305
purple acid phytases (PAPhy) 261, 263

r
refined, bleached, and deodorized (RBD) 342
refining process 323, 331, 345
Reichstein–Grüssner synthesis 12
renewable resources 73
renneting enzyme 147
rennets 145
repetitive batch 87
residence time 88, 89, 92, 275, 339
reversed micelles 82, 83
Rhizomucor miehei 149, 154
Rhizomucor pusillus 149
Rhizopus arrhius 12
ribosome binding site (RBS) 56
ripening enzymes
  enzyme modified cheese 153
  lipases/esterases 154
  proteases/peptidases 153–154
Röhm, Otto 4, 18
rotating bed reactor (rbr) 89, 90
Royal Society 5, 6

s
Saccharomyces cerevisiae 49, 50, 55–58, 105, 151, 233, 238, 355
St. Anthony's Fire 220
Sankyo Pharmaceutical Company of Tokyo 16
Schwann, Theodor 6
secreted recombinant protein 62
seed oil processing 325
selection markers (SM) 51–54, 56, 58, 59, 186
Selection of the fittest 31
selectivity, enzyme 29
self-cloning 60
sexual mutation 31
Shine–Dalgarno (SD) sequence 56
sialic acid (*N*-acetylneuraminic acid; Neu5Ac) 180
sialyl lacto-*N*-neotetraose (LST-c) 19–97, 181
sialyl lacto-*N*-tetraose (LST-a) 195–197
sialyltransferase (ST) activity 187, 193
single site mutation libraries 35, 36

slushing enzyme 290, 317–319
solid enzyme 79, 80
solid/liquid mixtures 84–87
space time yields (STY) 48, 58, 92, 391
sponge and dough process 101, 103
Sporobolomyces singularis 160
spray drying 175
starch processing industry 8, 9, 16, 19, 290
statin drugs 387, 388
stereoselective alcohol incorporation 377
Stickies 290, 301–305, 308
straight dough process 101
subcellular organelles 49
substrate channeling 75
substrate scope 26, 28, 29, 34, 35, 37, 354, 400
subtilisin 6, 9, 34, 128
sucrose phosphorylase (SP) 167, 168, 171
sugar cellobiose 168
sugar factory 5
suicide substrate 60
sunflower seeds 325
surfactants 82, 83
suspensions 78, 84, 85, 87, 138, 306, 311
Swiss fruit juice industry 18
synergistic effects 36, 104, 128, 264, 265
synthesis units (SU) 173

t

Taka-diastase 15, 206
Takamine, Jokichi 4, 15, 16, 206
tartaric acid 4, 10
therapeutic digestive enzymes 207
therapeutic protein production 47
thermomyces 63, 103, 108
Tolerase® G 135, 138
traditional rennet 147–148
transglucosidase (TG) 211
transglutaminase 7, 138, 162
transglycosylases (TGs) 184, 188
tricarballylic acid (TCA) 228
Trichoderma reesei 48–51, 53, 58–63, 239, 258
triglyceride oils 334, 343–344
triose phosphate isomerase (TPI1) system 55, 394
turnover frequency (TOF) 91
turnover number (TON) 91, 269
two-liquid-phase system (2LPS) 80, 81

u

UltraGI Replenish® 184
universal production host 49
5′ untranslated region (5′-UTR) 52, 57
UV-light or chemical mutagenesis 60

v

vegetable oil
  biodiesel 344
  enzymatic degumming
    in industrial practice 33–42
    phospholipase A enzymes 336–337
    phospholipase C enzymes 330, 331, 335, 336
  enzyme-assisted decoloring 344–345
  phospholipids 325–327
  production of 323
  triglyceride oils 343–344
  water-degumming process 327
vinegar production 10, 12
vital force 10
Vivinal GOS 160

w

Wafers 116, 117
wastewater treatment enzymes 316
water-degumming process 327, 330, 339
water-extractable arabinoxylan (WE-AX) 103
water-unextractable (WU-AX) 103, 115
Western world malt production 4
wheat flour quality factors 99
whey protein concentrates (WPC) 131
whey protein hydrolysate (WPH) 134

whey protein isolate (WPI) 131
whole-cell biocatalyst 12, 13, 396
whole cell catalysis 72
whole-cell microbial fermentation 185–186
wholemeal bread 117
wild-type enzymes 25–30, 193
wild-type promoter 58
wine production 4
wobble primers 228

x
xylanases 20, 58, 63, 100, 103, 104, 109, 110, 112, 117, 291, 313, 315

y
yeast 4, 10, 12, 29, 48–50, 52, 53, 55, 56, 97–101, 105, 107, 109, 111, 112, 118, 129, 131–133, 138, 151, 154, 207, 208, 222, 223, 227, 230, 257, 355, 399

z
zearalenone 220, 223, 227, 240, 241
Zeocin 53, 54
Zuo-Zhuan 4
zymase 10